Arthropod Diversity and Conservation in the Tropics and Sub-tropics

Akshay Kumar Chakravarthy
Shakunthala Sridhara
Editors

Arthropod Diversity and Conservation in the Tropics and Sub-tropics

Springer

Editors
Akshay Kumar Chakravarthy
Division of Entomology and Nematology
Indian Institute of Horticultural Research (IIHR)
Bengaluru, Karnataka, India

Shakunthala Sridhara (retired)
Department of Entomology
University of Agricultural Sciences
Gandhi Krishi Vignana Kendra (GKVK)
Bengaluru, Karnataka, India

ISBN 978-981-10-1517-5 ISBN 978-981-10-1518-2 (eBook)
DOI 10.1007/978-981-10-1518-2

Library of Congress Control Number: 2016955290

Printed on acid-free paper

This Springer imprint is published by Springer Nature
The registered company is Springer Science+Business Media Singapore Pte Ltd.

Foreword

The earth system is currently stressed severely due to human interventions. The ecological footprint of humanity has exceeded 1.6 times the biocapacity of biosphere. So much so that by 2050, twice as much food as today will be required to feed the world population. The future of mankind will be profoundly influenced by the way biological and natural resources are sustained, regulated and managed. This is because biodiversity is the foundation for sustainability and significantly contributes to quality of life of people and makes positive impact on the structure and functioning of the ecosystems. Arthropods constitute over 90 % of the animal kingdom. Bioecology of arthropods is intimately linked with global functioning and survival. However, documented literature and understanding on insects far exceed other classes of arthropods necessitating inputs on other classes of arthropods. This book titled *Arthropod Diversity and Conservation in the Tropics and Subtropics* brings together contributions of workers on arthropods. This compilation is an initiative at capturing the essentials of arthropod inventories, biology and conservation.

This edited compilation includes 26 chapters contributed by over 70 biologists on a wide range of topics embracing diversity, distribution, utility and conservation of arthropods and select groups of insect taxa. More importantly, the mechanism of sustaining the ecosystems of arthropods, their services and populations is also given in sufficient details. The use of modern biological tools such as molecular and genetic techniques regulating gene expression, as well as conventional, indigenous practices in arthropod conservation, are included. The contributors reiterate the importance of documenting and understanding the biology of arthropods with a holistic view before conservation issues at large are addressed. This is an important book for zoologists, entomologists, ecologists, conservation biologists, policy makers, teachers and students interested in the conservation and management of biological resources.

Madhav Gadgil

Goa University
Taleigão India
Centre for Ecological Sciences
Indian Institute of Science (IISc)
Bengaluru, India (previously)
December 2015

Preface

The trend in biological studies these days is towards biological diversity and conservation. Generally, many focus on vertebrates and flowering plants, although arthropods constitute more than 80 % of the total biota on planet Earth. Arthropods provide a multitude of ecosystem services and these are crucial for human sustenance and welfare. The concern of entomologists and other arthropod workers is to ensure sustenance of undisturbed populations, free from human interferences, by providing and maintaining safer and quality habitats for arthropods. In view of the importance and lack of published information on arthropod diversity and conservation, the current book, *Arthropod Diversity and Conservation in the Tropics and Subtropics*, was planned.

This book highlights concerns of not only entomologists but also other biologists studying arthropods to emphasize the importance of biological conservation of different communities of arthropods across the globe. Life forms dealt in this book cover arthropods as a community because despite such an important and vital group of animals, documented information on them with the exception of insects is scanty. These include scorpions, shrimps, prawns, crabs, ticks, mites, spiders, centipedes and millipedes that are unique in forms and distribution playing diversified roles in multiple biotopes. The crux of the problem for the conservation of arthropods is that the common public do not distinguish them as separate entities, recognize them as an independent group and appreciate their roles. So a comprehensive understanding on the roles of all the arthropod entities should be developed than insects alone.

Arthropods play an important role in maintaining the health of ecosystems, provide livelihoods and nutrition to human communities and are indicators of environmental change. Yet the population trend of several arthropod species is showing a decline. Arthropods evolved about 350 million years ago and constitute a dominant group with 1.2 million species influencing earth's biodiversity of animals, with insects predominant having about one million species. They are intimately associated with living and nonliving entities, and hence, ecosystem services offered by them are crucial. Arthropod conservation requires integrating conservation science and policy, with long-term planning and action plans. An attempt has been made here to indicate the mechanisms by which arthropod populations can be sustained in ecosystems and certain problematic species be managed without creating environmental side effects and economic burden. The plan for the conservation of arthropods and ecosystems should include a mixture of strategies like protecting key habitats and genomic studies to formulate relevant policies for *in situ* and *ex situ* conservation.

This book is a step forward in monitoring and encouraging people for their continued support in saving planet's earth-limited biological resources and their potential habitats and increasing awareness in generating appropriate technologies for their management. Implementing nature-friendly technologies to mitigate adverse impacts on arthropod populations is the major concern aimed to be reinforced in the minds of the public and its reinforcement amongst researchers in biological sciences. We hope that this book will serve as an additional step in this direction.

Bengaluru, India
2015

Akshay Kumar Chakravarthy
Shakunthala Sridhara

Prelude

Arthropod Diversity and Conservation in the Tropics and Subtropics

Knowledge available on arthropods in the tropics and subtropics does not do justice to their diversity and abundance. Detailed knowledge on arthropod diversity is required to develop integrated pest management and conservation strategies. Studies on arthropods from undisturbed and wild habitat patches are equally important in these tasks. However, it is not possible to collate information on all arthropod taxa, but it is possible to extrapolate from documented information to arrive at decisions on management/conservation issues, be it cropping systems or wild, urban or peri-urban habitats.

The book, *Arthropod Diversity and Conservation in the Tropics and Subtropics*, deals with select taxa of arthropods, their diversity, distribution, inventory and conservation issues. Currently, arthropod community is impacted by global warming which forms one of the most important factors influencing arthropods across landscapes as it does the biodiversity across the globe. Secondly, arthropods can serve as indicators of climate change. Two chapters in the book document data on this aspect. Prasannakumar and Kumar deal with salient features of arthropods affected by climate change. Arthropod-based molecules may hold promise for the new emerging disease and maladies. With this hope, Anudita, Varunrajan and Deepa have gained insights into the peculiar and interesting biomolecules from arthropods that have practical implications in biological control of pests and health care. The workers have emphasized the need to conserve arthropods. It will be interesting to know how ecologically important are the parasites on the arthropods. Jayashankar and his team have highlighted the role of arthropod parasites in regulating arthropod populations. Scorpions are venomous creatures, and their looks are frightening. Shakunthala Sridhara and co-workers have elucidated the evolutionary significance of venom in the life history of scorpion. Researchers have also highlighted ecological and biological aspects of other arthropod taxa namely crabs, Cladocera, millipedes, spiders and mites. Similarly, two species of wasps, both parasitoid and select phytophagous are dealt with. Ankita Gupta explores in an elegant way the distribution and diversity of braconid wasps which are parasitoids of pest species. Preethi and Lakshmi Devi chose chalcid wasps that include parasitoid species and few phytophagous species to study their distribution and diversity in Kerala, South India. Baaby Job and Olakkengill discuss at length diversity and distribution of sphecoid wasps which serve well as bioindicators of habitat change in South India.

Ants are other important bioelements in tri-trophic interactions. Presty John and Karmaly together have contributed a chapter on distribution and diversity of *Camponotus* ants. Sirisena and three co-workers from Sri Lanka have described in detail soft scales from the island country and unravel their peculiar characters. Fruit flies are key pests of tropical fruits. Laskar and co-workers document fruit flies from sub-Himalayan region to provide baseline data for their management.

Mrinmoy Das and others recorded faunal composition of scarab beetles in Assam, East India. The scarab beetles are notorious pests of roots of many cultivated plants, and they also serve as prey for a number of predators. Abrol and Uma Shankar have been studying honeybees over two decades and have pinpointed factors responsible for honeybee population decline particularly in tropics and subtropics. Khanjani from Iran determined factors for fluctuations in brown mite population in cherry orchards, Iran. Although some of these studies are not as exhaustive as insects and mites, they are expected to enlarge the scope of the book and underlie the importance of looking arthropods in a holistic manner. Often biologists focus on one specific taxon, ignoring others. However, it is desirable, while developing baseline data of a locality, to consider as many taxa as possible for the conservation and management of arthropods.

Mary and others have drawn attention of readers to giant moths that are specialists requiring unique combination of niche and food plants. Rapid population decline of bees in the wild is of concern to biologists today. Ritesh Inamke et al. discuss at length synthesis of management approaches for conserving insects. The approaches may also apply to conservation of other arthropod taxa. Prabhulinga and his team have worked on biological characteristics of a group of minute wasps called *Trichogramma* which are essentially important parasites of pests of cultivated crops because they kill the pests at egg stage and hence vital for the management of these pests that destroy reproductive parts of crops. Subba Reddy Palli proposes that by utilizing RNA interference technology, agricultural pests can be managed which would obviate the need for pesticides and substantially contribute to conservation of not only arthropods but natural resources. This is a revolutionary approach for pest control and a huge step in reducing/preventing pesticide contamination of environment.

Petersen and Reddy have drawn attention to methods that sustain *Apis dorsata* in tropical environments, and they also have made concerted efforts in preserving bee trees as a heritage site. Chandrika Mohan and co-workers worked out the possibilities of using coconut water as a culture media for mass rearing mite which is a predator on coconut mites. Rahul Ameen and his team provide evidence for using a native *Bacillus thuringiensis* isolate against polyphagous, defoliator pest, *Spodoptera litura* effectively.

Akshay Kumar Chakravarthy
Shakunthala Sridhara

Acknowledgements

First and foremost the editors thank the contributors for their patience as the manuscripts had to be revised several times before finalization for publication. Secondly, a number of youngsters doing PhD, senior research fellows, research associates, research assistants and young professionals helped in the final preparation of the manuscripts. It is difficult to name them all here, but Mr. Vasudev Kammar, Kumar K P, Chandrashekaraiah, Nitin K S, Rajendra Prasad, Nethravathi C J, Jayashankar M, Vijeth Arya, Madhu T N, Yatish K R, Savitha T, Sindhu, Vinesh L S, Rani A T, Onkara Naik, Prabhulinga T, Vijaya Kumar L, Subhash B Kandakoor, Basana Gowda G and Prasannakumar N R helped at various stages of the book preparation. For a book of this nature, it was often needed that we get highly specific information, which is not easily found in one or few publications as the information is highly scattered in the literature.

We profoundly thank the authorities of the University of Agricultural Sciences, GKVK, Bengaluru, and the Indian Institute of Horticultural Research (IIHR), Hesaraghatta, Bengaluru, for their support and encouragement. The staff and faculty of the Department of Entomology, UAS, GKVK, Bengaluru, and Division of Entomology and Nematology, IIHR, Bengaluru, lend support to the book-writing work, and some of the faculties have contributed chapters for the book.

Dr. Abraham Verghese, Dr. C T Ashok Kumar and Dr. N E Thyagaraj, authorities of the Indian Council of Agricultural Research (ICAR, New Delhi), in particular, Dr. N K Krishna Kumar, DDG (Horticultural Sciences), Dr. S Balwinder Singh, Ramesh Arora, P K Mehta and others, were instrumental in organizing the International Conference on Insect Sciences (ICIS) 2013 successfully. The editors profusely thank them for their help and contributions. We are thankful to Dr. S Sithanatham, Sun Agro Biotech, Chennai and Dr. H C Sharma, ICRISAT, Hyderabad, for their support and participation in ICIS 2013.

The editors thank immensely the International Springer Group, New Delhi, for their deep interest and great support for this publication.

Akshay Kumar Chakravarthy
Shakunthala Sridhara

Contents

Contributors

D.P. Abrol Division of Entomology, Sher-e-Kashmir University of Agricultural Sciences & Technology FOA, Jammu, Jammu and Kashmir, India

Periasamy Alagesan Post Graduate and Research Department of Zoology, Yadava College, Madurai, Tamil Nadu, India

Rahul Amin Department of Microbiology, Anand Agricultural University, Anand, Gujarat, India

D. Usha Anandhi Department of Zoology, Bangalore University, Bangalore, Karnataka, India

Anudita Department of Zoology, Bangalore University, Bengaluru, Karnataka, India

Badal Bhattacharya All India Network Project on White grubs and Other Soil Arthropods, Department of Entomology, Assam Agricultural University, Jorhat, Assam, India

Akshay Kumar Chakravarthy Division of Entomology and Nematology, Indian Institute of Horticultural Research (IIHR), Bengaluru, Karnataka, India

Hirak Chatterjee Palli Siksha Bhavana (Institute of Agriculture) Visva-Bharati, Birbhum, West Bengal, India

Mrinmoy Das All India Network Project on Soil Arthropod Pests, Department of Entomology, Assam Agricultural University, Jorhat, Assam, India

B.M. Deepa Department of Genetics, Indian Academy Degree College, Bengaluru, Karnataka, India

B. Doddabasappa Department of Entomology, College of Horticulture, Kolar, Karnataka, India

G. Basana Gowda ICAR-National Rice Research Institute, Cuttack, Odisha, India

Ankita Gupta Insect Systematics Division, National Bureau of Agricultural Insect Resources, Bangalore, Karnataka, India

Priyanka Hadole Department of Zoology, Sant Gadge Baba Amravati University, Amravati, Maharashtra, India

Gautam Handique All India Network Project on Soil Arthropod Pests, Department of Entomology, Assam Agricultural University, Jorhat, Assam, India

Tapan Kumar Hath Palli Siksha Bhavana (Institute of Agriculture) Visva-Bharati, Birbhum, West Bengal, India

K.S. Hemachandra Department of Agriculture Biology, Faculty of Agriculture, University of Peradeniya, Peradeniya, Sri Lanka

Ritesh Suresh Inamke Biological Control Research Laboratory, Anand Agricultural University, Anand, Gujarat, India

S.K. Jalali Division of Entomology, National Bureau of Agricultural Insect Resources (NBAIR), Bengaluru, Karnataka, India

Janardan Jani Biological Control Research Laboratory, Anand Agricultural University, Anand, Gujarat, India

M. Jayashankar Division of Entomology, Indian Institute of Horticultural Research, Bengaluru, India

Baaby Job Department of Zoology, St. Thomas College, Thrissur, Kerala, India

Presty John Department of Zoology, S.H. College, Cochin, Kerala, India

Current address: Kerala Forest Research Institute, Peechi, Kerala, India

A. Josephrajkumar Entomology Section, ICAR-Central Plantation Crops Research Institute, Alappuzha, Kerala, India

V. Kalarani Department of Biotechnology, Sri Padmavati Mahila Visvavidyalayam (Women's University), Tirupati, Andhra Pradesh, India

Vasudev Kammar Department of Entomology, University of Agricultural Sciences (UAS), Gandhi Krishi Vignana Kendra (GKVK), Bengaluru, Karnataka, India

Subhash B. Kandakoor AICRP on Potato, University of Agricultural Sciences, Dharwad, Karnataka, India

K.A. Karmaly St. Xaviers College for Women, Aluva, Kerala, India

M. Karuthapandi Freshwater Biology Regional Centre, Zoological Survey of India, Hyderabad, Andhra Pradesh, India

Mohammad Khanjani Department of Plant Protection, College of Agriculture, Bu-Ali Sina University, Hamedan, Iran

Kuldeep Khatri Biological Control Research Laboratory, Anand Agricultural University, Anand, Gujarat, India

K.P. Kumar Department of Entomology, University of Agricultural Sciences, Bengaluru, Karnataka, India

Nripendra Laskar Department of Agricultural Entomology, Uttar Banga Krishi Viswavidyalaya, Cooch Behar, West Bengal, India

P.P. Mary Department of Forestry, North Eastern Regional Institute of Sciences and Technology (NERIST) Deemed University, Nirjuli, Arunachal Pradesh, India

P. Lakshmi Devi Menon Department of Zoology, S.H. College, Cochin, Kerala, India

Chandrika Mohan Entomology Section, ICAR-Central Plantation Crops Research Institute, Alappuzha, Kerala, India

S. Murugan Department of Zoology, Bangalore University, Bangalore, Karnataka, India

K.S. Nitin Division of Entomology and Nematology, Indian Institute of Horticultural Research (IIHR), Bengaluru, Karnataka, India

J.L. Olakkengill Department of Zoology, St. Thomas College, Thrissur, Kerala, India

Subba Reddy Palli Department of Entomology, College of Agriculture, University of Kentucky, Lexington, KY, USA

Deepak Panpatte Biological Control Research Laboratory, Anand Agricultural University, Anand, Gujarat, India

Noushad Parvez Biological Control Research Laboratory, Anand Agricultural University, Anand, Gujarat, India

Ankit Patel Biological Control Research Laboratory, Anand Agricultural University, Anand, Gujarat, India

Leena Pathak Biological Control Research Laboratory, Anand Agricultural University, Anand, Gujarat, India

Stephen Petersen Toklat Apiaries, Fairbanks, AK, USA

T. Prabhulinga Division of Entomology, Central Institute for Cotton Research (CICR), Nagpur, India

S.R. Amulya Prasad Department of Zoology, Bangalore University, Bengaluru, Karnataka, India

N.R. Prasannakumar Division of Entomology and Nematology, Indian Institute of Horticultural Research, Bengaluru, Karnataka, India

N. Preethi Department of Zoology, NSS College, Malappuram, Kerala, India

D. Pujari All India Network Project on Soil Arthropod Pests, Department of Entomology, Assam Agricultural University, Jorhat, Assam, India

M.K. Radhika Entomology Section, ICAR-Central Plantation Crops Research Institute, Alappuzha, Kerala, India

P. Rajan Entomology Section, ICAR-Central Plantation Crops Research Institute, Alappuzha, Kerala, India

A.T. Rani Department of Entomology, University of Agricultural Sciences (UAS), Gandhi Krishi Vignana Kendra (GKVK), Bengaluru, Karnataka, India

D.V. Rao Freshwater Biology Regional Centre, Zoological Survey of India, Hyderabad, Andhra Pradesh, India

D. Chandrasekhar Reddy Department of Zoology, SV University, Tirupati, Andhra Pradesh, India

Muniswamyreddy Shankar Reddy Department of Zoology, Bangalore University, Bengaluru, India

Uma Shankar Division of Entomology, Sher-e-Kashmir University of Agricultural Sciences & Technology FOA, Jammu, Jammu & Kashmir, India

Harsha Shelat Department of Microbiology, Anand Agricultural University, Anand, Gujarat, India

Dipak Kumar Sinha Palli Siksha Bhavana (Institute of Agriculture) Visva-Bharati, Birbhum, West Bengal, India

U.G.A.I. Sirisena Post Graduate Institute of Agriculture, University of Peradeniya, Peradeniya, Sri Lanka

Shakunthala Sridhara Department of Entomology, University of Agricultural Sciences, Gandhi Krishi Vignana Kendra (GKVK), Bengaluru, Karnataka, India

Sridevi V Department of Biotechnology, Sri Padmavati Mahila Visvavidyalayam (Women's University), Tirupati, Andhra Pradesh, India

Ganesh Vankhede Department of Zoology, Sant Gadge Baba Amravati University, Amravati, Maharashtra, India

V. Varunrajan Division of Entomology and Nematology, Indian Institute of Horticultural Research, Bengaluru, Karnataka, India

Sumathi Vempalli Department of Biotechnology, Sri Padmavati Mahila Visvavidyalayam (Women's University), Tirupati, Andhra Pradesh, India

Rajababu Vyas Department of Microbiology, Anand Agricultural University, Anand, Gujarat, India

G.W. Watson California Department of Food & Agriculture, Plant Pest Diagnostic Center, Sacramento, CA, USA

H.N.P. Wijayagunasekara Department of Agriculture Biology, Faculty of Agriculture, University of Peradeniya, Peradeniya, Sri Lanka

K.R. Yathish Division of Entomology and Nematology, Indian Institute of Horticultural Research, Bengaluru, India

About the Editors

Akshay Kumar Chakravarthy is the Head and Principal Scientist at the Indian Institute of Horticultural Research (IIHR), Hesaraghatta, Bengaluru, India. With over three decades of experience in teaching, research and extension, Dr. Chakravarthy has been the investigator for over 30 research projects and has guided more than 25 postgraduate students. He received PhD from Punjab Agricultural University, Ludhiana, and Post Doctoral Fellowship from Indian Agricultural Research Institute, New Delhi. He is a fellow of the American Chemical Society, USA, member of several national and international scientific academia, advisor, panellist, referee, reviewer and editor associated with publication of over 30 national and international journals worldwide. He has 400 publications in the form of books, chapters, monographs, bulletins, papers, short notes, commentaries, letters and meeting and project reports. A field-oriented, widely travelled biologist, he is actively working on novel approaches in integrated pest management, host-plant interaction, vertebrate pest management, biodiversity and environmental conservation. Currently, he has initiatives on nanotechnology too.

Shakunthala Sridhara after obtaining her PhD in Animal Physiology from Bangalore University, joined a Ford Foundation project on vertebrate pest management in the University of Agricultural Sciences, Bangaluru, India, in 1973. Over the past 33 years, she has been researching on vertebrate pest management, specially the control of rodents in the agricultural context. She has researched extensively on the ecology, population dynamics, food selection and feeding behaviour of rodents, toxicology of rodenticides and adoption of rodent pest management at village level. Her studies on behaviour relevant to management of avian and mammalian pests are pioneering in the Indian context and well acknowledged culminating in adaptable technologies for their management. She has visited and interacted with specialists in the field across America and Europe several times. Keenly interested in animal behaviour studies and its application in pest and wildlife management, biodiversity conservation and animal produce, she is a member of several national and international scientific bodies including the presidentship of Ethological Society of India, Indian representative in the International Council of Ethologists, IUCN species specialist group on rodents, etc. She has retired as Professor and Head of Vertebrate Biology (Rodent Control) in the University of Agricultural Sciences, Bangalore followed by a stint as Emeritus Scientist of Indian Council of Agricultural Research for two years (2007–2009).

Impact of Climate Change on Arthropod Diversity

1

N.R. Prasannakumar and K.P. Kumar

Abstract

Today, understanding the consequences of changes in biological diversity due to global warming has become critically important. Climate change has profound influence on arthropod diversity. Principal effects of climate change on arthropod diversity include decrease in decomposers and predators and parasites and increased herbivory which affects the structure and function of ecosystems. With changes in arthropod communities, farmers are experiencing decreases in soil fertility and crop productivity. Amelioration includes organic farming, cultivation of climate-resilient crops, the use of water energy, wind energy, maintaining water bodies, reducing vehicular emissions, etc.

Keywords

Arthropod diversity • Climate change • Global warming

1.1 Introduction

Diversity of arthropods is defined as the existing variations within the arthropod communities from all the sources such as, inter alia, terrestrial, marine, and other aquatic ecosystems. It also includes ecological interactions, within and between

N.R. Prasannakumar (✉)
Division of Entomology and Nematology, Indian Institute of Horticultural Research, Hessarghatta, Bengaluru 560089, Karnataka, India
e-mail: prasannaent@gmail.com

K.P. Kumar
Department of Entomology, University of Agricultural Sciences, Bengaluru 560065, Karnataka, India
e-mail: kumarkp4830@gmail.com

A.K. Chakravarthy, S. Sridhara (eds.), *Arthropod Diversity and Conservation in the Tropics and Sub-tropics*, DOI 10.1007/978-981-10-1518-2_1

species and ecosystems (Mandal and Neenu 2012). Arthropods not only have a wide range of diversified ecological roles in natural resources but also economic benefits in agriculture (Wilson 1987; Isaacs et al. 2009). Their diverse functions include as prey resources for many taxa, as important predators, and as pollinators and seed dispersers (Bond and Slingsby 1984; Wilson 1987; Isaacs et al. 2009). Arthropods are also important for decomposition and nutrient cycling, creating nutrient-rich fertile soils for plants, community interactions, and food webs (Wilson 1987; Losey and Vaughan 2006; Seastedt and Crossley 1984).

Climate change, a current global concern, may be due to natural phenomenon or as a result of anthropogenic activity and exerts multitude of threats to all life forms. The Inter governmental Panel on Climate Change (IPCC 2001) reported that increased anthropogenic activities have been responsible for global warming, which is being observed over the last five decades. Mean surface temperature rose by 0.6 ± 0.2 °C during the twentieth century all over the world. It has been predicted to increase by 1.4–5.8 °C from 1990 to 2100. The possible consequences of global warming would not only be raised temperature but more crucially the rate of temperature raise (Root and Schneider 1993). Besides, increase in atmospheric CO_2 concentrations and the precipitation patterns are likely to have profound effects on overall growth of the plants (Reddy et al. 2010). The preindustrial levels of carbon in atmosphere rose from 285 μmol l^{-1}(600 gigatonnes (Gt)) to the current level of 384 μmol l^{-1}(800 Gt), which is predicted to rise to 1000 Gt by the end of 2050 (IPCC 2007). Such an abnormal rise in the levels of atmospheric CO_2 would result in direct and indirect global climate changes. Climate change will also have prolonged influence on arthropod populations. It will affect arthropods (e.g., herbivore insects) directly through effects on growth and development and indirectly by influencing host plants. On the other hand, CO_2 would affect arthropod herbivores indirectly through host plants.

Anthropogenic activities that have been identified are major contributors to global and regional climate change (IPCC 2001). The increasing frequency of climatic disasters is depicted in Fig. 1.1. To understand climate change, it is important to assess the cause by variety of factors from both humans and nature.

1.1.1 Impact of Climate Change on Arthropod Diversity

Climate change triggers considerable changes in flora and faunal diversity, their population distribution and interactions with other biotic and abiotic elements (Figs. 1.2 and 1.3). Distribution of arthropods will be influenced by changes in the cropping patterns due to climate change. Key insect pests such as cereal borers (*Chilo*, *Sesamia*, and *Scirpophaga*) and pod borers (*Helicoverpa*, *Maruca*, and *Spodoptera*) and sucking pests such as aphids and whiteflies may emigrate to cooler regions, leading to higher damage in field and horticultural crops. Global warming will also reduce the effectiveness of host plant resistance, transgenic plants, natural enemies, biopesticides, and synthetic chemicals for pest management (Sharma 2010a, b).

Sabine Nooten et al. (2014) investigated potential effects of warmer climate on insect fauna composition and structure through a multispecies transplant

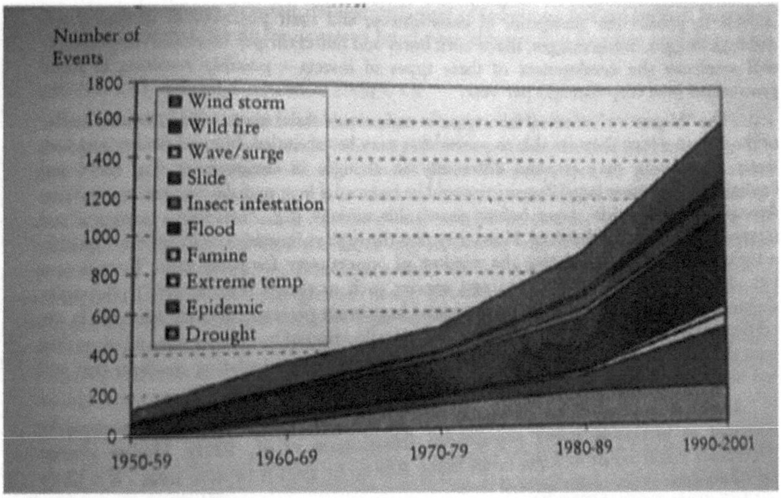

Fig. 1.1 Frequency of climate disasters (IPCC 2001)

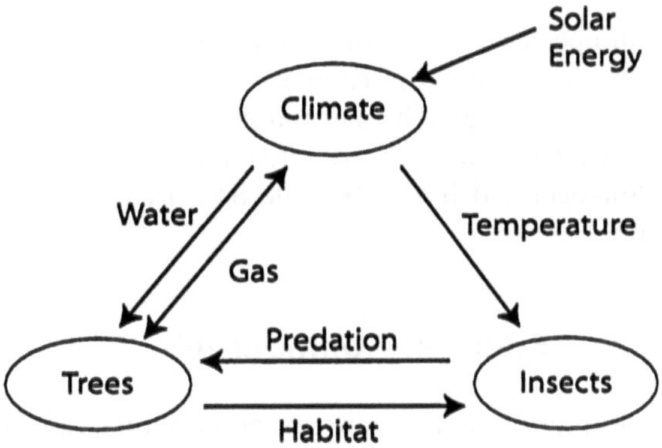

Fig. 1.2 Direct effect of climate change on arthropod pests (Raymond 2014)

experiment. Eight native Australian plant species were transplanted into sites approximately 2.5 °C (mean annual temperature) warmer than their native range. Incidence of insect infestation was monitored for the next 12 months. The morpho-species composition of Coleoptera and Hemiptera diversity differed distinctly between transplants at the control compared to the warmer sites. Community structure of feeding was also differed between the controls and transplants. There were considerable differences in community composition and feeding guild structure, for phytophagus and non-phytophagus, between transplants and congenerics at the warm sites. These results suggest that as the climate warms, considerable changes in the composition of insect communities occur (Nooten et al. 2014).

It is very vital to understand the causes and consequences of climate change on a wide range of biotic diversity (Chapin et al. 2000). Changes in vegetation and soil characteristics resulting from climate change may especially be detrimental for

Fig. 1.3 Indirect effect of climate change on arthropod pests (Raymond 2014)

arthropods because of their limited mobility, and many species require specific host plants (Wilson 1987; Kremen et al. 1993; Burghardt et al. 2008). Soil moisture, temperature, sunlight, and soil physical and chemical properties are also important determinants of the distribution and reproductive success of many arthropod taxa (Antvogel and Bonn 2001; de Souza and de Souza Modena 2004; Lassau et al. 2005).

1.2 Effect on Trophic Level

The trophic cascade concept, proposed by Hairston et al. (1960), has provided a valuable conceptual framework for community ecology. The trophic levels are in equilibrium because of interaction among biotic and abiotic factors in each and every level. Increased biotic potential and decreased environmental resistance usually result in outbreak of a species, whereas increase in environmental resistance and decreased biotic potential result in species extinction. Any alterations in trophic levels owe changes in climate over time.

1.2.1 Effect of Rainfall on Arthropod Diversity

Under the scenario of global warming, worldwide changes in rainfall patterns have been observed in recent decades (Karl and Trenberth 2003). For instance, regions like Australia, New Zealand (Plummer et al. 1999), Southern Brazil, and Eastern Argentina (Collischonn et al. 2001; Berbery et al. 2006) have received significantly more annual rainfall. Rainfall plays key role in keeping the population dynamics of any species in equilibrium by acting upon biotic potential and environmental resistance (Lima et al. 2002; Sala 2006). Consequently, changes in rainfall pattern may affect the entire services of the ecosystem (Martin 2001; Duffy 2003). However, based on currently available information, it is generated from arid and semiarid ecosystems, where water availability plays a significant role on primary productivity (Jaksic 2001; Letnic et al. 2005; Holmgren et al. 2006). In such areas, productivity of pulses increases in rainy season, followed by bottom-up effects of climate that occur with a lag through the food web (Ostfeld and Keesing 2000; Arim et al. 2007; Farias and Jaksic 2007). Nevertheless, in ecosystems where plants did not often experience problem of water deficit (e.g., wetlands), heavy rainfall led to floods which lead to changes in the flora which in turn destroyed terrestrial habitat and the biodiversity therein (Canepuccia et al. 2007, 2008).

A wide range of changes in precipitation are attributed to climate change which in turn will have a major effect on the abundance and diversity of fauna. Extreme climatic events such as drought are likely to decrease diversity as several trophic levels and alteration in composition of arthropod fauna, which subsequently affects the other associated taxa. In the forest ecosystems, chronic stress significantly altered intra- and interspecies composition, and the trees growing under high stress supported one-tenth the number of arthropods compared to trees growing under more favorable conditions (Talbot Trotter et al. 2008). Increasing tree stress was also correlated with an eight- to tenfold decline in arthropod species richness and abundance. Arthropod richness and abundance on individual trees were positively correlated with the tree's radial growth during drought. Stone et al. (2010) suggested that tree ring analysis could be used as a predictor of arthropod diversity.

Climate change aberrations alter the relative abundance of different insect species rapidly, and the species failing to tolerate the stresses gets extinct (Jump and Penuelas 2005; Thomas et al. 2004). The current extinction rates are 100–1000 times greater than the previous extinction rate, and nearly 45–275 species are becoming extinct every day. Effect of prognosticated rainfall in Austria is predicted to reduce abundance of spiders (Araneae, −47 %), cicadas, and leafhoppers (Auchenorrhyncha, −39 %), beetles (Coleoptera, −52 %), ground beetles (Carabidae, −41 %), leaf beetles (Chrysomelidae, −64 %), springtails (Collembola, −58 %), true flies (Diptera, −73 %), and lacewings (Neuroptera, −73 %) but increase the abundance of snails (Gastropoda, +69 %) (Zaller et al. 2014). Extreme rainfall in the Pampas region of Argentina heavily affected the species migration and limited their ability to expand their distribution range. The most affected among the beetle species are the habitat specialists (Xannepuccia et al. 2009). Loss of habitat has been observed at higher trophic levels, indirectly causing disturbances in food chain and food webs of the ecosystem.

1.3 Impact of Temperature on Arthropods

Arthropods are poikilothermic with temperature of their body being approximately the same as that of the environmental temperature. Therefore, temperature is probably the single most important factor influencing overall growth, development, and behavior of the arthropods (Das et al. 2008). The effect of temperature on arthropods overwhelms the effects of other environmental factors (Bale et al. 2002). Assessment made with every 2 °C rise in temperature on insect life cycle resulted in one to five additional life cycles per season (Yamamura and Kiritani 1998). Therefore, climate change resulting in increased temperature impacts crop pests in several ways. Although in certain cases, temperature effects might tend to slow down insect populations. However, warmer temperatures in cooler region result in more diversified and abundant populations of insects (Neumeister 2010).

Millipede, *Polydesmus angustus*, was significantly affected when temperature rose by 3.3 °C. Development was faster, with more number of individuals reaching adult stage resulting in earlier reproduction in spring. The positive effect of temperature helped the millipede females from overcoming post-diapause quiescence in late winter at Paris, and they started laying eggs as soon as temperature conditions improved (David et al. 2003). On the other hand, species that need a period of coolness to end winter diapause, resume development and reproduction, e.g., like the xystodesmid, *Parafontarialaminata armigera* (Fujiyama 1996) affected negatively by warming. Likewise, due to warming acceleration of the exhaustion of metabolic reserves in species that overwinter for long periods without feeding may be affected negatively (David 2009).

The effect of temperature on locomotor activity is ecologically important for arthropods, as it affects predator-prey interactions and activities such as sprinting (Bauwens et al. 1995), flying (Machin et al. 1962), swimming (Turner et al. 1985), and striking (Webb and Shine 1998). Studies by Carlson and Rowe (2009), on influence of temperature on striped bark scorpions, *Centruroides vittatus*, revealed that scorpions had significantly higher sprint speeds (defensive repertoire) at warmer temperatures. Males were significantly faster than females. At lower temperatures sting latency was longer and sting rate lower. Intriguingly, females appear capable of stinging at a higher rate than males. Desiccation enabled the scorpions to run faster than control (hydrated) scorpions, probably due to weight loss.

Increased temperatures could also significantly affect arthropod survival, development, distribution, and population size (Porter et al. 1991). Temperature influences insect physiology and development directly or indirectly through the physiology of host plants. It can exert different effects on insect species depending upon its development "strategy" (Bale et al. 2002). Insects with longer life cycle tend to adopt to temperature changes over the course of their life history, while crop pests with "stop and go" development strategy vis-a-vis temperature develop more rapidly during periods of suitable environment. Increased temperatures hasten the development of insects resulting in more generations and eventually leading to more crop damage (Diku and Mucak 2010).

Reduction in parasitism occurs due to altered host-parasitoid interactions owing to altered temperature (Lewis 1997). So is the gender ratio of thrips. Life cycle of subterranean insects might be more gradually influenced by temperature than terrestrial pests because soil provides an insulating medium against buffer temperature changes more than the air (Bale et al. 2002).

Although lower winter mortality of insects due to increases in winter temperatures occurs, their outbreaks also increases (Harrington et al. 2001). Higher average temperature might create favorable environment for extension of range of some crops to further north resulting in at least some of the insect pests of those crops to shift to the expanded crop areas. Insect species diversity in an area tends to decrease with higher latitude and altitude (Andrew and Hughes 2004). Rising temperatures of temperate climates could result in more insect species attacking more crops. Based on fossil records, it has been concluded that the diversity of insect species and rate of their feeding had increased historically with increasing temperature (Bale et al. 2002).

Increase in temperature may discourage the farmers to cultivate the host crop that can result in decrease in populations of insect pests specific to those crops. Besides, the factors that affect insect pests can also affect their natural enemies as well as entomopathogenic organisms, thereby resulting in increased mortality of insect populations. At higher temperatures, aphids have been shown to be less responsive to alarm pheromone which results in greater predation (Awmack et al. 1997).

Rice was damaged more by bug outbreaks and their poleward geographic emigration due to increase in the mean surface temperature by 1.0 °C over the last 40 years in Japan (Kiritani 2006). The winter mortality of adults of *Nezara viridula* and *Halyomorpha halys* was found reduced by 15 % by each rise of 1 °C. More than 50 species of butterflies showed northward movements and 10 species of previously migrant butterflies were found established on Nansei Islands during 1966–1987. It has also been reported that due to global warming, there has been a decline in abundance of *Plutella xylostella* and increase in natural enemies of *H. armigera* and *Trichoplusia ni* due to increase in number of generations.

Global warming increased infestation by *Heliothis zea* in North America (EPA 1989) and *H. armigera* in North India (Sharma 2010a, b), resulting in increased crop loss. The possible extension of overwintering period in *H. zea* due to climate change is shown in Fig. 1.4. An increase of 2 °C will reduce the number of generation of cherry aphid, *Rhopalosiphum padi* (L.), depending on the variations in mean temperature.

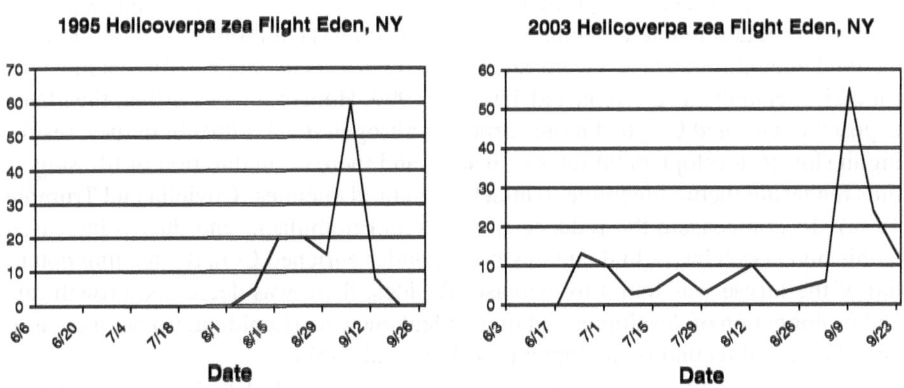

Fig. 1.4 Traps catch data indicating possible overwintering of corn earworm in Western New York (Petzoldt and Seaman 2007)

1.4 Impact of Elevated CO_2 on Arthropod Pests

Carbon dioxide emissions, the main contributor to global warming, are set to rise in 2014 reaching a record high of 40 billion tonnes. Carbon project in the UK has projected a rise of 2.5 % in burning fossil fuels globally. The world oceans have already become increasingly acidic as they absorb atmospheric carbon dioxide. Water reacts with CO_2 to form carbonic acid lowering pH of oceans. In fact, ocean pH has dropped from 8.0 to 7.7 by the twenty-first century. Mussels draw bicarbonate ions from seawater and use proteins in their bodies to make crystals of calcium carbonate to form their two-layer shells. In more acidic water, there are less bicarbonate ions available to make shells. Shells become more brittle in acidic water and break. This means in the wild in the future, mussels become vulnerable to the attack of predators as well as to the effect of ocean forces. It could also have effect on yields of mussels available for human consumption (Times of India 2014).

Understanding alterations in insect feeding behavior by increased CO_2 and O_3 will be important for predicting crop productivity as well as identifying minor pests assuming the status of major pests in future (Baker et al. 2000). Elevated CO_2 have an indirect effect on insect communities as changing host plant phenology creates imbalance in the insect-plant relationships and tritrophic interactions. Both warmer weather and elevated CO_2 alter the bionomics of insect pests (Hulle et al. 2008). Phytophagous insects are directly affected by alteration in the quality of plant parts exposed to elevated CO_2. The carbon/nitrogen (C/N) ratio generally decreases in plants grown under elevated CO_2, and this reduces their nutritive value. In order to compensate, insect tends to feed more leaf. Gypsy moth (*Lymantria dispar*) larvae fed on quaking aspen (*Populus tremuloides*) leaves exposed to high CO_2 performed poorer than larvae that were fed on leaves grown in ambient CO_2 (Cannon 1998). However, the response of larvae to red oak (*Quercus rubra*) leaves illustrated that this phenomenon was host plant dependent. Soybean grown in higher CO_2 conditions suffered more damage from insects like Japanese beetle, potato leafhopper, western corn rootworm, and Mexican bean beetle than at ambient atmosphere (Mahal and Agarwal 2010), perhaps due to increased level of simple sugars in soybean leaves that might have stimulated insect feeding (Hamilton et al. 2005). Leaves having lower nitrogen levels were eaten more by insects to obtain sufficient nitrogen for their metabolism (Coviella and Trumble 1999; Hunter 2001). Further, the plants exposed to elevated CO_2 had more carbon to nitrogen (C/N) ratios in tissues, resulting in slower developmental rates of insects and increase in duration of life stages, thus rendering them vulnerable to attack by natural enemies (Coviella and Trumble 1999). To compensate the reduced nitrogen and tremulacin and due to increased tannin and starch levels in the plants grown under enriched CO_2, the feeding potential of the larvae was found to increase. Besides, there was decreased growth rate and prolongation of developmental time in leaf chewers (Lepidoptera) but increased abundance and fecundity in Homoptera (Rao et al. 2006).

The elevated CO_2 produced more plant size and canopy density with high nutritional quality foliage and microclimate more favorable to pests and diseases. Insect abundance has been shown strongly correlated with plant biomass and height as larger plants increased structural complexity and greater range of resources that herbivores utilize (Lawton 1995).

Plant C/N ratios, condensed tannins, and gossypol were significantly higher, while concentration of the nitrogen was considerably lower in plants grown under elevated CO_2 levels compared to those grown in ambient CO_2 (Gao et al. 2008). Cotton aphid survival significantly increased with high CO_2 levels, whereas there are no significant differences in survival and lifetime fecundity of *Papilio japonica*. However, stage-specific larval durations of the Japanese lady beetle were significantly longer when fed on aphids from enriched CO_2 levels. It was opined that *Aphis gossypii* may assume serious pest status under high CO_2 conditions because of increased chance of survival of aphid and longer development duration of lady beetle. In case of castor semilooper larvae, foliage from elevated CO_2 was consumed more. The relative consumption rate of larvae increased, whereas the efficiency parameters, viz., efficiency of conversion of ingested food (ECI), efficiency of conversion of digested food (ECD), and relative growth rate (RGR), decreased in larvae fed upon enriched CO_2 foliage. The consumption and weight gain of the larvae were negatively and significantly influenced by leaf nitrogen, which was found to be the most important factor affecting consumption and growth of larvae (Rao et al. 2006).

Generally larger biomass, leaf area, and C/N ratios were observed in the host plants grown at enriched CO_2 than those grown at ambient CO_2. Plants with aphids had stunted growth resulting in lower biomass and leaf area than those without aphids (Hughes and Bazzaz 2001). The response of aphid populations to elevated CO_2 was species specific with one species being increased (*M. persicae*), other decreased (*Acyrthosiphon pisum*), and another three being unaffected. The C/N ratio would also be influenced by the N content of the soil and the ability of plants and organisms to fix atmospheric N (Hughes and Bazzaz 2001). Elevated CO_2 concentration increased the C/N ratio of plants, and a high C/N ratio negatively influenced leaf chewers but did not influence phloem feeders. Many species of phytophagous insects would encounter less nutritious host plants that would induce both longer larval developmental periods and higher mortality (Coviella and Trumble 1999).

Leaf-chewing insects consumed more foliage grown under high CO_2 levels than those grown at ambient CO_2. At elevated CO_2 clover shoot weight, lamina weight, and laminar area were greater than at ambient CO_2. Thrips population was not appreciably affected by CO_2, but laminar area scraped by pest feeding was significantly greater at elevated than at ambient CO_2 (Heagle 2003).

Ambient CO_2 levels did not affect above ground net primary productivity for positively dominant plant species in the understory community; C/N ratios of leaf tissue for four of the positively dominant understory taxa, amounts of herbivory, and arthropod abundance and richness across all trophic groups did not differ between ambient and elevated CO_2 plots (Sanders et al. 2004).

Elevated CO_2 levels had narcoleptic and behavioral changes in insects, and projected concentrations of atmospheric CO_2 up to 1000 ppm are unlikely to affect insects directly (Ziska and Runion 2007). Secondary CO_2-induced effects through crop hosts such as digestibility, chemical defenses, and canopy microclimate affect insect fecundity, or aphid populations could increase with increased CO_2 attributed to more fecundity (Awmack et al. 1996) and take more time for settling (Smith 1996). The potato aphid, *Aulacorthum solani*, responded differently to elevate CO_2 on bean (*Vicia faba*) and on tansy (Awmack et al. 1997). When plants were exposed to enriched CO_2, litter quality did not decline as consistently as tissue quality (Norby and Cotrufo 1998). Cotrufo et al. (1998) grew rooted cuttings of ash in ambient and CO_2-enriched atmosphere and fed leaf litter from these plants to *Oniscus asellus* individuals. High concentrations of CO_2 significantly lowered the adult weight of aphid, *Myzus persicae*, compared to aphids fed on plants grown under ambient CO_2 (Himanen et al. (2008).

Development, oviposition, and longevity of whiteflies were closely related to temperature, and there was more delay in developmental period from eggs to adults at low temperatures (Butler et al. 1983; Wang and Tsai 1996). It was also demonstrated that oviposition rate was influenced by the environmental conditions and the quality of the host plants (Byrne and Bellows 1991). However, Coll and Hughes (2008) found that whitefly female oviposition or adult longevity was not affected by elevated CO_2.

A decline in leaf miner population under elevated CO_2 compared to ambient CO_2 was also documented (Stiling and Cornelissen 2007). Further, increase in insect consumption of leaves was detected when caterpillars were exposed to elevated CO_2 (Schadler et al. 2007). Lincoln et al. (1986) observed that larvae of the soybean looper had higher feeding rates on plants grown under elevated carbon dioxide atmospheres than ambient CO_2.

1.5 Effect of Climate Change on Species Interactions

Arthropod-host plant interactions change vis-a-vis to the effects of CO_2 on nutritional properties and secondary metabolites of the host plants as discussed above (Sharma 2010a, b). Elevated levels of CO_2 induce more plant growth but may also persuade more damage from phytophagous insects (Gregory et al. 2009). Higher concentration of CO_2 in atmosphere is expected in the next century; many species of herbivorous insects will encounter less nutritious host plants that may encourage both prolonged life stages and higher mortality (Coviella and Trumble 1999). The positive or negative effect of elevated CO_2 on phytophagous herbivores will not only be species specific but also specific to each insect-plant system. Increased CO_2 enhances more plant growth rates: concurrently the drought stress will probably result in slower plant growth (Coley and Markham 1998). Thus climate change might modify interactions between phytophagous insects and host plants in more than one way (Bale et al. 2002; Sharma et al. 2010).

Increased CO_2 caused a slight decrease in nitrogen-based defenses and a slight increase in carbon-based defenses like tannins (Gore 2006). Acidification of water bodies by carbonic acid due to elevated CO_2 will affect the aquatic floral and faunal diversity. Lower foliar nitrogen content due to CO_2 would cause an increase in food consumption by the herbivores. Additionally severe drought may also increase the damage by insect species such as spotted stem borer, *Chilo partellus* on *Sorghum* (Sharma 2010a, b). Endophytes which play an important role in conferring tolerance to abiotic and biotic stresses in grasses may also undergo a change due to disturbance in the soil brought about by global warming (Newton et al. 2009).

Effects of ambient temperature on pollination success of alpine plants varied depending on the flowering time of individual populations that showed seasonal difference in thermal sensitivity and life cycle of pollinating insects (Kundo and Hirao 2006). Evans et al. (2002) reported that climate change was likely to alter balance between insects and their natural enemies and host interaction and thereby affecting insect pest development too. Thomas et al. (2004) observed that the overall performance of natural enemies altered vis-a-vis changes in herbivore quality and size induced by elevated temperature and CO_2 effects on plants. The encountering of pests by predators and parasitoids decreases due to production of additional plant foliage or altered timing of herbivore life cycles in response to plant phenological changes. Performance of natural enemies in controlling pests decreased if pest emigrate into other regions where the former is usually absent in those areas. In such a scenario, a new community of enemies might then provide some level of control.

Butterflies with strong habitat specificity and lower mobility had reduced distributions and fared worse under changing climate conditions than generalists occurring in the same geographic range (Warren et al. 2001). This led to conclusion that climate change will render habitat domination by mobile generalists. Voigt et al. (2003) reported that species in higher trophic levels were sensitive to climatic change due to the combined indirect effects of climate change on lower trophic levels and greater sensitivity to abiotic stress. Changes in top trophic levels caused trophic cascades and loss of predators, thereby disrupting interactions between lower trophic levels including herbivores and primary producers.

Life history traits such as resource specificity, geographical locations, trophic level, and dispersal ability are potentially good predictors of the magnitude and direction of the response of insect species to climate change (Pelini et al. 2009). The effect of climate change on species distribution and abundance involves not only direct effect on each species individually in an ecosystem but also on species interactions. Rapeseed-mustard is infested by two aphid species, *Lipaphis erysimi* and *M. persicae*, the former being dominant during severe winters and the latter during mild winters (Chander and Phadke 1994).

1.5.1 Effects on Farmers

Farmers are likely to encounter extensive effect on pest management strategies with changes in climate. Entomologists expect that insects may expand their distribution to newer areas and increase reproduction rates and overwintering success. This means that it is likely that farmers will have more diversified insect pests to manage. Based on current comparisons of insecticide usage between more southern states and northern states, this is likely to mean more insecticide use and expenses for northeastern farmers.

Entomologists predict extra generations of important pests in India like brown plant hopper, leafhoppers, aphids, thrips, etc. as a result of higher temperatures, probably necessitating more insecticide applications to maintain low damage thresholds. The number of cultural practices used by the farmers could be altered as per the changes in climate, although it is not clear whether these practices would be helped, hindered, or not affected by the anticipated changes. Crop rotation as an insect management strategy could be less effective with earlier insect immigration or increased overwintering of insects.

1.5.1.1 What Farmers Can Do?

Farmers should understand and remember that climate change is likely to be a gradual process. It is not precisely understood how climate change affects crops, arthropods, diseases, and their relationships. If climate is warmer, it does increase in crop yield and offset losses to pests. Clearly it will be important for farmers to be aware of crop pest dynamics and be flexible in choosing both their management methods and the crops they cultivate. Farmers should closely monitor the occurrence of pests in their fields and should keep a close watch on severity and frequency, and cost of managing pests over time will help in making decisions about whether it will be economical to continue to grow same crop or alternative crop or use a certain pest management technique. The farmers who make the best use of basics of integrated pest management (IPM) such as monitoring, pest forecasting, record keeping, and choosing economically and environmentally sound control measures are be most likely to be successful in dealing with effects of climate change.

1.6 The Paris Agreement

A conglomeration of 195 countries met to adopt measures to mitigate adverse effects of climatic change in Paris on December 12, 2015. These countries agreed to the goal of keeping the increase in the global average temperature to "well below 2 °C and pursuing efforts to limit the temperature increase to 1.5 °C above preindustrial levels." They also pursued a goal of zero net emissions – removing as much greenhouse gases from the atmosphere as is being added to it, by the second half of the century. United Nations Framework Convention on Climate Change (UNFCCC), which convened the Paris talks, felt the necessity of cooperation to tackle climate change. The UNFCCC, which dates from the 1992 Rio Earth Summit, called on nations to act "in accordance with their common but differentiated responsibilities." The new agreement requires a flow of 100 billion a year from developed countries to developing ones by 2020 to realize these goals. The pledges from China and India alone could double the world's wind and solar capacity within 15 years. Limiting warming to 1.5 °C will require reforestation in large landscapes. But to soak up really large amounts of carbon will require technologies capable of storing carbon dioxide deep underground. One idea is to grow plants – those absorb carbon dioxide out of the atmosphere, burning them in power stations, and burying the resulting burned-off carbon. But doing this at the scale required would mean growing fuel over millions of square kilometers. The world does not have that much land to spare. An alternative would be some sort of industrial process. This could be the beginning of the end of the fossil fuel era. Coal firms, however, are clinging to the hope that developing countries will provide enough new demand to persuade investors not to dump their shares. India, for example, generates 71 % of its electricity from coal. Its Intended Nationally Determined Contributions makes no commitment to cut total emissions. Its pledge to install 100 gigawatts (GW) of solar power capacity by 2022, up from just 5 GW now, would require reforms to its electricity sector that stretches credulity. Energy markets may respond only when national governments are seen to be serious at home about the environmental pledges they have made abroad (The Indian Express, December 18, 2015).

1.7 Conclusions

Climate changes have serious positive and negative impacts on arthropod diversity and abundance. Prediction of changes in geographical distribution and population dynamics of arthropods will be useful to mitigate the negative effects of climate change. For instance, pest outbreaks might occur more frequently, particularly during extended periods of drought, followed by erratic rainfall. Likewise, temperature directly affects insects, while CO_2 affects them through host plants. The probable effects of climate change on pests may include expansion of their distribution ranges, change in population growth rates, increased period of activity, alteration in crop pest synchrony and natural enemy-pest interaction, changes in interaction among pest species, and increased risk of invasion by migratory pests. Thus, adverse effects of climate change on the activity and effectiveness of natural enemies and other beneficial will be a major concern in future pest management programs. Therefore, there is a need to have a concerted look at the likely effects of climate change on crop protection and devise appropriate measures to mitigate the effects of climate change on food security on regional scale keeping in mind the soil, water availability, fertility, cropping patterns, and environmental factors, both existing and predicted.

References

Andrew NR, Hughes L (2004) Species diversity and structure of phytophagous beetle assemblages along a latitudinal gradient: predicting the potential impacts of climate change. Ecol Entomol 29:527–542. doi:10.1111/j.0307-6946.2004.00639.x

Antvogel H, Bonn A (2001) Environmental parameters and microspatial distribution of insects: a case study of carabids in an alluvial forest. Ecography 24:470–482

Arim M, Marquet PA, Jaksic FM (2007) On the relationship between productivity and food chain length at different ecological levels. Am Nat 169:62–72

Awmack CS, Harrington R, Leather SR, Lawton JH (1996) The impacts of elevated CO_2 on aphid-plant interactions. Asp Appl Biol 45:317–322

Awmack CS, Harrington R, Leather SR (1997) Host plant effects on the performance of the aphid *Aulacorthum solani* (Kalt.) (Homoptera: Aphididae) at ambient and elevated CO_2. Glob Chang Biol 3:545–549

Baker RHA, Sansford CE, Jarvis CH, Cannon RJC, MacLeod A, Walters KFA (2000) The role of climatic mapping in predicting the potential geographical distribution of non-indigenous pests under current and future climates. Agric Ecosyst Environ 82:57–71

Bale JS, Masters GJ, Hodkinson ID, Awmack C, Bezemer TM, Brown VK, Butterfield J, Buse A, Coulson JC, Farrar J, Good JEG, Harrington R, Hartley S, Jones TH, Lindroth RL, Press MC, Smyrnioudis I, Watt AD, Whittaker JB (2002) Herbivory in global climate change research: direct effects of rising temperature on insect herbivores. Glob Chang Biol 8:1–16

Bauwens D, Garland JT, Castilla AM, Van Damme R (1995) Evolution of sprint speed in lacertid lizards: morphological, physiological, and behavioural covariation. Evolution 49:848–863

Berbery EH, Doyle M, Barros V (2006) Regional precipitation trends Cap. V. In: Barros V, Clarke R, Silva Dias P (eds) Climate change in the la Plata basin: Consejo Nacional de Investigaciones Científicas y Técnicas. Conicet Press, Buenos Aires, pp 61–72

Bond W, Slingsby P (1984) Collapse of an ant-plant mutualism: the Argentine ant (*Iridomyrmex humilis*) and *Myrmecochorous proteaceae*. Ecology 65:1031–1037

Burghardt KT, Tallamy DW, Shriver WG (2008) Impact of native plants on bird and butterfly biodiversity in suburban landscapes. Conserv Biol 23:219–224

Butler GD, Henneberry TJ, Clayton TE (1983) *Bemisia tabaci* (Homoptera, Aleyrodidae) – development, oviposition and longevity in relation to temperature. Ann Entomol Soc Am 76:310–313

Byrne DN, Bellows TS Jr (1991) Whitefly biology. Annu Rev Entomol 36:431–457

Canepuccia AD, Isacch JP, Gagliardini DA, Escalante AH, Iribarne OO (2007) Water bird response to changes in habitat area and diversity generated by rainfall in a SW Atlantic coastal lagoon. Waterbirds 30:541–553

Canepuccia AD, Farias AA, Escalante AH, Iribarne OO, Novaro A, Isacch JP (2008) Differential responses of marsh predators to rainfall-induced habitat loss and subsequent variations in prey availability. Can J Zool 86:407–418

Cannon RJC (1998) The implications of predicted climate change for insect pests in the UK, with emphasis on non-indigenous species. Glob Chang Biol 4:785–796

Carlson BE, Rowe MP (2009) Temperature and desiccation effects on the antipredator behaviour of *Centruroides vittatus* (Scorpiones: Buthidae). J Arachnol 37:321–330

Chander S, Phadke KG (1994) Incidence of mustard aphid, *Lipaphis erysimi* and potato aphid, *Myzus persicae* on rapeseed crop. Annu Agric Res 15(3):385–387

Chapin FS, Zavaleta ES, Eviner VT, Naylor RL, Vitousek PM, Reynolds HL, Hooper DU, Lavorel S, Sala OE, Hobbie SE (2000) Consequences of changing biodiversity. Nature 405:234–242

Coley PD, Markham A (1998) Possible effects of climate change on plant/herbivore interactions in moist tropical forests. Clim Change 39:455–472

Coll M, Hughes L (2008) Effects of elevated CO_2 on an insect omnivore: a test for nutritional effects mediated by host plants and prey. Agr Ecosyst Environ 123:271–279

Collischonn W, Tucci CEM, Clarke RT (2001) Further evidence of changes in the hydrological regime of the River Paraguay: part of a wider phenomenon of climate change? J Hydrol 245:218–238

Cotrufo MF, Phil Ineson Scott Y (1998) Elevated CO_2 reduces the nitrogen concentration of plant tissues. Glob Chang Biol 4(1):43–54

Coviella CE, Trumble JT (1999) Effects of elevated atmospheric carbon dioxide on insect-plant interactions. Conserv Biol 13:700–712

Das DK, Behera KS, Dhandapani A, Trivedi TP, Chona N, Bhandari P (2008) Development of forewarning systems of rice pests for their management. In: Prakash A, Sasmal A, Rao J, Tewari SN, Behera KS, Singh SK, Nandagopal V (eds) Rice pest management. Applied Zoologists Research Association, Cuttack, pp 187–200

David JF (2009) Ecology of millipedes (Diplopoda) in the context of global change. Soil Org 81(3):719–733

David JF, Geoffroy JJ, Celerier ML (2003) First evidence for photoperiodic regulation of the life cycle in a millipede species, *Polydesmus angustus* (Diplopoda, Polydesmidae). J Zool 260:111–116
De Souza ALT, de Souza Modena E (2004) Distribution of spiders on different types of inflorescences in the Brazilian pantanal. J Arachnol 32:345–348
Diku A, Mucak L (2010) Identification and implementation of adaptation response measures to Drini – Mati River Deltas. Report on expected climate change impacts on agriculture & livestock and their influence in the other economic sectors in the DMRD.UNDP Climate Change Program p 34
Duffy JE (2003) Biodiversity loss, trophic skew and ecosystem functioning. Ecol Lett 6:680–687
Environment Protection Agency (EPA) (1989) The potential effects of global climate change on the United States. Vol 2: National Studies. Review of the Report to Congress, US Environmental Protection Agency, Washington DC, p 261
Evans H, Straw N, Watt A (2002) Climate change: implications for insect pests. For Comm Bull 125:99–118
Farias AA, Jaksic FM (2007) Effects of functional constraints and opportunism on the functional structure of a vertebrate predator assemblage. J Anim Ecol 76:246–257
Fujiyama S (1996) Annual thermos period regulating an eight-year life-cycle of a periodical diplopod, *Parafontaria laminata armigera* Verhoeff (Diplopoda). Pedobiologia 40:541–547
Gao F, Zhu SR, Sun YC, Du L, Parajulee M, Kang L, Ge F (2008) Interactive effects of elevated CO_2 and cotton cultivar on tri-trophic interaction of *Gossypium hirsutum*, *Aphis gossypii* and *Propylaea japonica*. Environ Entomol 37(1):29–37
Gore A (2006) An inconvenient truth: the planetary emergency of global warming and what we can do about it. Rodale Publisher, Emmaus
Gregory PJ, Johnson SN, Newton AC, Ingram JSI (2009) Integrating pests and pathogens into the climate change/food security debate. J Exp Bot 60:2827–2838
Hairston NG, Smith FE, Slobodkin LB (1960) Community structure, population control, and competition. Am Nat 94(879):421–425
Hamilton JG, Dermondy O, Aldea M, Zangerl AR, Rogers A, Berenbaum MR, Delucia E (2005) Anthropogenic changes in tropospheric composition increase susceptibility of soybean to insect herbivory. Environ Entomol 34:479–485
Harrington R, Fleming R, Woiwood IP (2001) Climate change impacts on insect management and conservation in temperate regions: can they be predicted? Agric For Entomol 3:233–240
Heagle AS (2003) Influence of elevated carbon dioxide on interactions between *Frankliniella occidentalis* and *Trifolium repens*. Environ Entomol 32(3):421–424
Himanen SJ, Nissinen A, Dong WX, Nerg AM, Stewart CN, Poppy GM, Holopainen JK (2008) Interactions of elevated carbon dioxide and temperature with aphid feeding on transgenic oilseed rape: are *Bacillus thuringiensis* (*Bt*) plants more susceptible to non-target herbivores in future climate? Glob Chang Biol 14:1437–1454
Holmgren M, Stapp P, Dickman CR, Gracia C, Graham S, Gutiérrez JR, Hice C (2006) Extreme climatic events shape arid and semiarid ecosystems. Front Ecol Environ 4:87–95
Hughes L, Bazzaz FA (2001) Effects of elevated CO_2 on five plant – aphid interactions. Entomologia Experimentaliset Appl 99(1):87–96
Hulle M, Bonhomme J, Maurice D, Simon JC (2008) Is the life cycle of high arctic aphids adapted to climate change? Polar Biol 31:1037–1042
Hunter MD (2001) Effects of elevated atmospheric carbon dioxide on insect-plant interactions. Agric For Entomol 3:153–159
IPCC (2001) Climate Change 2001: the scientific basis, report from working group I. Intergovernmental panel on Climate Change, Geneva
IPCC (2007) Summary for policy makers. In: Solomon S, Qin D, Manning M, Chen Z, Marquis M, Averty KB, Tignor M, Miller HL (eds) Climate change 2007: the physical science basis, Contribution of working group I to the IV assessment report of the Intergovernmental panel on Climate Change. Cambridge University Press, Cambridge, pp 1–18

Isaacs R, Tuell J, Fiedler A, Gardiner M, Landis D (2009) Maximizing arthropod-mediated ecosystem services in agricultural landscapes: the role of native plants. Front Ecol Environ 7(4):196–203

Jaksic FM (2001) Ecological effects of El Niño in terrestrial ecosystems of western South America. Ecography 24:241–250

Jump AS, Penuelas J (2005) Running to stand still: adaptation and the response of plants to rapid climate change. Ecol Lett 8:1010–1020

Karl TR, Trenberth KE (2003) Modern global climate change. Science 302:1719–1723

Kiritani K (2006) Predicting impacts of global warming on population dynamics and distribution of arthropods in Japan. Popul Ecol 48:5–12

Kremen C, Colwell RK, Erwin TL, Murphy DD, Noss RF, Sanjayan MA (1993) Terrestrial arthropod assemblages: their use in conservation planning. Conserv Biol 7:796–808

Kundo G, Hirao AS (2006) Habitat-specific responses in the flowering phenology and seed set of alpine plants to climate variation: implications for global-change impacts. Popul Ecol 48:49–58

Lassau SA, Hochuli DF, Cassis G, Reid CAM (2005) Effects of habitat complexity on forest beetle diversity: do functional groups respond consistently? Divers Distrib 11:73–82

Lawton JH (1995) Population dynamic principles. In: Lawton JH, May RM (eds) Extinction rates. Oxford University Press, Oxford, pp 147–163

Letnic M, Tamayo B, Dickman CR (2005) The responses of mammals to La Nina (El Niño Southern Oscillation) associated rainfall, predation, and wildfire in central Australia. J Mammal 86:689–703

Lewis T (1997) Thrips as crop pests. CAB International, University Press, Cambridge, p 740

Lima M, Nils C, Stenseth NC, Jaksic FM (2002) Food web structure and climate effects on the dynamics of small mammals and owls in semi-arid Chile. Ecol Lett 5:273–284

Lincoln DE, Couvet D, Sionet N (1986) Response of an insect herbivore to host plants grown in carbon dioxide enriched atmospheres. Oecologia 69:556–560

Losey JE, Vaughan M (2006) The economic value of ecological services provided by insects. Bioscience 56:311–323

Machin KE, Pringle JWS, Tamasige M (1962) The physiology of insect fibrillar muscle. IV. The effect of temperature on a beetle flight muscle. Proc R Soc Lond Ser B 155:493–499

Mahal MS, Agarwal N (2010) Impact of global climate change on arthropod fauna, pp. 50–56. In: Souvenier: National symposium on perspectives and challenges of integrated pest management for sustainable agriculture. Dr. Y.S Parmer university of Agriculture and forestry, Nauni, Solan, 19–21 Nov 2010

Mandal A, Neenu S (2012) Impact of climate change on soil biodiversity: a review. Agric Rev 33(4):283–292

Martin EM (2001) Abiotic vs. biotic influences on habitat selection of coexisting species: climate change impact? Ecology 82:175–188

Neumeister L (2010) Climate change and crop protection: anything can happen. Published by PAN Asia and the Pacific, Nov 2010, p 41

Newton AC, Begg G, Swanston JS (2009) Deployment of diversity for enhanced crop function. Ann Appl Biol 154:309–322

Nooten SS, Andrew NR, Hughes L (2014) Potential impacts of climate change on insect communities: a transplant experiment. PLoS One 9(1):e85987. doi:10.1371/journal.pone.0085987

Norby RJ, Cotrufo MF (1998) A question of litter quality. Nature 396:17–18

Ostfeld RS, Keesing F (2000) Pulsed resources and community dynamics of consumers in terrestrial ecosystems. Trends Ecol Evol 15:232–237

Pelini SL, Prior KM, Parker DJ, Dzurisin JDK, Lindroth RL, Hellmann JJ (2009) Climate change and temporal and spatial mismatches in insect communities. In: Letcher T (ed) Climate change: observed impacts on planet earth. Elsevier B.V. limited, Dordrecht, pp 215–230

Petzoldt C, Seaman A (2007) Climate change effects on insects and pathogens. climate change and agriculture: promoting practical and profitable responses. http://www.panna.org/sites/default/files/CC%20insects&pests.pdf

Plummer N, Salinger MJ, Nicholls N, Suppiah R, Hennessy KJ, Leighton RM (1999) Changes in climate extremes over the Australian region and New Zealand during the twentieth century. Clim Change 42:183–202

Porter JH, Parry ML, Carter TR (1991) The potential effects of climate change on agricultural insect pests. Agric For Meteorol 57:221–240

Rao MS, Khan MAM, Srinivas K, Vanaja M, Rao GGSN, Ramakrishna YS (2006) Effects of elevated carbon dioxide and temperature on insect-plant interactions-a review. Agric Rev 27(3):200–207

Raymond AC (2014) How climate change may affect arthropod pests http://www.amerinursery.com/growing/how-climate-change-may-affect-arthropod-pests/

Reddy AR, Rasineni GK, Raghavendra AS (2010) The impact of global elevated CO_2 concentration on photosynthesis and plant productivity. Curr Sci 99(10):46–57

Root TL, Schneider SH (1993) Can large-scale climatic models be linked with multi scale ecological studies? Conserv Biol 7:256–270

Sanders NJ, Belote RT, Weltzin JF (2004) Multitrophic effects of elevated atmospheric CO_2 on understory plant and arthropod communities. Environ Entomol 33:1609–1616

Schadler M, Roeder M, Brandal R (2007) Interacting effects of elevated CO_2, nutrient availability and plant species on a generalist invertebrate herbivore. Glob Chang Biol 13:1005–1015

Seastedt TR, Crossley DA Jr (1984) The influence of arthropods on ecosystems. Bioscience 34:157–161

Sharma HC (2010a) Effect of climate change on IPM in grain legumes. In: 5th International Food Legumes Research Conference (IFLRC V), and the 7th European conference on grain legumes (AEP VII), 26–30th Apr 2010, Anatalaya

Sharma HC (2010b) Global warming and climate change: impact on arthropod biodiversity, Pest management and Food security. National symposium on perspective and challenges of integrated pest management for sustainable agriculture, 19–21 Nov 2010, Solan

Sharma HC, Srivastava CP, Durairaj C, Gowda CLL (2010) Pest management in grain legumes and climate change. In: Yadav SS, McNeil DL, Redden R, Patil SA (eds) Climate change and management of cool season grain legume crops. Springer Science + Business Media, Dordrecht, pp 115–140

Smith H (1996) The effects of elevated CO2 on aphids. Antenna 20:109–111

Stiling P, Cornelissen T (2007) How does elevated carbon dioxide (CO_2) affect plant-herbivore interactions? A field experiment and meta-analysis of CO_2-mediated changes on plant chemistry and herbivore performance. Glob Chang Biol 13:1823–1842

Stone AC, Gehring CA, Whitham TG (2010) Drought negatively affects communities on a foundation tree: growth rings predict diversity. Oekologia 164:751–761

Talbot Trotter R III, Cobb NS, Whitham TG (2008) Arthropod community diversity and trophic structure: a comparison between extremes of plant stress. Ecol Entomol 33:1–11

The Indian Express. The Paris agreement on climate change. Dec 2015, p 20

Thomas CD, Cameron A, Green RE, Bakkenes M, Beaumont LJ, Collingham YC, Erasmus BFN, Siqueira MFD, Grainger A, Hannah L (2004) Extinction risk from climate change. Nature 427(6970):145–148

Times of India (2014) Climate change threat to world mussels population. Times of India, 25th Dec, p 10

Turner JS, Tracy CR, Weigler B, Baynes T (1985) Burst swimming of alligators and the effect of temperature. J Herpetol 19:450–458

Voigt W, Perner J, Davis AJ, Eggers T, Schumacher J, Bahram R, Fabian B, Heinrich W, Kohler G, Lichter D, Marstaller R, Sander FW (2003) Impact of climate change on crop-pest and pest-natural enemy interactions. Ecology 84:2444–2453

Wang KH, Tsai JH (1996) Temperature effect on development and reproduction of silver leaf whitefly (Homoptera: Aleyrodidae). Ann Entomol Soc Am 89:375–384

Warren MS, Hill JK, Thomas JA, Asher J, Fox R, Huntley B, Roy DB, Telfer MG, Jeffcoate S, Harding P, Jeffcoate G, Willis SG, Davies GJN, Moss D, Thomas CD (2001) Impact of global warming on butterflies distributions. Nature 41:65–69

Webb JK, Shine R (1998) Thermoregulation by a nocturnal elapid snake (*Hoplocephalus bungaroides*) in southeastern Australia. Physiol Zool 71:680–692

Wilson EO (1987) The little things that run the world (the importance and conservation of invertebrates). Conserv Biol 1:344–346

Xannepuccia AD, Ciccihino A, Escalante A, Navaro A, Isaach JP (2009) Differential responses of marsh arthropods to rainfall-induced habitat loss. Zool Stud 48:173–183

XAVIER SALAI- MARTIN (2006) The world distribution of income: falling poverty and convergence period. Q J Econ CXX(2):351–397

Yamamura K, Kiritani K (1998) A simple method to estimate the potential increase in the number of generations under global warming in temperate zones. Appl Entomol Zool 33:289–298

Zaller JG, Simmer L, Santer N, Tataw JT, Formayer H, Murer E, Hosch J, Baumgarten A (2014) Future rainfall variations reduce abundances of above ground arthropods in model agroecosystems with different soil types. Front Environ Sci 2:1–12

Ziska LS, Runion GR (2007) Future weed, pest and disease problems for plants. In: Newton CD, Carran RA, Edwards GR, Niklaus PA (eds) Agroecosystem in a changing climate. Taylor and Francis, Boca Raton, pp 261–273

Prospecting Arthropod Biomolecules for Medicinal and Therapeutic Use: Recent Breakthroughs

2

Anudita, V. Varunrajan, and B.M. Deepa

Abstract

Arthropods, particularly insects, are endowed with the richest chemical diversity which needs to be harnessed. Arthropods are reservoirs of neurotoxins, peptides, terpenoids, saponins, sugars, etc., that have the potential to cure modern human ailments like HIV, dengue, chikungunya, Ebola virus, Japanese encephalitis, and other syndromes. Basic knowledge of chemical interactions among and between organisms is important to understand the nature of disease treatment, controlling pest and food production, to name a few examples. In order to achieve improved livelihoods, it is necessary to use the chemistry of nature and to use it in a sustainable manner. Isolation, characterization of the molecule, and mass production need a holistic understanding of the chemical and physical properties of the substances arthropods produce. Pharmacognosy is still naive and needs to be further explored.

Keywords

Biomolecules • Entomotherapy • Pharmacognosy • Venom

Anudita (✉)
Department of Zoology, Bangalore University, Bengaluru 560056, Karnataka, India
e-mail: anuchamoli80@gmail.com

V. Varunrajan
Division of Entomology and Nematology, Indian Institute of Horticultural Research,
Hessaraghatta Lake P.O, Bengaluru 560 089, Karnataka, India
e-mail: varundivine@gmail.com

B.M. Deepa
Department of Genetics, Indian Academy Degree College,
Hennur Cross, Bangalore 560043, Karnataka, India
e-mail: deepagene@gmail.com

A.K. Chakravarthy, S. Sridhara (eds.), *Arthropod Diversity and Conservation in the Tropics and Sub-tropics*, DOI 10.1007/978-981-10-1518-2_2

2.1 Introduction

Arthropods use a wide variety of chemical substances for ecological adaptations, and this is the encircling idea of the field of chemical ecology. Progress in the field has advanced rapidly. There are enormous potentials for discovering natural products with numerous medicinal values from members of the phylum Arthropoda. Insects represent 80–90 % of the world's biodiversity, and hence, it is the largest and the most diverse group on earth. So far, 950,000 species of insects have been described, with some authors estimating that there are approximately 4,000,000 insect species on earth. Like all other organisms, insects and related arthropods mainly utilize chemistry to adapt to environments such as defense against predation or infection, communication and socialization, life cycle development, and surviving. The sheer number of insect species, to say nothing of the diversity of niches they inhabit or the huge variety of ways they interact with other species and their environment through chemistry, is awesome.

2.2 Entomotherapy and Ethnoentomology

The use of insects to treat medical ailments is termed as entomotherapy. Larval therapy is a method wherein the damaged and decaying tissues of the wound are allowed to be fed by the maggots of *Lucilia sericata* fly to heal the wound. Such practices were used by traditional medicine followers in countries such as Australia, Asia, and South America (Horobin et al. 2003; Wolff and Hansson 2003). Costa-Neto (2005) summarizes a list of different arthropods, both aquatic and terrestrial forms which include 14 orders spanning across insects, centipedes, and arachnids, which are used by different communities around the globe. The commercial value of the products based on insects was estimated to be worth more than hundred million US dollar. Studying the nature of insects was one of the most important aspects of chemical ecology. The exodus began for the isolation of "bombycol" from silk moth by Butenandt et al. (1959); since then the beneficial aspects of insects were explored for exploitation in the field of medicine, food, and industry. The early record of using bee sting and bee venom for arthritis treatment was described by Hippocrates (460–377 BC), and elsewhere leeches were used to remove "bad blood."

Costa-Neto (2005) describes in his review how people of different cultures use insects. The Arawak community of Guiana allow their babies to be stung by *Paraponera clavata* to facilitate them to walk early. The adults tolerate the sting so that they can gain immunity to the future stings. Marahna, a tribe from Brazil has a different way of using a related species of ant (*Pachycondyla commutata*). The women of the community undergo the stinging, and as a ritual these ants were woven into strings and tied across the forehead and chest by the subject. A South American tribe uses insects in an indirect way. They gather poisonous dart frogs (Dendrobatidae) for the poison secreted in their skin. The frogs produce poison as result of ingesting toxic arthropods which accumulate on their skin. This toxin was used to poison their arrows to kill during hunting. Asian medicine even today has exotic ingredients such as insects. From the early literatures of Chinese Materia

Medica by Read and Shennong Pharmacopoeia (100–200 AD), a useful list of medicinal arthropods is available, viz., bees, wasps, hornets, silkworm, caterpillars, mantises, flies, stink bugs, cicadas, mole crickets, silverfish, cockroaches, dragonflies, locusts, and lice; and extending the list are other arthropods such as spiders and scorpions. A few more to the list was provided in Compendium Materia Medica (1578). Both Chinese and European medicines use the blood of Spanish fly which contains cantharidin, is a potential blistering agent.

2.3 Venom and Neurotoxins

The class insecta embraces numerous poisonous insects. Among the arthropods, more poisonous species are represented by centipedes, chelicerates, and hymenopteran insects. Exclusively venomous arthropod groups are represented among scorpions and spiders (King and Hardy 2013). Therefore, arthropod's venom is a rich source of biologically active molecules in the natural world (Cherniack and Cherniack 2011; Pimenta and de Lima 2005; Schwartz et al. 2012).

The most popular venom to be explored is from spiders. Spiders, although a small group and less diverse among the arthropods, possess diverse neurotoxins to paralyze the prey. Some toxic compounds known from spiders include acylpolyamines, toxic peptides, sulfated nucleosides, polyamine compounds, and polyamine neurotoxins. Estrada et al. (2007) compiled a list of acylpolyamines, toxic peptides from different groups of spiders which have numerous potentias to treat pain and diseases related to malfunctioning of central nervous system such as Alzheimer, Parkinson, and Huntington disease. A study conducted by Schroeder et al. (2008) has identified novel sulfated nucleosides and polyamine compounds using NMR technology from more than 70 different spider species. Common pests such as the violin spider (*Loxosceles reclusa*) and hobo spider (*Tegenaria agrestis*) which were known for their toxicity for many years have such compounds. An array of polyamine neurotoxins from spiders are mostly found on the sensory receptors, which can be explored for drug discovery.

The venom of the spiders has different modes of action. Argiotoxin, isolated from the venom of *Agelenopsis aperta,* was effective against seizures which were audiogenic and N-methyl-D-aspartate (NMDA) induced. The most extensively studied spider for more than two decades, *Phoneutria nigriventer,* has some unique neurotoxins such as PhTx-3 (Tx-3) with an inhibitory effect on Ca^{2+}-dependent glutamate release. On the other hand, the colonial spider *Parawixia bistriata* and Chilean giant pink tarantula *Grammostola spatulata* have different modes of action; the latter uses a combination of toxins to kill its prey. The toxic mixture acts on the ion channels and inhibits their functions (Laurent et al. 2005). Voltage-gated calcium channels like N, P, and Q type were targeted by grammotoxin SIA, a neurotoxin isolated from the venom of spider *G. spatulata*; because of the targeted site of action, there are lots of avenues for potential research. There are few compounds with structural and functional similarity between some neurotoxic compounds found in spiders and other arthropods; one such compound is philanthotoxin which was first discovered from a predatory wasp species *Philanthus triangulum.*

Later such compounds were found in spider venom as well. This suggests a plausible convergent evolutionary relationship between these two neurotoxins in bringing down the prey. Philanthotoxin acts at the nerve junctions as a noncompetitive antagonist of both glutamate receptors and nicotinic acetylcholine receptors.

Many insect-related compounds have been developed as drugs. Structure–activity relationship (SAR) studies have used a chemical approach of producing over 100 analogs of philanthotoxin to improve specificity. Photolabile analogs are used to probe receptor structure and to design more selective antagonists of specific human receptors, with the goal of generating more useful compounds with therapeutic potential. Neurotoxins from invertebrates are diverse, and compounds such as philanthotoxin demonstrate numerous potentials and hence valuable in basic neuroscience research. Insects, like many arachnids, have diverse compounds in their venom such as poneratoxin, pederin, and philanthotoxin, to name a few. Poneratoxin is produced by bullet ant (*Paraponera clavata*). Bite by this ant is most painful, and the protein poneratoxin acts on the voltage-dependent ion channels in the insects. Further exploiting this phenomenon in control of insects was proposed using expression of this protein in insects using viral vectors. The suitability of using this compound and its analogs from other ant species needs to be explored. Pederin and philanthotoxin from rove beetle *Paederus fuscipes* and digger wasp *Philanthus triangulum,* respectively, have high potential as compounds for human medicine (Estrada et al.2007; Schroeder et al. 2008; Schwartz et al. 2012; King and Hardy 2013).

2.4 Honey

Honey has rich tradition in human civilization, be it for therapeutic or cultural reasons. Honey is produced by few genera of hymenopteran insects which include genera *Apis*, *Partamona*, *Tetragonisca*, *Plebeia,* and *Melipona.* Meda et al. (2004) list a few uses of honey in medical ailments such as topical applicator for treatment of burns, cold, sore throats, flu, irritable bowel syndrome, and tuberculosis. Honey in combination with lemon juice is used in the western cultures commonly against sore throats, flu, and colds. Efem et al. (1992) suggested honey for wound dressing owning to its antibacterial effects against both primary and secondary infections. Honeybees in addition to honey collect propolis, a waxy material from flowers which they use to seal the gaps in the comb. In addition to its structural function, it also has antimicrobial, antimutagenic, and antioxidant properties (Wang et al. 2002). Antimicrobial and antioxidant properties of honey and also four phenolic compounds, chrysin, pinocembrin, p-hydroxybenzoic acid, and naringenin, depend on the food source the bees forage.Hence, the antimicrobial and antioxidant properties of the honey differ among species of bees and plants (Basson and Grobler 2008). Honey is used in the treatment of skin disorders in combination with beeswax and olive oil. Royal jelly, a food source for developing queen larva, contains many compounds, of which 24-methylenecholesterol, 10-hydroxy-trans-2-decenoic acid, and 10-hydroxydecanoic acid have gained importance in traditional medicine and are widely used (Figs. 2.1 and 2.2) (Suzuki et al. 2008; Leung et al. 1997; Miyata 2007).

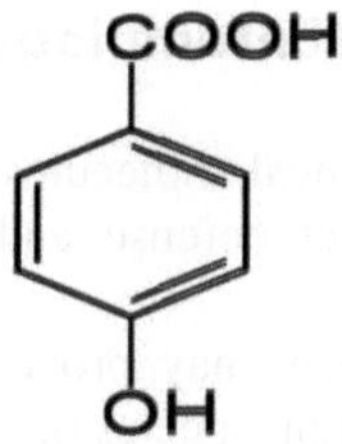

P-hydroxybenzoic acid

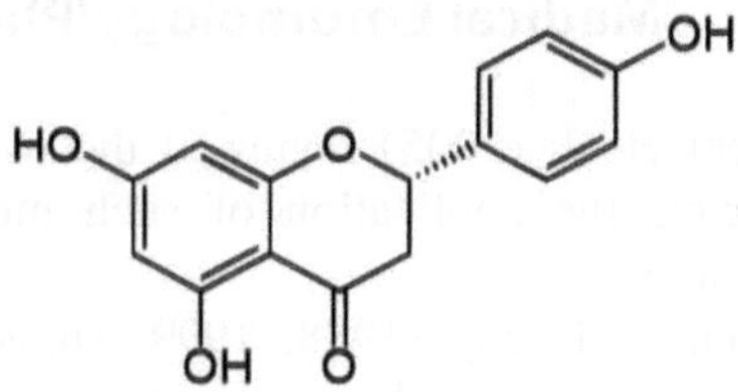

Naringenin

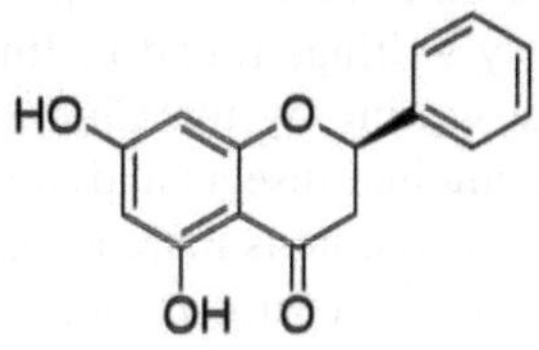

Pinocembrin

Fig. 2.1 Phenolic compounds in honey (Noda et al. 2005)

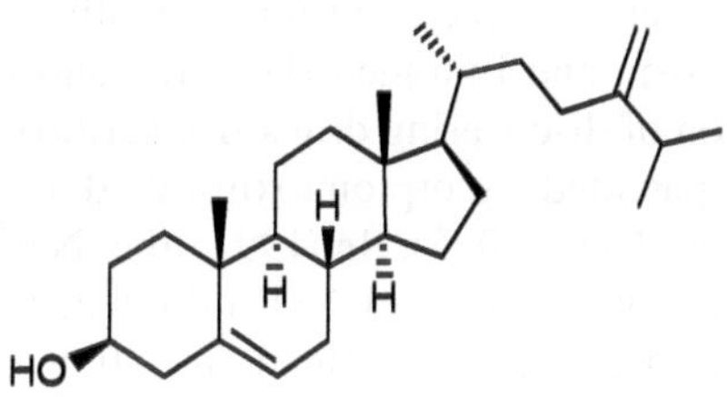

24-methylenecholesterol

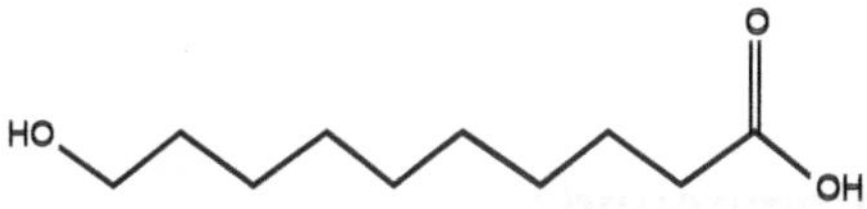

10-hydroxydecanoic acid

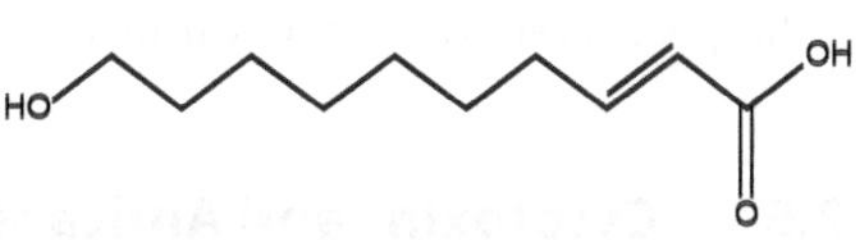

10-hydroxy-Trans-2-decenoic acid

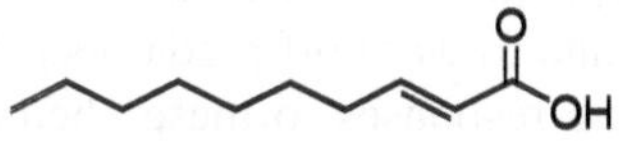

Trans-2-decenoic acid

Fig. 2.2 Royal jelly constituents (Noda et al. 2005)

2.5 Medical Entomology/Pharmaceutical Entomology

Laurent et al. (2005) reviewed the literature on chemical molecules in insects, and gave the application of such molecules in insect defense and medicinal application.

Ramos-Elorduy (1988, 1999) suggested that "insects may prove a valuable source of prototype drugs." For instance, there are different species of cockroaches found in the world; among them particularly species such as *Eupolyphaga sinensis* and *Opisthoplatia orientalis* are mass cultured in China to be used as medicine for "trauma and vulnerary." On the other hand, other species of cockroaches are used for internal feverish chills. Early writings found in Jing Shi Zheng Lei Da Guang Ben Cao, a traditional medical writing by the Chinese during the Song Dynasty (1280 AD), identify a list of medicinal insects and their uses (Namba et al. 1988). The practice of using insects as medicine is increasing nowadays. In the orient, 43 species of insects have been identified for use as medicine. They belong to 16 families and 6 orders.

Only two large-scale surveys of arthropods have been conducted to date: one in the 1960s, which continues today in the group led by Pettit, and a second group led by. Arthropod-based drug discovery by the Trowell group since 2002 at Australia's Commonwealth Scientific and Industrial Research Organization (CSIRO) and spun off into the company Entocosm. Those studies began with over 1,000 arthropod species, of which 80 % were insects that included representatives from 17 different orders of insects. Both Entocosm and Entomed (Dimarcq and Hunneywell 2003) were founded with the mission of discovering drugs from arthropods. The pharmaceutical company Merck has patented a scorpion toxin called margatoxin for use as an immunosuppressant (Costa-Neto 2005). In 1991, the National Biodiversity Institute of Costa Rica (INBio) entered into a US$1 million agreement with Merck to test insect extracts for their efficacy against infections, AIDS, cancer, and inflammatory conditions. New advances in analytical chemistry, new disease models for more efficient and high-throughput drug screening, and epidemics of emerging pathogens and other diseases provide significant rationale for deeper exploration of arthropods as sources of new drugs.

2.6 Cytotoxins and Anticancer Compounds

Arthropods have different groups of chemicals in their venoms and hence are desirable candidates for drug development. These chemicals are often used to paralyze prey by virtue of their toxicity and to ward off predators. Most often even humans are known to have physiological responses to these chemicals (Fox and Serrano 2007). The compounds with cytotoxic properties are often examined for their potential as anticancer chemotherapeutics. Farnesyl lamin, a cytotoxic compound, has the ability to inhibit farnesyl protein transferase; pharmacological properties of this compound were studied even before it was discovered from the natural source, an ant species *Monomorium fieldi* from Australia. Oldfield (1984, 1989) reported that among the 800 terrestrial arthropods studied which include centipedes, millipedes, crustaceans, insects, and spiders, around 4 % of the extracts were found to have at least some anticancer properties. To treat snake bites and headache, centepedes were ground into paste and used.

Maggot therapy, in addition to helping in the healing of bone tissue, also cleans the dead tissues, by the action of endogenous antibiotics and the secreted ammonia, and calcium carbonate provides a quick healing platform. Cantharidin is an example of a cytotoxic compound from blister beetles (Meloidae) (Figs. 2.3a, b and 2.4) and is being explored for potential medical application. Cantharidin, which is terpenoid in nature, is derived from *Lytta vesicatoria* (Spanish fly), *M. cichorii* (Chinese blister beetles), and the bodies of blister beetle, *Mylabris phalerata,* and is used in traditional Chinese and Vietnamese medicine as an ingredient to treat esophageal cancer, hepatoma, and skin diseases, and it has been found to inhibit the cancer cell proliferation in cancer cell lines (Moed et al. 2001; Rauh et al. 2007; Liu and Chen 2009). Blister beetles when disturbed produce droplets of blistering blood poured out from their leg joints so that it rubs off on the attacker. Blister beetles traditionally have been used to remove warts and for cancer treatment, and Greeks used it for enhancing sexual libido. Spanish fly is currently restricted for use only in animal husbandry and wart treatment in the United States. Unlike most chemotherapeutic

Fig. 2.3 Maggots of *Lucilia sericata*

Fig. 2.4 (**a**) Blister beetle (*Epicauta* sp., family Meloidae) deploying its typical defensive secretion of blood (hemolymph) enriched in the blistering agent. (**b**) Cantharidin (Source: Aaron 2010)

agents, cantharidin is found to act on leukemia progenitor and stem cells (Dorn et al. 2009). Commercial products synthesized from cantharid will be available in the supermarkets soon (Moed et al. 2001; Dorn et al. 2009).

Several fatty acids from insects also possess anticancer properties. Yoo et al. (2007) reported that numerous compounds were isolated from the flower beetle *Protaetia brevitarsis*. The scarab grubs were extracted in dichloromethane and further fractionated on silica gel. The fraction which contained anticancer activity was further analyzed by both NMR and GC-MS, which predominantly had two fatty acids: palmitic acid and oleic acid. They have also reported that an authentic standard of palmitic acid induces apoptosis in colon cancer cells. At high concentrations, fatty acids cause cell death by apoptosis or necrosis. Thus palmitic acid induces apoptosis in cancer cell lines, especially in colon cancer (Yoo et al. 2007) (Figs. 2.5, 2.6, and 2.7).

Pettit's (1977) anticancer activity compounds are from insects, namely, Asian rhino beetle (*Trypoxylus* (*Allomyrina*) *dichotoma*), the Asian butterfly *Catopsilia crocale,* and the wasp *Vespula pensylvanica*. Few current examples include the Texas lubber grasshopper *Brachystola magna* and the Asian butterfly *Byasa polyeuctes termessus*. From grasshoppers (*Brachystola magna*) collected from Texas in 1967 and preserved in isopropanol, three antineoplastic agents were isolated. New cytotoxic substances from butterfly extracts were identified by Pettit et al. (1991).

Fig. 2.5 Chinese red-headed centipede (Source: https://www.sciencenews.org/article/centipede-venom-fights-pain)

Fig. 2.6 The forest or wood scorpion (*Cercophonius squama*) (Source: http://www.australiangeographic.com.au/news/2013/12/venom-of-deadly-scorpions-has-medical-use)

Fig. 2.7 Leech, *Hirudo medicinalis* (https://commons.wikimedia.org/wiki/File:Hirudo_medicinalis.jpg)

Ethanolic extracts of *Byasa polyeuctes termessus* were subjected to activity-guided fractionation and were tested against the mouse-derived leukemia model P388. This resulted in "papilistatin," a new cancer cell growth inhibitor with the structure of aristolochic acid. This is present in other Asian butterflies that feed on food plants of the same genus (Aristolochia) as *B. polyeuctes termessus* (Pettit et al. 1991).

The nests of wasps is a store of anticancerous substances. "Paper wasps" (family Vespidae) build their nests with cellulosic plant material collected from different sources. Fujiwara et al. (2008) reported the first anticancer quinine 7, 8-seco-para-ferruginone from nests of the social wasp *Vespa simillima*. 7, 8-Seco-para-ferruginone and precursor ferruginol occur in the bark of Japanese cedar (*Cryptomeria japonica*) plants and is collected by the wasps for nest construction. The polyketide derivative pederin is a compound discovered in the hemolymph of *Paederus fuscipes* (Family: Staphylinidae). It was discovered in 1953 by Pavan and Bo from the extract of over 25 million field-collected beetles and this is another potential anticancercompound. Pederin is the primary component responsible for anticancerous property. It is produced by an internal symbiont present in the beetle, *Pseudomonas aeruginosa*, and it functions as a chemical defense, effective against spiders.

2.7 Antibiotics

Antimicrobial substances are important tools in the innate immune system for many animals such as insects, which are likely to function as a defense against microbial attack and infection. One such substance is the insect antimicrobial peptide. Cecropins were the first antimicrobial peptides to be discovered, whereas defensins are another common class of antimicrobial peptides which are found in a wide variety of organisms from plants to insects to humans, and their biological activity against bacterial and fungal pathogens has been reported. Melittin, a major component in the sting venom of honeybees and cecropins, is effective against Gram-negative bacteria. Genes encoding insect with antimicrobial peptides have even been identified for their potential use in transgenic plant production. An antimicrobial factor, lucifensin from the maggot species *Lucilia sericata*, is used clinically for better and faster wound healing (Opletalová et al. 2012). This peptide was found in different tissues of the insect as well as its secretions and also found to be effective against *Micrococcus luteus*.

Fly larvae (maggots) have been used for centuries to aid wound healing by removing dead tissue quickly while simultaneously protecting against infection. The isolation of antimicrobial substances from fly larvae was evident from several recent studies (Horobin et al. 2003). In addition to the previously mentioned lucifensin, several substances isolated from larvae of *Lucilia sericata* have been shown effective against a range of bacterial pathogens including methicillin-resistant *Staphylococcus aureus* (MRSA). Similarly an antimicrobial lipid, 1-lysophosphatidylethanolamine (C16:1), has been isolated from the common house fly (*Musca domestica*) (Meylaers et al. 2004). Another insect antimicrobial substance is 5-S-GA, which is effective against both Gram-positive (*Micrococcus luteus*) and Gram-negative (*Escherichia coli* and *Staphylococcus aureus*) bacteria (Leem et al. 1996).

Aquatic beetles produce p-hydroxybenzoic acid, probably to prevent microbe attachment to beetles, which would disrupt their ability to repel water. A similar compound, p-hydroxycinnamaldehyde, is an antimicrobial isolated from another Dipteran, the sawfly *Acantholyda parki*. Additionally to this cyclic proline dimer, other dipeptides isolated from flesh flies, *Sarcophaga peregrina* (Leem et al. 1996) and *Neobellieria bullata* (Meylaers et al. 2004) have been shown to possess antimicrobial properties. Apart from Lepidoptera and Diptera, antimicrobial properties have also been discovered from Coleoptera and Hymenoptera. Glandular secretions present in the nests of ants have also been shown to possess antifungal properties. Other social insects, such as social bees, also use antimicrobial substances to construct their nests.

2.8 Antiviral

Melittin and cecropins have been reported for anti-HIV activity. Melittin (from the sting venom of honeybees) and its analogs have also been shown effective against other viruses such as herpes simplex virus (Baghian et al. 1997; Matanic and Castilla 2004) and Junin virus (Matanic and Castilla 2004). Alloferons are another group of insect-derived antiviral peptides, originally discovered in the hemolymph of experimentally infected blowflies (*Calliphora vicina*). These peptides have been shown effective against influenza and herpes simplex virus. A detailed study found that one likely mode of action for alloferons against viruses could be activation of the NF-kB signaling pathway, which is found in nearly all animal cell types and is involved with cell stress, free radicals, and antigens from viruses and bacteria.

Various modern bioassays with silkworm powder have been tested and shown to inhibit absorption of glucose in human intestinal epithelium cells and reduce vasopressin expression in the hypothalamus of diabetic mice. The constituents responsible for the antidiabetic activity of silkworm powder are likely to include sugar-mimetic α-glucosidase-inhibiting alkaloids such as 1-deoxynojirimycin (DNJ) and other sugar-mimetic alkaloids including 1,4-dideoxy-1,4-imino-D-arabinitol (D-AB1) (67) and 1,4-dideoxy-1,4-imino-D-ribitol (Han et al. 2007; Konno et al. 2006).

References

Aaron TD (2010) Insects and their chemical weaponry: new potential for drug discovery. Nat Prod Rep 27:1737–1757

Baghian A, Jaynes J, Enright F, Kousoulas KG (1997) An amphipathic α-helical synthetic peptide analogue of melittin inhibits herpes simplex virus-1 (HSV-1)-induced cell fusion and virus spread. Peptides 18(2):177–183

Basson NJ, Grobler SR (2008) Antimicrobial activity of two South African honeys produced from indigenous *Leucospermum cordifolium* and *Erica* species on selected micro-organisms. BMC Complement Altern Med 8:41–67

Butenandt A, Beckmann R, Stamn D, Hecker E (1959) Uber den Sexuallockstoff des seidenspinners *Bombyx mori*. Reindarstellung und Konstitution. Z Naturforsch 14b:283–284

Cherniack EP, Cherniack EP (2011) Bugs as drugs, part two: worms, leeches, scorpions, snails, ticks, centipedes, and spiders. Altern Med Rev 16:50–58

Costa-Neto EM (2005) Entomotherapy or the medicinal use of insects. J Ethnobiol 25(1):93–114

Dimarcq JL, Hunneyball (2003) Pharma-entomology: when bugs become drugs. Drug Discov Today 8(3):107–110

Dorn DC, Kou CA, Png KJ, Moore MA (2009) The effect of cantharidins on leukemic stem cells. Int J Cancer 124:2186–2199

Efem SE, Udoh KT, Iwara CI (1992) The antimicrobial spectrum of honey and its clinical significance. Infection 20(4):227–229

Estrada G, Villegas E, Corzo (2007) Spider venoms: a rich source of acylpolyamines and peptides as new leads for CNS drugs. Nat Prod Rep 24(1):145–161

Fox JW, Serrano SMT (2007) Approaching the golden age of natural product pharmaceuticals from venom libraries: an overview of toxins and toxin-derivatives currently involved in therapeutic or diagnostic applications. Curr Pharm Des 13(28):2927–2934

Fujiwara Y, Mangetsu M, Yang P, Kofujita H, Suzuki K, Ohfune Y, Shinada T (2008) A quinone isolated from the nest of *Vespa simillima* and its growth inhibitory effect on rat liver cancer cell. Biol Pharm Bull 31(4):722–725

Han J, Inoue S, Isoda H (2007) Effects of silkworm powder on glucose absorption by human intestinal epithelial cell line Caco-2. J Nat Med 1(4):387–390

Horobin AJ, Shakesheff KM, Woodrow S, Pritchard DI (2003) Maggots and wound healing: an investigation of the effects of secretions from *Lucilia sericata* larvae upon interactions between human dermal fibroblasts and extracellular matrix components. Br J Dermatol 148:923–933

King GF, Hardy MC (2013) Spider-venom peptides: structure, pharmacology, and potential for control of insect pests. Annu Rev Entomol 58:475–496

Konno K, Ono H, Nakamura M, Tateishi K, Hirayama C, Tamura Y, Hattori M, Koyama A, Kohno K (2006) Mulberry latex rich in antidiabetic sugar-mimic alkaloids forces dieting on caterpillars. Proc Natl Acad Sci U S A 103(5):1337–1341

Laurent P, Braekman JC, Daloze D (2005) Insect chemical defense. Top Curr Chem 240:167229
Leem JY, Nishimura C, Kurata S, Shimada I, Kobayashi A, Natori S (1996) Purification and characterization of N-b-Alanyl-5-S-glutathionyl-3,4-dihydroxyphenylalanine, a novel antibacterial substance of *Sarcophaga peregrina* (flesh fly). J Biol Chem 271(23):13573–13577
Leung R, Ho A, Chan J, Choy D, Lai CK (1997) Royal jelly consumption and hypersensitivity in the community. Clin Exp Allergy 27:333–336
Liu D, Chen Z (2009) The effects of cantharidin and cantharidin derivates on tumour cells. Anticancer Agents Med Chem 9:392–396
Matanic VCA, Castilla V (2004) Antiviral activity of antimicrobial cationic peptides against Junin virus and herpes simplex virus. Int J Antimicrob Agents 23(4):382–389
Meda A, Lamien CE, Millogo J, Romito M, Nacoulma OG (2004) Therapeutic uses of honey and honey- bee larvae in central Burkina Faso. J Ethnopharmacol 95:103–107
Meylaers K, Clynen E, Daloze D, DeLoof A, Schoofs L (2004) Identification of 1-lysophosphatidylethanolamine (C(16:1)) as an antimicrobial compound in the housefly, Musca domestica. Insect Biochem Mol Biol 34(1):43–49
Miyata T (2007) Pharmacological basis of traditional medicines and health supplements as curatives. J Pharmacol Sci 103:127–131
Moed L, Shwayder TA, Chang MW (2001) Cantharidin revisited: a blistering defense of an ancient medicine. Arch Dermatol 137:1357–1360
Namba T, Ma YH, Inagaki K (1988) Insect-derived crude drugs in the Chinese Song dynasty. J Ethnopharmacol 24(2–3):247–285
Noda N, Umebayashi K, Nakatani T, Miyahara K, Ishiyama K (2005) Isolation and characterization of some hydroxy fatty and phosphoric acid esters of 10-hydroxy-2-decenoic acid from the royal jelly of honeybees (*Apis mellifera*). Lipids 40(8):833–838
Oldfield ML (1984) The value of conserving genetic resources. U.S Department of the Interior National Park Service, Washington, DC
Oldfield ML (1989) The value of conserving genetic resources. Sinauer Associates, Sunderland, pp 379
Opletalová K, Blaizot X, Mourgeon B (2012) Maggot therapy for wound debridement: a randomized multicenter trial. Arch Dermatol 148(4):432–438
Pavan M, Bo G (1953) Pederin, toxic principles obtained in the crystalline state from the beetle *Paederus fuscipes* Curt. Physiol Comp Oecologia 3:307–312
Pettit GR (1977) Biosynthetic products for cancer chemotherapy. Springer US, Boston. ISBN 978-1-4684-7232-5
Pettit GR, Sengupta D, Herald CL, Sharkey NA, Blumberg PM (1991) Synthetic conversion of bryostatin 2 to bryostatin 1 and related bryopyrans. Can J Chem 69:856–860
Pimenta AMC, de Lima ME (2005) Small peptides, big world: biotechnological potential in neglected bioactive peptides from arthropod venoms. J Pept Sci 11:670–676
Ramos-Elorduy C, Moreno Pino JM (1988) The utilization of insects in the empirical medicine of ancient. Mex J Ethnobiol 8(2):195–202
Ramos-Elorduy J, Motte-Florae E, Pino JM, Andary C (1999) Les insectesutilisesen medicine traditionnelle au Mexique: Perspectives. In Healing: yesterday today and tomorrow (CD ROM). Erga Ediziones, Geneva, pp 271–290
Rauh R, Kahl S, Boechzelt H, Bauer R, Kaina B, Efferth T (2007) Molecular biology of cantharidin in cancer cells. Chin Med 2:8–14
Schroeder FC, Taggi AE, Gronquist M, Malik RU, Grant JB, Eisner T, Meinwald J (2008) NMR-spectroscopic screening of spider venom reveals sulfated nucleosides as major components for the brown recluse and related species. Proc Natl Acad Sci U S A 105(38):14283–14287
Schwartz EF, Mourão CB, Moreira KG, Camargos TS, Mortari MR (2012) Arthropod venoms: a vast arsenal of insecticidal neuropeptides. Biopolymers 98:385–405

Suzuki KM, Isohama Y, Maruyama H, Yamada Y, Narita Y, Ohta S, Araki Y, Miyata T, Mishima S (2008) Estrogenic activities of fatty acids and a sterol isolated from royal jelly. Evid Based Complement Alternat Med 5:295–302
Wang XH, Andrae L, Engeseth NJ (2002) Antimutagenic effect of various honeys and sugars against Trp-p-1. J Agric Food Chem 50:6923–6928
Wolff H, Hansson C (2003) Larval therapy – an effective method of ulcer debridement. Clin Exp Dermatol 28:134–137
Yoo YC, Shin BH, Hong JH, Lee, Chee HY, Song KS, Lee KB (2007) Isolation of fatty acids with anticancer activity from *Protaetia brevitarsis* larva. Arch Pharm Res 30(3):361–365

Conservation of Arthropod Parasites: Restoring Crucial Ecological Linkages

3

M. Jayashankar, S.R. Amulya Prasad, and Subhash B. Kandakoor

Abstract

Parasitic biodiversity is focussed as a key component of conservation targets based on their ecological roles in the present review. Arthropods adopt parasitism as one of the common strategies for survival. However, some parasites are threatened by not only direct factors such as environmental conditions but also by indirect ones such as the effect on their hosts/prey. Conservation of parasites would help to sustain evenness in arthropod communities. Diptera, Hymenoptera, and Siphonaptera are orders of insects, mites, and ticks implicated as vectors in transmission of diseases in human populations and agricultural ecosystem of tropics and subtropics. Recently, outbreak of Zika viral disease has been reported in over 12 countries. This comprehensive review will be of value to scientists, students, and policy makers for biodiversity management.

Keywords

Conservation • Ecology • Parasites • Vector • Zika virus

M. Jayashankar (✉)
Division of Entomology, Indian Institute of Horticultural Research, Bengaluru 560 089, India
e-mail: jay81zoology@gmail.com

S.R.A. Prasad
Department of Zoology, Bangalore University, Bengaluru 560 056, Karnataka, India
e-mail: rajnarayan.ammu@gmail.com

S.B. Kandakoor
AICRP on Potato, University of Agricultural Sciences, Dharwad 580 005, Karnataka, India

A.K. Chakravarthy, S. Sridhara (eds.), *Arthropod Diversity and Conservation in the Tropics and Sub-tropics*, DOI 10.1007/978-981-10-1518-2_3

3.1 Introduction

Arthropods are evolved from wormlike coelomic animals, much earlier to most of other animals. There are parasites that exploit the host and some serve as hosts for organisms. Arthropods are generally ectoparasitic on, or in the skin of vertebrate hosts. Many arthropod species are hematophagous (rely on the blood), but others can be histophagous (rely on tissues). Parasitic infestations are transmitted from one host to another either by direct contact or by free-living larvae (Marquardt et al. 2000).

Mites and ticks are arthropods belonging to subclass Acari and class Arachnida. Because of the small size, they largely go unnoticed. A huge number of mites live as parasites on plants or animals and some on molds. To date nearly 30,000 mite species have been identified. One family of mites, Pyroglyphidae or west mites, live primarily in the nests of birds and other animals. These mites are primarily, largely parasitic and consume the blood, skin, and keratin of hosts (Gómez et al. 2012).

Ticks parasitise mainly terrestrial vertebrates living on land. Parasite infections are transmitted by bacteria, virus, rickettsia, protozoa, and other pathogens. They feed mainly on the blood, and their mouthparts are armed with small backward-facing teeth to aid in attachment. Two major tick families are identified based on morphological features: the Ixodidae (hard ticks with a tough cuticle and a large anterodorsal scutum) with some 650 species that attack reptiles, birds, and mammals and the Argasidae (soft ticks) with 160 species that infest and mainly attack higher vertebrates.

Fleas are wingless insects, have flattened body with hind limbs swollen, so that they can jump. The larvae are not parasitic but feed on debris associated mainly with bedding, den, or nest material, whereas the adult stages are parasitic and feed on host blood. There are about 2500 species of flea; most are parasitic on mammals (especially rodents) and some on birds. Flies and mosquitoes are winged insects with two pairs of wings attached to the thorax and a well-developed head with sensory and feeding organs. Species of fleas vary in their feeding habits, both as adults (parasitic or free living) and larvae (parasitic or free living). There are over 120,000 species belonging to 140 families (Rendell and Verbeek 1996).

3.2 Why Study Parasites?

Parasites have a huge impact on the health and well-being of the human population, particularly in the developing world. Parasites exploit resources of host to feed and reproduce. This lowers host fitness to considerable extent. Parasites can negatively impact human and animal health, food production, economic trade, and biodiversity. They can be difficult to study and have been regarded as having little or no influence on ecosystem functioning. Not surprisingly, parasitic biodiversity has not been considered as an important group by biologists and conservationists (Gompper and Williams 1998; Dunn et al. 2009; Griffith 2012). The stated goal of the field of conservation biology is to maintain biodiversity, including the evolutionary processes that drive and sustain it (Meffe et al. 2006). Yet to ignore the conservation of parasites is to ignore the conservation status of the majority of life on Earth, as parasitism represents the most common consumer strategy on the planet (Poulin and Morand 2000; Dobson et al. 2008). It also means neglecting a fundamental biological relationship, as infection is fundamental to the ecological and evolutionary drivers of biological diversity and ecosystem organization (Marcogliese 2004). Recent research has shown that basic assumptions about parasites, their ubiquity, and relevance need to be reexamined. Parasitism may be the most widespread trophic strategy in the animal kingdom. Also, documented literature reveals that parasites are diverse, have key roles in ecological and evolutionary processes, and that infection/parasitism may paradoxically result in ecosystem services of direct human relevance. Of the 42 broadly recognized classes, 31 are predominantly parasitic and most others have multiple parasitic groups (Poulin and Morand 2000; de Meeûs and Renaud 2002). Nest-dwelling ectoparasites feeding on the blood of nestlings and adults constitute an important selective force affecting avian life history evolution (Moller 1997). These ectoparasites may reduce the reproductive success of hosts by reducing nestling growth (Merino and Potti 1995) or inducing nest desertion (Oppliger et al. 1994) and may even affect parental health (Tomás et al. 2007). For hole-nesters, fleas (Siphonaptera), flies (Diptera), and mites (Acarina) constitute the most important class of nest-dwelling ectoparasites (Merino and Potti 1995; Rendell and Verbeek 1996). Ectoparasitic faunas in bird nests vary depending on host species, even in conditions of sympatry (Bennett and Whitworth 1992; Bauchau 1998). These differences may depend on interspecific variation in host defenses based on parental behavior, nestling immunity, or nest properties. Animal parasites and vectors help veterinarians, fishery biologists, induce diseases of farmed animals (including cultured fish), wild caught animals used for food or at risk from disease, and forensic scientists. Parasites are ubiquitous in the lives of animalism in the wild and constitute a major element of floral and faunal elements (Price 1980).

3.2.1 Major Ecological and Evolutionary Effects of Parasites

Parasites as vectors of disease causing pathogens can affect species distribution and population (Nichols and Gómez 2011). By exploiting resources from hosts, parasites can change animals' energy needs that affect growth, reproduction, competition, and survival. Some parasites even directly alter their hosts' behavior and force them to do their bidding. Parasites also drive adaptation and evolution as they and their hosts engage in "evolutionary arms races." Current investigations of food webs indicate that ~75 % of the links in food webs involve a parasitic effect; these links are keys of host population and for reducing the effect of toxicant. This means that extinctions of parasite may have unpredictable effects that impact the population of a wide range of living organisms. The parasitic fauna of any host species reflects its interaction with the host's feeding niche, latitudinal range, and social system.

3.2.2 Coexistence and Evolution of Parasites and Humans

Using a Bayesian coalescent modeling approach, it is estimated that clothing lice, *Pediculus humanus humanus* (Fig. 3.1), have evolved from head louse, *Pediculus humanus capitis*, ancestors at least by 83,000 and may be 170,000 years ago (Toups et al. 2011). Because clothing lice descended from head louse ancestors once humans took to clothing. The emergence of clothing lice provides data on proximate dates of the period when man started using clothes. Sucking lice (Phthiraptera: Anoplura) are obligate, permanent ectoparasites of eutherian mammals, parasitizing 12 of the 29 recognized mammalian orders and 20 % of all mammalian species. These bloodsucking insects are morphologically adapted for life on mammals: they are dorsoventrally flattened, are wingless with tibiotarsal claws for clinging to host hair, and have piercing mouthparts. In the fore literature, more than 540 described species of Anoplura have been documented. All modern mammal orders are supposed to have diverged by 75 Ma, giving scope suitable habitat for the establishment and speciation in sucking lice.

Although there is concern among experts on timing of diversification events in the association between anoplurans and mammals, there is considerable disagreement between the host and parasite phylogenies. These contrasting views may be due to a complex history of host switching and extinction events that occurred throughout the evolutionary association between sucking lice and mammalian hosts. The second example is related to the Inca Empire of western South America. Livestock levels over time are being studied because ancient Peruvian cultures relied on domesticated animals (mainly llamas) for food, wool, fuel, fertilizer, and transport. Changes in livestock densities affect the amount of dung and, therefore, nutrients deposited. Variation in habitat enrichment influences the number and species of mites present. Mites, e.g., *Hydrozetes* sp. (Figs. 3.2 and 3.3), are being

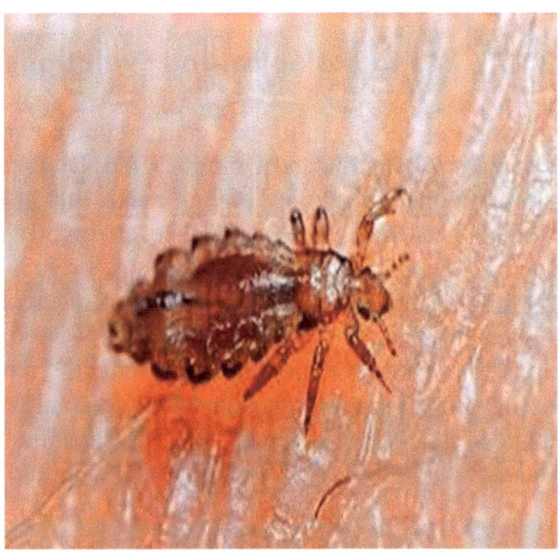

Fig. 3.1 *Pediculus humanus humanus*

Fig. 3.2 *Microterys flavus* (http://boutique.crisop.fr/microterys-flavus-10)

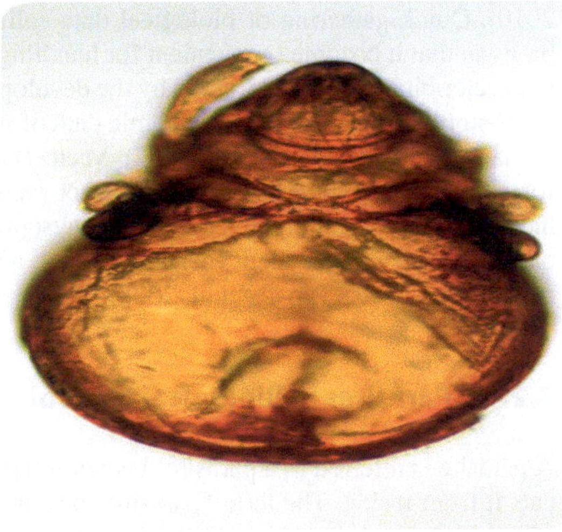

Fig. 3.3 *Hydrozetes* sp. (http://www.nhm.ac.uk/research-curation/life-sciences/terrestrial-invertebrates/research/mite-research/archaeological-indicators/index.html)

extracted from 1200 years of sediment samples from the in-filled lake basin of Marcacocha in Peru to assess mites as indicators of changing livestock densities and therefore agricultural activity, leading up to the end of the Inca Empire. The site is an area of pasture on an ancient trans-Andean trading route where llama caravans would have grazed and watered. Identifying the mites will shed light on their habitat and food preferences (http://www.nhm.ac.uk/research-curation/life-sciences/terrestrial invertebrates/research/mite-research/archaeological-indicators/index.html).

3.2.3 Taxonomic Revision

Parasites should not be excluded while making decisions on conservation (Gomez et al. 2012); rather they need to be included, because parasitism is perhaps the most prevalent life scape on planet Earth. Of all the animal phyla, 9 are entirely parasitic, 22 are predominantly parasitic, and the rest have parasitic groups scattered throughout them. There are an estimated 75,000–300,000 species that parasitize vertebrates, excepting the ones that choose spineless hosts. According to Gómez et al. (2012), numerically parasites may "outnumber free-living biodiversity by as much as 50 percent." Comparing the phylogenetics of the parasite, host, and vector is leading to better understanding of coevolution, co-speciation, and interactions between host-parasite relationships.

Conventional taxonomic descriptions, revisionary systematics, and faunal surveys are vital to the growth and development of collections and the preparation of keys and to contribute to parasite diversity. A revision in taxonomic procedure is in progress that will make taxonomy an even more reliable source of information for ecologists, cybertaxonomy (cyber-enabled taxonomy), which is able to produce results faster and better than ever before (Wheeler 2008, Wheeler and Valdecasa 2010). Quick gathering of biological data calls for development of suitable databases so that it becomes convenient for handling, retrieving, and analyzing the data. On vector biology, it is addressed by the development of VectorBase, a database that gives storage and analysis for genomic data of insect and other arthropod vectors of human diseases (Lawson et al. 2007). VectorBase contains information on genome for *Anopheles gambiae* and for other insect vectors. In addition, genomic sequences have initiated a community annotation system, a modern microarray expression data repository, controlled vocabularies for addressing anatomy, and insecticide resistance aspects (Topalis et al. 2008).

3.2.4 Parasites in Biological Control

A parasite is referred as a parasitoid when it fails to reproduce in its host but merely eats it from inside. The term "parasite" means insects that parasitize other insects. Many biological control workers, however, point to significant biological differences between such insects and other types of parasitic arthropods. "True" parasites are usually much smaller than their host, have a shorter life cycle than their host, and usually do not kill their host; examples include tapeworms and ticks. Recognizing this distinction, Reuter in 1913 coined the term "parasitoid" for insects that parasitize other insects. Some biological control workers prefer to use "parasite," whereas others use "parasitoid."

Many of the agricultural, horticultural, and forestry pests are attacked by one or more parasitoid species. It is estimated that, on a worldwide basis, there are about 68,000 species of parasitoids that are described. This forms about 10 % of all known insect species. Entomologists believe that about 10 % of all insects living on Earth are known to science and as many as 800,000 species of parasitoids actually live.

Most parasitoids belong to two major groups of insects: the orders Diptera (the true flies and their relatives), with about 15,000 known species of parasitoids, and within Diptera, the Tachinidae which is specially important for the biological control of agricultural, horticultural, and forestry pests. But the order Hymenoptera comprises the largest group of parasitoids and forms the most important group of natural enemies used in biological control programs. However, the parasitoids exist in several other groups as well, and an estimated 3000 species form the other major group of insects. The families of beetles that contain parasitoids are the carabids, one of the largest families of beetles. Many predators are important in biological control. About 500 species of ground beetles are parasitoids, primarily of soil-dwelling insects. These are ectoparasites, anchoring to the outside of their primary host. Several of these are in the genus *Lebia* that embraces parasites of Chrysomelidae. *Lebia grandis* parasitizes pupae of Colorado potato beetle. Many species of rove beetles (family Staphylinidae) are predators, and many of these are crucial for biological control; about 500 species are known to be parasitoids. The cedar beetles comprise a small family, with only about a half dozen species. Members of the genus *Sandalus* parasitize the nymphs of cicadas that are the soil borne, but are rare. The larvae of select blister beetles (family Meloidae) attack ground-nesting bees; some others attack the egg cases of grasshoppers. As egg cases carry several eggs, entomologists consider this group of blister beetles to be predators. Because each larva consumes only an egg case, others consider these to be egg-case parasitoids. Most of the parasitoids attack the pupae of flies, such as scavenging flies (houseflies, blowflies, flesh flies, etc.).

One species, *Aleochara bilineata*, is a common parasite of cabbage maggot. Family Ripiphoridae is a small family of about 400 species that parasitizes larvae of bees and wasps. Limited species of *rhipidius* are parasites of cockroaches. The wingless, larviform female of the parasite lays eggs on the ground in areas frequented by the hosts. The first instar attaches itself to a passing cockroach and enters the body, completing its larval life as an endoparasite. The order Strepsiptera is a small group of uncommon parasites closely related to the beetles. Bees, wasps, leafhoppers, and plant hoppers are the most common hosts, but a few species attack grasshoppers or bristletails. The order Neuroptera, which are predators, consists of the lacewings and ant lions. Adult mantidflies (family Mantispidae) are predators and have the front legs strongly modified for catching and holding the prey, similar to the modifications of legs found in praying mantids. The larvae are all parasitic within the egg sacs of spiders, which are ground dwelling.

3.2.5 In Agri-Horticultural Ecosystems

The egg parasitoids of the genus *Trichogramma* (Fig. 3.4) are the most frequently produced natural enemies in the world (Li 1994). Commercially produced species of *Trichogramma* parasitize the eggs of many lepidopteran pest species and have been used as biological control agents against several agricultural pests. The short generation time of *Trichogramma* wasps, and the fact that they can be reared on

Fig. 3.4 *Trichogramma* sp. (http://www.ecured.cu/index.php/Trichogramma_spp.)

factitious hosts, allows these wasps to be produced rapidly and affordably (Li 1994; Smith 1996). Despite the user-friendliness of *Trichogramma* species, there have been at least as many failures of biological control using *Trichogramma* as there have been successes (Smith 1996), and the quality of mass-reared wasps is one potential factor of variability in the success of biological control programs (Bigler 1989; Cerutti and Bigler 1995; Kuhlamann and Mills 1999; Losey and Calvin 1995; O'Neil et al. 1998). One parameter of natural enemy quality important to biological control programs is whether fitness of commercially produced wasps is sustained for long periods. Releasing the natural enemies to establish the population before pest densities have begun to increase has an advantage. The natural enemy reproduces on the target pest or an alternate host, and its population increases to levels sufficient to control the target pest later in the season. In China, inoculative releases of *Trichogramma* produce wasps which later in the season move into adjacent fields to suppress pests of cotton (Li 1994). Conservation as a biological control method includes crop management measures that encourage natural enemies and beneficials and increase their impact on pests. Interplanting ryegrass in seed corn production fields lowered soil temperatures which otherwise would be lethal to released *Trichogramma* (Orr et al. 1997). *Trichogramma* species commonly parasitize bollworms (corn earworm) on corn, sorghum, and other crop pests. These crops serve as an important source of adult wasps that disperse to cotton (Oatman 1966).

Parasitic wasps form important natural enemies of aphids – categorized under Aphelinidae and Braconidae. *Aphelinus flavipes* on cotton aphid is similar to *A. semiflavus. A. semiflavus* parasitizes several aphid species, including the green peach aphid *A. colemani.* This braconid wasp is cosmopolitan and occurs in many parts of the world. It reproduces well on cotton aphid, green peach aphid, and other species. This cosmopolitan species parasitizes numerous aphids in diverse cultivated ecosystems. In greenhouse, they parasitize larger aphids than does *A. colemani*, particularly potato aphid and foxglove aphid, *A. matricariae.*

Most of the parasites of leaf miners parasitize the larvae and pupae. This European braconid wasp, *Dacnusa sibirica*, is a solitary endoparasite of all instars of *Liriomyza bryoniae* and *L. trifolii*, and the chrysanthemum leaf miner. *Diglyphus isaea* parasitizes *L. trifolii*, *L. bryoniae*, and chrysanthemum leaf miner. *D. pulchripes* parasitizes the vegetable leaf miner. The eucoilid wasp, *Ganaspidium utilis* native to subtropical areas of North America, is a larval-pupal parasite of *L. trifolii*. The American braconid wasp, *Opius dimidiatus*, is one of the most abundant larval-pupal parasites of the vegetable leaf miner and *L. trifolii*. The braconid wasp *Opius dissitus* is a parasite of the vegetable leaf miner in Florida which also parasitizes *L.trifolii*. *Opius pallipes*, European braconid wasp, is a solitary endoparasite of all instars of *L. bryoniae* and chrysanthemum leaf miner larvae.

The family Encyrtidae includes parasitic wasps that are important natural enemies of soft scales. The cosmopolitan wasp, *Coccophagus lycimnia*, parasitizes over 47 species of soft and armored scales and has proven effective against brown soft scales in citrus and ornamental plants. *Metaphycus helvolus*, a small encyrtid wasp from South Africa, attacks young nymphal stages of several species of soft scales; *M. helvolus* readily attacks black and hemispherical scales, as well as brown soft scale; *Microterys flavus* (Fig. 3.2) is another internal parasite of brown soft scale. This parasite is also effective against other soft scales and commercially available for release in citrus orchards.

Encyrtus infelix and *Encyrtus lecaniorum* have been successfully utilized alone and in combination with *Metaphycus helvolus* for suppressing scales in French greenhouses. *E. lecaniorum* parasitized hemispherical and soft brown scales more effectively than *M. helvolus* or *C. lycimnia*. *E. lecaniorum* mimics ants that tend the scales, thereby preventing the other parasites from attacking the scales. The ectoparasitic wasp *Aphytis melinus* (Figs. 3.5, 3.6, 3.7, and 3.8) from India and Pakistan parasitizes select species of armored scales, including California red scale, ivy scale, San Jose scale, and oleander scale.

The Mediterranean wasp *Anagyrus pseudococci* (Fig. 3.9) looks like *Leptomastix dactylopii*, but it attacks half-grown to full-grown citrus mealybug nymphs. *Leptomastidea abnormis* attacks only young, second-instar citrus mealybugs and develops as a solitary endoparasitoid. The Brazilian wasp, *L. dactylopii*, is the most frequently used parasite for suppressing citrus mealybug. *Pauridia* (=*Hungariella*) *peregrine* is a solitary endoparasitoid of the long-tailed mealybug attacking the first

Fig. 3.5 *Aphelinus semiflavus* (http://ponent.atspace.org/fauna/ins/fam/aphelinidae/aphelinus_aph.htm)

Fig. 3.6 *Aphidius colemani* (http://www.arbico-organics.com/category/pest-solver-guide-aphids)

Fig. 3.7 *Aphidius ervi* (http://www.arbico-organics.com/category/pest-solver-guide-aphids)

and second instars. *H. peregrine* has been used successfully in cultivated ecosystems in subtropical areas and in greenhouses on ferns.

Select species of trichogrammatid and mymarid wasps parasitize eggs of thrips. Most of the parasitic wasps that have been found to attack thrips larvae are tropical or subtropical and belong to *Eulophidae. Ceranisus* (=*Thripoctenus*) spp. are eulophid wasps that develop inside thrips larvae. After female wasps lay single eggs in young thrips larvae, the thrips continue to appear and behave normally while the wasps develop inside them. *Ceranisus* (=*Thripoctenus*) spp. are eulophid wasps that develop inside thrips larvae. *C. menes* is a solitary endoparasitoid that attacks the first instar of western flower thrips. It occurs worldwide and has been found parasitizing western flower thrips on alfalfa and roses in California. *Thripobius semiluteus* is a parthenogenic wasp that was introduced in 1987 from Brazil and Australia as a possible control agent for greenhouse thrips in avocado orchards in southern California.

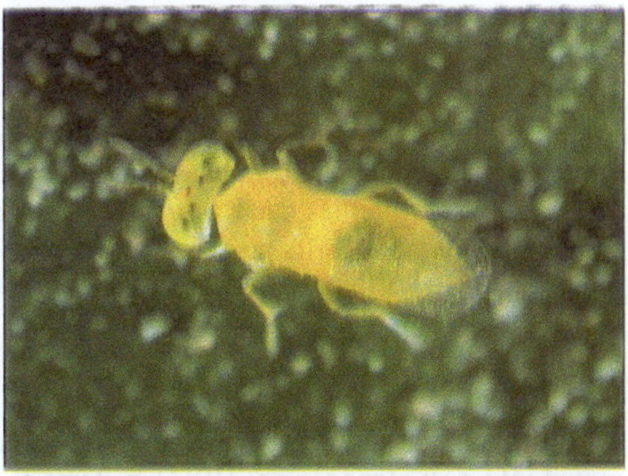

Fig. 3.8 *Aphytis melinus* (http://californiaagriculture.ucanr.edu/landingpage.cfm?article=ca.v047n01p16&fulltext=yes)

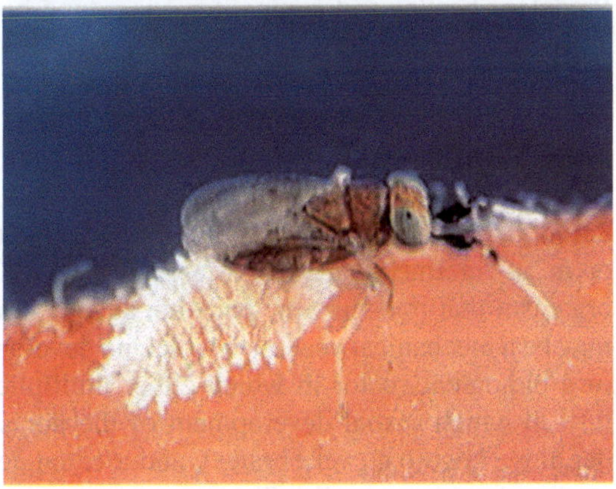

Fig. 3.9 *Anagyrus pseudococci* (http://bio-bee.ru/wp-content/uploads/2013/01/anagirus-400-266.jpg)

Encarsia formosa is an effective parasite under greenhouse conditions in semi-tropical areas of the New World. It has been used since the 1920s to suppress the greenhouse whitefly in greenhouses. *E. formosa* also parasitizes sweet potato white-fly in greenhouses, but this species is not a suitable host for this wasp, so pest suppression is not satisfactory. *Encarsia luteola* resembles *E. formosa* but is slightly smaller and lighter brown, and males are regularly produced. In the field it is usually present on upper leaves and parasitizes greenhouse whitefly. *Encarsia pergandiella*, an aphelinid wasp, native to North America, parasitizes greenhouse, sweet potato, and banded-wing whiteflies. *E. inaron* is a Mediterranean species parasitizing the pupae of greenhouse and sweet potato; *E. lutea* parasitizes sweet potato whitefly pupae; *E. meritoria* from California parasitizes banded-wing, greenhouse, and sweet potato whiteflies but prefers other species of chalcids, such as the Iris whitefly (*Aleyrodes spiraeoides*). The aphelinid wasp *Eretmocerus eremicus* (=*californicus*), indigenous to deserts in California and Arizona, parasitizes banded-wing silverleaf and sweet potato whiteflies. *Eretmocerus haldemani* is primarily a parasite of sweet potato and banded-wing whiteflies, although sometimes it will parasitize greenhouse whiteflies. *Eretmocerus mundus*, a Mediterranean parasite of sweet potato whitefly, oviposits on nymphal whitefly stage, although it likes second and third instars. *Encarsia formosa* is the most widely used parasite for suppressing whiteflies. It has been extensively utilized for several years in commercial vegetable greenhouses all over the world. The parasite can control whiteflies on vegetables like cucumber and tomatoes and ornamental plants, such as poinsettia. On vegetable crops that can tolerate a few whiteflies such as tomato, a single inoculative release of *Encarsia formosa* may satisfactorily suppress.

3.2.6 In Animal Husbandry

Most of the parasites belong to Hymenoptera, and they have been frequently integrated into pest management (IPM) systems to control houseflies and stable flies. Predatory beetles and competitor, viz., dung beetles, have been used for the management of horn flies (*Haematobia irritans*) and other insects of veterinary importance.

3.2.6.1 Transmission via Vectors

Parasites have both direct (the parasite is passed from one host to another through air, food, or water) and indirect life cycle (parasite develops or multiplies in a vector or in an intermediate host) (Marquardt et al. 2000). A vector is an invertebrate organism that transmits the parasitic agent from one vertebrate host to the next. It may be a mechanical vector (in which no development or multiplication takes place) or a biological vector in which either multiplication or development occurs. The host in which sexual development of the parasite (trematodes) takes place is the definitive host (e.g., vertebrates), and asexual development takes place in the intermediate host (e.g., snails). Transport or paratenic hosts are those in which a parasite does not develop or multiply but is carried to the next host usually through ingestion of the transport host.

Rapid and widespread urbanization brings an imbalance between development and quality of natural habitats. This has resulted in the increase of urban pest populations like cockroaches, ants, termites, rats, crows, etc. Vector-borne diseases like dengue, malaria, and filariasis are responsible for the prevalence of morbidity and mortality across tropical countries. In epidemiology studies, any agent that carries and transmits an infectious pathogen into another living organism is considered as a vector. Further it is postulated that parasites cause population declines in threatened mammals, have wide host ranges, and are transmitted via biting and bloodsucking arthropods, contaminated food, and water (Pedersen et al. 2007). They are generally ectoparasitic on the skin of vertebrate hosts. Many species are hematophagous (suck blood), while others are histophagous (tissue feeders) and bite or burrow in dermal tissues causing pain, infection and reactions that irritate the skin. Infestations are transmitted from one host to another by direct contact or free-living larvae or adults seek hosts for parasitizing.

Arthropods are involved in multiple parasitic relationships, either as parasites or as vectors for other arthropods or microorganisms. Arthropod vectors causing diseases include arachnid vectors such as soft ticks, hard ticks, and mites and insect vectors such as mosquitoes, flies (tsetse flies, sand flies, black flies, horseflies, biting midges), lice (head or body lice, crab lice), flea (human flea, rat flea, cat flea, jigger flea, bedbugs), and triatomine bugs (Arcari and Azzali 2000). Arthropods particularly mosquitoes, flies, and ticks act as vectors for many of the neglected tropical diseases affecting humans and domestic animals (Meyer 2012). Man continues to be a victim of arthropod borne diseases such as malaria, dengue, dengue hemorrhagic fever, Japanese encephalitis, and plague (Artsob 2000). Larvae or nymphs or adults may cross from one host to another to cause infections, while eggs or pupae may contaminate existing natural resources. Insects (fleas, flies, and lice) and arachnids (ticks and mites) depend on close contact between hosts. Many adults actively seek hosts to feed or oviposit. Winged insects (mosquitoes, flies) fly to new hosts to feed, while fleas jump onto passing hosts. Some adult flies (botflies) do not feed on hosts but oviposit eggs from which larvae emerge and feed on host tissues and fluids. Tick larvae actively seek hosts by climbing vegetation and questing for passing hosts. Some species complete life cycle on the same host (one-host ticks), while others detach after feeding and drop to the ground to molt before seeking new hosts and complete life cycle.

Transmission of diseases is the strategy used by protozoan parasites which inhabit the blood or internal tissues inside its host. This strategy implicates a hematophagous arthropod providing as an intermediary between successive vertebrate hosts. Many diseases caused by protozoa are transmitted by a volley of arthropod vectors. The vectors are not simply "flying syringes" but constitute a second host for the protozoan parasite. So, the pathogenicity life history of vector-transmitted diseases involves complex interactions and protozoan-vector interaction analogous to the complex human-protozoan interactions. The triatomine bugs defecate during feeding, and the excrement that contains the parasites inadvertently gets into the open wound by the host responding to pain and irritation from the bite.

Crustaceans that are parasitic serve as both hosts and vectors of viruses and parasites and other microbial pathogenic organisms. Important groups of parasitic crustaceans affecting commercial aquaculture species are external parasites: the Branchiura, Copepoda, and Isopoda. Crustacean parasites are numerous and have a worldwide distribution in marine and brackish water aquaculture systems. Copepods comprise the largest group of crustacean parasites on fish causing economical loss. Parasitic crustaceans are abundant and are globally distributed in fresh, brackish, and saline waters (Jithendran et al. 2008).

Vector transmission probably evolved multiple times. Vector transmission embraces complicated complex interactions between humans and vectors. This involves bioecology of human-arthropod interactions and bioecological cognizances so vector-transmitted parasites exhibit complicated life history involving interactions between humans, arthropods, and other organisms. An arthropod borne pathogen is transmitted mainly in two ways i.e., mechanical transmission and biological transmission. In mechanical transmission, no reproduction or development of pathogen takes place in the arthropod vector. But in the other type, the pathogen undergoes some type of biological development in the system of arthropod vector to complete life cycle (Gubler 2009). The bioecology of vector species and their associations with man serve as possible template for controlling the transmission of diseases to humans.

3.3 Concerns and Threats: Parasite Paradox

Parasites like hosts are not immune to the threats that normally affect wild species, and the present biodiversity crisis may well be primarily represented by the loss of affiliate species (Dunn et al. 2009). Emergence of pandemics and emerging disease pinpoint one of the consequences of global change but do not preclude that several parasites are threatened by it. It is proven that it creates risks for parasite perpetuation (Hudson et al. 2006; Lafferty 2012). For instance, land-use variation and contamination can both reduce the abundance and diversity of parasite species (Lafferty 1997; Huspeni and Lafferty 2004; Bradley and Altizer 2007). Climate change restricts parasite transmission (Afrane et al. 2012) resulting in phenological mismatches between parasites and hosts (Rohr et al. 2011). Parasites are threatened also by deliberate actions by humans to suppress or eliminate them. In certain situations, the extinction of parasites of public health or veterinary importance can be undoubtedly a gain, but suppression efforts affect populations beyond those initially targeted (Kristensen and Brown 1999). In other situations, daily veterinary practices can have the nontarget effect of eliminating intermediate hosts and interrupt enzootic transmission cycles in species other than those receiving the treatment (Spratt 1997; Wardhaugh et al. 2001). Parasites and taxa are threatened not only by direct environmental changes but also are adversely affected by the threats acting upon their hosts (Colwell et al. 2012). Parasites' dependense on host populations implies that they are endangered and co-extinct when hosts reduce. Several parasites need a threshold host population size for effective transmission. Few species of parasites will be endangered much before this decline which is irreversible (Altizer et al. 2007; Powell 2011). However, extinctions in dependent taxa are likely to represent the majority of extinction events in the current age of unprecedented biodiversity loss (Koh et al. 2004; Dunn et al. 2009); discrepancies remain between the number of documented and expected co-extinctions (Dunn et al. 2009). However, the threat of co-extinction must be carefully analyzed in many parasite conservation programs. Estimating the extent of the co-extinction threat for specific parasites will depend on the understanding of host and parasite behavior, natural history, phylogeny, and vital attributes such as host specificity and bioecology (Gómez et al. 2012).

Characterization of parasite species in wild animals is still in its infancy. Also, complicated man-made and environmental change makes wildlife disease problematic to forecast. Speeding-up changes in external factors, in addition to translocation of hosts and parasites, act synergistically to produce hard-to-predict disease outcomes in different ecosystems. These outputs are further complicated by the intimate connections between diseases in wild animals and diseases in man and domestic animals. Hence, it is important to unravel the interactions of parasites in wildlife, their response to environmental change, emerging diseases, and the role of humans and domestic animals to parasitic infections.

3.3.1 Ticks

Ticks (Ixodidae) can affect considerable economic loss in livestock production, and it becomes essential to control ticks, and it is virtually impossible to raise livestock economically in several parts of the world. The factors considered in the monetary losses in livestock production caused by ticks are unknown. However, results in research programs show the effects of tick control as it is related to mineral supplements, nutrition level, and cattle breed. Nevertheless data on control costs (inclusive of labor, equipment, and chemicals) and the benefits derived from these control programs are wanting. Ticks are the known vectors of agents that cause many economically important diseases in domestic livestock (Hoogstraal 1970; Neitz 1956). Some of the most important are anaplasmosis, east coast fever, theileriosis, babesiosis, equine encephalomyelitis virus, and Q fever virus (Berge and Lennette 1953; Stoker and Marmion 1955).

Ticks directly cause poor health and production losses to hosts by several species of parasites. Ticks transmit numerous kinds of viruses, bacteria, and protozoa among domestic animals. These microbes infect diseases that can seriously devitalize or fatal to domestic animals and humans. Ticks are particularly important to domestic animals in tropical and subtropical regions, where the higher temperatures enable several species to multiply. Also, the large populations of wild animals in warm countries provide a reservoir of ticks and infective microbes that spread to domestic animals. Farmers of livestock regulate several methods to control ticks. Veterinarians and animal health agencies work at private, national, and international levels to decrease the harm caused by ticks and diseases transmitted by them.

Range of Ticks Affecting Domestic Animals

Rhipicephalus (*Boophilus*) *annulatus* engorged female

Ticks are related to spiders in Arthropoda. Ticks are in the subclass Acari which comprises of several orders of mites and one tick order, the Ixodidae. There are also parasitic mites, but all ticks are parasitic feeders on the blood. Few mite species are mistaken for larval ticks at infestations, but their feeding habits are unique. All ticks have an incomplete or partial metamorphosis: after the egg hatches, a series of similar stages (=instars) develops from a six-legged larva to an eight-legged nymph and then a sexually developed, eight-legged adult. Ixodidae have the important genera *Amblyomma* (Fig. 3.10), *Dermacentor*, *Haemaphysalis*, *Hyalomma*, *Ixodes*, *Margaropus*, and *Rhipicephalus*. The important boophilid ticks, formerly of the genus *Boophilus*, are now classified under a subgenus within the genus *Rhipicephalus*. These genera are known as hard ticks because their outer surfaces contain hard sclerites. Within ten genera are about 100 species of significance to domestic animals and humans (Taylor 2007).

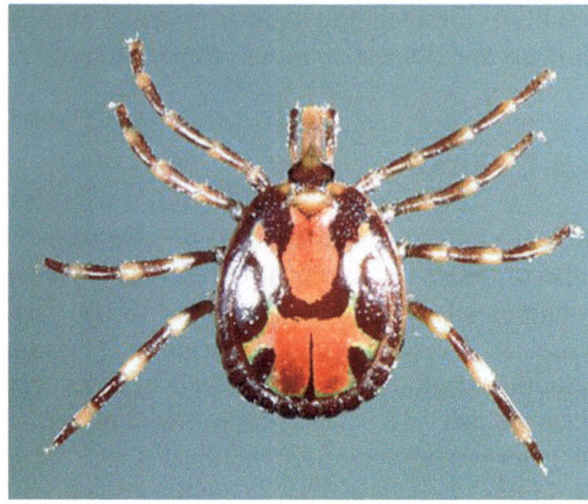

Fig. 3.10 Adult male bont tick, *Amblyomma variegatum* (Fabricius 1794)

3.3.2 Mites

The most important groups of arthropods transmitting etiological agents pathogenic to livestock are blood feeding and are inherently involved in transmission of diseases. Ticks, mosquitoes, tsetse flies, and biting midges, for instance, have leading roles in the transmission of pathogens causing severe livestock and poultry diseases. Of lesser importance are hematophagous arthropod groups that mechanically transmit pathogens. Horseflies, deer flies, stable flies, horn flies, and others have been implicated in disease transmission although not by continuous feeding.

There are arthropod groups in which several species are not bloodsucking. They are muscoid flies, grasshoppers, and beetles but, which by contact, transfer pathogens and serve as intermediate hosts of helminths. Meanwhile, instances can also be cited for a range of transmission modes and cycles within each of the major class of vectors.

The mite, *Psorergates bovis*, causes pruritus, but does little harm to cattle (Geevarghese et al. 1997). But *Psorergates ovis* feeding on sheep induces inflammatory and hypersensitive responses in the epidermis, resulting in intense pruritus and formation of scabs. Further spread to the skin and fleece of sheep happens when the sheep groom.

3.3.3 Mosquitoes

Mosquitoes are proven notorious as transmitters of severe human diseases. There is little to document the adverse effects on public health due to malaria, yellow fever, filariasis, and several mosquito-borne diseases of arboviral etiology. Rift Valley fever and the equine encephalitides are prime diseases of livestock spread by mosquitoes. Although over 2500 species of mosquitoes have been described worldwide

Table 3.1 List of diseases transmitted by arthropod parasites

Diseases	Causal organisms	Mode of transmission
Colorado tick fever	*Coltivirus*	Transmitted to humans by ticks
Fever	*Rickettsia rickettsii*	
Lyme disease	*Borrelia burgdorferi*	
Psoroptic skin disease	*Psoroptes ovis*	Transmitted to animals by mites
Psorergatic skin disease	*Psorergates bovis*	
Sarcoptic skin disease	*Sarcoptes*	
Mosquito-borne encephalitis	Mosquito-borne virus	Transmitted by mosquito
Plague	*Yersinia pestis*	Transmitted to human by infected fleas

in 18 genera and subgenera, species of the greatest significance as vectors of pathogenic agents are detected in the genera *Aedes*, *Culex*, *Anopheles*, and others.

The southern cattle tick, *Boophilus microplus*, is a vector of cattle babesiosis, bovine anaplasmosis, and benign bovine theileriosis. This tick is present in the warmer, humid regions of the West Indies, Mexico, Central America, South America, Africa, Australia, the Orient, and Micronesia. Earlier it has been found in Southern Florida and Southern Texas and is found in Puerto Rico and St. Croix, US Virgin Islands. A closely related species, *B. annulatus*, the cattle tick, was the most significant external parasite of cattle in the United States. It is a common vector of bovine babesiosis and has also been implicated in the spread of bovine anaplasmosis, bovine theileriosis, and spirochetosis of cattle, goats, sheep, and horses (Table 3.1).

3.3.4 Mosquito-Borne Encephalitis

Encephalitis is a disease caused by mosquito-borne viruses (arboviruses) adversely influencing the central nervous system. Infections vary from un apparent to mild, nonspecific illnesses (fever, headache, musculoskeletal pain, and malaise) to occasionally severe illness of the central nervous system resulting in permanent neurologic damage and death.

These viruses naturally infect a wide range of birds and mammals and are transmitted between animals by mosquito vectors. Occasionally, infected mosquitoes will attack man or cattle that are "dead ends" for the viruses, with little or no chance of subsequent transmission to other mosquitoes. Often viral infections result in severe illness or death of hosts like man or horses (EEE and WEE).

No commercial vaccine is available for humans, but vaccines for WEE and EEE are readily available for horses. The best protective practices are protection from mosquito bites, particularly during early evening hours, and repelling them away. Mosquito populations can be suppressed by eliminating breeding sites for vector species. Killing mosquitoes with area-wide applications of insecticides has been most effective in preventing epidemics.

Diseases like plague caused by bacteria *Yersinia pestis* are acute. Humans get infected by the bites of infected fleas but also directly from exposure to tissues or body fluids from diseased animals, especially when skinning animals. The symptoms of disease are sudden fever and shivering. Swollen and painful lymph nodes (buboes) developed in the armpits, groin, and other areas 2–6 days following contact. In addition to causing itchy bump, septicemic infection may develop involving other organs. Secondary infection of the lungs results in plague pneumonia that can be transmitted by a person to another by aerosol. The disease initially may be mild and short-lived but often advances to a severe form.

One sixth of the world's human population suffers from one or the other tropical diseases, and children are often the most harmed. There are presently 14 listed neglected tropical diseases (NTDs). Most can be prevented but the scale of the problem is huge. Further these diseases are now established on the global health agenda. They have a plethora of hosts, a multitude of developmental stages, that often bear little resemblance from one stage to another, and most forms remain hidden in other animal guts and tissues. Arthropods are responsible for hundreds of millions of cases of disease in man and animals every year. Over the past 30 years, there has been a global reemergence of infectious diseases in humans and animals and vector-borne diseases in particular, with an increased frequency of epidemic transmission and expanding distribution on Earth. The major concern is that important vector-borne diseases are found in the tropics and subtropics, in the areas where resources are most limited and surveillance is poor or absent. With highly increased human and animal mobility, globalization and trade has made these diseases not problems of the tropics alone but of the world. This underscores the need for physicians and veterinarians in non-endemic areas to be aware of vector-borne diseases and knowledge about them. New, mode-of-action chemistries are urgently required to update vector management practices in tropical countries where arthropod-borne diseases are endemic, especially where vector populations have acquired insecticide resistance.

3.4 Conservation Initiatives

The bioecology and control of parasite and vector interactions are vital to fully understand epidemiology before effective control strategies can be developed. Similarly, basic research on pathogen and blood-feeding insect vector biology along with drug discovery programs, clinical trials of vaccines, and field-based studies on vaccine efficacy are required.

Conservation of bloodsuckers and disease-causing agents is to explicitly weed out death and disease. Under natural conditions, this is a normal part of a functioning ecosystem, but for human world, it is not easily acceptable. Parasites should be recognized as "meaningful conservation targets no less relevant than their hosts." Only one parasite is listed in the IUCN Red List as endangered because of the rarity of its host, the world's smallest pygmy hog-sucking louse (*Haematopinus oliveri*) (Whiteman and Parker 2005). It is estimated that there are between 75,000 and

300,000 helminth species parasitizing the vertebrates. Of them 3–5 % are threatened in the coming 50–100 years. Pandemics and emerging disease illustrate one of the consequences of global climatic change. Similarly many parasite species are also threatened by extinction.

3.4.1 Co-extinction of Species

The concept of co-extinction (parasites, mutualists, and commensals becoming extinct alongside their hosts) has been around since the 1990s. In general, it is expected that inefficiently transmitted parasites are lost initially; later efficiently transmitted species with low host specificity will persist at low host densities. Host variety of a parasite will reduce as each potential host is lost or declines in host range and population. This indicates that parasitic species will decline at a rapid rate than their host species. Populations of Galapagos mockingbirds, *Nesomimus melanotis*, have dropped dramatically since they were first described by Charles Darwin in 1835 (http://www.nhm.ac.uk/research-curation/lifesciences/terrestrialinvertebrates/research/mite-research/mites-galapogos/index.html). As parasites are directly affected by the number of mockingbirds, genetic data obtained from research may help to conserve and reintroduce mockingbird populations. The trematode *Pleurogonius malaclemys* infects snails only when the endangered diamondback terrapin, the single host for the trematode (Byers et al. 2011), and when a diamondback terrapin population declines; it takes its parasites too.

3.4.2 Regulation of Host Populations

Parasites often serve as modulators of host population, which in generalist pathogens may result in strong frequency-dependent control over relative abundance across the host community. When the black-footed ferret (*Mustela nigripes*) was deloused during captive breeding after extinction in the wild, the ferret louse (*Neotrichodectes* sp.) may become extinct. All species are important, and adding back only the cute and charming ones undercuts the notion that reintroduction programs are about ecosystems, not just aesthetics. Determining the part played by parasites in regulating natural ecosystems still remains a major challenge for biologists. If the core job of conservation biologists is to sustain functional food webs and food chains, then it is crucial that parasites as an important and essential component of biodiversity are sustained.

Parasites are also biotic entities of natural selection altering a range of host characteristics, from phenotypic polymorphism and secondary sexual features to sexual reproduction (Wegner et al. 2003; Lively et al. 2004; Blanchet et al. 2009). These effects ultimately lead to biological diversification by changing host reproductive isolation and speciation (Summers et al. 2003). Common conservation strategies for hosts such as captive management, reintroduction, and translocation include wide-spectrum veterinary treatments to limit or prevent parasite transmission (Phillips and Scheck 1991; Moir et al. 2012). By maintaining disease-free host populations, the likelihood of conservation intervention success may be increased at the cost of parasite decline, particularly, for parasites of endangered, rare, or spatially restricted hosts (Gómez et al. 2012). For instance, the extinction of the louse *Colpocephalum californici* is suspected to be associated with the *ex-situ* veterinary treatment of California condors (Koh et al. 2004). However, such interventions can lead to unanticipated and negative impacts for hosts, including increased susceptibility of hosts to infection following reintroduction or translocation (Gompper and Williams 1998; Almberg et al. 2012). This indicates that maintenance of host-parasite relationships in managed wildlife populations can be beneficial and highlight vital play of the vertebrate parasitologists in conservation programs.

3.4.3 Modulators of Competitive Interactions

The differential effects of infection of generalist parasites can regulate competitive interactions. For instance, parapox virus-mediated competition can explain the ecological success of introduced gray squirrels (*Sciurus carolinensis*) in Europe (Tompkins et al. 2001). Infection can also affect reproductive behaviors and output, for example, causing abortion or sterility. In the most extreme case, parasitic castrators divert the host's metabolism for their own reproductive success, driving changes in host density and maturation rates (Lafferty and Kuris 2009).

3.4.4 Conservation vs Control

Parasitic agents which normally develop in hosts other than humans but can infect humans if given the opportunity are called *zoonotic agents*. Modern research has revealed that emerging diseases in humans have a zoonotic reservoir, that reservoirs are often wildlife species populations (Jones et al. 2008), and that anthropogenic influence is usually associated with human and wildlife disease emergence patterns (Daszak et al. 2000).

Associations among competitors, predators, and preys have historically been seen as the foundation of community structure. Parasites – long ignored in community ecology – are now recognized as playing an important role in affecting species interactions and consequently affecting ecosystem function. Parasitism can interact with other ecological drivers, resulting in both detrimental and beneficial effects on ecosystem services and biodiversity. Species interactions including parasites are also vital to understanding several biological invasions and emerging infectious diseases. This review combines community ecology and epidemiology to develop a wide-ranging monitor of how parasites and pathogens alter different aspects of communities, enabling the new emergence of ecologists to include parasites as a prime consideration in future programs.

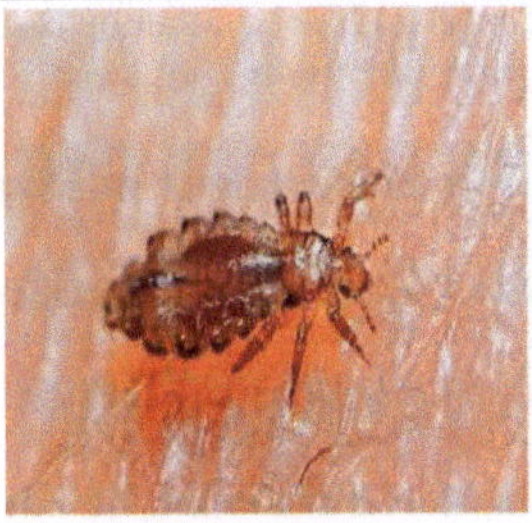		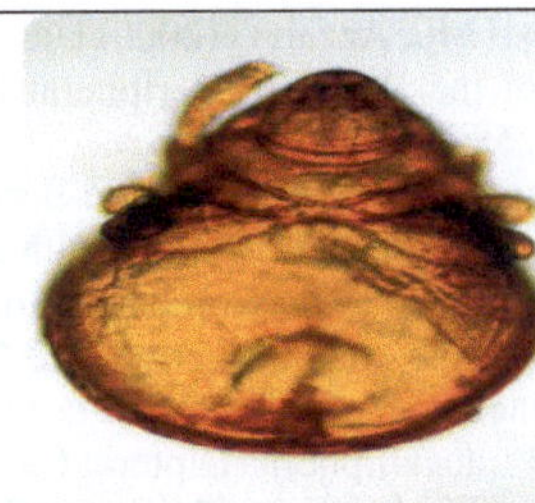
Fig.1 *Pediculus humanus humanus*	**Fig.2** ***Microterys flavus*** http://boutique.crisop.fr/microterys-flavus-10	**Fig.3** *Hydrozetes sp.* http://www.nhm.ac.uk/research-curation/life-sciences/terrestrial-invertebrates/research/mite-research/archaeological-indicators/index.html

Fig.4 *Trichogrammma* sp. (http://www.ecured.cu/index.php/Trichogramma_spp.)	**Fig.5** *Aphelinus semiflavus* http://ponent.atspace.org/fauna/ins/fam/aphelinidae/aphelinus_aph.htm	**Fig.6** *Aphidius colemani* http://www.arbico-organics.com/category/pest-solver-guide-aphids
	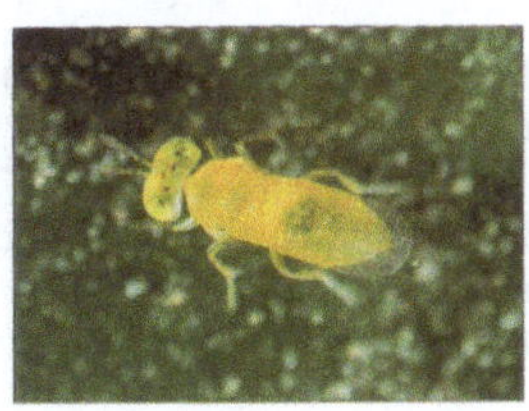	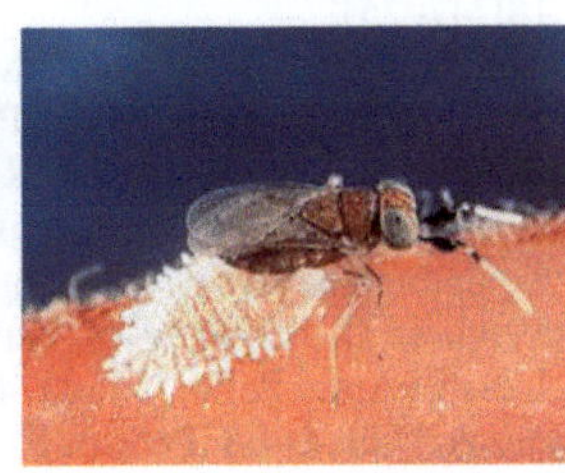
Fig.7 *Aphidiu servi* http://www.arbico-organics.com/category/pest-solver-guide-aphids	**Fig.8** *Aphytis melinus* http://californiaagriculture.ucanr.edu/landingpage.cfm?article=ca.v047n01p16&fulltext=yes	**Fig.9** *Anagyrus pseudococci* http://bio-bee.ru/wp-content/uploads/2013/01/anagirus-400-266.jpg

Plate 3.1 List of select important parasitoids

References

Afrane YA, Githeko AK, Yan G (2012) The ecology of Anopheles mosquitoes under climate change: case studies from the effects of deforestation in East African highlands. Ann N Y Acad Sci 1249:204–210

Almberg ES, Cross PC, Dobson AP, Smith DW, Hudson PJ (2012) Parasite invasion following host reintroduction: a case study of Yellowstone's wolves. Proc R Soc Lond Ser B Biol Sci 367:2840–2851

Altizer S, Nunn CL, Lindenfors P (2007) Do threatened hosts have fewer parasites? A comparative study in primates. J Anim Ecol 76:304–314

Arcari ML, Azzali G (2000) Ultrastructural and three-dimensional aspects of the lymphatic vessels of the absorbing peripheral lymphatic apparatus in Peyer's patches of the rabbit. Anat Rec 258:71–79

Artsob H (2000) Arthropod-borne disease in Canada: a clinician's perspective from the 'Cold zone'. Paediatr Child Health 5(4):206–212

Bauchau V (1998) Comparison of parasitism level in two sympatric passerines: the pied flycatcher and the great tit. Écoscience 5:164–171

Bennett GF, Whitworth TL (1992) Host, nest, and ecological relationships of species of Protocalliphora (Diptera: Calliphoridae). Can J Zool 70:51–61

Berge TO, Lennette EH (1953) World distribution of Q fever: human, animal and arthropod infection. Am J Hyg 57(1):25–43

Bigler F (1989) Quality assessment and control in entomophagus insects used for biological control. J Appl Entomol 108:390–400

Blanchet S, Rey O, Berthier P, Lek S, Loot G (2009) Evidence of parasite mediated disruptive selection on genetic diversity in a wild fish population. Mol Ecol 18:1112–1123

Bradley CA, Altizer S (2007) Urbanization and the ecology of wildlife diseases. Trends Ecol Evol 22:95–102

Byers JE, Altman J, Grosse AM, Huspeni TC, Maerz JC (2011) Using parasitic trematode larvae to quantity an elusive vertebrate host. Conserv Biol 25:85–93

Cerutti F, Bigler F (1995) Quality assessment of *Trichogramma brassicae* in the laboratory. Entomologia Experimentaliset Applicata 75:19–26

Colwell RK, Dunn RR, Harris NC (2012) Coextinction and persistence of dependent species in a changing world. Annu Rev Ecol Evol Syst 43:183–203

Daszak P, Cunningham AA, Hyatt AD (2000) Emerging infectious diseases of wildlife – threats to biodiversity and human health. Science 287:443–449

deMeeûs T, Renaud F (2002) Parasites within the new phylogeny of eukaryotes. Trends Parasitol 18:247–251

Dobson A, Lafferty KD, Kuris AM, Hechinger RF, Jetz W (2008) Homage to Linnaeus: how many parasites? How many hosts? Proc Natl Acad Sci U S A 105:11482–11489

Dunn RR, Harris NC, Colwell RK, Koh LP, Sodhi NS (2009) The sixth mass coextinction: are most endangered species parasites and mutualists. Proc R Soc Lond 276:3037–3045, Series B, Containing papers of a Biological Character

Geevarghese G, Fernandes S, Kulkarni SM (1997) A checklist of Indian ticks (Acari: Ixodidae). Indian J Anim Sci 67:566–574

Gómez A, Nichols ES, Perkins S (2012) Parasite conservation, conservation medicine, and ecosystem health. In: Aguirre AA, Daszak P, Ostfeld RS (eds) Conservation medicine: applied cases of ecological health. Oxford University Press, New York, pp 67–81

Gompper ME, Williams ES (1998) Parasite conservation and the black-footed ferret recovery program. Conserv Biol 12:730–732

Griffith GW (2012) Do we need a global strategy for microbial conservation? Trends Ecol Evol 27:1–2

Gubler DJ (2009) Vector-borne diseases. Rev Scientifiqueet Tech (Int Off Epizootics) 28(2):583–588

Hoogstraal H (1970) Symposium on current world research on tick-disease relationships and control of tick vectors of disease. Entomol Soc Am Misc Publ 6:327–372

Hudson PJ, Dobson AP, Lafferty KD (2006) Is a healthy ecosystem one that is rich in parasites? Trends Ecol Evol 21:381–385

Huspeni TC, Lafferty KD (2004) Using larval trematodes that parasitize snails to evaluate a salt marsh restoration project. Ecol Appl 14:795–804

Jithendran KP, Natarajan M, Azad IS (2008) Crustacean parasites and their management in brackishwater finfish culture. Mar Finfish Aquac Netw 13:47–50

Jones KE, Patel NG, Levy MA, Storeygard A, Balk D, Gittleman JL, Daszak P (2008) Global trends in emerging infectious diseases. Nature 451:990–993

Koh LP, Dunn RR, Sodhi NS, Colwell RK, Proctor HC, Smith VS (2004) Species co-extinctions and the biodiversity crisis. Science 305:1632–1634
Kristensen TK, Brown DS (1999) Control of intermediate host snails for parasitic diseases – a threat to biodiversity in African freshwaters? Malacologia 41:379–391
Kuhlamann U, Mills NS (1999) Comparative analysis of the reproductive attributes of three commercially produced trichogramma species (Hymenoptera: Trichogrammatidae). Biocontrol Sci Technol 9:335–346
Lafferty KD (1997) Environmental parasitology: what can parasites tell us about human impacts on the environment? Parasitol Today 13:251–255
Lafferty KD (2012) Biodiversity loss decreases parasite diversity: theory and patterns. Philos Trans R Soc Lond B Biol Sci 367:2814–2827
Lafferty KD, Kuris AM (2009) Parasitic castration: the evolution and ecology of body snatchers. Trends Parasitol 25:564–572
Lawson D, Arensburger P, Atkinson P, Besansky NJ, Bruggner RV, Butler R, Campbell KS, Christophides GK, Christley S, Dialynas E (2007) VectorBase: a home for invertebrate vectors of human pathogens. Nucleic Acids Res 35(Database issue):D503–D505
Li YL (1994) Worldwide use of Trichogramma for biological control on different crops a survey. In: Wajnberg E, Hassan SA (eds) Biological control with egg parasitoids. Cab International, Oxon, pp 37–55
Lively CM, Dybdahl MF, Jokela J, Osnas EE, Delph LF (2004) Host sex and local adaptation by parasites in a snail-trematode interaction. Am Nat 164:S6–S18
Losey JE, Calvin DD (1995) Quality assessment of four commercially available species of trichogramma (Hymenoptera: Trichogrammatidae). J Econ Entomol 88:1243–1250
Marcogliese DJ (2004) Parasites: small players with crucial roles in the ecological theatre. Ecohealth 1:151–164
Marquardt WC, Demaree RS, Grieve RB (2000) Parasitology and vector biology, 2nd edn. Academic, San Diego, pp 1–13
Meffe GK, Carroll RC, Groom MJ (2006) What is conservation biology? In: Groom MJ, Meffe GK, Carroll RC (eds) Principles of conservation biology. Sinauer Associates, Sunderland, pp 3–25
Merino S, Potti J (1995) Mites and blowflies decrease growth and survival in nestling pied flycatchers. Oikos 73:95–103
Meyer M (2012) A high coverage genome sequence from an archaic Denisovan individual. Science 338:222
Moir ML, Vesk PA, Brennan KEC, Poulin R, Hughes L, Keith DA, McCarthy MA, Coates D (2012) Considering extinction of dependent species during translocation, ex situ conservation, and assisted migration of threatened hosts. Conserv Biol 26:199–207
Moller AP (1997) Parasitism and the evolution of host life history. In: Clayton DH, Moore J (eds) Host–parasite evolution: general principles and avian models. Oxford University Press, Oxford, pp 105–127
Neitz WO (1956) A consolidation of our knowledge of the transmission of tick-borne diseases. Onderstepoort 1. Vet Res 27:115–163
Nichols E, Gómez A (2011) Conservation education needs more parasites. Biol Conserv 144:937–941
O'Neil RJ, Giles KL, Obryeki JJ, Mahr DL, Legaspi JC, Katorich K (1998) Evaluation of the quality of four commercially available natural enemies. Biol Control 11:1–8
Oatman ER (1966) Parasitization of Corn earworm eggs on sweet corn silk in southern California, with notes on larval infestations and Predators. J Econ Entomol 59:830–835
Oppliger A, Richner H, Christe P (1994) Effect of an ectoparasite on lay date, nest-site choice, and hatching success in the great tit (*Parus major*). Behav Ecol 5:130–134
Orr DB, Landis DA, Mutch DR, Manley GV, Stuby SA, King RL (1997) Ground cover influence on microclimate and trichogramma (Hymenoptera: Trichogrammatidae) augmentation in seed corn production. Environ Entomol 26:433–438

Pedersen AB, Jones KE, Nunn CL, Altizer S (2007) Infectious diseases and extinction risk in wild mammals. Conserv Biol 21:1269–1279
Phillips M, Scheck J (1991) Parasitism in captive and reintroduced red wolves. J Wildl Dis 27:498–501
Poulin R, Morand S (2000) The diversity of parasites. Q Rev Biol 75:277–293
Powell FA (2011) Can early loss of affiliates explain the coextinction paradox? An example from Acacia-inhabiting psyllids (Hemiptera: Psylloidea). Biodivers Conserv 20:1533–1544
Price PW (1980) Evolutionary biology of parasites. Princeton University Press, Princeton, p 237
Rendell WB, Verbeek NAM (1996) Are avian ectoparasites more numerous in nest boxes with old nest material? Can J Zool 74:1819–1825
Reuter OM (1913) Lebensgewohnheiten und Instinkte der Insekten. Friendlander, Berlin
Rohr JR, Dobson AP, Johnson PTJ, Kilpatrick AM, Paull SH, Raffel TR, Ruiz- Moreno D, Thomas MB (2011) Frontiers in climate change-disease research. Trends Ecol Evol 26:270–277
Smith SM (1996) Biological control with *Trichogramma*: advances, successes and potential of their use. Annu Rev Entomol 41:375–406
Spratt DM (1997) Endoparasite control strategies: implications for biodiversity of native fauna. Int J Parasitol 27:173–180
Stoker MGP, Marmion BP (1955) The spread of Q fever from animals to man. The natural history of a rickettsial disease. Bull WHO 13(78):1–806
Summers K, McKeon S, Sellars J, Keusenkothen M, Morris J, Gloeckner D, Pressley C, Price B, Snow H (2003) Parasitic exploitation as an engine of diversity. Biol Rev 78:639–675
Tomás G, Merino S, Moreno J, Morales J (2007) Consequences of nest reuse for parasite burden and female health and condition in blue tits, *Cyanistes caeruleus*. Anim Behav 73:805–814
Tompkins DM, Greenman JV, Hudson PJ (2001) Differential impact of a shared nematode parasite on two game bird hosts: implications for apparent competition. Parasitology 122:187–193
Topalis P, Lawson D, Collins FH, Louis C (2008) How can ontologies help vector biology? Trends Parasitol 24:249–252
Toups MA, Kitchen A, Light JE, Reed DL (2011) Origin of clothing lice indicates early clothing use by anatomically modern humans in Africa. Mol Biol Evol 28:29–32
Wardhaugh KG, Holter P, Longstaff B (2001) The development and survival of three species of coprophagous insect after feeding on the faeces of sheep treated with controlled-release formulations of ivermectin or albendazole. Aust Vet J 79:125–132
Wegner KM, Reusch TBH, Kalbe M (2003) Multiple parasites are driving major histocompatibility complex polymorphism in the wild. J Evol Biol 16:224–232
Wheeler QD (Ed) (2008) The new taxonomy, systematics association special publication. CRC, Boca Raton, pp 76:33–53
Wheeler QD, Valdecasa AG (2010) Taxonomy: add a human touch too. Nature 467:788
Whiteman NK, Parker PG (2005) Using parasites to infer host population history: a new rationale for parasite conservation. Anim Conserv 8:175–181

Diversity and Ecology of Scorpions: Evolutionary Success Through Venom

4

Shakunthala Sridhara, Akshay Kumar Chakravarthy, V. Kalarani, and D. Chandrasekhar Reddy

Abstract

The scorpions are lowly evolved but highly successful group of animals. This success is discussed with relation to size, morphological diversity, efficient utility of primitive sense organs, efficient foraging, feeding and digestion system as well as life history strategies of utilizing venom for defense and offence, parthenogenesis and increase in instars. In fact their life history strategies are unique and adaptive to warm, arid and semiarid conditions. Research on scorpions in India is two fold: Discovery and identification of new species, and Metabolic studies and Physiological adaptations including rhythmicity. The chapter after giving a brief classification of the order highlights the biology, the survival strategies and the epidemiology of scorpion venom and also an account of species diversity and distribution across the world. Brazil, Colombia, Peru, Panama, the Amazonian Andes, Mexico, Iran, India, South Africa and the USA have maximum density and diversity of scorpions with high degree of endemism in Mexico,

S. Sridhara (✉)
Department of Entomology, University of Agricultural Sciences, Gandhi Krishi Vignana Kendra (GKVK), Bengaluru 560065, Karnataka, India
e-mail: shakunthallasridhara@yahoo.com

A.K. Chakravarthy
Division of Entomology and Nematology, Indian Institute of Horticultural Research (IIHR), Hessaraghatta, Bengaluru 560089, Karnataka, India
e-mail: chakravarthyakshay@gmail.com

V. Kalarani
Department of Biotechnology, Sri Padmavati Mahila Vishwavidyalam (Women's University), Tirupati 517502, Andra Pradesh, India

D.C. Reddy
Department of Zoology, SV University, Tirupati 517502, Andhra Pradesh, India

A.K. Chakravarthy, S. Sridhara (eds.), *Arthropod Diversity and Conservation in the Tropics and Sub-tropics*, DOI 10.1007/978-981-10-1518-2_4

Iran, India and Brazil. However, all over the habitats of scorpions in developing countries, development works have a serious impact on both density and species composition.

Keywords
Adaptations • Evolution • Scorpions • Venom • Endemism

4.1 Introduction

Scorpions are one of the oldest known terrestrial arthropods with their fossils ageing 430 million years but interestingly appear much similar to the present-day species. These venomous arthropods belong to class Arachnida, order Scorpiones, and are distributed across several extant families and subfamilies and number about 1400 species globally. They have an elongated body and a segmented tail terminating in a venomous bulby tip. There are four pairs of legs, and the head has a pair of pedipalps ending in plier like pincers which help in grasping (Polis 1990).

Scorpions are relatively big arthropods averaging a size of 6 cm. The males are usually slender with longer tails than females. The black emperor scorpion, *Pandinus imperator*, found in Guinea is the biggest scorpion, with a body length of about 18 cm, and weighs around 60 g. *Hadogenes troglodytes*, the rock scorpion from South Africa, is the longest scorpion whose female attains a length of 21 cm. The smallest scorpion, *Microtityus fundorai*, measures just 12 mm and is an inhabitant of the Caribbean. Some ancestral forms were giants reaching a length of 35–100 cm. Species from deserts and arid zones are usually yellowish to light brown in colour compared to the brownish to black species inhabiting moist areas, mountains and agricultural fields (Bastawade 2006; Ayal 2007).

4.2 Scorpion Classification

The classification of scorpions has undergone revisions as newer specimens are collected across the globe. Stockwell (1989) proposed the creation of superfamilies leading to several changes in the existing number of families, promoting some subfamilies to families, and revision of genera leading to newer families. Analysing the various taxonomic publications including that of Stockwell and his own work, Lourenco et al. (2001) proposed six Super families and gave their distribution.

4.2.1 Superfamily: Bothriuroidea Simon, 1880 (With One Family)

Distribution: South America, Australia and India (Himalayas).

4.2.2 Superfamily: Buthoidea C.L. Koch, 1837 (With Two Families)

Distribution: all continents except Antarctica, in tropical, subtropical including India and to some extent temperate regions.

4.2.3 Superfamily: Chaeriloidea Pocock, 1893 (With Two Families)

Distribution: Asia-Bangladesh, India, Indonesia, Malaysia, Nepal, Singapore, Sri Lanka and Vietnam.

4.2.4 Superfamily: Chactoidea Pocock, 1893 (With Five Families)

Distribution: North Africa, Southern Europe, Middle East, Southern Central Asia including India, Pakistan, Indonesia, Malaysiia, South and Central America, Mexico (Baja California) and Guatemala.

4.2.5 Superfamily: Vaejovoidea Thorell, 1876 (With Two Families)

Distribution: Mexico and the South/Southern USA, Asia (Turkey) and Europe (Greece).

4.2.6 Superfamily: Scorpionoidea Latreille, 1802 (With Eight Families)

Distribution: Africa, South and Southeast Asia, Indonesia, North and Central America, Northern South America, the Caribbean and Middle East.

4.3 Ecology and Behaviour

Scorpions inhabit diverse habitats although the greatest species diversity and greater abundance are usually found in arid environments and deserts. They live successfully in grasslands, savannas, deciduous forests, montane pine forests, rain forests and caves. They have also been found under snow-covered rocks at an altitude of 4000–4200 m in the Himalayas in Northern India. In contrast, some species occur at a depth of more than 800 m (Polis 1990).

Scorpions are basically nocturnal and hide in their burrows and natural crevices or lie under rocks and beneath loose tree barks. They can fluoresce under ultraviolet

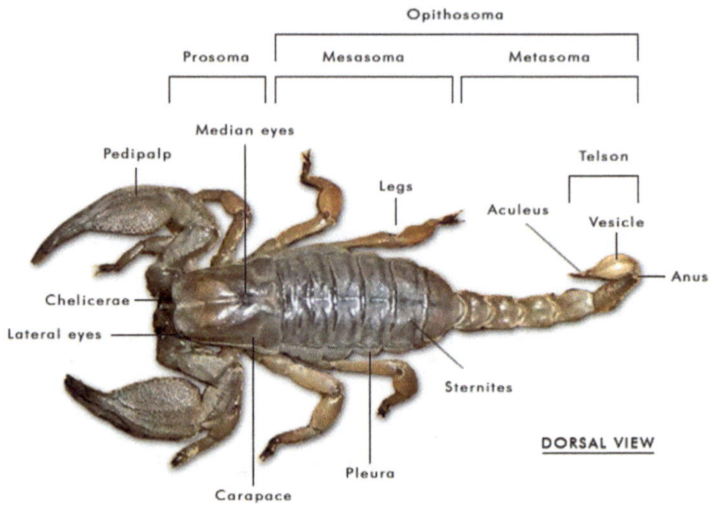

Fig. 4.1 External morphology of a typical scorpion (Source: Polis 1990)

light, a character utilised by biologists to study them during night using ultraviolet light. They show great adaptability to live successfully even under extreme conditions like sandy areas with minimal water, hard rocky substances, cooler interiors of caves, high desert temperatures up to 47 °C and intertidal zones and can tolerate submersion under water for a day or 2 (Hadley 1974).

The body is composed of an anterior cephalothorax and an abdomen. They have mouth parts called chelicerae, a pair of pedipalps which with their pincer-like function aid in prey capture and defense. They are also used in excavating burrows, during courtship and are covered with sensory hairs. They have four pairs of legs. The cephalothorax is covered by a carapace that has a pair of median eyes and two to five pairs of lateral eyes (Fig. 4.1).

The abdomen has 12 distinct segments. The first seven segments form the mesosoma (body) and the posterior five segments form the tail or metasoma. The tip of the tail bears the telson which is a bulb-like structure containing the venom gland ending in a curved sting to inject venom (Fig. 4.1). The ventral side of the body has a pair of comb-like sense organs called pectins. They are used to sense the texture and vibration of surfaces. Males have larger pectins than females with more teeth. They also serve as chemoreceptors and detect pheromones (Polis and Yamashita 1991).

Scorpions perceive the environment through visual, tactile and chemical sense organs. Although incapable of forming sharp images, their central eyes are one of the most sensitive light receptors in the animal kingdom. Using the shadows cast by the stars, they easily navigate during night. The lateral eyes called ocelli can sense only changes in light intensity and help in establishing daily rhythms of activity (Smith 1995). Some species are known to possess light receptors in the tail.

Scorpions detect and intercept vibrations through both air and ground. Long thin hairs, situated on the pedipalps, are deflected by all vibrations originating from the air. By detecting the direction of air movement, they help in the capture of prey, detect predators and navigate. The thin cuticle covering the slits in the tarsal leg segments detects ground vibrations produced by prey, predators and mates.

Chemoreceptors are situated in the oral cavity and perceive taste. They are also present in the pectins and help to locate prey in some species. The pectins of male scorpions have more and bigger teeth and are used to detect the pheromones emitted by the receptive females. Scorpions are nocturnal and carnivorous, often digging out stationary prey, insect eggs and termites. But most of the time, they are opportunistic feeders waiting for prey to come close rather than hunt them. They feed mostly on insects, beetles, crickets, flies, wasps, spiders, centipedes, other scorpions, snails, tadpoles, small snakes and rodents (Cloudsley-Thomson 1991). They can detect vibrations from prey even at a distance of 20 cm. Once they catch the prey, it is either killed or paralyzed with a powerful, venomous sting. The two pairs of chelicerae are employed to position the prey in the mouth to chew. As it masticates the prey, digestive fluids secreted from the mid gut pour over it converting the prey into an easily digestible matter. The liquefied portion is sucked into the stomach by a pumping action. The indigestible matter, which is now a ball-shaped solid, is cast aside. The external, partial digestion helps the scorpion to derive the maximum nutrients from its prey, minimizing energy expenditure for the process of digestion. Capture of prey, external digestion or suction is a slow process and can take several hours. Scorpions are terrestrial foragers and, similar to other predators, return to the same feeding area every night. When their territory is disturbed by natural calamities or by anthropogenic activities, they enter human habitat particularly homes and buildings (Polis 1990, 1993).

4.4 Life Cycle

Breeding is seasonal, mostly taking place during warmer months of spring. Males are known to travel hundreds of metres to reach the receptive female attracted by the sexual pheromone. During courtship, the male using pedipalps grasps the pedipalps of females and leads her into a courtship dance and deposits his spermatophore on a solid surface and then pulls the female over it. She, in turn, draws the spermatophore into her genital pore which is on the ventral side of the body at the anterior portion. Fertilization is internal. Gestation is prolonged extending from few months to over a year or 2 (Golizadeh et al. 2015).

Although scorpions are ovoviviparous, embryonic development is of two types: *apokogenic* embryos develop from ova containing plenty of yolk, whereas *katoikogenic* embryos develop from yolkless ova. They receive nourishment through a specialised feeding apparatus, the *diverticulum*, which extends from the mother's ovariuterus. This "teat" receives and transports nutrients from the hepatopancreas of the mother to the embryo. Some species reproduce parthenogenetically.

Fig. 4.2 Mother with babies on her back (Source: Dawn et al. 2001)

The young are born alive and immediately climb onto the mother's back. In most species the mother folds her legs to form a "birth basket" to catch the young as they are born. Averages of 25–35 young are born at a time. The process of giving birth may last from several hours to several days. The young remain on the back of the mother till the first instar (Fig. 4.2). After a few days, the translucent baby scorpions climb down, moult and climb back to the mother's back; stay there for another 4–5 days, and then leave to pursue their own individual life. The period from birth to leaving the mother takes place about 1–3 weeks depending on the species. After moving away from the mother, the young moult 5–6 times more to reach maturity. Scorpion's average life span is about 3–5 years although some species may live up to 50 years (Polis and McCormick 1987; Polis 1990).

4.5 Metabolism and Physiology

Scorpions are highly flexible in their metabolism. The metabolic activities can be slowed down to such a level that they can survive on as little as one insect in a year. This capacity is of extreme survival value when food becomes scarce or absent or during natural disasters. Not just slowing down, they are also capable of recovering normal metabolism instantly at the sight of a prey and charge at them when conditions become normal. They are also capable of consuming unlimited food when available with the excess food being stored as fat in hepatic tissue. The capacity to lower metabolism to the minimal levels coupled with almost nil activity during adverse conditions enables scorpions to thrive even in extreme, harsh environments (Polis 1990; Cloudsley-Thomson 1991). Their water requirements are mostly met from their prey and also by drinking from small pools, puddles and water bodies. The baby scorpions stay on the back of their mothers till their first moult absorbs moisture from the mother.

4.6 Antipredator Behaviour

Despite themselves being venomous predators, because of their relative larger bodies and abundance, scorpions are readily available to predators like owls, lizards, small snakes, bats and small rodents. A few vertebrates are known to be specialised predator of scorpions, at least seasonally. A few larger arthropods such as spiders and centipedes devour scorpions. Predation by other animals and cannibalistic activities act as limiting factors on scorpion abundance and distribution.

Scorpions too have evolved certain anti predatory strategies, the most effective and obvious one being the venom injecting. Venom is both offensive and defensive. One of its components is toxic to invertebrate predators, while the other acts as a deterent to vertebrate predators. Nocturnal life and living beneath solid objects are also anti predatory adaptations. Even during night their activities are limited to a few hours. On the other hand, some of their vertebrate predators have learnt to bite off their tail, and a few others are immune to scorpion venom (Polis and McCormick 1987; Polis 1990).

4.7 Social Life

The social life of scorpions is poor and almost non-existent with very few exceptions. A vast majority of them are solitary living. Interactions with conspecifics are limited to few stages of life, namely, at birth, during courtship and during cannibalism. In contrast, most inter- and intraspecies interactions are highly aggressive with few exceptions and occur for territorial defence and for accessing mates. Very few intraspecies interactions come close to social behaviour such as mother-offspring associations till the first moult of scorpion babies (Fig. 4.2) which in some cases may extend up to few years. In extreme cases the offspring even after attaining adulthood remain with adults forming family group; some of these family groups cooperate in prey capture. Such behaviour is seen in emperor scorpions. Extended families indulge in burrow digging and also share burrows. The bark scorpions in Arizona overwinter in aggregates of 20–30 (Fig. 4.3). Bark scorpions are one of the few scorpions that climb vertical walls but fail to cross horizontal ceilings (Polis 1990).

Fig. 4.3 Winter aggregation of bark scorpions (Source: Dawn et al. 2001)

4.7.1 Survival Strategies

4.7.1.1 Parthenogenesis

Of the several theories advocated to explain regional abundance of scorpions geographically, parthenogenetic reproduction in at least one species, *Tityus serrulatus*, in man-made environments in Brazil contributing to such regional, huge populations makes an interesting theory (Lourenco and Cuellar 1995). Depending on life history strategies and their habitat types, scorpions are divided into two broad categories: *equilibrium* and *opportunistic* (Polis 1990). Equilibrium species inhabit stable, natural environments, produce single-egg clutches, do not store sperm, exhibit long life spans, have low population density and population peaks, have weak mobility and are highly endemic (Lourenco 1991a, b; Polis 1990). Contrastingly invasion of disturbed environments, production of multiple clutches from a single insemination, elaborate sperm storage capacities, brief embryonic diet, short life spans, high population density, rapid mobility and wide distribution characterize opportunistic species as in genera *Centruroides*, *Tityus* and *Isometrus* (Kovoor et al. 1987).

The opportunistic scorpion species, *Tityus serrulatus*, of Brazil is highly toxic and was originally restricted to small areas of natural savannas. It has recently invaded the expanding human settlements in Western Brazil and is a serious public health hazard. Its population was characterized by the absence of males probably due to parthenogenetic mode of reproduction resulting in all female populations (Matthiesen 1962). Either the bisexuals in the populations were eliminated during evolution or the species was parthenogenetic even before invading newer habitats where bisexuals were also absent. Lourenco and Cuellar (1995) base their theory of parthenogenesis as a life history strategy to explain the extraordinary proliferation of *T. serrulatus* in Brasilia, the capital of Brazil which had earlier *Tityus fasciolatus* as the dominant species (93 %) followed by *Bothriurus araguayae* and *Ananteris balzani* during 1971–1975 (Lourenco 1981). Since its first appearance in the city in

1980, *T. serrulatus* has reproduced profusely forming 70 % of total scorpion population displacing *I. fasciolatus* as the dominant species. The life history traits of the successful species are density-independent population fluctuation and parthenogenetic reproduction compared to bisexual, density-dependent population with 1:3 sex ratio of *T. fasciolatus*. Similarly displacements of local species by *T. serrulatus* are occurring in other cities of Brazil, Colombia and Panama (Lourenco et al. 1994a, b).

Parthenogenesis is reported in four more species of scorpions, but their populations have both unisexual and bisexuals (Lourenco and Cuellar 1994). In natural environments, having sexually heterogeneous populations, bisexuals tend to be dominant and occupy more favourable areas of the habitat, pushing unisexuals to the periphery (Lourenco 1991c). A similar colonization of favourable valleys with subsoil water by *T columbianus* and invasion of human-altered environments by *T. serrulatus* are seen in Colombia (Lourenco and Cuellar 1994).

4.7.1.2 Polymorphism

Thirty-two species of scorpions whose sexually mature instars have been determined (Francke and Sissom 1984) have exhibited a seventh moult. Of these 15 species which exhibited more than 1 adult instar, 3 were males, 6 were females and 6 belonged to both sexes. A detailed study by Benton (1991) on the life history of *Euscorpius flavicaudis* in Southern England indicated two maturation strategies in this species: (i) maturing at sixth instar (normal) and be small or (ii) delay maturing till the seventh instar and be larger, 1.2 times in males and 1.21 in females, the bigger size bestowing both ecological and behavioural advantages. Large males have bigger chelae, the weapons of offence and defense which help the male to dominate smaller males in the fight for females during mating season and also aid the males to "persuade" unwilling females to mate (Benton, in press). Thus, the seventh instar males have a higher preseason reproduction success. The large females reproduce heavier offspring as they consume more food and can easily capture relatively bigger prey (Polis and McCormick 1987).

4.7.1.3 Natural Threats

The major threats to scorpion populations are predators. Their location and other prey are also important because birds not only predate on scorpions but also on centipedes which occupy habitats similar to scorpions. Bird's keen eyesight and swift movements make them a formidable predator. The mammalian predators are small rats and mice, while in deserts they are prone to be consumed by snakes and lizards. Bats, notably the pallid bat, *Antrozeus pallidus*, an inhabitant of Western USA, are immune to scorpion sting and are capable of detecting UV lights reflected by scorpions and sweep down on them easily during the night (Fig. 4.4; Polis 1990).

Life of scorpions is also threatened by other cannibalistic scorpions. But this phenomenon is rare and limited to certain inter- and intraspecific encounters such as the female devouring the exhausted male after mating in some species if the male is not quick enough to move away after mating. Starvation and fights over territory also lead to cannibalism. Other mortality factors are parasites and pathogens.

Fig. 4.4 The pallid bat with scorpion in its mouth (Source: http://blog.nature.org/science/2013/08/12/scorpion-eating-bat/)

By far the most dangerous predator of scorpion is human being. In some parts of the world, mainly China, scorpions are a delicacy eaten in vegetable and fruit preparations. However, the major threat is by anthropogenic activities such as expanding agriculture, hydroelectric projects, road construction and land conversion for industrial and agricultural purposes, all of which either destroy or fragment their natural habitat. A classical example of such a tragedy is reported from Afghanistan (Stewart 2012). A 40-ha montane scrubland near the city of Ghazni, which had a sizable population of *Mesobuthus caucasicus*, was converted into a military base during 2004–2006. The conversion process led to 50-fold increase in hard land surface and 20-fold increase in human population. Concrete walls, barriers and sandbags on gravel substrate not only destroyed the habitat of scorpions but also raised ambient temperature by 2.3 °C. Despite the presence of *M. caucasicus* in Ghazni city, Stewart (2012) could not locate a single scorpion in the military base. He postulated that the rapid anthropogenic activities at the base surpassed the scorpion's capacity to disperse to safer, alternate environments possibly leading to their extinction the base.

Other factors detrimental to scorpions include the tendency of humans to kill them as exotic pests and research on scorpion venom to exploit its medicinal properties. Despite predation, diseases, habitat destruction or fragmentation and detrimental human activities, scorpions are still thriving successfully all over the world, and none of the species is at the risk of extinction. On the contrary, new species are being discovered quite frequently.

4.7.1.4 Scorpion Venom

Venom of scorpions has two primary functions, that of killing the prey and self-defense. Additionally the sting may be used to subdue the female prior to mating. Although all scorpions are venomous, their basic instinct is to escape and hide in threatening situations. They are capable of controlling venom flow; as a result some stings can be venom less or only mildly lethal. Of the nearly 1400 species, only 25 are known to have venom potent enough to kill human beings. The genera that are notoriously lethal are *Androctonus*, *Buthus*, *Hottentotta* and *Leiurus* in North Africa and Middle East; *Tityus* in South America; *Mesobuthus* throughout Asia including India; *Centruroides* in Mexico, Southern USA, Central America and the Caribbean and *Parabuthus* in Western and Southern Africa. The three most dangerous scorpions of the world are:

1. The Indian red scorpion, *Hottentotta tamulus*
2. The death stalker scorpion, *Leiurus quinquestriatus*
3. The Arabian fat-tailed scorpion, *Androctonus crassicauda*

4.7.1.5 Epidemiology of Scorpion Bites

In some of the underdeveloped and developing countries, scorpion stings may cause severe deaths, but reliable data is not available as most victims do not seek medical treatment preferring to go to local practitioners of traditional medicine. Some studies suggest mortality rates of up to 4 % in hospitals. Scorpion stings pose significant public health threat in Sahara Africa, South Africa (Lamoral 1998), Iran (Mohseni et al. 2013), and India (Bawaskar and Bawaskar 2012).

Scorpion bites have been reported from the Middle East, South India, Mexico and Southern Latin America (Chippaux and Goyffon 2008). Tunisia is the most affected with reported cases of 4000 stings per year (Mansour 2007). In Khuzestan in South-Western Iran, scorpion stings are reported to cause significant human mortality due to stings of *Hemiscorpius lepturus* (Pipelzaden 2007). An epidemiological study over a period of eight years (2006–2012) in scorpion sting-prone areas of Ramshir district in the province of Khuzestan in South-Western Iran indicated high prevalence during June–October each year. More than 50 % of the victims were in the age group of 15–34 years and were stung on legs and hands. The species involved were *Mesobuthus phillipsii*, *Hemiscorpius lepturus*, *Orthochirus iranus*, *Androctonus crassicaudata* and *Scorpio maurus* (Gholizadeh et al. 2015). In another study, most victims were stung by *H. lepturus* although the number of collected species was comparatively less. This paradox was attributed to the ecology, distribution of scorpion and their more frequent confrontations with humans (Mohseni et al. 2013).

In Asia, India is the most affected with a reported incidence of 0.6 % (Chippaux and Goyffon 2008). People from endemic areas of Western Maharashtra, Karnataka, Andhra Pradesh, Sourashtra in the state of Gujrat and Tamil Nadu are prone to scorpion stings during the hot periods of April–June and September–October. Higher incidence of stings correlating with higher ambient temperature was attributed to increased agricultural activities prior to monsoon rains (Bawaskar and Bawaskar 2012).

Scorpion venom is a complex mixture of mucopolysaccharides, hyaluronidase, phospholipase, serotonin, histamine, enzyme inhibitors and toxins which are mostly neurotoxic. They affect the function of Na^+, K^+ and other ion channels. On the positive side of venom, it is extensively used in physiological and pharmaceutical research. The therapeutic properties of scorpion venom include anticancer, antimicrobial, antiepileptic, analgesic, antimalarial, pesticidal and insecticidal effects. They are reported to modulate cardiovascular function and autoimmune disease (Joseph and George 2012).

4.7.1.6 Clinical Manifestations

In severely affected people convulsions, paralysis and cardiovascular irregularities precede death. Generally, the effects of venom depend on species of scorpion and venom dose/body weight. Changes in body temperature are known to increase the sensitivity towards venom and influence the action of the venom (Murthy and Zave 1998). In India, Israel, Brazil and Mexico, cardiac manifestations are common. The other pathological results are tissue necrosis and haemolysis (Iran), neurological effects (South Africa and the USA) and acute pancreatitis in Tunisia. Nowadays, antivenoms are available against most lethal species bringing down death rates (Bawaskar and Bawaskar 2012).

The venom of less hazardous scorpions produces hemotoxins which cause mild to strong local effects. These consist of oedema, discolouration and pain. In fact the sting is less painful than of a bee sting, and complete recovery occurs within minutes or hours or days depending on the species of scorpion and the mental status of the victim. Early hospitalization and immediate administration of accurate doses of scorpion antivenom are important actions to be taken when stung.

4.8 Venom Protein as Fingerprinting Tool for Taxonomy

Biotopic variations of the scorpion venom peptides were successfully used to supplement species identification in India for *Mesobuthus tumulus* (Newton et al. 2007) Later Salama and Sharshar (2013) measured total proteins and protein profiles using SDS-PAGE bands of eight species of venomous scorpions collected from different localities in Egypt. The calculated similarity coefficients of the protein bands of these species were used to establish species-specific patterns in *Androctonus bicolor*, *Androctonus australis*, *Androctonus amoreuxi*, *Androctonus crassicaudata*, *Leiurus quinquestriatus*, *Buthus arenicola* and *Orthochirus innesi* (all belonging to family Buthidae) and *Scorpio maurus palmatus* (family Scorpionidae). The identification starting from family to species level through genera tallied with morphologically based classification. The electrophoresis patterns indicated that there were inter-family, inter-genus and inter-species variations between species studied. The study proposes to use variations in the venom protein both qualitatively and quantitatively as supplementary evidence to morphological characteristics in identifying scorpion species.

4.9 Scorpion Diversity and Distribution

The impressive diversity and abundance of scorpions in tropical, Neotropical and arid zones indicate that they contribute significantly to the biomass of animal communities of the habitats (Polis 1993). Their role as both predators and prey makes them an important link in the food web. Although their low metabolism limits role in energy and nutrient recycling, their relatively high biomass and role in the food webs make them an important component of community structure (Ayal 2007).

Scorpions successfully live in all terrestrial habitats, the exception being tundra, high-latitude taiga and very high mountain tops (Fig. 4.5). The taxonomical richness of scorpions exhibit two types. One type is similar to other fauna and flora, with maximum diversity prevailing in humid tropics and neotropics decreasing towards poles and equator (Laurenco 1994). There are several reasons for greater biodiversity in tropics. The foremost is the existence of large climatically similar surface areas. Other factors include relatively small fluctuations in temperature, high degrees of speciation and low levels of extinction. The second reason could be historical-geographical with large areas of earth surface being tropical and subtropical during the tertiary geological period. Thus, both evolutionary and geological processes of the past seem to have led to the high diversity of scorpions in tropics. Thirdly, tropical areas receive more solar radiation than those in the temperate zone leading to higher productivity and density. The higher temperatures can also shorten generation times and increase maturation rates accelerating the speciation process in the tropics (Lourenco 1994).

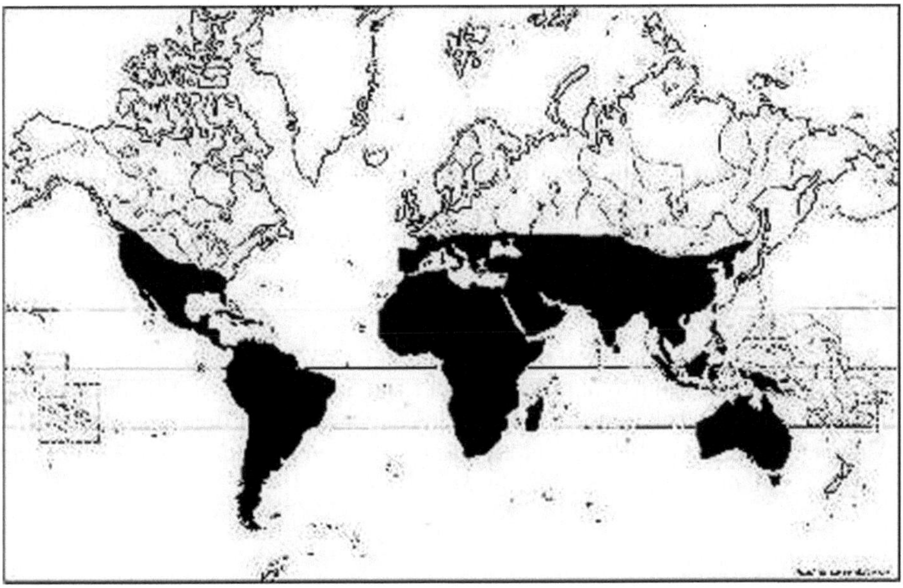

Fig. 4.5 World distribution of scorpions (Lourenco 2001)

The second interesting feature of scorpion diversity and distributions is their successful colonization of arid areas. Unlike most animal groups which have rich biotic representation in tropics compared to deserts and sandy areas, the most diverse communities of scorpions are found in arid regions (Hadley 1974). Their remarkable adaptation to such extreme ecosystems includes resistance to high temperatures (Cloudsley-Thompson 1962), metabolic and behavioural adaptations, ability to conserve water for prolonged periods even under very low humid conditions and living inside burrows deep enough not to feel the high ambient temperature (Hadley 1974).

Of the factors influencing scorpion diversity, rainfall, temperature and possibly competitions are important, but not vegetation according to Koch (1981). Later Smith (1995) argued that most species resistance will ultimately depend on long-term effects of chronic disturbances such as wind and water erosion, nutrient drifts, etc. which in turn depend on human's management and exploitation of resources. These results were based on a long and extensive study on the scorpions of Western Australian wheat belt. Later, Eric et al. (2010) again based on studies in Australia suggested that ruggedness and edaphic factors such as soil depth, texture and nutrient status were strongly correlated with the pattern of scorpion species richness and distribution in arid areas. Prendini (2005) attributed the species richness and endemism of South African scorpions to historical changes in geomorphology, climate and their ecological requirements. Summarily the available data showed that globally, regional species diversity varied from 1 to 13, with most areas having 3–7 species, and deserts averaging seven (Pollis and Yamashita 1991).

The following compilation is indicative of the adaptive radiation of scorpions in their ecological niche and medical importance of their venom which is the extreme evolution of self-defense. Countries having records of high venomation also have great species richness; one probable reason could be more research being carried out on the venom and its antidote and hence greater collection and sampling of scorpions. In other words, given more attention to sampling and taxonomic categorization, the number of scorpion species in many countries will be much greater than the present figures. Thus, scorpions represent a silent, simple evolutionary success despite the lack of sexual colouration, ornaments or vocal abilities, not changing much morphologically from millions of years. Probably intensive studies on their metabolism and behaviour may shed light on the secrets behind their spectacular success. Following is their country wise distribution in tropics and neotropics:

4.9.1 Algeria

Three species belonging to families Buthidae and Scorpionidae were collected from the pine, cedar and oak forests situated in the national park of Belezma in Northeast Algeria (Sadine et al. 2012). Buthidae was represented by *Androctonus bicolor* and *Buthus accinatus* and Scorpionidae by *Scorpio maurus*.

4.9.2 Australia

Scorpion diversity in **Australia** is low, only equalled by the former USSR which had only 14 species recorded (Fet 2010). Altogether 43 species occurred belonging to four families—Buthidae (12 species), Bothiuridae (6 species), Liochelidae (5 species) and Urodidae (20 species).

A later exhaustive study (Eric et al. 2010) across 179,000 km^2 in the arid zones of North-Western Australia revealed richer species diversity. Temperature and soil characteristics such as soil depth, texture and nutrient status were strongly correlated with patterns of scorpion species richness and distribution. However, there was a weak correlation between species diversity and sandiness of the soil.

4.9.3 Brazil

So far 150 species have been described in **Brazil**. The endemic taxa in this region are very high, i.e. more than 80 %. At least 3 families, 6 genera and 14 species were found in Rio Negro region, of which 93 % were endemic. The families were Buthidae, Liochelidae and Chactidae. Buthidae was better represented than Chactidae (Lourenco 1994, 1997). A new species, *Opisthacanthus* (family Liochelidae) was reported in 2003 by Lourenco and Fe. In 2005 two new species of Buthidae belonging to genus *Tityus* were described from Barcelona and São Gabriel da Cachoeira, respectively, in central and upper Rio Negro (Lourenco 2005), namely, *Tityus syloe* and *Tityus nelsini*. These two species resemble two other species of *Tityus*, namely, *T. strandi* and *T. raquelae*, distributed in Amazonas and Para in Brazilian Amazonia.

The scorpions of **Equador** are one of the least studied in South Africa. It is only since 1980 frequent studies have been carried out and several new species have been described. The country has 47 species, 8 genera and 5 families (Brito and Borges 2015).

South America is the most studied area with high endemism of scorpions. The belt comprising of Southern Colombia, Ecuador, north of Peru, Guyana, Imataca, Manaus and Santa Marta has a record number of 126 species, 17 genera and 6 families. Lourenco (1994) considers this region as the richest area in number of scorpion species and more diverse than even Baja California Desert, the area with the highest scorpion diversity and abundance in the world reported hitherto. With future studies and discovery of newer species, this part of South America may have the highest scorpion diversity in the world (Lourenco 1994).

Panama is inhabited by 3 families, five genera and 14 species. *Tittyus tayvona* is the dominant species.

4.9.4 China

The scorpion fauna of **China** consisted of 53 species and subspecies belonging to 12 genera under five families. Of these, 33 species and 1 genus are endemic. The topography and geography of east-west and north-south regions are distinct. Accordingly the distribution and species richness of scorpion vary between these localities. Only four species are recorded from east and Chinese islands. These are *Isometrus maculatus*, *Lychas mucronatus*, *Liocheles australasiae* and *Mesobuthus* *martensii martensii*. Three species seem to be extinct or invalid, namely, *Isometrus heinansis* (Lourenco et al. 2005), *Lychas scutilus* and *Mesobuthus martensii martensii* (Birula 1897). The genera *Chaerilus*, *Euscorpius* and *Mesobuthus* live in the Western region of China, namely, Tibet, Yunnan and Xinjiang. These are the areas of the richest scorpion diversity in China. Northern China has only two genera, namely, *Mesobuthus* (eight spp.) and *Rozanus* (one sp.).

4.9.5 India

Of the nearly 120 species of scorpions reported in **India**, none are lethal to human beings under normal circumstances. Despite the diversity and abundance, scorpions in India are poorly understood in terms of their ecology, biology and behaviour. Most studies are in the nature of discovery of newer species. The first major compilation of Indian scorpions was by the British zoologist, Innes Pocock, in 1894. After a century, Tikader and Bastawade (1983) published a book detailing the available information and also describing several new species. The book still remains as the reference guide for Indian scorpions. Of late there is a great surge of research on surveying newer species of scorpions by three major teams: that of Lourenco in Tamil Nadu, Pondicherry and Himalayas, Pande and his colleagues concentrating in Maharashtra, and Zambre and Mirza reporting several species from Western Ghats and also Andhra Pradesh and Kerala.

The Indian scorpions are classified into five main families, namely, Buthidae, Chaerilidae, Euscorpidae (Vaejovidae), Liochilidae and Scorpionidae (Tikader and Bastawade 1983). Some of the recently discovered scorpions in India are *Euscorpiops kamengensis* from Arunachal Pradesh (Bastawade 2006), *Euscorpiops bhutanensis*, *Euscorpiops* species either *beccaloniae* or a new taxon, *Euscorpiops asthenurus*, *Scorpiops leptochirus*, *Chaerilus pictus*, *Orthochirus krishnai* (Zambre and Bastawade 2009), *Chaerilus annapurna* from the high plateaus of the Himalayas (Lourenco and Duhem 2010) and 14 other species.

Because of the diverse climate, geographic conditions, land use patterns, habitats, lifestyles and cultural variability, scorpions in India present a unique assemblage of species with much variations. Although there is a lacuna on research on Indian scorpions, new species are being discovered, and select aspects of physiology on *Heterometrus fulvipes* have been documented. Based on the available date, scorpions are categorised into burrowing, rock dwelling and arboreal. The burrowing scorpions include several genera of buthids such as *Odontobuthus*, *Androctonus*, *Orthochirus*, *Compsobuthus* and *Mesobuthus*, and chaerilids like *Heterometrus* and Ischerids are true burrowing species. *Heterometrus* do not dig deep burrows only pits to take shelter. They are just 6″–9″ found beneath boulders. Ischerids excavate deep burrows (1′–2′) which terminate in a bulb-like structure where the scorpions dwell. Some buthids live in colonies in the same burrow. Vaejovids live inside cracks and crevices of the rocks. Vaejovids and chaerilids live at high altitude; the former occur both in the Himalayas and Deccan, while the chaerilids are known only from the Himalayas.

4.9.5.1 Species Diversity Studies

In a lone study on species diversity and microhabitat preferences in a heterogeneous habitat of cultivated fields, mango orchards, boulders, rocks, different soil conditions and scrubland, Pande et al. (2012) found five genera of three families, namely, Buthidae, Scorpionidae and Euscorpidae, represented by eight species as measured by quadrat sampling in Saswad-Jejuri in Western India. The overall species richness was 0.7, Shannon diversity index was 1.1, and evenness index was 0.6. *Hottentotta tamulus* was the dominant species and *Heterometrus phipsoni* was found in lowest numbers. *Hottentotta xanthopus* was a co-dominant species. *Hottentotta pachyurus* was found only under tree barks, whereas *Isometrus rigidus* was seen only under heaped stone rubble. Both these species were microhabitat specific. *H. phipsoni* was found only in scrub-land with stones but shared it with *H. tamulus*. The authors suggest that these microhabitats need to be protected. There was species overlap in the occupancy of other microhabitats, namely, loam and stones in hilltops, veldt with stones, red and black soil in cropland, grassy hilltops with stones, black soil in mango orchards, beneath tree barks, hill slopes with boulders and eucalyptus plantations. These were grouped into two-clade microhabitats that were arid and xeric, the second one with some vegetation. Wasteland like the current study area is under the threat of developmental work, and hence scorpion fauna living here is prone to destruction and migration if they survive. To save the scorpions, the authors advocate impact assessment studies before initiating major land changes by both public and private sector.

A second study, strictly speaking, was not on species diversity, but the researchers could capture four species of scorpions for their venom extraction work, thereby identifying the scorpion habitats in the districts of Bellary and Chitradurga in Karnataka (Nagraj et al. 2015). They were *Hottentotta rugiscutis*, *Hottentotta tamulus*, *Lychas tricarinatus* (found underneath stone and dead wood situated in damp soil) and *Heterometrus swammerdami* (burrow living at a depth of (5^{1}–10^{1}).

4.9.5.2 Social Behaviour

A lone publication (Shivashankar 1994) indicated the presence of the mother and young of *Heterometrus fulvipes* in the same burrow even after the completion of the final moult. The group exhibited division of labour in foraging, burrow extension activities and shared food.

4.9.5.3 Breeding and Reproduction

Heterometrus fulvipes exhibited definite breeding season in July–August with a gestation period of 11 months in South India (Subbaram and Reddy 1976). In *Heterometrus phipsoni*, gestation was 71 days, parturition occurred in the month of June, and the number of young born was only 7 (Mirza and Sanap 2009). Testicular development of male *H. fulvipes* was identified to have four stages of development occurring seasonally. These stages were accompanied by changes in protein content, glycogen and glucose levels in the testicular and pancreatic tissues. Seasonal changes in the histology of the cephalothorax accompanied by changes in the biochemistry of different tissues including testes indicated a possible role of neurosecretory cells in the courtship and mating behaviour of *H. fulvipes* (Soba et al. 2007).

4.9.5.4 Laboratory Studies

In contrast to meagre studies on ecology and behaviour on scorpions in India as also occurring elsewhere, significant research has been carried out on most aspects of physiology of *Heterometrus fulvipes* mostly at Sri Venkateswara University, Tirupati, in the early 1960s; later at Bangalore University, Bangalore, in the late 1960s; and at Nagarjuna University, in Andhra Pradesh presently. The work initiated by the late Prof. Pampapathi Rao diverged in two mainstreams—biochemistry and physiology on one hand and neurophysiology including receptors employing electrophysiological techniques on the other hand. In fact it was for the first time electrophysiological techniques were introduced to study the activity of the nervous system and sense organs particularly in scorpions in India by Pampapathi Rao. Briefly the investigation was in the following lines:

(i) *Cardiac metabolism and haemolymph*: The heart of male *Heterometrus fulvipes* was morphologically bigger than that of female, was carbohydrate rich and had greater rates of fatty acid conversion to citrate compared to smaller hearts of females which were lipid rich, a biochemical status suited to higher metabolic and physical activities of males (Padmanabhanaidu et al. 1984).

(ii) *Haemolymph*: Three tests, namely, agglutination test, clearance test and enzyme immunoassay tests, carried out on *Heterometrus fulvipes* indicated a well-integrated internal defense system in the form of haemocytes and lymph glands (Amarnath 1996).

(iii) *Activity rhythms*: Diurnal activity was demonstrated in *H. fulvipes* with peaks of activity occurring during the first four hours of the night (Kasiah 1989). Organs involved in locomotion, namely, leg, tail and pedipalpal muscles, exhibited diurnal rhythmicity. Rhythmicity in physiological parameters such as blood glucose and liver glycogen (Changal Raju et al. 1973), heartbeat, muscle dehydrogenase (Venkateshwara Rao and Govindappa 1967), acid and alkaline phosphatase activities (Chandrashekara Reddy and Padmanabhanaidu 1997), lipid levels (Mastaniah et al. 1978) and acetyl cholinesterase activity (Uttaman and Srinivasa Reddy 1980, 1985) were reported for *H. fulvipes*.

(iv) *Neurophysiology and sensory organs*: Exhaustive laboratory investigations were carried out, and significant information was generated on rhythmicity in neural activity, vision,photo receptors, extraocular photo reception, cuticular sense organs, giant fibres, neuro humoral secretions and biochemical studies on sub oesophageal gland (Habibulla 1971, 2004; Ramakrishna and Pampapathi Rao 1970; Sanjeeva Reddy and Pampapathi Rao 1970; Geethabali and Papmpapathi Rao 1973; Geethabali 1976; Venkatachari 1971; Vasantha et al. 1977; Yellamma et al. 1980).

(v) *Thermal acclimation*: Acclimation to both low and high temperatures in *H. fulvipes* was studied in the laboratory (Vijayalakshmi 1964; Kalarani et al. 1991a, b, 1992). Detailed investigations on oxygen consumption by the whole animal, pedipalp and heart muscles, levels of glycogen, lactate, pyruvate in the pedipalp and heart muscles, changes in the oxygen consumption, alpha phosphorylase activity and isocitrate dehydrogenase activity in the hepatopancreas of animals acclimated to temperatures below and above habitat temperature led to the conclusion that adaptation to low temperatures is more stressful.

4.9.6 Iran

Due to the high incidence of sting, scorpion fauna of **Iran** particularly in South West has been extensively sampled. An earlier report (Mirsha et al. 2011) recorded 51 species, 18 genera and 4 families, namely, Buthidae, Scorpionidae, Hemiscorpidae and Diplocentridae. Buthidae was the most diverse. The greatest diversity was seen in South-Western Iran probably due to more sampling. The authors suspect the presence of many more species.

4.9.7 Israel

Israel is inhabited by three families, namely, Buthidae, Scorpionidae and Diplocentridae, spread across 9 genera and 19 species in its arid and mesic areas (Levy and Amitai 1980; Levy and Alon 1983).

4.9.8 Malaysia

A preliminary checklist of scorpions at Kuala Lompat, Krau Wildlife Reserve, **Malaysia**, revealed three species, namely, *Heterometrus longimanus*, *Isometrus zideki* and *Liocheles australasiae*, belonging to three families: Scorpionidae, Buthidae and Liochelidae (Izzat-Husna et al. 2014).

4.9.9 Mexico

According to the latest review of scorpion fauna of **Mexico** by Santibanez-Lopez et al. (2016), the country harbours the highest diversity of scorpions in the world including some of the world's most medically important species. The 281 diverse species of scorpions in Mexico represent 12 % global diversity. There are 9 families and 38 genera. These include Buthidae, Euscorpidae, Superstitionidae, Typhlochactidae, Vaejovidae, Diplocentridae, Chactidae, Caraboctonoidae.

Baja California Sur, coast of Guerrero, the Isthmus of Tehuantepec, Central Valley in Oaxaca, the valley of Cuicatlan-Tehuacan (Southern Mexico) and the Yucatan Peninsula (Eastern Mexico) are the hotspots of scorpion biodiversity in Mexico (Williams 1980).

Apart from taxonomical, geographical and medicinal aspects, studies also have been carried out on litter size, burrow size, density, biomass, life cycle, courtship, mating plug and aspects of ecology to some extent.

4.9.10 South Africa

Prendini (2005) in his classical study compiled and analysed 140 scorpion species of **South African countries** using a geographical information system. The taxonomic composition of scorpions of South Africa was distinctly different in the Western and Eastern regions. The greatest species richness and endemism characterized arid regions with rugged topography, complex geology or substratal heterogeneity. The speciation, species richness and endemism were attributed to historical changes in geomorphology, climate and specific ecological requirements. The scorpion fauna of Southern Africa is very rich, and their composition differs in north-eastern and north-western parts of the subregion. Most Southern species were primitive, while those of north were more evolved (Hewit 1925). Prendini (2001) very strongly argues for an ecological explanation within a historical context for these regional differences for which he analysed the collected scorpions using a geological information system.

4.9.10.1 Endemism

Forty-six percent of the genera and 96 % of the species of South Africa are endemic to the subregion. Three genera, namely, buthids (*Afroisometrus*, *Karasbergia* and *Pseudolychas*), buthriurids (*Brandbergia* and *Lisposoma*) and theliochelids (*Cheloctonus*), are endemic to this region.

4.9.10.2 Spatial Distribution

Several genera are restricted to the Western region—*Brandbergia*, *Lisposoma* and *Karasbergia*—while the Eastern region is inhabited by species belonging to the genera *Afroisometrus*, *Lychas*, *Pseudolychas*, *Cheloctonus* and *Opisthocanthus*. Three genera, viz. *Hottentotta*, *Uroplectes* and *Hadogenes*, are evenly distributed between Eastern and Western regions. The two largest genera, *Opistophthalmus* and *Parabuthus* dominate the arid Western half, but several endemic species of the two genera inhabit both Eastern and Western parts of the regions. The species are also associated with geomorphologic features of landscape such as sand systems, sandy areas, mountain ranges with rugged topography, plains and alongside the river Orange. Prendini (2001) attributes the high species richness and endemism to the cumulative effects of geomorphology and palaeoclimatic changes.

4.10 Conclusion

In spite of occupying habitats less favourable compared to other animals and not having undergone much evolutionary changes, scorpions have thrived and are living successfully in all habitats except the extreme snowy, mountainous and aquatic mediums. It appears that both behaviour and physiology have played a vital role in their successful existence preventing them from reaching alarming situations, not at least immediately. However, the fast pace of development occurring in countries where it has the greatest diversity, such as Brazil and other South American countries, Mexico and India, is definitely going to threaten their survival, specially some of the endemic species which did not have the necessity to evolve and migrate hitherto. When projects seek clearance from the environment departments, keeping scorpions also in the picture is necessary to preserve the diversity and abundance of scorpions. The high demand of pharmaceutical industry for research may affect their population levels. Effects to breed them successfully in the laboratory may alleviate this threat.

Further, it is not heartening to observe that most studies on scorpions have focused on taxonomic diversity and historical biogeography in many species-rich and not so rich countries, namely, Brazil, Andean Neotropical countries, Iran and the USA (California). One exception is Prendini's (2005) classical review on the scorpions of South Africa who not only identified areas of species richness and endemism but also related these to broad-scale land elements, phylogeny and historical geographical changes. To some extent exhaustive studies have been carried out on the physiology of one species (*Heterometrus fulvipes*) in India which is an encouraging trend. Such studies across countries with similar or different ecosystem elements may throw more light on the evolution, success and preservation of scorpions.

These remarkably successful groups of animals deserve and need to be studied in detail—their ecology, biology, behaviour and other aspects—so as to understand the secret of their abundance and distribution in various ecosystems of the earth apart from their well understood venom.

References

Amarnath (1996) Effect of Xenogenic implantation on the scorpion, *Heterometrus fulvipes.* Thesis submitted to Sri Venkateswara University for Doctoral degree, Tirupati, India

Ayal Y (2007) Trophic structure and the role of predation in shaping hot desert communities. J Arid Environ 68:171–187

Bastawade DB (2006) Arachnida: Scorpionida, Uropygi, Schizomida and Oncopodid Opiliones (Chelicerata) Arunachal Pradesh State Fauna Ser 13(2):449–465

Bawaskar HS, Bawaskar PH (2012) Scorpion sting: update. J Assoc Physicians India 60:46–55

Benton TG (1991) The life history of *Euscorpius flavicaudus* (Scorpiones, Chactidae). J Arachnol 19:105–110

Birula AA (1897) Miscellanea scorpiologica II Zur synonymie der russischen scorpione. Annuri Muse's Zool L'Acad Impe'rialedes Sci, St Petersburg 9:20–27

Brito G, Borges A (2015) A checklist of the scorpions of Ecuador (Arachnida: Scorpion) with notes on the distribution and medical significance of some species. J Venom Anim Toxins 21:23. doi:10.1186/40409-015-0023

Chengal Raju D, Basha Mohideen MD, Narasimham C (1973) Circadian rhythm in blood glucose and liver glycogen levels of scorpion, *Heterometrus fulvipes*. Experientia 29:964-965

Chippaux JP, Goyffon M (2008) Epidemiology of scorpionism: a global appraisal. Acta Trop 107:71–79

Cloudsley-Thompson JL (1962) Lethal temperatures of some arthropods and the mechanism of heat death. Exp Appl. 5:276–280

Cloudsley-Thompson JL (1991) Ecophysiology of desert arthropods and reptiles. Springer, Berlin

Dawn HG, Kirk AS, Carl Olson, Paul Baker (2001) Scorpions, AZ1223, April 2001. An AZ cooperative extension bulletin. University of Arizona Cooperative Extension, pp 2

Dyanand Y (1996) Neuroethology of the scorpion. *Heterometrus fulvipes* and *Buthus tamulus* with special reference to the cuticular sense organs of the walking legs. Thesis submitted to Sri Venkateswara University, Tirupati for Doctoral degree

Eric S, Volschenk, Burbridge AH, Durrant BJ, Harvey MS (2010) Spatial distribution patterns of scorpions (Scorpiones) in the arid Pilbara region of Western Australia. Rec West Aust Mus Suppl 78:271–284

Francke OF, Sissom WD (1984) Comparative review of the methods used to determine the number of molts to maturity in scorpions (Arachnida), with analysis of the post-birth development of *Vaejovis coahuilae* Williams (Vaejovidae). J Arachnol 12:1–20

Geethabali (1976) Central course of photic input in the ventral nerve cord of scorpion (*Heterometrus fulvipes*). Experientia 32(3):345–347

Geethabali, Pampapathi Rao K (1973) A metasomatic neural photoreceptor in the scorpion. J Exp Biol 58:189–196
Gholizadeh S, Lalehzari E, Buvani MM, Hosseini A, Khalkhali HR, Rafinejad J (2015) Bioecology and scorpion envenomation in Ramshir district, Khuzestan Province, South Western Iran. Appl Entomol Zool. doi:10.1007/s13355-015-0367-2
Habibulla M (1971) Neurosecretion in the brain of a scorpion *Heterometrus swammerdami* – a histochemical study. Gen Comp Endocrinol 17:253–255
Habibulla M (2004) Studies on the brain modulation of circadian rhythm in the scorpion *Heterometrus swammerdami.* Proceedings of the 28th annual meeting of American Arachnological Society Norman, Oklahoma
Hadley NF (1974) Adaptation biology of desert scorpions. J Arch 2:11–28
Hewitt J (1925) Facts and theories on the distribution of scorpions in South Africa. Trans R Soc South Afr 12:249–276. http://blog.nature.org/science/2013/08/12/scorpion-eating-bat/
Izzat- Husna M, Riza SA, Amirudin BA (2014) Notes on scorpion fauna in Kuala Lompat, Krau Wildlife Reserve, Pahang, Malaysia. J Wildl Parks 28:121–129
Joseph B, George J (2012) Scorpion toxins and its applications. Int J Toxicol Pharmacol Res 4(3):57–61
Kalarani V, Reddy DC, Naidu BP, Davis RW (1991a) Time course of thermal acclimation in the tropical scorpion, *Heterometrus fulvipes*. J Therm Biol 16(3):135–139
Kalarani V, Reddy DC, Naidu BP, Davis RW (1991b) Acclimatory response in enzymatic activity and metabolite concentrations in the tropical scorpion, *Heterometrus fulvipes*. J Therm Biol 16(6):327–332
Kalarani V, Murali Mohan P, Davis RW (1991c) Thermal acclimation and metabolism of the hepatopancreas in the tropical scorpion, *Heterometrus fulvipes*. J Therm Biol 17(3):141–146
Kassiah A (1989) Neuroethology of the scorpion, Heterometrus fulvipes: habitat, cuticular sense organs and behaviour. Ph.D. thesis, Sri Venkateswara University, Tirupati
Koch LE (1981) The scorpion of Australia: aspects of their ecology and zoogeography. In: Keast A (ed) Ecological biogeography of Australia. W. Junk, The Hague, pp 873–884
Kovoor J, Lourenco WR, Munoz-Cuevas A (1987) Concentration des spermatozoids dans les voies genitals des females et biologic de la reproduction des scorpions (Chelicerates). C R Acad Sci Paris 304:259–264
Lamoral BH (1998) The scorpions of South West Africa. Ph. D thesis, University of Natal, Pietermaritzburg
Levy G, Alon A (eds) (1983) Scorpions, plants and animals of the land of Israel, an illustrated encyclopedia 2. Ministry of Defense, The Publishing House, Society for the Protection of Nature, Israel, pp 17–28
Levy G, Amitai P (1980) Fauna Palaestino, Arachnida I. Scorpions. The Israel Acad Sci and Humanities, Jerusalem
Lourenco WR (1981) Sur l'ecologie du scorpion Buthidae: *Tityus trivittatus fasciolatus* pessoa, 1935. Vie Millieu 31:71–76
Lourenco WR (1991a) Biogeographiqu evalutie, ecologie et les strategies biodemographiques chez les Scorpions neotropicaux. C R Soc Biogeogr 67:171–190
Lourenco WR (1991b) La province biogeographique guayanaise: etude de la biodiverseities et des centres endemisme en rue de la conservation des patrimoines genetique. C R Scoc Biogeogr 67: 113–131
Lourenco RW (1991c) Parthenogenesis in the scorpion *Tityus columbianus* Thorell (Scorpiones: Buthidae). Bull Br Arachnol Soc 8:274–276
Lourenco WR (1994) Diversity and Endemism in tropical versus temperate scorpion communities. Biogeographica (Paris) 70(3):155–160
Lourenco WR (1997) Synopsis de la faune de scorpions de Colombie, avec des considerations sur le systematique et al biogeographic des especes. Rev Suisse Zool 164:61–94
Lourenco WR (2001) The scorpion families and their geographical distribution. J Venom Anim Toxins 7(11):3, on-line version ISSN 1678-4936
Lourenco WR (2005) Scorpion diversity and extremism in the Kio-Negro Region of Brazilian Amazonia with the description of two new species of *Tityus* CL Koch (Scorpions, Buthidae). Amazonia 18(34):203–231

Lourenco WR, Cuellar EO (1994) Notes on the geography of parthenogenetic scorpions. Biogeographica 70:19–23
Lourenco RW, Cuellar EO (1995) Scorpions, Scorpionism, life history strategies and parthenogenesis. J Venom Anim Toxins 1(2). http://dx.doi.org/10.1590/50104-79301995000200002
Lourenco WR, Duhem B (2010) The genus *Chaelius simon* and description of a new species. Zookeys 37:13–15
Lourenco WR, Qi JY, Zhu MS (2005) Description of a new species of scorpion from China (Tibet) belonging to the genus *Mesobuthus* Vachon (Buthidae) and *Heterometrus* Ehrenberg (Scorpionidae). Zootaxa 998:1–16
Lourenco WR, Knox MB, Yoshizawa MAC (1994a) L'invasion d'une communaute a le stade initial d'une succession secondaire parune espece parthenogenetiquede scorpion. Biogeographica 70:77–91
Lourenco WR, Kox MB, Yoshizawa M (1994b) L'invasion d'une communaute e le stade initial d'une succession secondaire parune initial d'une succession secondaire par use espece parthenogentiquede scorpion. Biogeographica 70:77–76
Mansour N (2007) Delay and characterization of scorpion management in Thisisi-Bouzid region. Arch Inst Pasteur Tunis 78:25–31
Mastanaiah S, Doreswamy Reddy V, Chengala Raju D, Swamiy KS (1978) Circadian rhythmic activity of lipase in the scorpion, *Heterometrus fulvipes* (C.L. Koch). Curr Sci 47:130–132
Matthieson FA (1962) Parthenogenesis in scorpions. Evolution 16:255–256
Mirsha MSI, Sari A, Hosseinlie S (2011) History of study and checklist of the scorpion fauna (Arachnida: Scorpiones) of Iran. Prog Bio Sci 1(2):16–28
Mirza ZA, Sanap R (2009) Notes on the reproductive biology of *Heterometrus phipsoni* pocock 1893 (Scorpiones: Scorpionidae). J Threatened Taxa 1(9):488–490
Mohseni A, Vazirianzadeh B, Hossienzadeh M, Salehchea M, Moradi A, Moravvej SA (2013) The roles of some scorpions, *Hemiscorpius lepturus* and *Androctonus crassicauda*, in a scorpionism focus in Ramhormorz, South Western Iran. J Insect Sci 13:59
Murthy KRK, Zave MA (1998) Effect of Indian red scorpion on thyroxine, triiodothryronine in experimental acute myocarditis and its reversal by species specific antivenom. Indian J Exp Biol 30:16–21
Nagraj SK, Dattatreya P, Boranmurthi TN (2015) Indian scorpions collected in Karnataka: maintenance in captivity, venom extraction and toxicity studies. J Ven Anim Toxins Including Trop Dis 21:51–54
Newton KA, Clench MR, Deshmukh R, Jayasheelan K, Strong PN (2007) Mass finger printing of toxic fractions from the venom of the Indian red scorpion, Mesobuthus tamulus: biotopic specific variations in the expression of venom peptides. Rapid Comm Mass Spectrum 21:3467–3476
Padmanabha Naidu B, Jayaram V, Kalarani V, Chandrashekara Reddy D (1984) Studies on the cardiac metabolism of scorpion: some physical and metabolic parameters. Indian J Comp Anim Physiol 2(1):58–63
Pande S, Bastawade D, Padhyye A, Pawashe A (2012) Diversity of scorpion fauna of Saswad-Jejuri, Pune district, Maharashtra, Western India. J Threatened Taxa 4(2):2381–2389
Patil AE, Shah UH (2011) Types of hemocytes in scorpion *Mesobuthus tamulus tamulus*. Bioscan 6(40):597–599
Pipelzaden MH (2007) An epidemiological and a clinical study on the Iranian scorpion, *Hemiscorpius lepturus*. Toxicon 50:548–592
Pocock RI (1894) Scorpions and their geographical distributions. Nat Sci 4:353–364
Polis GA (1990) The biology of scorpion. Stanford University Press, Stanford, p 585
Polis GA (1993) Scorpions as model vehicle to advance theories of population and community ecology: the role of scorpions in desert communities. Memories Queensland Mus 33:401–410
Polis AG, McCormick SJ (1987) Intraguild predation and competition among scorpions. Ecology 68:332–343
Polis GA, Yamashita T (1991) The ecology and importance of predaceous arthropods in desert communities. University of Arizona Press, Tuscon, pp 186–222
Prendini L (2001) Substratum specialization and speciation in Southern African scorpions: the effect hypothesis revisited. In: Memorium Gary A, Polis V, Fet V, Selden PA (eds) Scorpion 2001. British Arachnological Society, Burnham Beeches, Bucks, pp 113–138

Prendini L (2005) Scorpion diversity and distribution in Southern Africa: pattern and process. In: BA Huber, BJ Sinclair, Kwlampe (eds) African biodiversity: molecules, organisms, ecosystems: 5th international symposium on tropical biology proceedings. Bonn, Germany, pp 25–68

Ramakrishna T Pampapathi Rao K (1970) Nonspecific sensory input from the eyes of the scorpion, *Heterometrus fulvipes*. Proc Indian Acad Sci 73(4):202–207

Reddy DC Padmanabhanaidu B (1997) Diurnal rhythmicity of alkaline and acid phosphatases in the scorpion, *Heterometrus fulvipes* (CL Koch). J Comput Physiol Ecol 2:217–219

Sadine SE Youcef A Chenchouni H (2012) First data on scorpion diversity and ecological distribution in the national park of Belzma, Northeast Africa. Serker 13(12):27–37

Salama W Sharshar KR (2013) Surveillance studies of scorpion species in Egypt and compositions of their crude venom protein profiles. J Basic Appl Zool 66(2):76–86

Santi-Banez Lopex CE Francke OE, Ureta C Popanu LD (2016) Scorpions from Mexico: from species diversity to venom complexity. Toxins 9(2):2–18

Shivashankar TV (1994) Advanced sub social behaviour in the scorpion, *Heterometrus fulvipes*. Brunner Arachnida J Biosci 1(10):81–90

Smith GT (1995) Species richness, habitat and conservation of scorpion in the Western Australian wheat belt. Rec West Aust Mus Suppl 52:55–66

Snjeeva RP, Pampapathi RK (1970) The central course of the hair afferents and the pattern of contralateral activation in the central nervous system of the scorpion. Heterometrus Fulvipes J Exp Biol 53:165–169

Sobha K Veeraiah K, Gopal Rao N (2007) Cytological study of neurosecretion in relation to reproduction in the male scorpion, *Heterometrus fulvipes* (Koch) Braz. J Morphol 24(2):137–147

Stewart AK (2012) Military base growth in Afghanistan: a threat to scorpion population? J Arachnology 40(2):245–248

Stockwell SA (1989) Revision of the phylogeny and higher classification of scorpions (Chelicerata). Ph.D thesis, University of California, Berkley

Subburam V Reddy TG (1976) Breeding season and gestation period of the scorpion *Heterometrus fulvipes*. J Bombay Nat Hist Soc 75(2):513–514

Tikader BK Bastawade (1983) Fauna of Indian scorpions: Scorpionida: Arachnida- vol-III. Zoological Survey of India, Calcutta, pp 671

Uthaman M Srinivasa Reddy Y (1980) Neural functions in the regulation of oxygen consumption rhythm in the tissues of the scorpion, *Heterometrus fulvipes* (Koch CL). Chronbiology 7:533–537

Uthaman M Srinivasa Reddy Y (1985) Rhythmic changes in the activity of neurosecretory cells of the scorpion, *Heterometrus fulvipes* (Koch C). Chronobiology 12:331–338

Vasantha N Venkatachari SAT Murali Mohan P Sasira Babu K (1977) On the possible mode of action of neurohormones on cholinesterase activity in the ventral nerve cord of the scorpion, *Heterometrus fulvipes*. Experientia 33(2):238–239

Venkatachari SAT (1971) Analysis of spontaneity in the nervous system of the scorpion, *Heterometrus fulvipes*. Indian J Exp Biol 9:338–344

Venkatehwara Rao P Govindappa S (1967) Dehydrogenase activity and its diurnal variations in different muscles of the scorpion, *Heterometrus fulvipes*. Proc Indian Acad Sci 66:243–249

Vijayalakshmi C (1964) Studies on energy metabolism of scorpion with special reference to thermal acclimation. Ph.D thesis, Sri Venkateswara University, Tirupati, India

Williams SC (1980) Scorpions of Baja California, Mexico and adjacent islands. Occas Pap Calif Acad Sci 135:1–127

Yellamma K Mohan M, Sasira Babu K (1980) Morphology and physiology of giant fibres in the seventh abdominal ganglion of the scorpion *Heterometrus fulvipes*. J Compar Physiol 59(3):286–298

Zambre AM, Bastwade DB (2009) Description of male *Orthochirus krishnai* (Scorpiones: Buthidae) from India with comment on its taxonomic status. J Threatened Taxa 1(12):621–623

Zambre A Sanap RV, Mirza ZA (2014) A new high elevation scorpion species of the genus Scorpions peters, 1861 (Scorpiones: Euscorpiidae: Scorpiopinae) From the Himalayas, India. CR Biologies (http://dx.dor.org/146:10.1016/j.cvi.2014)

An Overview of Crustacean Diversity in Mangrove Ecosystem

5

S. Murugan and D. Usha Anandhi

Abstract

Mangrove forests are the protective barrier of coastal inhabitants against natural calamities and prevent soil erosion protecting the land behind. These typical ecosystems provide areas for breeding, nesting, foraging and shelter for economically important organisms. Hence, most of the growing population in the world live within easy reach of coastal areas. Crustaceans like lobsters, crabs, crayfish, shrimps, barnacles, etc. are very important in the nutrient recycling and are most crucial in human economy. Crustaceans are unique source of nutrients like proteins, fats and minerals to aquatic life as well as to human beings. As mangroves act as nursery, high juvenile abundance of many aquatic organisms is seen. Today these economically important ecosystems are among the most threatened habitats with 30–50 % of global loss due to natural and anthropogenic disturbances. Awareness in public through media, Non Government Organisations (NGOs), Government organisations and educational institutes who must come together for restoration, rehabilitation and conservation of this delicate and precious ecosystem.

Keywords

Crustaceans • Conservation • Ecosystem • Mangrove

S. Murugan (✉) • D.U. Anandhi
Department of Zoology, Bangalore University, Bangalore 560056, Karnataka, India
e-mail: muru3986@gmail.com

A.K. Chakravarthy, S. Sridhara (eds.), *Arthropod Diversity and Conservation in the Tropics and Sub-tropics*, DOI 10.1007/978-981-10-1518-2_5

5.1 Introduction

Mangrove ecosystems are the woody halophyte-dominated forests located in muddy, loose and wet soils at the junction of terrestrial and marine ecosystem rich in productivity with commercially important fishes, crustaceans and molluscs. Mangrove forests act as a protective barrier to the coastal inhabitants against the natural calamities like flood, tsunami hurricanes and prevent soil erosion protecting the land behind. A number of studies have shown that mangrove ecosystems act as a natural green belt of the world (IUCN 2005).

In India, mangroves are diverse with 125 species, of which 39 species are mangroves and 86 species are mangrove associates, amounting to 56 % of the world's mangrove species with equally diverse fauna of birds, reptiles, fishes, crustaceans, molluscs and insects including both resident and visiting fauna. Reptiles like snakes, turtles, estuarine crocodiles and monitor lizards are common in mangroves which survive the range of salinity. Mammals like various species of monkeys, otters, deers, fishing cats and wild pigs are most common in mangrove forests (http://www.indian-ocean.org/bioinformatics/mangroves/mangcd/fact.htm).

5.2 Mangrove

Mangrove forests are usually dominated by red mangrove (*Rhizophora* sp.), black mangrove (*Avicennia* sp.) and white mangrove (*Laguncularia* sp.) The red mangrove being the tallest with a maximum height of 25 m and acting as a basis for detrital food chain in estuary has been well documented (Hutchings and Saenger 1987). Black mangrove is the second tallest species reaching a height of 20 m, and the white mangrove is the smallest of the three mangroves with a maximum height of 15 m. Mangrove forest are the most productive ecosystems (Por 1984), and they hold an enormous carbon stock per unit area (Twilley et al. 1992). The mangrove has aerial root system, either prop roots or stilt roots (Fig. 5.1) which are important organs of breathing at high tides and help in anchoring of the plant which is a characteristic of *Rhizophora*. They also occur in *Bruguiera*, *Ceriops*, *Avicennia alba* and *A. officinalis*. These roots provide an important nursery ground for crustaceans, molluscs and fishes (Odum and McIvor 1990). Mangrove plants along with the root system help in sedimentation (Krauss et al. 2003) by slowing water velocity (Mazda et al. 1997).

Globally tropical mangroves are the most productive natural ecosystem (Alongi 2009), and their distribution is divided as the Atlantic East Pacific hemisphere and the Indo West Pacific hemisphere, where the former one has less species diversity than the latter (Fig. 5.3). Mangroves are globally distributed in four ecotypes: fringe, riverine, basin and scrub forests (Twilley 1998) with key faunal species richnesses (Lee 2008). The Matang Mangrove Forest Reserve that consists of roughly 40,151 ha of pure and mixed stands of *Rhizophora* and *Bruguiera* had been established in Perak, Malaysia (Gan 1995). The mangrove forest covers about 360,000 ha in India (Govindasamy and Kannan 1996), which is the world's fourth largest mangrove

Fig. 5.1 Prop roots or stilt roots

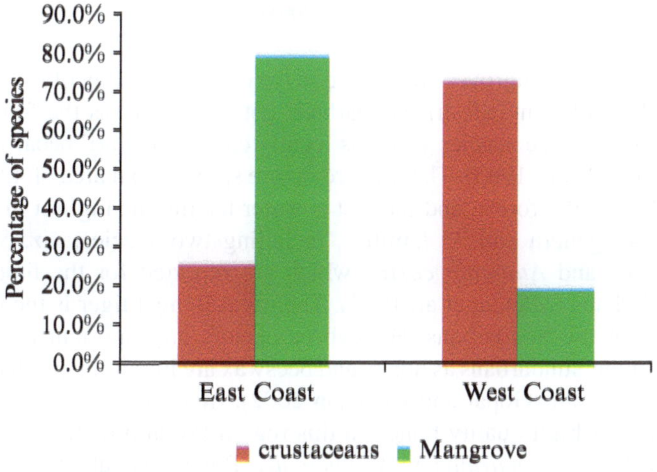

Fig. 5.2 Distribution of crustaceans and mangroves in Indian coasts

area (Mandal et al. 1995), of which 80 % are along the east coast and 20 % on the west coast (Fig. 5.2).

The Forest Survey of India (1997) reported that the total mangrove area of India is 4822 km^2. The major part of mangrove is found in states like West Bengal, Andaman and Nicobar Islands, Orissa, Gujarat, Andhra Pradesh, Tamil Nadu, Kerala, Goa, Maharashtra and Karnataka. The largest area of mangrove occurs at the Sundarbans, West Bengal, followed by Andaman and Nicobar Islands. The mangrove species diversity is highest in Orissa with 101 species, followed by 92 species in West Bengal and in Andaman and Nicobar Islands and the lowest of 40 species in Gujarat (Kathiresan 2000) (Fig. 5.3).

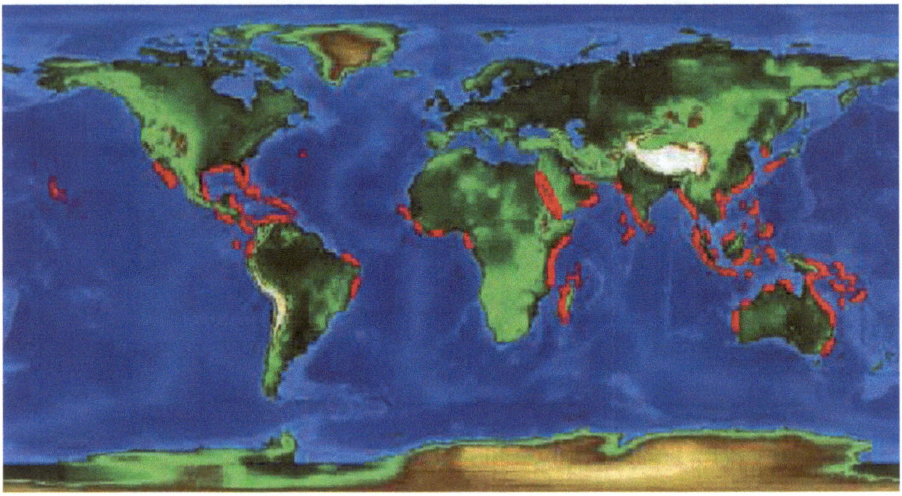

Fig. 5.3 World atlas of mangroves (Source: FAO 2008)

The Sundarbans mangroves are located in the coast of the Bay of Bengal (sharing between India and Bangladesh). Sundarbans got this name, as the forest is dominated by *Heritiera* tree species which is locally called 'sundari' because of its elegance (Jain and Sastry 1983). The mangroves are spread to an area of 4262 km^2, but only 2320 km^2 is the forest, and the rest is water having the highest diversity with 69 species, 49 genera and 35 families, including two species, viz. *Scyphiphora hydrophyllacea* and *Atalantia correa* which are reported for the first time from Indian Sundarbans (Mandal et al. 1995). The royal Bengal tiger is the unique resident species of the Sundarbans. Honeybees occupy an important position in the mangroves of the Sundarbans as honey and beeswax are important products obtained from this ecosystem (Gopal and Chauhan 2002). *Aegiceras corniculatum* flowers form a source of high-quality honey in this region (Wealth of India, vol. I. 1985) followed by *Avicennia marina* and *A. officinalis* (Banerjee et al. 1989).

Pichavaram mangrove forest is located in the Southeast Coast, in the Cuddalore district of Tamil Nadu State, which receives brackish water from Vellar estuary and Coleroon estuary. Pichavaram mangrove ecosystem harbours 35 species of 26 genera and 20 families, including one new species *Rhizophora annamalayana* Kathir (Kathiresan 1995). The dominant mangrove species in Pichavaram is *Avicennia marina* followed by *Avicennia officinalis*, *Excoecaria agallocha*, *Rhizophora apiculata* and *Rhizophora mucronata* (Jyoti Srivastava et al. 2012). Pichavaram mangroves harbour 46 species of crabs from five different stations (Ravichandran and Kannupandi 2007). Being an ideal feeding and breeding ground for many commercially valuable faunal species like fishes, prawns, molluscs, etc., it attracts a large number of birds which in turn attract an appreciable number of tourists. Chandrasekaran and Natarajan (1992) reported an annual harvest of 245 tons of fish, prawn and crab from this mangrove ecosystem, and 3000 traditional poor

Fig. 5.4 Netravathi-Gurupur estuarine mangrove

fishermen living in 14 hamlets depend on this for their livelihood security. Eight prawn species are dominant in Pichavaram mangroves (Kathiresan and Bingham 2001) and finfishes like *Ambassis gymnocephalus*, *A. commersoni*, *Arius subrostratus*, *Chanos chanos*, *Etroplus suratensis*, *Gerres filamentosus*, *G. abbreviatus*, *Liza parsia*, *L. macrolepis*, *L. subviridis*, *Lates calcarifer*, *Lutjanus argentimaculatus*, *Mugil cephalus*, *Osteomugil cunnesius*, *Pomadasys Pondicherry kaakan*, *Plotosus canius*, *Scatophagus argus*, *Siganus javus*, *S. canaliculatus* and *Terapon jarbua* (Kathiresan 1999).

According to the Government of India report (1997), 6000 ha of mangrove cover is present in Karnataka which is spread in estuaries of Netravathi-Gurupur, Mulki-Pavanje, Udyavara-Pangala, Swarna-Sita-Kodi, Chakra-Haladi-Kollur, Baindur hole, Shiroor hole Venkatapur, Sharavathi, Aghanashini, Gangavali and Kali river estuarine complexes. Most mangroves are of the fringing type in linear formations along the river or estuarine banks. Some of the mangrove species present in Karnataka are *Acanthus ilicifolius*, *Lumnitzera racemosa*, *Excoecaria agallocha*, *Aegiceras corniculatum*, *Porteresia coarctata*, *Bruguiera cylindrica*, *Bruguiera gymnorrhiza*, *Kandelia candel*, *Rhizophora apiculata*, *Rhizophora mucronata*, *Sonneratia alba*, *Sonneratia caseolaris*, *Avicennia marina* and *Avicennia officinalis*. Netravathi-Gurupur estuarine mangrove complex is located near Mangalore joining the Arabia Sea at right angle, whereas the Gurupura River runs parallel to the coast before joining the Netravati River which together forms the estuarine mangrove complex (Fig. 5.4). The influence of the tide is felt to a distance of about 20 km inland. This estuarine complex harbours eight mangrove species and many associated plants along with molluscs, polychaetes, crustaceans (*Penaeus indicus*, *P. monodon*, *Metapenaeus monoceros*) and a number of finfishes. Finfish and shellfish fisheries and aquaculture are important livelihood with clam, prawn and crab collection activities. The benthic animals and planktons play a crucial role in these estuarine and mangrove ecosystems like pelagic and terrestrial communities.

5.3 Crustacean Diversity in Mangrove Ecosystem

Crustaceans are the most diverse arthropod group under water, and few have evolved to live on land, though they still require the damp area or an easy access of water to live. Crustaceans include penaeid prawns, non-penaeid prawns, lobsters, crabs, crayfishes, shrimps, barnacles, etc. which are important in human economy due to their role in the marine and terrestrial food chain. Crustaceans are also very important in the nutrient recycling. They possess hard and flexible exoskeleton, two pairs of antennae, a pair of mandibles, a pair of compound eyes, a pair of maxillae and a pair of appendages in each of the body segments. Most of them are dioecious and free living, few are sessile and few others are parasitic. Reproduction is either through the larval stage nauplius or miniature adults. The crustaceans and molluscs, being the dominant macrofauna of the ecosystem, provide a major contribution to the mangrove forest (Hutchings and Saenger 1987) as they make burrows helping other faunal diversity to get the necessary ground for feeding and to establish the boundary required for breeding. Some organisms like barnacles filter water through these burrows. Globally the high number of juvenile crustaceans is found in mangrove ecosystem. The main reason for this is the high abundance of food, lower predation pressure due to shallow-water, microhabitats with higher turbidity and reduced visibility (Beck et al. 2001) and their complex physical structure, for example, root modifications (Lee 2008; Nagelkerken 2009).

To understand the significance of faunal diversity in mangrove ecosystem, numerous studies have been carried out at the east and west coasts of India (Kumar 2001; Saravanakumar et al. 2007). Mud crabs, hermit crabs, lobsters, shrimps and prawns are predominately present in these ecosystems. As mangroves bloom, germinate and fruit, changes occur in invertebrate populations in response to change in food resource. The mangrove ecosystem is continuously being utilised by the vertebrates in these areas. Some vertebrates visit only during favourable periods. Over 90 % of commercial fishery and 70 % of sport fishery depend on the natural mangrove ecosystem for food and habitat (Lewis et al. 1985). Mangroves are dominated by marsh crabs near upper zones and fiddler crabs (Fig. 5.5) closer to shore. As fiddler crabs feed on detritus and micro-organisms living in the detritus, they play a crucial role in cycling of nutrients in the mangroves (Dev Roy and Sivaperuman 2012). This nutrient cycling from the sediment is due to the digging behaviour of the bottom strata and remineralisation action of the detritus in the ecosystem (Robertson 1991). Digestion of structural carbon aided by cellulase enzymes has been studied on terrestrial detritivorous and herbivorous animals (Linton and Greenaway 2004, 2007). Recent studies also revealed the same in many estuarine animals, particularly detritivorous grapsid crabs (Adachi et al. 2012). The diversity effects in the mangrove ecosystem differ between top-down and bottom-up trophic levels in

Fig. 5.5 Fiddler crabs

detritus-based systems (Srivastava et al. 2009; Kominoski et al. 2010). Most of the species are exclusively visitors including humans exploiting the same.

Comprehensive compiled lists of fishes, molluscs, crustaceans, echinoderms and algae of economic importance associated with mangroves that report 732 species in Western Central Pacific, 640 species in Eastern Indian Ocean and 654 species in Western Indian Ocean have been prepared by Matthes and Kapetsky (1988). The commercially important species among these are *Mugil cephalus*, *Scylla serrata*, *Penaeus* sp., *Crassostrea* sp. and *Meretrix* sp. All over the world, about 65 bird species have been listed as endangered or vulnerable. Saenger et al. (1983) have stated that about 150–250 mangrove bird species would be in each of the main biogeographical regions. Aveline (1980) reports that decapod crustaceans are the primary composition of mangrove forests' invertebrate fauna, and these play a crucial role in the mangrove ecosystem dynamics. Hamilton and Snedaker (1984) report that 60 % of commercial catch depends on mangroves in Fiji.

A total of 3111 mangrove-inhabiting faunal species are distributed in different states of India and include prawns, crabs, molluscs, fishes, insects, reptiles, amphibians and mammals (Kathiresan and Quasim 2005). Till now 26 lobsters, 162 hermit crabs, 705 brachyuran crabs and 84 prawns/shrimp species have been recorded in India (Venkataraman and Wafar 2005). In India 11 species of edible crabs have been identified (*Portunus sanguinolentus*, *P. pelagicus*, *Charybdis feriatus*, *C. hicifem*, *C. annulata*, *C. natator*, *Scylla tranquebarica*, *S. serrata*, *Matuta lunaris*, *Sesarma tetragonum* and *Varuna litterata*), inhabiting the coastal and brackish water environments and supporting the commercial fisheries. A total of 48 prawn species are reported from the mangrove ecosystem of India with 34 in the east coast, 16 in Bay Islands and 20 in the west coast (Kathiresan 2000). Samant (1986) recorded 121 species of birds along the Maharashtra coast. Approximately 60 % of commercially important fish species are associated with mangrove ecosystem (Untavale 1986). Molluscs are abundant in South Canara particularly clams during the non-monsoon period (Chatterji et al. 2002). The estimation of the Central Marine Fisheries Research Institute (CMFRI), South Canara, showed an increase of bivalves harvest with 4583 tons in 2006, compared to 905 tons in 1997. Boominath et al. (2008) estimate harvest of 22,000 tons of edible bivalves per year from Aghanashini estuary of Honnavar Forest Division worth of Rs. 66 crore at current market prices.

In mangrove ecosystem, copepods are the major planktonic component with zooplanktonic copepods comprising 60–80 % of biomass (López-Ibarra and Palomares-García 2006) and play the role as prey to many juveniles and immature adults (Sommer et al. 2002). These copepods transfer energy and organic matter from the primary producers to higher trophic levels of the aquatic system (Parsons et al. 1984) though the abundance and distribution are influenced by hydrographical conditions (Santhanam and Perumal 2003).

Barnacles are one of the most fouling organisms in the estuarine ecosystem, and there are more than 1000 known species of barnacles (Chan and Lee 2007) in the world. The barnacles in mangrove ecosystem attach to hard submerged object in the estuarine water or other organisms like crabs, molluscs, shell of turtles, etc. or roots and the lower region of mangrove plant (Grunbaum 2010; Madin 2010).

The mud crab *Scylla* sp. the most important large commercial crab species reaching up to 3 kg in weight is found in mangrove habitats in the world. FAO (2012) and Grubert et al. (2012) report 37,000 tons of *Scylla* sp. being caught globally excluding many countries that do not report the mud crab catches to FAO. This clearly shows that the global mud crab catch is much higher than the given figures of FAO. Joel et al. (1985) reported the distribution and zonation of 29 crab species in the Pulicat Lake. Chakraborthy and Chowdhury (1992) studied 18 species of crabs in the island of Sundarbans mangrove. Mangrove leaf litter decomposition by crabs, lobsters and sesarmids plays a key role between primary and secondary producers (Ajmal Khan et al. 2005). Mangrove trees shed about 7.5 tons of leaves per acre per year, and the energies stored in these leaves are utilised by the crabs before these are carried away by tides (Ashton 2002) and crucially contribute towards the secondary production through the coprophagous food chain (Gillikin et al. 2001). The crabs are a unique source of protein to aquatic life as well as to human being (Siddiqui and Zafar 2002). Sesarmids depend on the sediments of mangroves than the leaves for carbon assimilation (Bouillon et al. 2002; Skov and Hartnoll 2002). The other major commercially valuable crustaceans associated in mangrove are prawns though a large part of commercial harvest (Fig. 5.6) of prawn is in offshore sectors (Vance et al. 2002).

The major part of juvenile stage in prawns is dependent on the mangrove ecosystem as they supply all necessary environment and nutrients (Potter et al. 1986). These juveniles spend only few months in the estuaries and then migrate to offshore up to tens of kilometre which makes it difficult to quantify their nursery grounds. The prawn fishery catch varies to a wide extent as the distribution is dependent on the physical factors like salinity and temperature (Vance et al. 1985). Besides all these factors, a greater abundance of prawn juveniles are found in mangrove areas of estuaries than in the other parts (Primavera 1998).

Fig. 5.6 Commercial harvest of prawns

Some of the prawn and shrimp species reported in the Indian mangroves are *Metapenaeopsis coniger*, *Metapenaeus monoceros*, *M. affinis*, *M. brevicornis*, *M. dobsoni*, *M. lysianassa*, *M. kutchensis*, *M. moyebi*, *Parapenaeopsis longipes*, *P. sculptilis*, *P. stylifera*, *P. canaliculatus*, *P. indicus*, *P. japonicus*, *P. latisulcatus*, *P. merguiensis*, *P. monodon*, *P. semisulcatus*, *P. penicillatus*, *Acetes erythraeus*, *A. indicus*, *Solenocera crassicornis*, *Macrobrachium lamarrei*, *M. idella*, *M. idea*, *M. dayanum*, *M. javanicum*, *M. malcolmsonii*, *M. mirabile*, *M. rosenbergii*, *M. rude*, *M. scabriculum*, *Palaemon concinnus*, *P. debilis*, *P. styliferus*, *P. semmelinkii*, *P. tenuipes*, *Alpheus crassimanus*, *A. paludicola*, *Caridina gracilirostris*, *C. brachydactyla*, *Lucifer hanseni* and *Thalassina anomala* (Kathiresan and Rajendran 2005).

As the mangroves provide soft sediment for burrowing animals and hard rooted structure for the animals requiring firm substratum with adequate nutrients, these areas are abundantly inhabited by bivalves. Till date 5070 mollusc species have been recorded in India with 3370 marine species (Venkataraman and Wafar 2005), the rest are freshwater and terrestrial species. As the molluscs are rich in nutrients especially proteins, fats and minerals, mollusc (shell) fishery mainly includes bivalves like clams, mussels and oysters. High juvenile abundance of many aquatic organisms is seen in mangrove-associated areas like coral reefs and seagrasses (Mumby et al. 2004; Jelbart et al. 2007).

5.4 Ecological and Economic Significance

Mangrove ecosystem provides many goods and services though all of which are not quantified in terms of economic value. Mangroves, serving as a critical nursery for many commercially important organisms, play an important role in the health and the economic well-being of fisheries. Mangrove forests being the highly productive ecosystem have also been a potential contributor of timber, poles, food, medicines

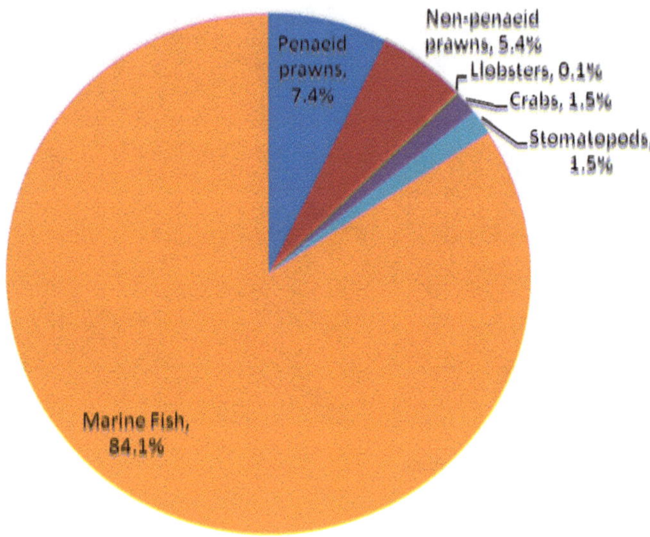

Fig. 5.7 Marine Fish Landing in India (Source: CMFRI 2013)

and a wide variety of other items, indicating conservation of this ecosystem to be a realistic global economic investment (Ronnback 1999). Timber from mangrove is widely used to produce charcoal, tannins and resins for dying and leather making, furniture, bridges, poles for fish cages and traps, alcohol, boats and many other products (Kathiresan and Bingham 2001). The major roles of mangroves and the macrofauna associated with the ecosystem are in providing food security, coastal protection, maintaining water quality, carbon dioxide fixation, source of medicine and salt-resistant genes, dyes, fodder, paper pulp, timber, firewood, manure, fibre, perfumes, recreation and tourism. The leaves, fruits, seeds and seedlings of *Avicennia marina* and vegetative parts of other mangrove species are consumed as vegetables, and some of the mangrove flora also has toxic compounds that are used for their antifungal, antibacterial and pesticidal properties (Bandaranayake 1998).

Mangroves are among the most productive ecosystems as they provide support both directly and indirectly to all life forms. The rich diversity in the mangrove ecosystem influences primary producer abundance, nutrient input, tidal action, annual production and many other factors in the ecosystem. In the past decade, global production of shellfish has doubled in the coastal zone (Naylor et al. 2000). Inhabitants of the estuaries and mangrove forest derive the major direct benefits (Hamilton and Snedaker 1984) as these ecosystems are the important contributors for coastal economy. The economy that depends on mangrove swamps, estuaries, grass beds and coral reefs are severely affected worldwide as these areas are being degraded or destroyed (Wilkinson 1992).

FAO (2012) reports a global consumption of 78.9 million tons of fish, crustaceans and molluscs from oceans contributing to 16.6 % of global protein intake by human in 2011. India has exclusive economic zone of 2.02 million sq. km and 0.53 million sq. km of continental shelf. CMFRI (2013) reports that India has an average annual crustacean resource landing of about 4.36 lakh tons (Fig. 5.7) which is 15.9 % of marine fish landing that included 7.4 % penaeid prawns, 5.4 % non-penaeid prawns, 0.1 % lobsters, 1.5 % crabs and 1.5 % stomatopods. In India the east coast contributes 26.4 % and the west coast 73.6 % (Fig. 5.2) of the total crustacean landing. The demand for fish, crustaceans, molluscs and their products has seen a dramatic increase in the past decades due to the economic importance and provides employment and livelihood to millions of people. Crustaceans and molluscs take a crucial role in ecological functioning of mangrove ecosystem (Lee 1999). Thus, functioning of mangrove ecosystems is reflected by the richness and abundance of species as they become an indicator of change in mangrove ecosystem. Mangrove swamps, estuaries and lagoons and their grass beds act as 'nurseries' and provide nutrients and shelter for economically important fishes, crustaceans and molluscs, and these areas are among the world's most ecologically critical and threatened resources (Pauly and Ingles 1999). Fishery resources from the mangrove ecosystem are regarded more valuable than the wood and other natural and agricultural goods (Lacerda 1993). Man gets fish, crabs, shellfish, reptile skins, honey and other products from most of the world's mangrove forests (Nurkin 1994). Thus, these ecosystems are an important economic resource and ecological asset to every county. As numbers of molluscs are soft bodied, their shells are used in preparation of a number of medicinal oils and medicines to treat many diseases. Apart from these, molluscs like cuttlefish, squids and octopods are used as fish bait. Mangroves also help greatly towards the recharge of underground water table.

5.5 Loss of Diversity

Hostile habitat and human abuse of mangroves are resulting in endangering this unique ecosystem (Kathiresan and Ravikumar 1995), and today these ecosystems are one among the most threatened ecosystems in the world (Kelleher et al. 1995). Many studies have shown that these ecosystems are being destructed twice faster than the well-publicised destruction of tropical rain forests. The consumption of macro consumers like grapsid crabs and gastropods would significantly reduce the detrital carbon stock in tropical mangroves (Kristensen et al. 2008; Lee 2008). Natural and anthropogenic disturbances at various levels (Wolanski et al. 2000) are serious threats to mangrove ecosystem functioning (Osborn and Polsenberg 1996). Degradation and loss of mangrove are mostly due to salt production, wood extraction, coastal industrialisation and urbanisation, coastal aquaculture and other anthropogenic activities (Macintosh 1996). These activities adversely affect the crustacean fauna and their resources worldwide (Suseelan and Nair 1994). Degradation or deforestation of mangroves has resulted in 1–2 % loss annually, and thus approximately 30–50 % of global mangroves were lost in the past half century (Polidoro et al. 2010). IPCC (2007) predicts that if the present rate of loss continues, 30–40 % of coastal wetlands and 100 % of mangrove forests (Duke et al. 2007) could be lost from the world within the next century. Globally only 6.9 % mangrove area is covered under the existing protected area network (Giri et al. 2011). The Government of India (1987) reports a loss of 40 % mangrove area in the last century from India. Mangrove area of about 191,300 ha was lost during 1987–1997 all over India. A total of 20 faunal species has been reported to be threatened from Indian mangrove ecosystem (Dev Roy and Sivaperuman 2012).

The brackish water ecosystems that include estuaries, backwaters, saltwater lakes, mangroves, etc., harbouring immense wealth of juvenile prawns, are reported with considerable shrinkage due to an authorised and unauthorised reclamation in the past few decades. Mangrove reforestation is limited to two to three species due to economic reasons of the developing countries (Gan 1995). Overfishing has resulted in depletion in number of fish, crustacean and molluscs. Some popular species, like whales, dolphins and sea turtles, are on the verge of extinction. The recent global mangrove forest estimate reveals that the loss is more than half of what it was once (Spalding et al. 1997) and the remaining is in degraded condition (MAP 2005). The survival of mangrove seedlings has been suggested to be affected by the attachment of the barnacles in many areas (Havanon et al. 1995; Rawangkul et al. 1995). This degradation will result in the reduction or loss of goods and services provided by mangrove ecosystem (Duke et al. 2007). Vietnam, Mexico, Singapore, the Philippines and Thailand have lost most of their mangrove as a result of urbanisation and exploitation (Spalding et al. 1997). All mangrove species are susceptible to herbicide stress (Odum et al. 1982), and stress of single defoliation would kill the entire plant.

Data on carbon storage in mangrove wood or sediments are very less, but the available data suggests that mangrove forests can accumulate enormous carbon (Twilley et al. 1992; Alongi et al. 2000). In Sawi Bay, carbon in mangrove sediments is derived from land and from phytoplankton stimulated by inorganic nutrients (Ayukai and Alongi 2000). As mangroves are an important carbon sink which get converted to biomass, the loss of these crucial ecosystems has resulted in a huge loss of carbon, stored as mangrove biomass (Cebrian 2002). Human beings cut and fill mangroves for agricultural and industrial development and urbanisation (Linden and Jernelov 1980), and these mangroves are converted as salt flats to produce salt (Nurkin 1994). The changed pattern of fishing creates biological damages to important shrimp species, besides it has also generated social problems in many parts of the world. Many countries do not report the actual figures of the resources that are caught from these ecosystems despite producing thousands of tons of resources which makes it difficult to assess the global loss. With the limited data that are available from different countries, FAO has assessed the loss of resources to an alarming level requiring immediate attention.

5.6 Conservation

Natural conservation is the best way of conservation where it is being left alone without interruption, as they are self-sustaining ecosystems. The best effort by human in conservation could be by reducing and preventing future damages and taking steps to remove the past damages caused. The most common measures adopted in tropical fisheries are restriction of fishing, closed seasons for fishing, catch quotas allotment, mesh size regulation and restriction of juveniles capturing at nursery grounds. These measures are drafted keeping in view of different breeding periods and to have sustainable resource. Implementation of the above measures is only possible by public understanding and participation. Governments and people must understand the social and economic importance of mangrove ecosystem and action to be taken towards the conservation and improvisation of these resources. Following the Charter by the International Society for Mangrove Ecosystems (Field 1995) would be a great step towards the conservation. Restoration and rehabilitation of mangrove plants should also be initiated, which are already being carried out in few of coastal countries, though restoration and natural regrowth are slow (Sherman et al. 2000). Countries like Papua New Guinea, Australia and Belize show no substantial change, and in few countries like Cuba, the restoration projects have helped to regain mangrove forests (Field 2000). But due to the lack of proper planting techniques and adequate site selection, there is a mixed outcome in most restoration and rehabilitation projects (Ellison and Farnsworth 2000).

The Indian Constitution Amendment (1976) states that *it shall be the duty of every citizen of India to protect and improve the natural environment including forests, lakes, rivers and wildlife.* To advise the Indian Government about mangrove conservation and development, a National Mangrove Committee under the Ministry of Environment and Forests was set up by the Government of India in 1976. The Environment (Protection) Act (1986) declares a Coastal Regulation Zone (CRZ) in which industrial and other activities such as discharge of untreated water and effluents, dumping of waste, land reclamation and bunding are restricted in order to protect the coastal environment. Mangroves are included in the most ecologically sensitive category of coastal stretches.

As most rehabilitation projects adopt the monoculture or low diversity polyculture with an emphasis on silviculture to produce timber, wood chips, charcoal and fuel wood, it is highly difficult to restore fauna and ecosystem function of mangrove forests (Ellison and Farnsworth 2000). Even then, if the hydrological and geomorphological conditions provided at optimal level and seedlings inoculated with bacteria that promote nitrogen fixation and plant growth (Holguin et al. 2001), the rehabilitated forests can reach the biomass, structure and productivity in a long period of 20–25 years (McKee and Faulkner 2000) and can reduce the loss of mangroves worldwide to a level reducing not more than an annual global loss rate of about 1 % (Kaly and Jones 1998). To reach high success in the rehabilitation, less destructive mud crab cultivation has been trailed in most regions of the world (Keenan and Blackshaw 1999). Till date high level of mangrove rehabilitation projects are successful only in Pakistan, Cuba and Bangladesh (Spalding et al. 1997) wherein an impressive attempt along the large portion of tropical coastline has been observed (Saenger and Siddiqi 1993). Though it is well known that the prohibition is not possible due to socio-economic problems, attempts have to be made in controlling the destruction of the juvenile prawns, and alternation for protection of young prawns in the estuarine systems has to be considered as prawn fishery is mostly dependent on the emerging immature adults from the estuarine ecosystem.

Educating people by providing better vision on mangrove habitats, resources, relevant legislation, policies and conservation strategies by awareness campaigns along with magazines, films, posters, pamphlets, documentaries, exhibitions, study tours, incentives for protection of mangroves, etc. would help in successful restoration and rehabilitation of these ecosystems.

References

Adachi K, Toriyama K, Azekura T, Morioka K, Tongnunui P, Ikejima K (2012) Potent cellulase activity in the hepatopancreas of mangrove crabs. Fish Sci 78:1309–1314

Ajmal Khan S, Raffi SM, Lyla PS (2005) Brachyuran crab diversity in natural Pichavaram and artificially developed mangroves Vellar estuary. Curr Sci 88:1316–1324

Alongi DM (2009) Paradigm shifts in mangrove biology. In: Perillo GME, Wolanski E, Cahoon DR, Brinson MM (eds) Coastal wetlands: an integrated ecosystem approach. Elsevier, Amsterdam, pp 615–640

Alongi DM, Wattayakorn G, Ayukai T, Clough BF, Wolanski E, Brunskill GJ (2000) An organic carbon budget for mangrove-fringed Sawi Bay, southern Thailand. Phuket Mar Biol Cent Spec Publ 22:79–85

Ashton EC (2002) Mangrove sesarmid crab feeding experiments in Peninsular Malaysia. J Exp Mar Biol Ecol 273:97–119

Aveline LC (1980) Fauna de manguezais brasileiros. Rev Brasil Geogr 42(2):786–821

Ayukai T, Alongi DM (2000) Pelagic carbon fixation and heterotrophy in shallow coastal waters of Sawi Bay, southern Thailand. Phuket Mar Biol Cent Spec Publ 22:39–50

Bandaranayake WM (1998) Traditional and medicinal uses of mangroves. Mangroves Salt Marshes 2:133–148

Banerjee LK, Sastry ARK, Nayar MP (1989) Mangroves in India: identification manual. Botanical Survey of India, Calcutta

Beck MW, Heck KL, Able KW, Childers DL, Eggleston DB, Gillanders BM, Halpern B, Hays CG, Hoshino K, Minello TJ, Orth RJ, Sheridan PF, Weinstein MP (2001) The identification, conservation and management of estuarine and marine nurseries for fish and invertebrates. Bioscience 51:633–641

Boominath M, Chandran MDS, Ramachandra TV (2008) Economic valuation of the bivalves in the Aghanashini estuary, west coast, Karnataka. ENVIS Technical Report 30. Ministry of Science and Technology, Government of India

Bouillon S, Koedam N, Raman AV, Dehairs F (2002) Primary producers sustaining macro-invertebrate communities in intertidal mangrove forests. Oecologia 130:441–448

Cebrian J (2002) Variability and control of carbon consumption, export, and accumulation in marine communities. Limnol Oceanogr 47:11–22

Chakraborthy SK, Chowdhury A (1992) Community structure of macrobenthic polychaetes of intertidal region of Sagar Island, Hooghly estuary, Sundarbans, India. Trop Ecol 35(1):97–104

Chan BKK, Lee KH (2007) Barnacles of Taiwan: biodiversity and ecology. National Museum of Natural Science, Taiwan

Chandrasekaran VS, Natarajan R (1992) Small-scale fishery of Pichavaram mangrove swamp, Southeast India. Naga, lhe lCI ARM Qy_argr_ly, ICLARM, Manila 15:41–43

Chatterji A, Ansari ZA, Ingole BS, Bichurina MA, Sovetova, Boikov YA (2002) Indian marine bivalves: potential source of antiviral drugs. Curr Sci 82:1279–1282

CMFRI (2013) Crustacean fisheries division manual on taxonomy and identification of commercially important crustaceans of India. Compiled by Josileen Jose and Lakshmi Pillai S

Dev Roy MK, Sivaperuman C (2012) Fauna of ecosystem of India-Mangrove. Published by Director, Zoological Survey of India, Kolkata, pp 1–11

Duke NC, Meynecke JO, Dittmann S, Ellison AM, Anger K, Berger U, Cannicci S, Diele K, Ewel KC, Field CD, Koedam N, Lee SY, Marchand C, Nordhaus I, Dahdouh-Guebas F (2007) A world without mangroves? Science 317(5834):41–42

Ellison AM, Farnsworth EJ (2000) Mangrove communities. In: Bertness MD, Gaines SD, Hay ME (eds) Marine community ecology. Sinauer Associates, New York, pp 423–442

FAO (Food and Agriculture Organization of the United Nations) (2012) The state of world fisheries and aquaculture. Fisheries and Aquaculture Department, Food and Agriculture Organization of the United Nations, Rome

Field CD (1995) Journey amongst Mangroves. International Society for Mangrove Ecosystems, Okinawa, p 140

Field CD (2000) Mangroves. In: Sheppard CRC (ed) Seas at the millennium: an environmental evaluation, vol III, Global Issues and Processes. Pergamon Press, Amsterdam, pp 17–31

Forest Survey of India (1997) State of forest report 1997. Ministry of Environment and Forest, Dehradun

Gan BK (1995) A Working Plan for the Matang Mangrove Forest Reserve (fourth revision). Published by the State Forest Department of Perak Darul Ridzuan, Malaysia, pp 214

Gillikin DP, De Grave S, Tack JF (2001) The occurrence of the semi-terrestrial shrimp Merguia oligodon (De Man, 1888) in Neosarmatium smithi H. Milne Edwards, 1853 burrows in Kenyan mangroves. Crustaceana 74:505–508

Giri C, Ochieng E, Tieszen LL, Zhu Z, Singh A, Loveland T, Masek J, Duke N (2011) Status and distribution of mangrove forests of the world using earth observation satellite data. Glob Ecol Biogeogr 20:154–159

Gopal B, Chauhan M (2002) Biodiversity and its conservation in the Sundarban Mangrove Ecosystem. Aquat Sci 68(2006):338–354

Government of India (1987) Mangroves in India –status report. Ministry of Environment and Forests, New Delhi, pp 52–55

Government of India (1997) The state of forest report. Forest Survey of India, Ministry of Environment and Forests, New Delhi, pp 5–6, 38

Govindasamy C, Kannan L (1996) Ecology of the rotifers of the Pichavaram Mangroves, southeast coast of India. Indian Hydrobiol 1:69–76

Grubert M, Leslie M, Winning M (2012) Mud crab *Scylla serrata*, *Scylla olivacea*. In: Flood M, Stobutzki I, Andrews J, Begg G, Fletcher W, Gardner C, Kemp J, Moore A, O'Brien A, Quinn R (eds) Status of key Australian fish stocks reports 2012. Fisheries Research and Development Corporation, Canberra, pp 131–127

Grunbaum M (2010) How do barnacles attach to whales? http://www.scienceline.org/2010/03/how-do-barnacles-attach-to-whales/. Accessed 17 Jan

Hamilton LS, Snedaker SC (1984) Handbook for mangrove area management. IUCN/ UNESCO/ UNEP/East-West Centre, Honolulu, 123 pp

Havanon S, Chukwamdee J, Anunsiriwat A, Meepol W (1995) Study on mangrove forest structure at Samut Songkram Province. In: The ninth national seminar on mangrove ecology, mangrove conservation for Thai society in the next decade, National Research Council of Thailand, Bangkok; Pap No. 111-02

Holguin G, Vazquez P, Bashan Y (2001) The role of sediment microorganisms in the productivity, conservation, and rehabilitation of mangrove ecosystems: an overview. Biol Fertil Soils 33:265–278

Hutchings P, Saenger P (1987) Ecology of mangroves. University of Queensland Press, St Lucia, p 388

IPCC (2007) Summary for policymakers. Climate Change 2007: impacts, adaptation and vulnerability. In: Parry ML, Canziani OF, Palutikof JP, van der Linden PJ, Hanson CE (eds) Contribution of working group II to the fourth assessment report of the Intergovernmental Panel on Climate Change. Cambridge University Press, Cambridge, pp 7–22

IUCN (2005) Early observations of tsunami effects on mangroves and coastal forests. The World Conservation Union (IUCN). http://www.iucn.org/info_and_news/press/water-wetlands-statement

Jain SK, Sastry ARK (1983) Botany of some tiger habitats in India. Botanical Survey of India, Howrah, pp 40–44

Jelbart JE, Ross PM, Connolly RM (2007) Fish assemblages in seagrass beds are influenced by the proximity of mangrove forests. Mar Biol 150(5):993–1002

Joel DR, Sanjeeva Raj PJ, Raghavan R (1985) Distribution and zonation of Shore crabs in the Pulicat lake. Proc Indian Acad Anim Sci 95(4):437–445

Kaly U, Jones GP (1998) Mangrove restoration: a potential tool for coastal management in tropical developing countries. Ambio 27:656–661

Kathiresan K (1995) *Rhizophora annamalayana* – a new species of mangrove. Environ Ecol 13:240–241

Kathiresan K (1999) Impact of mangrove biodiversity on associated fishery resources and fishers' income. A project final report submitted to WWF, Washington, DC, pp 142

Kathiresan K (2000) A review of studies in Pichavaram Mangroves southeast coast of India. Hydrobiologia 430(1):185–205

Kathiresan K, Bingham BL (2001) Biology of mangroves and mangrove ecosystems. Adv Mar Biol 40:81–251

Kathiresan K, Quasim SZ (2005) Biodiversity of mangrove ecosystems. Hindustan Publishing Corporation, New Delhi, p 251

Kathiresan K, Rajendran N (2005) Coastal mangrove forests mitigated tsunami. Estuar Coast Shelf Sci 65:601–606

Kathiresan K, Ravikumar S (1995) Vegetative propagation through air layering in two species of mangroves. Aquat Bot 50(1):107–110

Keenan CP, Blackshaw A (1999) Mud crab aquaculture and biology. In: Proceedings of an International Scientific Forum held in Darwin, Australia, 21–24 April 1997. ACIAR Proceedings No. 78. Canberra, Australia: Australian Centre for International and Agricultural Research, pp 216

Kelleher G, Blcakley C, Wells S (1995) A global representative system of marine protected areas, vol 1, A joint publication of World Bank, The Great Barrier Reef Marine Park Authority and the World Conservation Union (WCN). The World Bank, Washington, DC

Kominoski JS, Hoellein TJ, Leroy CJ, Pringle CM, Swan CM (2010) Beyond species richness: expanding biodiversity ecosystem functioning theory in detritus-based streams. River Res Appl 26:67–75

Krauss KW, Allen JA, Cahoon DR (2003) Differential rates of vertical accretion and elevation change among aerial root types in Micronesian mangrove forests. Estuar Coast Shelf Sci 56:251–259

Kristensen E, Bouillon S, Dittmar T, Marchand C (2008) Organic carbon dynamics in mangrove ecosystems: a review. Aquat Bot 89:201–219

Kumar RS (2001) Intertidal zonation and seasonality of benthos in a tropical mangrove. Int J Ecol Environ Sci 27:199–208

Lacerda LD (1993) Conservation and sustainable utilization of Mangrove Forests and their present state of conservation in Latin America and Africa Regions, Part I-Latin America, Mangrove Ecosystems technical reports, vol 2. International Society for Mangrove Ecosystems, Okinawa, Japan, pp 272

Lee SY (1999) Tropical mangrove ecology: physical and biotic factors influencing ecosystem structure and function. Aust J Ecol 24:355–366

Lee SY (2008) Mangrove macrobenthos: assemblages, services and linkages. J Sea Res 59:16–29

Lewis RR, Gilmore RG, Crewz DW, Odum WE (1985) Mangrove habitat and fishery resources of Florida. In: Seaman W Jr (ed) Florida aquatic habitat and fishery resources. American Fisheries Society, Kissimmee, pp 281–336

Linden O, Jernelov A (1980) The mangrove swamps: an ecosystem in danger. Ambio 9:81–88

Linton SM, Greenaway P (2004) Presence and properties of cellulase and hemicellulase enzymes of the gecarcinid land crabs *Gecarcoidea natalis* and *Discoplax hirtipes*. J Exp Biol 207:4095–4104

Linton SM, Greenaway P (2007) A review of feeding and nutrition of herbivorous land crabs: adaptations to low quality plant diets. J Comp Physiol B Biochem Syst Environ Physiol 177:269–286

López-Ibarra GA, Palomares-García R (2006) Estructura de la comunidad de copépodos en Bahía Magdalena, México, durante El Niño 1997–1998. Rev Biol Mar Oceanogr 411:63–76

Macintosh DJ (1996) Mangroves and coastal aquaculture: doing something positive for the environment. Aquacult Asia I(2):3–8

Madin J (2010) Biofouling and short- term dynamics on fish cage netting in relation to fish rearing and environmental factor. PhD thesis, University of Malaya

Mandal RN, Ghosh DC, Sarkar AK, Naskar KR (1995) New distributional record of Scyphiphora hydrophyllacea Gaertn. f. and Atalantia corea M. Roem. from the inter tidal mangrove forests of the Indian Sundarbans. J Natl Bot Soc 49:71–74

MAP (Mangrove Action Project) (2005) Available at: http://mangroveactionproject.org/

Matthes H, Kapetsky JM (1988) Worldwide compendium of mangrove-associated aquatic species of economic importance. FAO Fish Circ 14:236

Mazda Y, Wolanski E, King B, Sase A, Ohtsuka D, Magi M (1997) Drag force due to vegetation in mangrove swamps. Mangroves Salt Marshes 1:193–199

McKee KL, Faulkner PL (2000) Biogeochemical functioning of restored and natural mangrove forests in southwest Florida, USA. Restor Ecol 8:247–259

Mumby PJ, Edwards AJ, Ernesto Arias-Gonzalez J, Lindeman KC, Blackwell PG, Gall A, Gorczynska MI, Harborne AR, Pescod CL, Renken H, Wabnitz CCC, Llewellyn G (2004) Mangroves enhance the biomass of coral reef fish communities in the Caribbean. Nature 427(6974):533–536

Nagelkerken I (2009) Evaluation of nursery function of mangroves and seagrass beds for tropical decapods and reef fishes: patterns and underlying mechanisms. In: Nagelkerken I (ed) Ecological connectivity among tropical coastal ecosystems. Springer, New York, pp 357–399

Naylor RL, Goldburg RJ, Primavera JH, Kautsky N, Beveridge MCM, Clay J, Folke C, Lubchenco J, Mooney H, Troell M (2000) Effect of aquaculture on world fish supplies. Nature 405:1017–1024

Nurkin B (1994) Degradation of mangrove forest in South Sulawesi, Indonesia. Hydrobiologia 285:271–276

Odum WE, McIvor CC (1990) Mangroves. In: Myers RL, Ewel JJ (eds) Ecosystems of Florida. University Press of Florida, Gainesville, pp 517–548

Odum WE, McIvor CC, Smith TS (1982) The ecology of mangroves of South Florida: a community profile. U.S. Fish and Wildlife Service, Office of Biological Services, Washington, DC, FWS/OBS – 81/24

Osborn JG, Polsenberg JF (1996) Meeting of the mangrovellers: the interface of biodiversity and ecosystem function. Trends Ecol Evol 11:354–356

Parsons TR, Takahashi M, Hargrave B (1984) Biological oceanographic processes. Pergamon Press, Oxford, p 332

Pauly D, Ingles J (1999) The relationship between shrimp yields and intertidal vegetation (mangrove) areas: a reassessment. In: A Yanez-Arancibia, Lara-Dominguez AL (eds) Mangrove ecosystems in tropical America. Instituto de Ecologia, A.C. Xalapa, Mexico; UICN/ORMA Costa Rica; NOAA/NMFS Silver Spring, Maryland, pp 311–316

Polidoro B, Carpenter K, Collins L, Duke N, Ellison A, Ellison KC, Farnsworth EJ, Fernando ES, Kathiresan K, Koedam NE, Livingstone SR, Miyagi T, Moore GE, Ong JE, Primavera JH, Salmo SG III, Sanciangco JC, Sukardjo S, Wang Y, Yong JWH (2010) The loss of species: mangrove extinction risk and geographic areas of global concern. PLoS ONE 5(4). doi:10.1371/journal.pone.0010095

Por FD (1984) The ecosystem of the mangal: general considerations. In: Por FD, Dor I (eds) Hydrology of the mangal. W junk Publ, The Hague, pp 1–14

Potter I, Penn J, Brooker K (1986) Life cycle of the western school prawn, Metapenaeus dalli Racek, in a Western Australian estuary. Mar Freshw Res 37(1):95–103

Primavera JH (1998) Mangroves as nurseries: shrimp populations in mangrove and non-mangrove habitats. Estuar Coast Shelf Sci 46(3):457–464

Ravichandran S, Kannupandi T (2007) Biodiversity of crabs in Pichavaram mangrove environment. Zoological Survey of India. National Symposium on Conservation and Valuation of marine Biodiversity, pp 331–340

Rawangkul S, Angsupanich S, Panitchart S (1995) Prelirninary study of barnacles damaging the mangrove plantation Rhizophora mucronata at Tha Phae canal, Nakorn Si Thammarat. In: The ninth national seminar on mangrove ecology, mangrove conservation for Thai Society in the next decade, National Research Council of Thailand Bangkok, Pap No. 111-06

Robertson AI (1991) Plant-animal interactions and the structure and function of mangrove forest ecosystems. Aust J Ecol 16:433–443

Ronnback P (1999) The ecological basis for economic value of sea food production supported by mangrove ecosystems. Ecol Econ 29:235–252

Saenger P, Siddiqi NA (1993) Land from the sea: the mangrove afforestation program of Bangladesh. Ocean Coast Manag 20:23–39

Saenger B, Hegerl EJ, Davie JD (1983) Global status of mangrove ecosystems. Inter Union Cons Natu Nat Res 3:9–18

Samant JS (1986) Avifauna of aningroves around Ratnagiri, Maharashtra. In: The Mangroves: Proc Nat Symp Biol Oil Cons Mangroves, pp 456–466

Santhanam P, Perumal P (2003) Diversity of zooplankton in Parangipettai coastal waters, southeast coast of India. J Mar Biol Ass India 45:144–151

Saravanakumar A, Sesh Serebiah J, Thivakaran GA, Rajkumar M (2007) Benthic macrofaunal assemblage in the arid zone mangroves of Gulf of Kachchh, Gujarat. J Oce Uni Chi 4(6):303–309

Sherman RE, Fahey TJ, Battles JJ (2000) Small-scale disturbance and regeneration dynamics in a neo-tropical mangrove forest. J Ecol 88:165–178

Siddiqui MZH, Zafar (2002) Crabs in the Chakaria Sundarban area of Bangladesh. J Noami 19:61–77

Skov MW, Hartnoll RG (2002) Paradoxical selective feeding on a low nutrient diet: why do mangrove crabs eat leaves? Oecologia 131:1–7

Sommer U, Stibor H, Katechakis A, Sommer F, Hansen T (2002) Pelagic food web configurations at different levels of nutrient richness and their implications for the ratio fish production: primary production. Hydrobiologia 484:11–20

Spalding MD, Blasco F, Field CD (1997) World mangrove Atlas. The International society for Mangrove Ecosystems (ISME), Okinawa, p 178

Srivastava DS, Cardinale BJ, Downing AL, Duffy JE, Jouseau C, Sankaran M, Wright J (2009) Diversity has stronger top-down than bottom-up effects on decomposition. Ecology 90:1073–1083

Srivastava J, Farooqui A, Hussain SM (2012) Vegetation history and salinity gradient during the last 3700 years in Pichavaram estuary, India. J Earth Syst Sci 121(5):1229–1237

Suseelan C, Nair KP (1994) Endangered, vulnerable and rare estuarine shellfishes of India. In: Threatened fishes of India. NATCON Publication 4, pp 237–251

Twilley RR (1998) Mangrove wetlands. In: Messina MG, Conner WH (eds) Southern forested wetlands. Ecology and management. Lewis Publishers, Boca Raton, pp 445–473

Twilley RR, Chen RH, Hargis T (1992) Carbon sinks in mangroves and their implications to carbon budget of tropical coastal ecosystems. Water Air Soil Pollut 64:265–288

Untavale AO (1986) Mangroves of India. In: Mangroves of Asia and the pacific-status and management. UNDP/UNESCO Project Research and Training Pilot Program on Mangrove ecosystems (RAS/79/002. UNOP/UNESCO, Manila)

Vance DJ, Staples DJ, Kerr JD (1985) Factors affecting year-to-year variation in the catch of banana prawns (*Penaeus merguiensis*) in the Gulf of Carpentaria, Australia. J Conseil 42(1):83–97

Vance D, Haywood M, Heales D, Kenyon R, Loneragan N, Pendreyne R (2002) Distribution of juvenile penaeid prawns in mangrove forests in a tropical Australian estuary, with particular reference to *Penaeus merguiensis*. Mar Ecol Prog Ser 228:165–177

Venkataraman K, Wafar M (2005) Coastal and marine biodiversity of India. Indian J Mar Sci 34(1):57–75

Wealth of India (Raw Materials) (1985) *Acacia ferruginea*. vol I (reprint 2003), CSIR, New Delhi, pp 32–33

Wilkinson CR (1992) Coral reefs of the world are facing widespread devastation: can we prevent this through sustainable management practices? In: Richmond RH (ed) Proceeding of the 7th International Coral Reef Swamp, Guam, 22–27 June 1992. vol 1. pp 11–21

Wolanski E, Spagnol S, Thomas S, Moore K, Alongi DM, Trott L, Davidson A (2000) Modelling and visualizing the fate of shrimp pond effluent in a mangrove-fringed tidal creek. Estuar Coast Shelf Sci 50:85–97

Phylogenetic Status, Diversity, Economic and Medicinal Importance of Crabs

6

Kalarani Varada, Sumathi Vempalli, Sridevi V, and Chandra Sekhara Reddy Dandala

Abstract

Crustaceans that include lobsters, crayfish, crabs, prawns, shrimps, and barnacles are the most important part of the macrobenthic fauna. Crabs belong to the infraorder Brachyura and are one of the dominating groups of decapod crustaceans. Indian mangroves alone are serving as a good habitat for over 138 brachyuran crab species. The total number of freshwater crab species forms 65 % of brachyuran species. But due to limitation in dispersal abilities, crabs have tended to be endemic to small areas; as a result, a large proportion of them are threatened to extinction. Culturing of crabs under captivity seems to alleviate the threat to some extent. Due to the multiple roles of crabs such as indicators of pollution, a good source of protein, an important stratum in the food web, and having direct and indirect socioeconomic, medicinal, and sociocultural values, more attention needs to be paid to explore the potential of different crab species for their economic and/or medicinal value. The present study investigates the antimicrobial and antifungal activity of the hemolymph of the South Indian field crab, *Oziotelphusa senex senex*.

Keywords

Brachyura • Crabs • Diversity • Hemolymph • Phylogenetic status

K. Varada (✉) • S. Vempalli • S. V
Department of Biotechnology, Sri Padmavati Mahila Visvavidyalayam (Women's University), Tirupati 517 502, Andhra Pradesh, India
e-mail: kala.dandala@gmail.com

C.S.R. Dandala
UGC - Emeritus Scientist, Department of Fishery Science and Aquaculture, Sri Venkateswara University, Tirupati 517 502, Andhra Pradesh, India

A.K. Chakravarthy, S. Sridhara (eds.), *Arthropod Diversity and Conservation in the Tropics and Sub-tropics*, DOI 10.1007/978-981-10-1518-2_6

6.1 Introduction

Crustaceans are the important part of macrobenthic fauna. Lobsters, crayfish, crabs, prawns, shrimps, and barnacles are the different groups of crustaceans. The majority of decapods occurring in tropical and subtropical regions of the world are exhibiting a significant decline in species number in the temperate and cold regions (Boschi 2000). Classification within the order Decapoda of phylum Arthropoda depends on the structure of the gills and legs and types of larval development, giving rise to two suborders: Dendrobranchiata and Pleocyemata. The Dendrobranchiata consists of prawns, including many species commonly referred to as "shrimp," while Pleocyemata includes the remaining groups, including the "true shrimp" (Elena Mente 2008). Those groups which usually walk rather than swim (Pleocyemata, excluding Stenopodidea and Caridea) form a clade called Reptantia (Scholtz and Richter 1995). Crabs, the decapod crustaceans, belong to the infraorder Brachyura, which typically have a very short projecting "tail" (abdomen), usually entirely hidden under the thorax. Crabs belong to 93 families and exist in an array of microhabitats across the oceans, freshwater, and even on land, and are found throughout the world's tropical and semitropical regions.

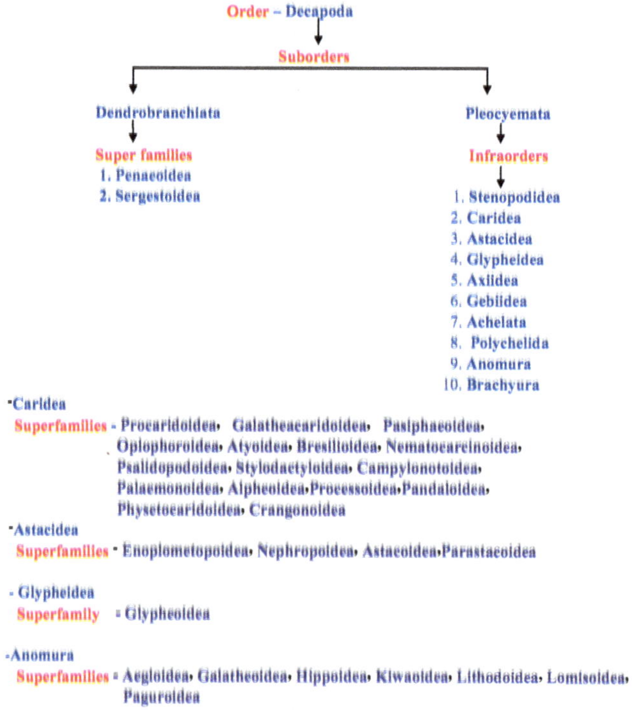

Fig. 6.1 Classification of order - Decapoda to the level of superfamilies (De Grave et al. 2009) (Fig. 6.1)

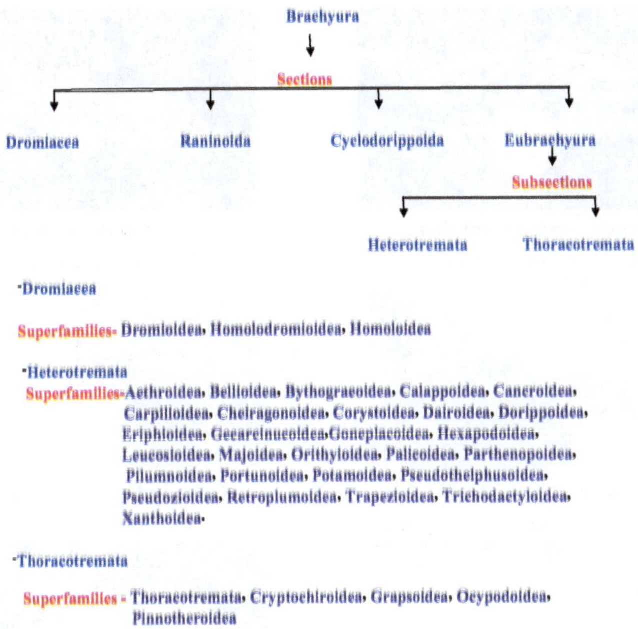

Fig. 6.1 (continued)

6.2 Distribution, Ecological and Economic Importance of Crabs

Brachyuran crabs are represented by 700 genera and up to 10,000 species worldwide (Kaestner 1970; Ng 1988; Melo 1996; Martin and Davis 2001; Ng et al. 2008; Yeo et al. 2008), out of which 2600 species are present in the Indo-West Pacific. In India alone, 705 brachyuran crab species belonging to 28 families and 270 genera have been reported by Venkataraman and Wafar in 2005. Maximum crab catches were found in the Gulf of Mannar, Palk Bay, Nagapattinam, and Pondicherry of Tamil Nadu coast (Rao et al. 1973), as well as Chennai coast of the same state. In Indian mangroves, 138 brachyuran crab species were reported (Kathiresan and Qasim 2005; Ng et al. 2008), with 19 species in the Gulf of Kutch (Gujarat state) (Trivedi et al. 2012), 15 species in the Pondicherry coast (Satheeshkumar and Khan 2011), and 54 crab species belonging to family Portunidae from Chennai coast (Krishnamoorthy 2007). In different areas of the Gulf of Mannar, the occurrence of 32 species in Manauli Island, 26 species in Appa Island, 22 species in Nallathanni Island, and 18 species in Karaichalli Island were reported (Jayabaskaran and Ajmal Khan et al. 2007). A total of 38 species of brachyuran crabs (Ravichandran and Kannupandi 2007) were reported from Pichavaram mangrove areas.

Brachyuran crabs, a diverse group of crustaceans, are found at a depth of 6000 m to seashore. They are dominant in many estuarine habitats where salinity and

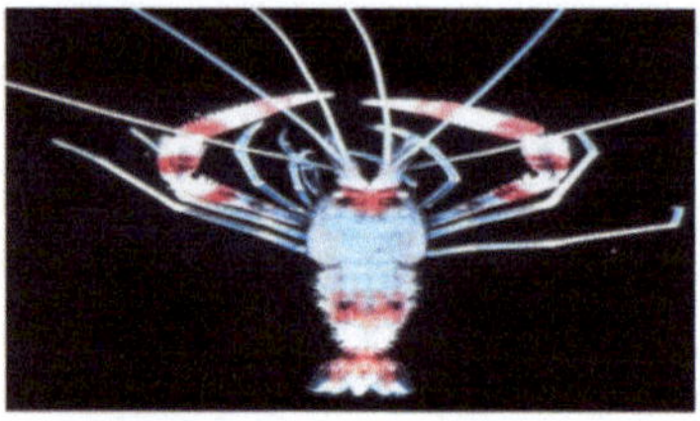

Infraorder - **Stenopodidea**
Ex: *Stenopus hispidus*

Infraorder - **Caridea**
Ex: *Heterocarpus ensifer*

Infraorder - Astacidea
Ex: *Austrotamobius pallipes*

Infraorder - Glypheidea
Ex: *Mecochirus longimanatus*

Infraorder - **Axiidea**
Ex: *Axius serratus*

Infraorder - **Gebiidea**
Ex: *Upogebia deltaura*

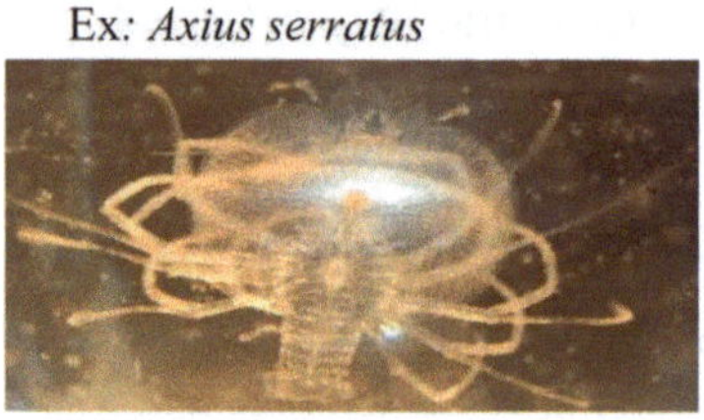

Infraorder - **Achelata**
Ex: *Phyllosoma*

Infraorder -**Polychelida**
Ex: *Stereomastis sculpta*

Infraorder -**Anomura**
Ex: *Dardanus megistos*

Infraorder - **Brachyura**
Ex: *Liocarcinus vernalis*

Fig. 6.2 Representative species of different infraorders of Decapoda temperatures can fluctuate dramatically daily (Ng et al. 2008). Tropical and subtropical regions have more number of crab species compared to temperate and cold regions (Fransozo and Negreiros-Fransozo 1996; Boschi 2000).

The total number of freshwater crab species (undescribed species) is thought to be actually 65 % higher, potentially up to 2155. All freshwater crabs belong to eight families, each with a limited distribution. Although various crabs from other families are also able to tolerate freshwater conditions (euryhaline) or are secondarily adapted to freshwater, the phylogenetic relationships between these families are still debateable, and it is not clear how frequently the freshwater lifestyle has evolved among the true crabs. Few fossils of freshwater crabs have also been found; the oldest fossil is *Tanzanonautes tuerkayi*, from the Oligocene epoch of East Africa. Freshwater crabs are known to exhibit direct development and maternal care of a small number of offspring. Marine crabs release thousands of larvae to the sea surface. This resulted in limited dispersal abilities of freshwater crabs and they tend to be endemic to small regions. Consequently, a large proportion of them are at the verge of extinction.

Being the predominant group of the mangrove ecosystem (Macintosh 1996), crabs are thought to play a significant ecological role in their structure and function (Lee 1999). They act as connecting links between the primary detritus at the base of the food web and consumers at higher trophic levels (Macintosh 1996). They play an important role in nutrient recycling. The feces of all the crabs consist of carbon, nitrogen, phosphorus, and trace metals which form a rich food for other consumers. Crabs and their larvae are consumed by many predators and omnivorous fishes. Thus they are of immense value in recycling nutrients. The burrowing activities of crabs are known to improve soil aeration (Smith et al. 1991), allow seawater penetration, and are responsible for nutrient exchange (Paphavasit et al. 1990). As a result, the topography and textural properties of mangrove soils are altered (Warren and Underwood 1986). Crabs along with other animals modify the mangrove sediment and thus have the potential to nourish mangrove vegetation, structure, and productivity.

The South Indian edible freshwater crab, *Oziotelphusa senex senex*, is an inhabitant of rice fields and irrigation canals. Although originally limited to freshwater, it can also survive in 100 % seawater (Reddy and Reddy 2006). Their ecological role through their important position in the food web in more than one level cannot be ignored. They provide prey for many invertebrates and vertebrates and in turn feed on a variety of plant materials competing with other small herbivores, small fishes, prawns, and invertebrates (Gherardi et al. 1989). Thus crabs among other invertebrates are considered as an essential shell fishery product (Nalan et al. 2003) and significant organisms inhabiting Southeast Asian freshwaters and play a key role in recycling nutrients.

Perhaps another significant role of crabs is the transmission of diseases. Crabs of the genera *Sudanonautes* and *Liberonautes* have been identified as secondary hosts of lung flukes (*Paragonimus* spp.), several species of which cause pulmonary paragonimiasis in the rain forest regions of the West and West-Central Africa (Voelker et al. 1975; Voelker and Sachs 1977; Cumberlidge 1999). The mode of transmission is through the consumption of partially cooked crab flesh (Ollivier et al. 1995). In Southern Nigeria, crabs are often eaten raw, particularly by children, resulting in high rates of infection (Udonsi 1987). Crabs also act indirectly as vectors of onchocerciasis (river blindness), a disease caused by a blood fluke spread by black flies (Diptera). Larvae of these flies have a phoretic association with crabs and other freshwater invertebrates, most notably mayflies whose larvae and pupae attach themselves to the carapace and leg bases of crabs.

6.3 Sociocultural Value

The association between crabs and religion dates back to the first century. Romans worshiped the crab as a sea god and named the crab as Neptune. The crab, *Charybdis feriatus*, has been called as symbol of the cross by Christians from the coastal areas of India. They do not eat this crab and release it to the sea when accidently caught along with fish. The abnormal growth of body cells due to cancer/tumor resembles the nesting ground of *Cancer* spp., so the disease's name denotes this killer disease. Even in astrology, crabs represent one of the 12 powerful zodiac signs.

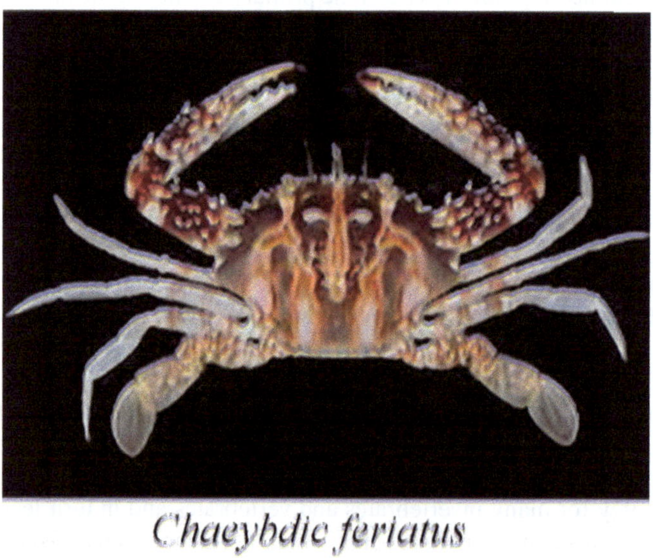

Charybdis feriatus

6.4 Socioeconomic Value

Crabs play a valuable role as indicators of pollution. *Potamonautes warreni*, found commonly along the Orange River, has a large body and is easy to capture with bait. The river drains much of the heavily polluted mining by products of the Northern

South Africa. The crab is investigated as a possible bioindicator of metals in these habitats (Van Eeden and Schoonbee 1991; Sanders et al. 1999; Schuwerack et al. 2001).

The smaller intertidal crabs are not economically valuable but can be used in the preparation of high energy-yielding, cheaper artificial pellet feeds for the farming of edible varieties of seafood. In some parts of the world, powdered crabs are fed to livestock and poultry as a supplement to their diet (Panning and Peter 1933; Agarwal and Kumar 1987). Crab's powder is also used as fertilizer in agricultural operations (Agarwal and Kumar 1987).

Crabs form an important constituent in the marine fish landings in the world. Crab meat, frozen or dried, and crab concentrate of non marine species are considered as table delicacies and are exported (Samuel 1968). Live crabs, cut crabs, and meat are exported from India to countries like Japan, the USA, France, Hong Kong, Malaysia, and Singapore. Crabs have become a very important component of the coastal and mangrove ecosystems worldwide because of their high consumer demand fetching high prices both in local and export markets (Chitravadivelu 1994). True brachyuran crab, *Ranina ranina*, fishery is also thriving commercially since World War II (Brown and Caputi 1985). The meat of *R. ranina* is considered a delicacy among the tourists, and its pegged price in Philippine Peso is 200–300/kg at the local market in Zamboanga City in the southern part of the Philippines (Tito and Alanano 2008). Although the mud crab fishery in the Philippines was started with the work of Arriola in 1940 with the study of the life history of *Scylla serrata*, the farming was started only 30 years earlier (FCI Act 1999). In Japan, Japanese mitten crab, *Eriocheir japonica*, is a very popular seafood so much so that to replenish the declining populations of this crab species, its restocking management programs such as seed production, releasing young crabs, etc. are being taken up to protect the natural population from extinction by overexploitation.

Although crabs are thought to be inferior to prawns and lobsters, they have a high demand in Vietnam (Yeo et al. 2008), Australia (Gardner et al. 1998), and the USA (Cohen and Carlton 1997). The ripening gonads and orange ovaries of matured females of the Chinese mitten crab are considered delicacies (Veldhuizen and Stanish 1999).

Freshwater crabs are eaten by human beings in many countries. The species conserved in Africa are *Sudanonautes aubryi* in Ivory Coast (Bertrand 1979) and *S. africanus* and *S. kagoroensis* in Nigeria (Okafor 1988; Cumberlidge 1991). Some tribes do not eat them (Cantrell 1980). Large pseudothelphusid freshwater crab is eaten by tribals in South America (Finkers 1986). In Thailand, large potamids and parathelphusids are occasionally eaten by locals (Ng 1988). Potamids are important in the diet of the rural and hill tribes of Northern Vietnam (Yeo and Ng 1998). *Liberonautes nanoides*, the dwarf river crab, is caught during the dry season using basket traps in Liberia (Sachs and Cumberlidge 1991). This small species is used as an ingredient in soups.

Apart from being a table delicacy and a good source of protein, crabs are of use in several other ways. They can tolerate a wide range of environmental variations and are considered as plastic and thus are always chosen to test tolerance levels of aquatic fauna. Brachyuran crabs have been used as tools in the elucidation of physiological mechanisms, fertilization, regeneration and cell association, and mechanisms of drug action. Crabs have recreational values, such as fishing large and small crabs, and are used esthetically too such as keeping colorful crabs in the aquarium. More colorful Indo-Chinese potamid crabs of the genus *Demanietta* have a good market as specimens for aquariums (Yeo et al. 2008). *Potamonautes lirrangensis* ("Malawi blue crab"), which inhabits Lake Malawi and rivers in the upper Congo catchment is sold frequently as an aquarium species locally (often under the name *P. orbitospinus*).

Bacteria, viruses, fungi, and several other pathogens influence the shell diseases among crustaceans. Shell infestation by biological agents such as bacteria (Sawyer 1991), fungi (Stentiford et al. 2003), dinoflagellates (Gottfried et al. 2003), and foulers (Heath 1976) is a major problem in aquaculture, causing heavy economic losses. Bacterial pathogens rank as the most common etiological agents of crustacean decapods, as they cause more diseases than all other agents (Roberts 2004). Gram-positive forms affect and cause serious diseases among crustaceans. Several shell diseases caused by *Vibrio*, *Pseudomonas*, and *Aerococcus* are major causes of economic loss (Prince et al. 1993). On the other hand, fungal growth on the surface of eggs, larvae, and shell of adult causes extensive mortalities. Some of these infections become systemic and invade all body organs making crabs nonedible.

6.5 Medical Importance of Crabs

The Central Institute of Fisheries Technology, Cochin, India, has been endeavoring to develop a variety of prawn/crab products having food, medicinal, and industrial value (Gopakumar 1993). Crab meat is believed to possess therapeutic qualities for colds, asthma, and wheezing. Freshwater crabs are known to have toxic and medicinal properties especially to heal injuries and stomach ailments (Dai 1999). According to the traditional belief, medicinal liquor prepared by baking the crab shell enriches it with calcium (Sriphuthorn 2000). Crab curry is used to treat cold, asthma, and typhoid and is given as tonic to convalescing patients (Anonymous 1972). The tonic derived by pounding the whole body of freshwater crabs in a mortar is used to detoxify the blood (Sriphuthorn 2000). This water is administered to women who have a history of miscarriages.

A very interesting behavior is displayed by *Potamonautes raybouldi*, the tree-hole crab of the East Usambara Mountains in Tanzania and the Shimba Hills in Kenya (Bayliss 2002; Cumberlidge and Vannini 2004). Not the crab itself, but the water from the tree hole in which it lived is reported to be medically important. The value of the water is enhanced by the behavior of the crab. The crab neutralizes the naturally acidic water in tree holes by crushing the shells of captured snails and adding them into the water, resulting in increased pH and levels of dissolved calcium (Bayliss 2002).

Many highly active biocompounds have recently been isolated from reef brachyurans associated with antimicrobial, antileukemic, anticoagulant, and cardioactive properties. The exoskeleton of many crustaceans including crabs is chiefly composed of chitin, an animal polysaccharide. It plays an important role in biochemical activities and chemical applications in medical and pharmaceutical industries (Murugan and Ramakrishna 2004; Yadav and Bhise 2004; Takeuchi et al. 2001; Kato et al. 2003). Chitin and its N-acetylated analogs are used for their antimicrobial, antitumor, and immuno-enhancing activities (Gohel et al. 2006). Similarly, chitinolytic enzymes are used extensively in agriculture, biological, and environmental fields. It is also believed that chitin-enriched crab's shells help to reduce cholesterol and triglyceride levels in the blood. Many physiologists, biologists, and researchers have been using crabs as a biological model due to their ready availability (Warner 1977; Burggren and Mc Mahon 1988; Reinecke et al. 2003).

The development of resistance to antibiotics in aqua farms has led to the search for alternate antimicrobial agents especially among arthropods. Such an activity was detected in the hemolymph and/or hemocytes especially those of crustaceans (Soderhall et al. 1996; Taylor et al. 1997). The antimicrobial activity and lysozyme activity of the hemolymph of some crustaceans such as *Heliothis virescens* (Lockey and Ourth 1996) and *Manduca sexta* (Mulnix and Dunn 1994), and antimicrobial peptides isolated from the plasma and hemocyte of crabs (Shao-Yang et al. 2006) provide scope for the exploration of the medical use of the hemolymph of some common crustaceans to control pathogen proliferation. It has been observed that in various invertebrate species, the hemolymph elicits the synthesis of a number of antimicrobial peptides and proteins after bacterial injection (Noga et al. 1996). It is believed that circulating hemocytes are playing an important role in the innate immune system of invertebrates, including being the storage reservoir of several immune components, such as lectins, coagulation factors, and protease inhibitors (Hoq et al. 2003). Decapod hemocytes are known to contain several immune effectors, and they play a major role in the cellular and humoral defense mechanisms of the host (Ravichandran et al. 2010).

We investigated the antibacterial and antifungal activity of the hemolymph of the South Indian edible freshwater/field crab, *Oziotelphusa senex senex*, a brachyuran belonging to the family of Gecarcinidae on clinical pathogenesis in this study.

6.6 Materials and Methods

Collection of animals: *Oziotelphusa senex senex*, commonly known as the freshwater field crab (Fabricius) (Fig. 6.3), was used for the present investigation. The male and female crabs were collected from rice fields and irrigation canals (free from pesticides and pollutants) in and around Tirupati (Andhra Pradesh, South India). Crabs were maintained in the laboratory at 28 ± 1 °C in tubs partially filled with freshwater. They were made to acclimatize to laboratory conditions (12:12 L: D) for at least 7 days before experiments began. Water in the tubs was changed on daily basis. Sheep meat ad libitum was provided as food for them. Crabs are starved for 1

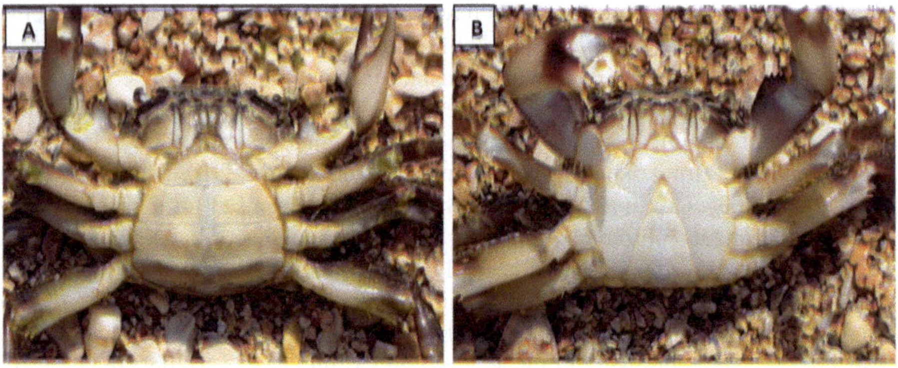

Fig. 6.3 The freshwater crab, *Oziotelphusa senex senex*. (**a**) Female and (**b**) male

day before the commencement of the experiment to avoid changes due to prandial activity.

Collection of hemolymph: Using the hypodermic syringe, the hemolymph was collected from each walking leg of the animal. Hemolymph was collected in the presence of sodium citrate buffer, pH 4.6 (2:1, V/V), to avoid hemocyte degranulation and coagulation. Equal volume of physiological saline (0.85 %, NaCl, w/v) was added to it. Hemolymph was centrifuged at 2000 rpm for 15 min at 4 °C to remove hemocytes. Supernatant was collected by aspiration and stored at 4 °C until use.

Microbial strains used: Antibacterial activity of crab hemolymph was determined against four pathogenic bacterial strains, viz., *Escherichia coli*, *Staphylococcus aureus*, *Klebsiella pneumoniae*, and *Enterococcus faecalis* (Fig. 6.4) and four pathogenic fungal strains, viz., *Aspergillus flavus*, *Aspergillus niger*, *Penicillium citrinum*, and *Rhizopus* spp. (Fig. 6.5).

Antibacterial assay: Antimicrobial activity of the hemolymph of the male and female was tested separately against *E. coli*, *S. aureus*, *K. pneumoniae*, and *E. faecalis* through agar disk diffusion method (Benkerroum et al. 1993). Sterile Luria agar plates were prepared, and 100 μl (10^5 CFU/ml) of each pathogenic bacterium was evenly spread on the surface of the plates. Disks measuring about 0.5 mm diameter dipped separately in the hemolymph of male and female crabs were placed in each of the plates containing one of the above pathogenic bacterial strains. All the plates were incubated at 37 °C for 24 h, and the zone of inhibition was observed against respective controls prepared using LB medium.

Antifungal assay: Antifungal activity of the hemolymph of the male and female was tested separately against *A. flavus*, *A. niger*, *P. citrinum, and Rhizopus* spp. through agar disk diffusion method (Benkerroum et al. 1993). Sterile potato dextrose agar plates were prepared, and 100 μl (10^5 CFU/ml) of each pathogenic fungus was evenly spread on the surface of the plates. Disks measuring about 0.5 mm diameter dipped separately in the hemolymph of male and female crabs were placed in each of the plates containing one of the above pathogenic fungal strains. All the plates were incubated at 37 °C for 24 h, and the zone of inhibition was observed against respective controls prepared using potato dextrose broth.

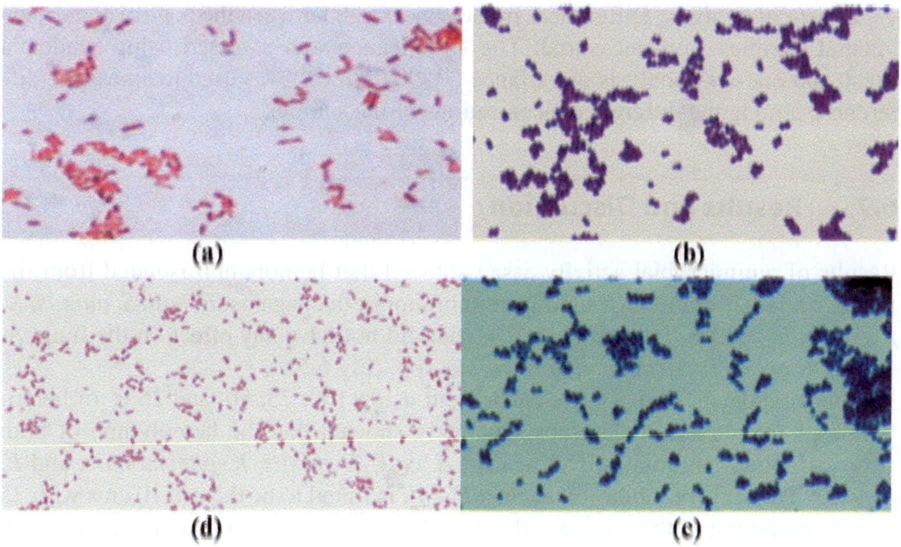

Fig. 6.4 Bacterial pathogens. (**a**) *Escherichia coli.* (**b**) *Staphylococcus aureus*. (**c**) *Enterococcus faecalis*. (**d**) *Klebsiella pneumoniae*

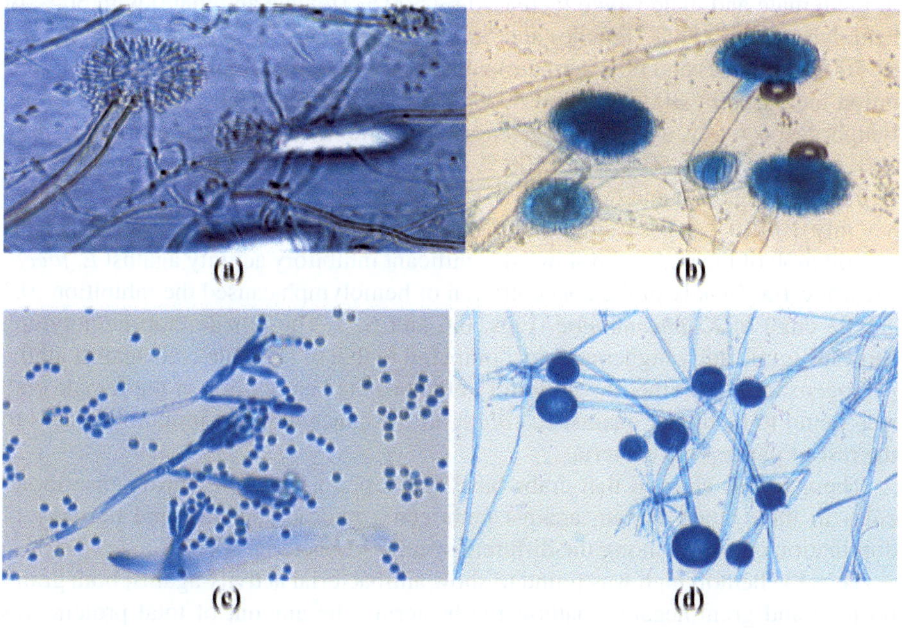

Fig. 6.5 Fungal pathogens. (a) *Aspergillus flavus*. (b) *Aspergillus niger*. (c) *Penicillium citrinum*. (d) *Rhizopus* spp

Statistical analysis: Differences in the zone width between the control and experimental samples were measured. The significance was assessed using Student's *t*-test, and one-way analysis of variance ($P < 0.05$ level) was used to assess the differences among the different concentrations of the sample.

6.7 Results and Discussion

Results of antimicrobial activity assay showed that hemolymph isolated from the female and male crabs have significant inhibitory effect against *E. coli*, *S. aureus*, *K. pneumoniae*, and *E. faecalis* (Fig. 6.4), but did not have any effect on the fungi *A. flavus*, *A. niger*, *P. citrinum, and Rhizopus* spp. (Fig.6.5).

The antagonistic activity test carried out using 100, 75, 50, and 25 % of hemolymph of female and male crabs separately showed that the hemolymph of both sexes exhibited inhibitory effect against *E. coli*, *S. aureus*, *K. pneumoniae*, and *E. faecalis*. The zone of inhibition against *E. coli* is found to be 0.5 ± 0.10 cms with 25 % of hemolymph of the female and male which increased by 60 % (female) and 20 % (male) with 50 % of hemolymph, by 80 % (female) and 40 % (male) with 75 % of hemolymph, and by 100 % (female) and 80 % (male) with 100 % of hemolymph. The zone of inhibition against *S. aureus* is 0.1 ± 0.10 cms with 25 % of hemolymph of the female and male which increased by 500 % (female and male) with 50 % of hemolymph, by 600 % (female) and 500 % (male) with 75 % of hemolymph, and by 700 % (female) and 600 % (male) with 100 % of hemolymph. The zone of inhibition increased by 300, 700, and 800 % with 50, 75, and 100 % of female hemolymph, respectively, against *K. pneumoniae*, while the zone of inhibition was 400, 500, and 700 % with 50, 75, and 100 % of female hemolymph, respectively, against *E. faecalis* (Figs. 6.6 and 6.7).

Only 100 (0.7 ± 0.14 cms) and 75 % (0.5 ± 0.12 cms) concentrations of the hemolymph of the male crab showed significant inhibitory activity against *E. faecalis*, while 100 % only of the concentration of hemolymph caused the inhibition (0.2 ± 0.03 cms) of *K. pneumoniae* (Figs. 6.6 and 6.7). The female crab hemolymph showed significantly higher activity compared to that of the male crab against all the pathogens tested. Further the inhibitory effect of the hemolymph of the female crab was found to be higher against *E. coli* at all the concentrations tested compared to the rest of pathogenic bacteria.

These results showed that crabs have developed a range of self-defense molecules in their blood system against pathogenic microorganisms, and the way of domination changes among the different species of bacterial species.

Thus the hemolymph was found to show antibacterial activity against both gram-positive and gram-negative pathogenic bacteria. The amount of total protein was noticed to be 150 μg/ml and 162 μg/ml in the female and male, respectively (Priya Rathna et al. 2014). Crabs are a wonderful resource of proteins with a wide range of antimicrobial properties which is highly supported in the hemolymph study of *Charybdis lucifera* (Stewart and Zwicker 1972), shore crabs (*Carcinus maenas*) (Chisholm and smith 1992), and penaeid shrimp (Destoumieux et al. 1997). The antimicrobial activity has been reported earlier in the hemolymph of the blue crab,

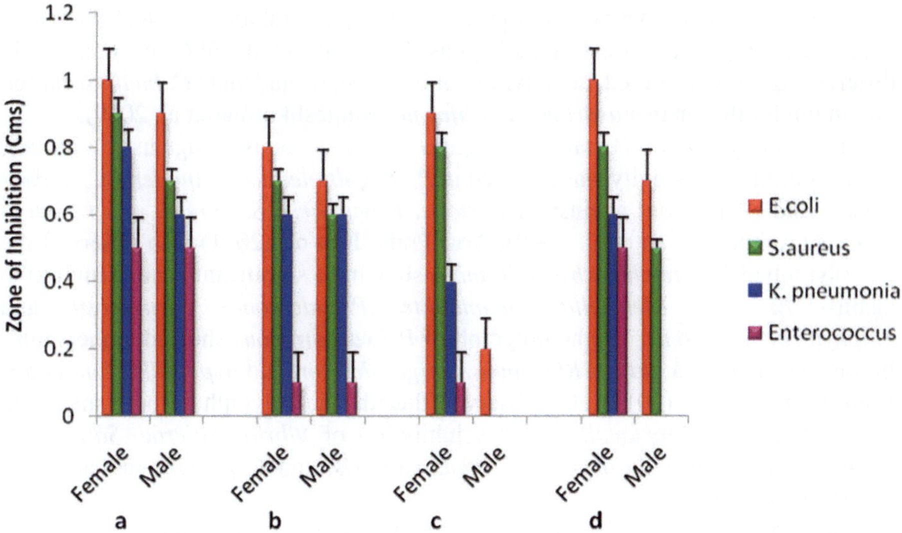

Fig. 6.6 Antimicrobial activity of (**a**) 100 %, (**b**) 75 %, (**c**) 50 %, and (**d**) 25 % hemolymph of female and male crabs against *E. coli*, *S. aureus*, *K. pneumoniae*, and *E. faecalis* (values are mean ± SD (*n* = 5) of five individual observations)

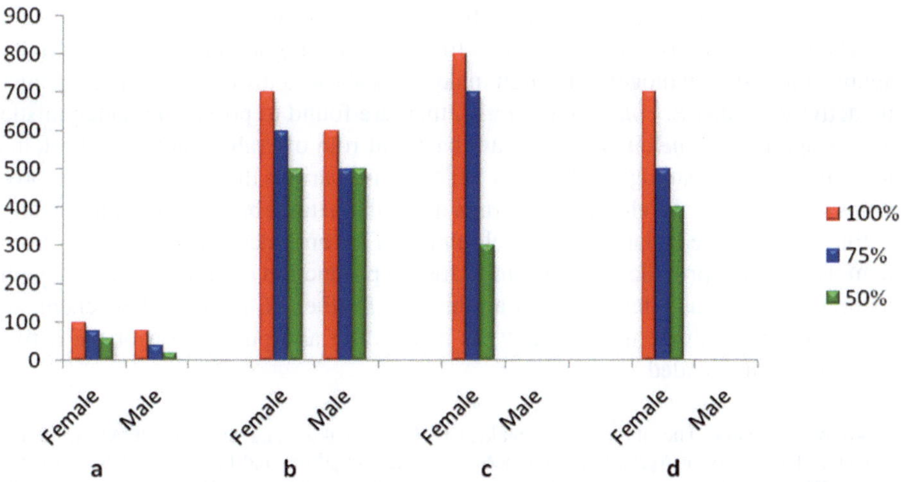

Fig. 6.7 Percent change in antimicrobial activity of (**a**) 100 %, (**b**) 75 %, (**c**) 50 %, and (**d**) 25 % hemolymph of female and male crabs against *E. coli*, *S. aureus*, *K. pneumoniae*, and *E. faecalis*

Callinectes sapidus (Edward et al. 1996), *Carcinus maenas* (Schnapp et al. 1996), mud crab *Scylla serrata* (Hoq et al. 2003), and *Ocypode macrocera* (Ravichandran et al. 2010). The hemolymph of male *C. lucifera* also showed significantly higher inhibition of *P. aeruginosa* compared to the hemolymph of the female crab (Rameshkurmar et al. 2009).

Earlier crab hemolymph was found to have lesser inhibitory effect on fungal pathogens compared to bacterial pathogens (Kawababa et al. 1996). Among the six different strains considered, hemolymph of both female and male *C. lucifera* caused maximum inhibition of *Fusarium moniliforme* (Rameshkurmar et al. 2009).

The hemolymph of *Portunus pelagicus* exhibited activity against *S. aureus*, whereas minimum activity was detected in *P. sanguinolentus*, *C. lucifera*, *C. amboinensis*, and *D. dehaani* against *K. oxytoca*, *L. vulgaris*, *S. aureus*, and *K. pneumoniae* (Anbuchezhian et al. 2009). Arul Prakash et al. (2011) also observed the hemolymph of *Paratelphusa hydrodromous* showing a significant zone of inhibition against *E. coli*, *Klebsiella pneumoniae*, *Pseudomonas aeruginosa*, and *Staphylococcus aureus*. The hemolymph of *P. hydrodromous* showed highest inhibition zone against *A. flavus*, *Rhizopus*, *A. niger*, *Mucor*, and *Aspergillus fumigatus*. Priya Rethna et al. (2014) also observed that the hemolymph of both male and female *Liagore rubromaculata* shows inhibition of *Vibrio cholerae*, *Salmonella typhi*, *E. coli*, *Klebsiella oxytoca*, *Proteus vulgaris*, and *E. faecalis* indicating its wider applications.

Thus fast growth rate, high meat yield, and excellent palatability and resistance to the pathogens have led the South Indian region to the development of rapidly increasing aquaculture industry of different species of crabs (Reddy and Reddy 2006). Thus the development of inland crab fishery is safety warranted under the aquaculture program for the economic development of the targeted people and the country as a whole without disturbing the natural ecosystem.

The hemolymph of a female crab is found to have higher antimicrobial activity against the tested pathogens. Though male crabs showed lower levels of antagonistic activity against *E. coli* and *S. aureus*, they are found to possess no antagonistic agents against *K. pneumoniae*. The antibacterial role of crab hemolymph proteins identified against pathogenic bacteria which cause various diseases in humans provides scope for the development of drugs using the natural biota in an economically viable manner. The isolation and development of the antimicrobial compounds from hemolymph will provide an opportunity for the production of new compounds with natural activities as alternatives to antibiotics. Further, to identify their chemical nature and to evaluate their potency as novel drug and purification of the active compounds are needed.

Acknowledgments The authors are thankful to Prof. Shakunthala Sridhara, ICAR Emeritus Scientist, University of Agricultural Sciences, GKVK, Bangalore, and Prof. A. K. Chakravarthy, Head, Division of Entomology and Nematology, Indian Institute of Horticultural Research, Bangalore, for providing the opportunity to contribute this chapter for the book.

References

Agarwal SK, Kumar S (1987) The bionomics and fishery prospects of a freshwater crab of Kumaun. Environment, Problems and Development, Western Himalaya, pp 275–280

Anbuchezhian RM, Ravichandran S, Rameshkumar G, Ajithkumar T (2009) Influence of crab haemolymph on clinical pathogens. Adv Biol Res 3(3–4):104–109

Anonymous (1972) Bharat ki Sampada (Raw materials), 2: Council of Scientific and Industrial Research, New Delhi

Antibacterial activity from the blue crab *Callinectes sapidus*. Fish Shellfish Immunol 6(6):403–412

Arul Prakash A, Balasubramanian S, Gunasekaran G, Prakash M, Senthil RP (2011) International scholarly research network, Isrn Pharmacology

Bayliss J (2002) The East Usambara tree-hole crab (Brachyura: Potamoidea: Potamonautidae) – a striking example of crustacean adaptation in closed canopy forest. Tanzania Afr J Ecol 40:26–34

Benkerroum N, Ghiuti Y, Sandine WE, Tantaoui- Elaraki A (1993) Methods to demonstrate the bactericidal activity of bacteriocins. Lett Appl Microbiol 17:80

Bertrand JY (1979) Observations sur les crabes *Sudanonautes africanus* (Milne-Edwards 1869) leurs terriers dune savane de Lamto (Côte d'Ivoire). Bulletin de la Societe Zoologique de France 104:27–36

Boschi EE (2000) Biodiversity of marine decapod brachyurans of the *Americas*. J Crustac Biol 20(2):337–342

Brown RS, Caputi N (1985) Factors affecting the growth of undersized western rock lobster, *Punulirus cygnus* George, returned by fisherman to the sea. Fish Bull 83:567–574

Burggren WW, Mc Mahon BR (1988) Biology of the land crabs: an introduction. In Burggren WW and Mc Mahon BR (eds) Biology of the land crabs. Cambridge University Press, Cambridge, pp 1–5

Cantrell MA (1980) On the terrestrial behaviour of the freshwater crab *Potamonautes choloensis* (Chace). Nyala 6:95–107

Chisholm JRS, Smith VJ (1992) Antibacterial activity in the hemocytes of the shore crab, *Carcinus maenas*. J Mar Biol Assoc UK 72:529–542

Chitravadivelu K (1994) Aspects of fishery and species composition of edible crabs in the Jaffna lagoon. J Natl Sci Counc Sri Lanka 21(1):43–55

Cohen AN, Carlton JT (1997) Transoceanic transport mechanisms: introduction of the Chinese mitten crab, Eriocheir sinensis, to California. Pac Sci 51(1):1–11

Cumberlidge N (1991) The respiratory system of *Globonautes macropus* (Rathbun, 1898), a terrestrial freshwater crab from Liberia (Gecarcinucoidea, Gecarcinucidae). Crustaceana 61:69–80

Cumberlidge N (1999) The freshwater crabs of West Africa: family Potamonautidae, vol 36. Institut de Recherche pour le Developpement, Collection Faune et Flore Tropicales, Paris, pp 1–382

Cumberlidge N, Vannini M (2004) A new species of tree-hole living freshwater crab (Brachyura: Potamoidea: Potamonautidae) from coastal East Africa. J Nat Hist (in press)

Dai AY (1999) Fauna Sinica (Arthropoda, Crustacea, Malacostraca, Decapoda, Parathelphusidae, Potamidae). Editorial Committee of Fauna Sinica, Academia Sinica Science Press, Beijing, p 501

De Grave Sammy N, Pentcheff D, Ahyong ST (2009) A classification of living and fossil genera of decapod crustaceans. Raffles Bull Zool Suppl 21:1–109

Destoumieux D, Bulet P, Loew D, Dorsselaer VA, Rodriguea L, Bachere E (1997) Penaedine, a new family or antimicrobial peptides isolated from the shrimp *penaeus vannamei* (Decapoda). J Biol Chem 72:28398–28406

Edward NJ, Arroll TA, Fan Z (1996) Specificity and some physiochemical characteristics of antibacterial activity from blue crab *Callinectes sapidus*. J Fish Shellfish Immunol 6:403–412

Elena Mente (2008) Reproductive biology of crustaceans: case studies of decapod crustaceans. Science Publishers, Enfield. ISBN 978-1-57808-529-3 pp 16

FCI Act (1999) Foster care independence act of 1999. Child Welfare League of America 1999-11-23

Finkers J (1986) Los Yanomami y su sistema alimenticio (Yanomami Nii Pe"). Vicariato Aposto'lico de PuertoAyacucho, Puerto Ayacucho. Monografıa 2:262

Fransozo A, Negreiros-Fransozo ML (1996) Brazilian coastal crustacea decapoda. In: de Bicudo CE and Menezes MNA (eds) Biodiversity in Brazil: a first approach. Proceedings of the workshop methods for the assessment of biodiversity in plants and animals, Campos do Jordão, São Paulo, Brazil, 26–30 May 1996, CNPq – Conselho Nacional de Desenvolvimento Cientı'fico e Tecnolo'gico, São Paulo, Brazil pp 275–287

Gardner C, Rush M, Bevilacqua T (1998) Nonlethal imaging techniques for crab spermathecae. J Crustac Biol 18:64–69

Gherardi F, Micheli F, Vanini M (1989) Movement patterns and dispersal of the hermit crab, *Clibanarius longitarsus* in a mangrove swamp. Mar Behav Physiol 16:209–223

Gohel V, Singh A, Maisuria V, Phdnis A, Chatpar HS (2006) Bioprospecting and antifungal potential of chitinolytic microorganisms. Afr J Biotechnol 5:54–72

Gopakumar K (1993) Marketing of value-added products. Seaf Export J XXV(5):25

Gottfried PP, Taylor DM, Hoenig JM, Shields JD, Pickavance R (2003) Monitoring the prevalence of the parasitic dinoflagellate *Hematodinium* sp. in snow crabs, *Chionoecetes opilio* from Conception Bay, Newfoundland. Dis Aquat Org 53:67–75

Heath DJ (1976) The distribution and orientation of epizoic barnacles on crabs. Zool J Linn Soc 59:59–67

Hoq MI, Seraj MU, Chowdhury S (2003) Isolation and characterization of antibacterial peptides from the mud crab, *Scylla serrata*. Pak J Biol Sci 6(11):1345–1353

Jayabaskaran R, Ajmal Khan S (2007) Diversity of brachyuran crabs in Gulf of Mannar (Southeast coast of India). In: Kannaiyan S, Venkataraman K (eds) Biodiversity conservation of Gulf of Mannar biosphere reserve. National Authority, Chennai, pp 68–82

Kaestner A (1970) Invertebrate zoology III. Crustacea. (trans and adopted from the second German edited by HW Levi). Intersciences Publishers, New York

Kathiresan K, Qasim SZ (2005) Biodiversity of mangrove ecosystems. Hindustan Publishing Corporation, New Delhi, p 251

Kato Y, Onishi H, Machida Y (2003) Application of chitin and chitosan derivatives in the pharmaceutical field. Curr Pharm Biotechnol 4(5):303–309

Kawababa SI, Nagayama R, Hirata M, Shigenaga T, Agarwala KL, Saito T (1996) Tachycitin, a small granular component in horseshoe crab *Hemocytes* is an antimicrobial protein with chitin-binding activity. J Biochem 120:1253–1260

Krishnamoorthy P (2007) Brachyura zoological survey of India, Fauna of Chennai coast. Ecosyst Ser 1:83–109

Lee SY (1999) The effect of mangrove leaf litter enrichment on macrobenthic colonization of defaunated sandy substrates. Estuar Coast Shelf Sci 49:703–712

Lockey TD, Ourth DD (1996) Purification and characterization of lysozyme from hemolymph of *Heliothis virescens* larvae. Biochem Biophys Res Comm 220:502–508

Macintosh DJ (1996) Mangroves and coastal aquaculture: doing something positive for the environment. Aquac Asia 1:3–8

Martin JW, Davis GE (2001) An updated classification of the recent crustacea. Nat Hist Mus Los Angel County Sci Ser 39:1–124

Melo GAS (1996) Manual de identificacao dos Brachyura (caranguejos e siris) do litoral brasileiro. Editora Pleiade, Sao Paulo

Mulnix AB, Dunn PE (1994) Structure and induction of a lysozyme gene from the tabacco hornworm, *Manduca sexta*. Insect Biochem Mol Biol 24:271–281

Murugan R, Ramakrishna S (2004) Bioresorbable composite bone paste using polysaccharide based nanohydroxyapatite. Biomaterials 25(17):3829–3835

Nalan G, Dlua K, Yerlikayaa P (2003) Determination of proximate composition and mineral contents of blue crab (*Callinectes sapidus*) and swim crab (*Portunus pelagicus*) caught off the gulf of Antalya. Food Chem 80:495–498

Ng PKL (1988) The freshwater crabs of peninsular Malaysia and Singapore. Department of Zoology, National University of Singapore, Shinglee Press, Singapore, pp 1–8, 1–156

Ng PKL, Guinot D, Davie PJF (2008) Systema Brachyurorum: part I. An annotated checklist of extant brachyuran crabs of the world. Raffles Bull Zool 17:1–286

Noga EJ, Arroll TA, Fan Z (1996) Specificity and some physicochemical characteristics of the antibacterial activity from blue crab Callinectes sapidus. Fish Shellfish Immunol 6:403–412

Okafor FC (1988) The ecology of *Sudanonautes (Sudanonautes) africanus* (H. Milne-Edwards) (Crustacea: Decapoda) in southeastern Nigeria. Trop Ecol 29:89–97

Ollivier G, Boussinesq M, Albaret JL, Cumberlidge N, Farhati K, Chipaux JP, Baysadde-Dufour C (1995) Etude épidémiologique d'une distomatose à Paragonimus sp. au sud-Cameroun. Bull Soc Pathol Exot 88:164–169

Panning A, Peter N (1933) Die Chinesische wollhandkrabbe (Eriocheir sinensis H. Milne-Edwards) Deutschland. Zool Anz Suppl 104:1–180

Paphavasit N, Dechaprompun S, Aumnuch E (1990) Physiological ecology of selected mangrove crabs: physiological tolerance limits. In Field CD (ed) Mangrove ecosystems occasional papers No. 5. UNDP/UNESCO Regional Mangroves Project RAS/86/120, pp 1–19

Prince DL, Bayer RC, Loughlin M (1993) Etiology and microscopy of shell disease in impounded American lobsters, *Homarus americanus*. Bull Aquac Assoc Can 93:87–89

Priya RE, Ravichandran S, Jawaharlal P (2014) Antimicrobial and antioxidant proteins from the crab, *Liagore rubromaculata* (de haan, 1835). World J Pharm Pharm Sci 3(10):533–541

Rameshkurmar GS, Ravichandran G, Kaliyavarathan T, Ajithkumar T (2009) Antimicrobial peptide from the crab, *Thalamita crenata* (Latreille, 1829). World J Fish Mar Sci 1(2):74–79

Rao PV, Vedavyasa M, Thomas MM (1973) Crab fishery resources of India. In: Proceeding symposium on living resources of the seas around India 581–591, India

Ravichandran S, Kannupandi T (2007) Biodiversity of crabs in Pichavaram mangrove environment. Zoological survey of India. National symposium on conservation and valuation of marine biodiversity India pp 331–340

Ravichandran SS, Wahidulla L, Souza D, Rameshkumar G (2010) Antimicrobial lipids from the hemolymph of brachyuran crabs. J Appl Biochem Biotechnol 162(4):1039–1051

Reddy RP, Reddy PS (2006) Isolation of peptide hormones with pleiotropic activities in the freshwater crab, *Oziotelphusa senex senex*. Aquaculture 259:424–431

ReineckeA J, Synman RG, Nei JAJ (2003) Uptake and distribution of lead (pb) and cadmium (cd) in the freshwater crab *Potamonautes perlatus* (Crustacea), in the Eerste river, South Africa. Water Air Soil Pollut 145:395–408

Roberts RJ (2004) The bacteriology of teleosts. In: Roberts RJ (ed) Fish pathology. W.B. Saunders, Philadelphia

Sachs R, Cumberlidge N (1991) The dwarf river crab *Liberonautes latidactylus* nanoides from Liberia – a new second intermediate host of *Paragonimus uterobilateralis*. Trop Med Parasitol 42:73–74

Samuel CT (1968) Marine fisheries of India. M/s. ST Reddiar & Sons, Cochin

Sanders MJ, Du Preez HH, Van Vuren JHJ (1999) Monitoring cadmium and zinc contamination in freshwater systems with the use of the freshwater crab, Potamonautes warreni. Water SA 25:91–98

Satheeshkumar P, Khan AB (2011) An annotated checklist of brachyuran crabs (Crustacea: Decapoda) from Pondicherry Mangroves, south east coast of India. World J Zool 6:312–317

Sawyer TK (1991) Shell disease in the Atlantic rock crab, *Cancer irroratus* SAY, 1817, from the northeastern United States. J Shellfish Res 10:495–497

Schnapp D, Kemp GD, Smith VJ (1996) Purification and characterization of a prolin rich antibacterial peptide, with sequence similarity to bactenecin-7, from the hemocytes of the shore crab, *Carcinus maenas*. Eur J Biochem 240:532–539

Scholtz G, Richter S (1995) Phylogenetic systematics of the reptantian Decapoda (Crustacea, Malacostraca). Zool J Linnean Soc 113(3):289–328

Schuwerack PM, Lewis JW, Jones P (2001) The potential use of the South African river crab, *Potamonautes warreni*, as a bioindicator species for heavy metal contamination. Ecotoxicology 10:159–166

Shao-Yang H, Jan-Hsiung H, Wei-Ting H, Yang-Hui Y, Mark HungChih C, Hong-Yi G, Tze-Ting C, Tzu-Hsuan Y, Thomas TC, JennKhan L, Jen-Leih W (2006) Structure and function of antimicrobial peptide penaeidin-5 from the black tiger shrimp, *Penaeus monodon*. Aquaculture 260:61–68

Smith TJ, Boto KG, Frusher SD, Giddins RL (1991) Keystone species and mangrove forest dynamics: the influence of burrowing by crabs on soil nutrient status and forest productivity. Estuar Coast Shelf Sci 33:419–432

Soderhall K, Iwanaga S, Vasta GR (1996) Crustacean immunity. Ann Rev Fish Dis 2:3–23

Sriphuthorn K (2000) Shell survivor. Bangkok Post, 14 Dec. http://www.ecologysia.com/NewArchives/Dec_2000/bangkokpost_141200_Outlook01.html. Assessed on 2/1/01

Stentiford GD, Evans M, Bateman K, Feist SW (2003) Coinfection by a yeast-like organism in *Hematodinium* infected European edible crabs *Cancer pagurus* and velvet swimming crabs *Necora puber* from the English Channel. Dis Aquat Org 54:195–202

Stewart JE, Zwicker BM (1972) Natural and induced bactericidal activities in the hemolymph of the lobster; Homarus americanus: products of hemocyte-plasma interactions. Can J Microbiol 18:1499–1509

Takeuchi H, Yamamoto H, Kawashima Y (2001) Mucoadhesive nanoparticulate systems for peptide drug delivery. Adv Drug Deliv Rev 47(1):39–54

Taylor SW, Kammerer B, Bayer E (1997) New induced lysis. Biochemistry 34:4393–4401

Tito OD, Alanano JP (2008) Some aspects of fisheries and biology of spanner crab (*Ranina ranina*, Linneaus) in Maluso, Basilan Province, Philippines. Memoirs of Faculty of Kagoshima University Special Issue. Published by Faculty of Fisheries, Kagoshima University, Kagoshima pp 40–48

Trivedi JN, Gadhavi MK, Vachhrajani KD (2012) Diversity and habitat preference of brachyuran crabs in gulf of Kutch, Gujarat, India. Arthropods 1:13–23

Udonsi JK (1987) Endemic Paragonimus infection in Upper Igwun Basin, Nigeria: a preliminary report on a renewed outbreak. Ann Trop Med Parasitol 81:57–62

Van Eeden PH, Schoonbee HJ (1991) Bio-accumulation of heavy metals by the freshwater crab *Potamonautes warreni* from a polluted wetland. S Afr J Wildl Res 21:103–108

Veldhuizen TC, Stanish S (1999) Overview of the life history, distribution, abundance, and impacts of the Chinese mitten crab, *Eriocheir sinensis*, Report prepared for the US Fish and Wildlife Service. California Department of Water Resources, Sacramento, p 26

Voelker J, Sachs R (1977) Über die Verbreitung von Lungenegeln (Paragonimus africanus und P. uterobilateralis) in West-Kamerun und Ost-Nigeria auf Grund von Untersuchungen an Süsswasserkrabben auf Befall mit Metazerkarien. Tropenmed Parasitol 28:129–133

Voelker J, Sachs R, Volkmer KL, Braband H (1975) On the epidemiology of paragonimiaisis in man and in animals in Nigeria, west Africa. Vet Med Rev 1–2:161–175

Warner GF (1997) The biology of crabs. First published in Great Britain, Paul Elek (Scientific books Ltd.) 54–58 Caledonian Road, London pp 202

Warren JH, Underwood AJ (1986) Effects of burrowing crabs on the topography of mangrove swamps in New South Wales. J Exp Mar Biol Ecol 102:223–235

Yadav AV, Bhise BB (2004) Chitosan a potential biomaterial effective against typhoid. Curr Sci 187(9):1176–1178

Yeo DCJ, Ng PKL (1998) Freshwater crabs of the *Potamon tannanti* species group (Crustacea, Decapoda, Brachyura, Potamidae) of northern Indochina. Raffles Bull Zool 46:627–650

Yeo DCJ, Ng PKL, N Cumberlidge, C Magalhaes, SR Daniels, Campos M (2008) A global assessment of freshwater crab diversity (Crustacea: Decapoda: Brachyura). In Balian EV, Lévequè C, Segers H, Martens M (eds) Freshwater animal diversity assessment. Hydrobiologia 595:275–286

Millipedes: Diversity, Distribution and Ecology

7

Periasamy Alagesan

Abstract

Millipedes, the ancient arthropods which belong to class Diplopoda, are the third most diverse class of terrestrial arthropods, following Insecta and Arachnida (Golovatch and Kime, Soil Org 81(3):565–597, 2009). They have a long and distinguished history on our planet and existed on earth more than 100 times longer than man, and their fossil records demonstrate their worldwide distribution (Almond, 1985; Hannibal, 1986). With an estimated total of more than 80,000 extant species, only about 12,000 millipede species have been formally described in 3005 genera, 145 families and 16 orders (Shelley, 2007; Sierwald and Bond, 2007). They also constitute one of the major groups of soil and litter fauna in temperate and tropical environments. Millipedes play an important role in energy flow as well as in the humification of soil and circulation of minerals in terrestrial ecosystems.

Keywords

Millipedes • Diversity • Distribution • Ecological significance

P. Alagesan (✉)
Research Department of Zoology, Yadava College, Madurai 625 014, Tamil Nadu, India
e-mail: maniragavi@rediffmail.com

A.K. Chakravarthy, S. Sridhara (eds.), *Arthropod Diversity and Conservation in the Tropics and Sub-tropics*, DOI 10.1007/978-981-10-1518-2_7

7.1 Introduction

Millipedes are a diverse group of invertebrates under phylum Arthropoda that are the most vital consumers of detritus across several terrestrial ecosystems. They are common litter and soil animals that occur in most parts of the globe's natural zones or belts except the snowy zones (Arctic and Antarctic) as well as the driest deserts (Milkov 1997). In most cultures, the millipedes are called by local names, as they have innumerable legs. For example, Miljoenpoot (Dutch), Tausendfüßler (German), Quilopodo (Portuguese), Jongvoo (Swahili), Songololo (Zulu), Kaki seribu (Indonesia) and Maravattai (Tamil). In Latin, the word 'milli' stands for thousand and 'pede' for foot, so this arthropod ought to be a thousand-footed animal. However, some millipedes have a lot of legs, but none actually have thousands. The record holder is *Illacme plenipes* which has an amazing 350 pairs of legs and the millipede *Polyxenus lagurus* with the least number of legs (24 legs or 12 pairs). But normally most millipedes have 100–300 legs (Fig. 7.1).

One of the most notable characteristics of millipedes is the beautifully coloured exoskeleton, compact, strong dorsal covering, which is fortified largely with calcium salts, making the millipede incompressible. The animal is round and cylindrical or hemispherical in shape. They have a relatively inflexible body composed of three units: the head, a variable number of trunk segments and a pygidial segment that contains the anus. The trunk following the head comprises a column or first trunk unit, three segments embracing six legs, a pair per segment and many similar leg-bearing ring segments, with four legs on each segment (Blower 1985). The body segments bear a number of thigmotactic hairs sensitive to touch (Lawrence 1984). The last leg-bearing segments are called telson, containing one or two apodous rings. The telson is frequently utilised for the identification of millipede species. If a moving millipede is observed from its side, it appears as a little gliding train, for which the coordination of many legs is required.

In the mouthparts of millipedes, mainly a pair of mandibles is present, with a few exceptions, the biting parts armed with blunt and rather clumsy 'teeth', which are

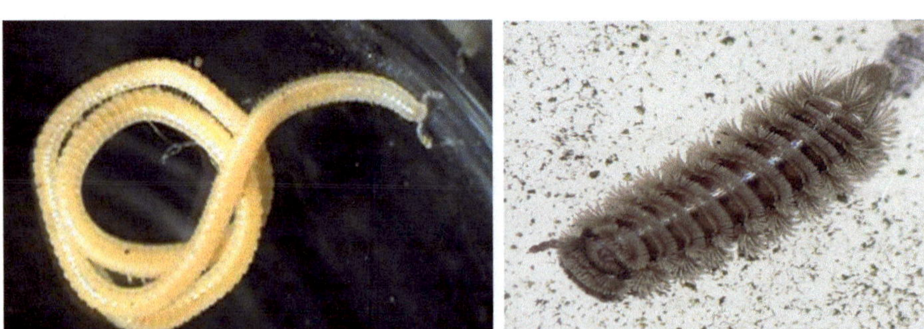

Fig. 7.1 Millipede species: longest millipede, *Illacme plenipes*, and smallest millipede, *Polyxenus lagurus*

Fig. 7.2 Millipede species: largest millipede, *Archispirostreptus gigas*, and pill millipede, *Glomeris marginata*

used to masticate and grind the bigger particles into smaller ones (Hopkin and Read 1992). The gnathochilarium, which serves as the floor of the buccal cavity, is rarely involved in chewing. Along the front edge of the gnathochilarium, the taste organs are found. As the food passes into the mouth, these organs contact the food initially. The shape of the gnathochilarium changes from one order to another order, and hence, it is also used in classification.

As millipedes live in dark areas, vision is not well developed. In the members of Polydesmida, eyes are wanting (Blower 1985), while in the majority of the superorder Juliformia, a group of ocelli, just above the antennae, is found. Millipedes exhibit a wide range of body lengths. The smallest millipede, *Polyxenus lagurus*, is 3–6 mm, whereas the largest millipede, *Archispirostreptus gigas*, is 275 mm in length when fully matured. Compared to insects and spiders, millipedes as a taxon are long-lived (Fig. 7.2). The pill millipede, *Glomeris marginata*, takes several years to mature sexually and can live up to 11 years (Fig. 7.2).

7.2 Reproduction and Life Cycle

Millipedes have separate sexes and are typically sexual, having the openings of the reproductive ducts of both male and female on the trunk's seventh segment ring. In all orders of Diplopoda, the female millipedes have a pair of oviducts, which separately opens through the vulvae, posterior to the second pair of legs. Lawrence (1984) noticed that the external male sex organs are often a helpful means to identify the species, whereas the females do not have these pads. After the process of insemination, the female stores the sperm in the so-called spermathecae. The eggs will only be fertilised when they leave the body at oviposition (Hopkin and Read 1992). Millipede eggs are usually laid in the soil in various forms and shapes. The number of eggs laid by a female millipede varies widely from as few as 3 or 4 up to 2,000 eggs in one clutch. As the egg is energetically a closed biological system, embryogenesis is largely dependent on the egg's nutritive reserve, the yolk.

Generally, the eggs of millipedes are typically very yolky and are nutritious to nurture the offspring until after emergence following the second moult because many immature millipedes remain in an egg capsule and do not feed till then. Except for the males of Platydesmida, there are reports of other millipedes guarding their eggs (Hopkin and Read 1992). Hatchlings are usually legless. They develop legs and add segments at each moult. Born with only three pairs of legs, the larva possesses four ring segments fully developed. It soon sheds the skin; after every moult, it initiates more ring segments as well as legs, appearing in the proliferation zone, near the posterior end of the trunk. This process of embryonic development is called as anamorphosis. The young larvae are pale in colour in comparison with the matured ones, and they take more than 12 months to reach full size (Blower 1985). Generally there are seven or eight moults after the six first-legged stadia. The skin shedding also continues from time to time during the adult life. The whole process of moulting lasts approximately 3 weeks, during which the millipedes are immobile. The moulted skin is often eaten, because the cuticle is a rich source of calcium salts, which is utilised for hardening the new cuticle (Lawrence 1984).

7.3 Natural Enemies and Defence

Diplopods are attacked by a variety of organisms, including vertebrates. The alimentary canal of a number of South African birds had millipedes; it is an indication that the birds can overcome the discouraging defensive glands. In South Africa, assassin bugs, *Lepidocoris elegans* (Heteroptera: Reduviidae), consume mostly larger species of the family Spirostreptidae (Fig. 7.3).

Similarly, parasitic mites attack *Doratogonus* in large numbers but rarely feed on *Chersastus* (Lawrence 1984). The Southeast Asian ant, *Probolomyrmex*, is a specialised predator of Polyxenida (Ito 1998). Some millipedes of the order Julida are often affected by *Laboulbeniales, the* ectoparasitic fungi (Rossi and Balazuc 1977).

Millipedes typically are slow-moving and non aggressive creatures. Even though they may appear defenceless, the majority of millipedes exhibit three kinds of defensive mechanisms against a number of predators.

First, most of the millipede species have a thick, hard exoskeleton that gives some protection.

Fig. 7.3 The predatory bug, *Lepidocoris elegans* (Heteroptera: Reduviidae) attacking the millipede *Xenobolus carnifex*

Second, most millipedes have the ability to curl into a spiral or ball that protects the sensitive head of the animals. For example, Eisner and Davis (1967) explained that an African pill millipede, Sphaerotheriida sp., is able to escape predation from a number of enemies because of its very hard exoskeleton and ability to form into a tight ball. However, this millipede is unable to escape predation from the banded mongoose, which hurls the millipede through its hind legs and smashes the millipedes against a rock or another hard surface and then eats them.

Third, many millipedes employ a defensive secretion that can cause stinging, irritation or sedation of potential predators. In many instances, the exudates contain hydrogen cyanide or quinones, but others discharge unusual molecules. Carrel and Eisner (1984) found that predators like wolf spiders, *Lycosa* sp., which attempted to prey upon quinazolinone-secreting millipedes, *Glomeris marginata*, were quickly deterred by the exuded fluid; those few that persisted and consumed some secretion were sedated for hours even if they had ingested less than one droplet of the secretion. On the other hand, Polyzoniidae millipedes such as *Polyzonium rosalbum* secrete ployzonimine, a well-known ant deterrent.

7.4 Pest Status

Though millipedes are mainly considered as saprophytes, some are treated as indoor nuisance pests, as they crawl into basements, first-floor rooms, living room, outside walls and ceilings of houses. Johns (1967) reported that the millipede, *Ommatoiulus moreletii*, caused significant nuisance in New Zealand. Urban infestations by the millipedes, *Oxidus*, have been reported in Tennessee, USA (O'Neil and Reichle 1970). Outbreaks of the train millipede, *Parafontaria laminata*, led to suspension of the train service in central Japan (Fig. 5; Niijima and Shinohara 1988). Interestingly, Alagesan and Muthukrishnan (2009) recorded a millipede, *Xenobolus carnifex*, as a nuisance household pest especially among the poor hut dwellers in many southern districts of Tamil Nadu, India. The millipede lives and infests the organic thatched roof of the huts and drastically decreases the durability of the thatch. The poor inhabitants have to endure a practically continuous shower of faecal pellets. During the rainy season, the adults and larvae descend on all the household articles, food and drinking water.

Alagesan and Ganga (1989) reported that the adults and juveniles of *Harpurostreptus* sp. feed on tender buds and roots of newly planted tapioca and cause stunted growth or death of the plant. Millipedes usually swallow the soft and easily digestible parts, like young shoots or fine roots (Hopkin and Read 1992). Studies have shown that sweet potato farmers of Northeastern Uganda considered the millipedes as serious arthropod pest next to sweet potato weevils (Ebregt et al. 2005). Furthermore, other important crops, such as groundnut, maize, beans (kidney beans and other grain legumes), sesame, sunflower, cotton, cabbage, cassava and banana pseudostems, were also affected by millipede species (Ebregt et al. 2007).

7.5 Millipede Evolution

The class Diplopoda is grouped together with Chilopoda (centipede), Pauropoda and Symphyla under the superclass Myriapoda of subphylum Mandibulata (Lawrence 1984; Blower 1985). The terrestrial millipedes were one of the first land animals, arising during the Ordovician period more than 450 million years ago (Wilson and Anderson 2004). According to the fossil record, diplopods became very diverse and exceedingly common in the forests during Devonian and Carboniferous periods (Shear and Kukalova-Peck 1990). Millipedes are the major components in the evolutionary systematics of the phylum Arthropoda.

7.5.1 First Fossil Diplopod

Pneumodesmus newmani is a fossil millipede that lived 428 million years ago, in the late Silurian period (Wilson and Anderson 2004). The oldest known creature to have lived on land and the first myriapod was discovered in a layer of sandstone rocks on the foreshore of Cowie, near Stonehaven in 2004 by Mike Newman, an amateur palaeontologist from Aberdeen. The species was later given the specific epithet '*newmani*' as an honour.

The single, 1 cm-long fragment of *P. newmani* depicts small paranota (keels) high on the body and long, slender legs. The fossil is important because its cuticle contains openings which are interpreted as spiracles, part of a gas exchange system that would only work in air. This makes *P. newmani* the earliest documented arthropod with a tracheal system and indeed the first known oxygen-breathing animal on land.

Not only fossil evidence but also the results of recent molecular, genetic and neuroembryological studies support the pancrustacean model in which millipedes diverged from the crustacean-insect clade (Telford and Thomas 1995). For example, hemocyanin proteins (Hc's) in insects and crustaceans are more similar in amino acid sequences to one another than they are to millipede Hc's (Hanger-Holler et al. 2004). Likewise, phylogenetic analysis using the structure of three nuclear protein-coding genes places the insect/hexapod clade deep within the crustaceans, far distant from the myriapods (Regier et al. 2005). Morphological data on neurogenesis in the euarthropod group does not support the Atelocerata model, but it is consistent with the pancrustacean model of arthropod phylogeny (Fanenbruck et al. 2004; Harzsch 2006).

The phylogeny of the class Diplopoda at the ordinal level was well established by Hoffman (1969). Extant millipedes are classified into sixteen taxonomic orders (Shelley 2007) (Table 7.1).

The orders are combined into two subclasses (Penicillata and Chilognatha) based on the presence or absence of a hard, calcified exoskeleton. The majority of millipedes belong to the subclass Chilognatha, which has a calcified exoskeleton; hence, the orders in this subclass are further grouped into two infraclasses, namely,

Table 7.1 Classification and diagnostic features of Diplopoda

Subclass	Infraclass	Order	Body shape	Rings	Body length
Penicillata		Polyxenida	Unique through the tufts of setae		Up to 4 mm
Chilognatha	Pentazonia	Glomeridesmida	Flat	22	
		Glomerida	Short, flat	12 (may look like 11)	2.5–20 mm
		Sphaerotheriida	Large, flat, tank-like	13	Up to 95 mm and 50 mm wide
	Helminthomorpha	Callipodida	Round, with longitudinal ridges	>30	Up to 100 mm
		Chordeumatida	Short, body tapering to end, long legs	26–32, most with 30	4–25 mm
		Julida	Snake-like	>30	10–120 mm
		Platydesmida	Flat	>30, up to 110	Up to 60 mm
		Polydesmida	In dorsal view forms without paranota, resembles mealworm larvae	19–21, most with 20	3–130 mm
		Polyzoniida	Leech-shaped, back domed or arched, belly side flat	>30	
		Siphoniulida	Nematode-like	>30	Less than 7 mm
		Siphonocryptida	Flat	>30	Up to 10 mm
		Siphonophorida	Long, worm-like and thin	>30, up to 192	Up to 36 mm
		Spirobolida	Snake-like	>30	Up to 200 mm
		Spirostreptida	Snake-like	>30, up to 90	6–300 mm
		Stemmiulida	Laterally compressed, higher than wide	>30	Up to 50 mm

Shelley (2007)

Pentazonia and Helminthomorpha. Pentazonia are relatively short, broad millipedes, whereas Helminthomorpha are elongated, worm-like millipedes.

Millipedes are useful components of invertebrate diversity assessments and also other biodiversity studies. Despite being conspicuous and abundant arthropods in most habitats, the group is under investigated in several aspects. Relative to their diversity in temperate ecosystems, there are few studies documenting aspects of the group's phylogeny, evolution, behaviour, physiology and ecology. Apart from ignoring readily available sources of diagnostic characters, millipede biologists specialised in systematics have neglected many parts of the world, namely, the tropics. In both the New and Old Worlds, millipede taxonomic endeavours have much focused on temperate and northern hemisphere taxa. The millipedes' diversity in Europe and North America is extensively studied compared to other parts of the world. This continues even today despite the acknowledgement among experts that a rich tropical millipede fauna remains woefully understudied (Sierwald and Bond 2007). The temperate areas of Australia and South American countries in the southern hemisphere have received very less attention in recent years. These understudied areas pave way for new exciting frontiers of diplopodology to young researchers.

7.6 Millipede Distributions

Since these animals are very slow in their movements, are cryptic in their habit and have limited power of dispersal, every patch of land has different faunas of its own, and it is so with tropical forests too. Millipedes with their long life cycles and large number of stadia with frequent moults have limited power of dispersal. This results in a high degree of speciation and endemism, which is very evident with the Indian fauna also (Bano 1998). Perhaps their obscure nature might not have evinced interest among the zoologists and entomologists, and thus there is a lack of expertise on the subject. The taxonomy of this group remains at alpha stage, and the biology is at infant stage, excepting for a small group of North European fauna (Hoffman 1990). Verhoeff (1936) reported a few species of millipedes from India. Carl (1941) brought out a monograph on the millipede fauna of South Peninsular India and Ceylon in which he reported several new genera and species.

The oldest millipede fossils, apparently detritovores (Crawford 1992), are documented from the mid-Silurian and Devonian periods. however, subsequently they diverged into different species. Since the Carboniferous, those early diplopods represented six extinct orders and are recognised as fully terrestrial, in view of trachea (spiracles). In spite of the long fossil record of Diplopoda, it remains so fragmentary that only few extinct species appear relevant to modern millipede distributions. In general, millipedes are typical of the forest floor, where they find not only enough food and shelter like leaf litter or dead wood but also a sufficiently high humidity. These animals are usually quite sensitive to water deficit, being meso- to hygrophiles, and are calciphiles.

7.6.1 Environments and Habitats

Typical habitat of the Diplopoda is temperate (especially deciduous), subtropical or tropical forests (in particular, humid ones). The most typical habitats are leaf litter, the litter/soil interface, the uppermost soil and dead wood. All kinds of milder environments and habitats can be considered as rather typical, such as boreal forest, compost, dung, greenhouses, etc. Millipedes, however, can also dwell high in the mountains (above timberline), grasslands, caves, deeper in the soil, under the bark of trees, desert and even in environments that can be considered as extreme.

Hopkin and Read (1992) concluded that there are five millipede life forms:

1. *Bulldozers* or *rammers*, represented chiefly by the long-bodied orders Julida, Spirobolida and Spirostreptida;
2. *Wedge types, or litter-splitters*, with short-bodied Polydesmida equipped with strong paraterga being the most typical order;
3. *Borers*, typified by the chiefly flat-bodied orders Chordeumatida, Polyzoniida, Platydesmida and Siphonophorida;
4. *Rollers*, represented by Glomerida and Sphaerotheriida capable of conglobation;
5. *Bark dwellers*, characterised by Polyxenida which is tiny, soft bodied and swift.

These diplopod morphotypes are distributed among the following universal terrestrial arthropod life forms:

(a) *Stratobionts*, restricted to litter and the uppermost soil, dominant in the Diplopoda and represented by all five morphotypes
(b) *Pedobionts* or *geobionts*, mainly restricted to mineral soil and represented by the smaller juloid, glomeroid and polydesmoid morphotypes, usually implying body miniaturisation or elongation, shortening of appendages, often also decolouration of the teguments and loss of eyes
(c) *Troglobionts*, likewise represented by the juloid, glomeroid and polydesmoid morphotypes, but usually implying a drastic elongation of the extremities, depigmentation of the teguments, blindness, sometimes mouthpart modifications, often also 'cave gigantism'
(d) Under-bark *xylobionts* or *subcorticoles*, represented by all five morphotypes, but either particularly flat bodied (polydesmoids, platydesmoids) or miniature (polyxenoids, glomeroids), often also especially thin (juloids)
(e) *Epiphytobionts*, again with all five morphotypes involved, but usually characterised by very small body sizes

7.6.2 Extreme Environments

In spite of the group's pronounced monotony, Diplopoda have managed, at least marginally, to populate an adequate number of individuals even in the upheaval habitat patches.

7.6.2.1 Marine Littoral

Seawater is regarded as an insurmountable barrier for millipedes. Still there are a few exceptions that can be termed as littoral species such as *Thalassisobates littoralis* living under stones and seaweeds, in rock crevices or shortly above the tidal zone, sometimes also in coastal caves around the Mediterranean, reported also from the coasts of Sweden and Eastern USA (Enghoff 1987).

No special studies have been conducted to reveal the mechanisms of salt tolerance in these diplopods, but all these species represent the juloid, polydesmoid or polyxenoid morphotype.

7.6.2.2 Freshwater Habitats

Millipede *Polydesmus denticulatus* can stay alive underwater for 75 days (Zulka 1996). It is noteworthy that all diplopods proven or suggested to be strongly associated with freshwater habitats appear to belong to the polydesmoid morphotype in the tropics and subtropics; apparently due to a geologically far longer and stable evolution, numerous terrestrial arthropods, including a few millipedes, have developed special adaptations to life underwater (Adis 1997).

Millipedes lacking any morphological and/or physiological adaptations must rely on vertical migrations along the tree trunks to survive the flood period (Minelli and Golovatch 2001).

7.6.2.3 Deserts

Deserticolous millipedes are also scarce, but some of them inhabit the globe's harshest deserts such as the Sonora and Kalahari. Perhaps the best-known example is *Orthoporus ornatus*, a species living from Southern USA to Northern Mexico. Among the adaptations to arid environments, this large species (up to 185 mm long) shows a low rate of water loss due to a rather impermeable cuticle, waxy epicuticle, ability of water uptake from unsaturated air, protection of eggs in pellets to prevent water loss and the use of cracks, burrows, stones and other shelters to buffer the wide temperature oscillations that occur in deserts.

In addition, the animals remain dormant deep in the soil most of their life, with the most active periods coinciding closely with periods of rain, when swarming is often observed. As a behavioural adaptation, individuals may thermoregulate by alternately basking and retreating into cool shelters (Crawford et al. 1987). Similar observations were made on the large spirostreptid *Archispirostreptus syriacus* in Israel where, depending on locality, the life span varies from 9 to 11 years (Bercovitz and Warburg 1985). Their large size seems to be an advantage as heat gain and loss is less rapid, as is water loss (Hopkin and Read 1992). The soil, however, must always be loose enough to penetrate, especially by active burrowers (Minelli and Golovatch 2001).

7.6.2.4 High Mountains

Many mountains of temperate and tropical lands are to be regarded as natural foci and refugia for millipedes. In the Himalayas, where the timberline lies well above

4000 m MSL, the diversity of Diplopoda is quite pronounced, with over half of the about 200 described species of Polydesmida.

Thus, among the 17 species of *Nepalmatoiulus* currently known from the Himalayas, *N. ivanloebli* occurs between 2200 and 4800 m MSL representing perhaps the highest altitude millipede recorded globally (Enghoff 1987).

7.6.3 Caves

Cavernicoles are common among Diplopoda. Few diplopods are endemic to lava caves and none to snow, sea or salt caves (Vandel 1964). Until recently, the caves of tropical countries, despite the rich fauna they had long been shown to contain, harboured only few millipedes.

The cave millipede *Amplaria muiri* was recorded by Shear and Krejca (2007) in the crystal caves of California, USA. The lightly pigmented individuals that form most of the cave population may be the result of growth and development in the absence of light.

7.6.3.1 Deeper Soil

Geobites is a unique species among several species of millipedes. Some information is available for the minute, parthenogenetic, calcicolous species, *Geoglomeris subterranea*, seen in France, Britain and Ireland and occurs not only in the uppermost soil, usually between 10 and 20 cm deep, but also under boulders, in ant nests and in old chalk and limestone quarries (Lee 2006). Beyond any life form approach, subsoils are used by most Diplopoda as shelter for aestivation, oviposition and development.

The soil properties are crucial for numerous millipedes, especially calcicoles that can tolerate neither increased acidity nor waterlogged habitats. Not surprisingly, juvenile stadia in most species of diplopods can only be encountered in special chambers more or less deep in the soil, whereas adults and certain advanced juvenile stadia are the only truly active stages. Sandy soils and friable loams are much more easily penetrated than heavy clay soils and tend to have higher populations of millipedes unless they are too dry (Golovatch and Kime 2009).

7.6.3.2 Other Extreme Habitats

Select species of diplopods are found in suspended soil (mainly of epiphytes); on vegetation; in bird, ant and termite nests; under the bark of trees; and in dung and compost. Many brightly coloured, apparently aposematic species of *Desmoxytes* and *Centrobolus* typically inhabit arboreal vegetation in Southeast Asia (Enghoff et al. 2007) and Southern Africa (Hoffman 1999), respectively. The millipede, *Pelmatojulus tigrinus*, is a large juliform that feeds voraciously on large nests of termites in Ivory Coast. The millipedes use termites both for shelter and as food, which are rich in calcium (Mahsberg 1997). Several dozen species of Polyxenida, Glomerida, Julida, Spirostreptida, Spirobolida and Polydesmida commonly occur in ant nests all over the globe (Stoev and Lapeva-Gjonova 2005).

The household millipede pest *Xenobolus carnifex* survives and reproduces successfully both in natural and urban habitats of southern districts of Tamil Nadu, India. It is pre adapted to survival in high ambient temperature and low relative humidity for nearly 7 months in a year and to breeding in an unfamiliar breeding site at 3–4 m elevation from the ground level. The other major adaptations found in this species are resistance to desiccation and ability to climb vertical walls rapidly to reach the roof of the huts (Alagesan and Muthukrishnan 2005).

7.7 Ecological Significance

7.7.1 Millipedes as Efficient Decomposers

Millipedes are appreciated for their ecological importance and have a crucial role in fragmentation and decomposition of leaf litter, thus resulting in nutrient cycling within the soil. Until recent past, the ecological role of the diplopods has been studied with an emphasis in temperate ecosystems but not in tropical forests, despite the fact that millipedes reach their maximum diversity and biomass in this ecosystem and are probably the biggest arthropods that occur in the soil and litter of this environment (Swift et al. 1979). Usually millipedes are found in habitats containing much leaf litter in order to feed, remain hidden and stay moist, all at the same time. McBrayer and Reichle (1971) reported that millipedes are litter fragmenters. Temperate species of millipedes tend to eat leaf litter about five times of their weight. They digest some of the plant materials themselves, particularly any proteins and simple sugars. Millipedes are often referred as 'litter transformers and ecosystem engineers' similar to earthworms (Kaneko and Ito 2004).

Dangerfield (1993) states that millipedes as detritus feeders may process up to 30 % of the annual dead organic matter input to most soils and affect the decomposition of this material through fragmentation, inoculation with microbial spores and physical disturbance. Kheirallah (1990) reported that soil macrofauna like millipedes are known to contribute significantly to the breakdown of leaf litter by fragmentation and to install bacteria and fungi thus increasing the available colonisation surface and hence may accelerate decomposition.

The role of millipedes in forest ecosystem is important, and they act as bulldozers or rammers, wedge type or litter-splitters, borers, typified rollers and bark dwellers (Manton 1977; Hopkin and Read 1992). The survival of every deciduous forest depends on millipedes, since diplopods are one of the prime mechanical decomposers of wood and leaf litter, especially in the tropics. Millipedes proved to be the important members of the detritus food web in the leaf litter-rich ecosystem (Martens et al. 2002; Wilson and Anderson 2004).

Many millipedes have to find natural shelters, such as stones, trunks of fallen trees and sometimes deserted termite mounds. In general they stay in the shallow topsoil layer, but during the hot dry season, they can burrow to greater depths. Though millipedes lack the necessary digging structure for burrowing, they are found overwintering in burrows (Dangerfield and Telford 1991). Hopkin and Read

(1992) reported that, the head capsular is usually heavily calcified to facilitate burrowing. Besides, the ventral origin of the legs is seen as an adaptation for burrowing (Manton 1985).

7.7.2 Effective Role in Nutrient Recycling

Millipedes are ecologically important in facilitating nutrient cycling through decomposition of dead plant tissues, perhaps much more so than is envisioned (Hattenschwiler and Gasser 2005). Since nutrient cycling is one of the key processes governing ecosystem dynamics (Tilman et al. 1996), millipedes may have tremendous impact on the ecosystem. In select tropical regions, millipedes are more important than earthworms in soil recycling (Costaneto 2007). Dangerfield and Milner (1996) reported millipedes as detritovores, apparently affecting nutrient cycling through the redistribution of organic material and, consequently, the release of chemical elements such as nitrogen (N) in the soil. Through this process, the millipedes improve the soil fertility and enhance the growth of bacterial and fungal species in soil. Warren and Zou (2002) observed a significant positive correlation between the biomass of millipedes and the nitrogen content of *Leucaena* in a plantation in Puerto Rico. *Parafontaria laminate* increased nitrogen mineralisation with discharge excretion as soil aggregates (Okai et al. 2008). Loranger et al. (2008) also demonstrated high feeding preference of nitrogen-rich leaf litter by three millipede species in semievergreen tropical forest of Guadeloupe.

Setala and Hunta (1990) revealed that soil fauna significantly increased the release of N in microcosms that stimulated conifers. Diplopods have been experimentally shown to have a significant effect on the mass loss of high C/N ratio of litter comparing to low C/N ratio litter (Hattenschwiler and Gasser 2005). It is evident that the drastic reduction of C/N ratio may be due to the occurrence of millipedes, and that favoured the growth of plants on forest floors. Diplopods are important sinks for calcium (Ca) and accumulate Ca and magnesium (Mg) up to fivefold higher than raw leaf litter (Krivollutzky and Pokarzhevsky 1977). Leaching of Ca, Mg and nitrogen increased in proportion to the number of millipedes. Ashwini and Sridhar (2006) also observed a positive correlation in pill millipede abundance and biomass to soil calcium content in Western Ghats. The presence of millipedes positively changes concentrations of many nutrients, not just ammonia, phosphorus, calcium and magnesium (Smit and van Aarde 2001).

The contribution of the millipedes to the ecosystem is significant as they transfer energy through their faeces and produce animal biomass for consumption by members in the next trophic level which are called as secondary consumers. The degree of alteration of food processed by these animals varies so widely according to the morphology and physiology of their digestive tract that the faeces produced by one group plays a different role in the food web of the ecosystem from that produced by other groups.

7.7.3 As Agents in Humification and Soil Enrichment

Soil formation is a complex process involving the participation of the edaphic macrofauna. Millipedes constitute dominant components of soil fauna, and they have prime ecological importance for litter breakdown within the decomposition cycle and humification in the soil (Schäfer 1990; Crawford 1992). Millipedes, woodlice, earthworms and slugs are the major soil macrofauna demonstrated to increase the nutrient leaching from dead leaf litter and enhance the humification of the soil (Morgan et al. 1989). Bano (1992) also reported that millipedes play an important role in the soil formation process. Categorised as predominantly saprophagous, millipedes contribute to improvement of the humic part of the soil and help to increase the microflora through their faecal pellets.

Litter fragmentation and passage through the gut of macro-arthropods, such as millipedes and isopods, favour the establishment of soil bacterial population (Tarjovorsky et al. 1991). These epigeic invertebrates greatly affect decomposition processes both directly through fragmentation of organic material and indirectly through control or stimulation of microbial population and dissemination of their propagules (Lavelle and Spain 2001). The composition of the symbionts swallowed during host nutrition depends strongly on the degree of destruction of the leaf fall consumed. Millipedes deprived of symbionts and own enzymes are largely characterised by low assimilation rates. Apparently to avoid this and get inoculated, these diplopods repeatedly eat their own faecal pellets (Sriganova 1980). The faecal pellet of millipedes also acts as an important food source for coprophagic insects and other saprophagous animals such as earthworms (Scheu and Wolters 1991).

The roles of the microorganisms associated with millipedes and the role of the millipedes themselves are complicated by the practice of coprophagy, the ingestion of faeces of one's own or of another. Communities of microbes present in faeces can be a source of nutrition for millipedes and/or a source of various enzymes which the millipedes cannot produce themselves.

Many millipedes depend on the microbial population in their gut to derive nutrients and energy from wood, foliage and detritus. Parle (1963) reported that the gut of humus-feeding animals provides a favourable environment for microbial growth. Symbiotic microflora in the gut of litter-feeding animal converts decaying organic matter into assimilable energy (Bano et al. 1976). Cellulose and hemicellulose degradation in a millipede gut due to a rich aerobic microbial population was observed by Taylor and Crawford (1982).

Millipedes are not well equipped with specialised enzymes to enable them to digest the plant litter itself. It is suspected that microorganisms in the alimentary canal play a crucial role in the digestion of food (Blower 1985) and indirectly influence the fluxes of nutrients (Anderson et al. 1985). Leaf litter feeding by millipedes affects decomposer soil microorganisms, which enhances the activity of the latter and often increases biomass after litter passage through the millipede gut (Anderson and Bignell 1980; Maraun and Scheu 1996). The digestive system of most millipedes is rich in enzymes. Diplopods also appear to support temporary microbial symbionts that destroy cellulose, pectin and many plant products such as protein and carbohydrates (Kozlovskaya 1976).

Anderson and Bignell (1980) found bacteria to be abundant in the gut of millipede and in the faecal matter than in leaf litter that they feed upon. The presence of cellulolytic enzymes in millipedes capable of cellulose degradation has been suggested (Sriganova 1980). Millipedes break up dead plant material into small pieces thus increasing the surface area. This is important because the surface is the only part the bacteria and microfungi can reach and they are the main agents of recycling in the soil. An American biologist calculated that millipedes in an unnamed forest in USA contributed 2 tons of manure per acre per year to the forest floor.

7.7.4 Millipedes as an Alternative to Earthworm in Organic Farming

It is one of the upcoming areas of research that millipedes can be also employed in organic farming. Like vermicomposting, 'milli-composting' is an ecofriendly technique involving no pollution and hence is the most suitable method for converting organic waste materials when compared to the conventional methods like landfilling, incineration, biogas production, etc. Martens et al. (2002) reported that earthworms and millipedes proved to be the important members of the detritus food web in the agricultural ecosystem and both use manure as the food source. Ashwini and Sridhar (2006) studied that the millipede, *Arthrosphaera magna*, efficiently composted the leaves of cocoa and acacia for organic farming. Since millipedes play a significant role in the composting of plant residues, they should be considered as economically significant producers of compost in organic farming, besides enriching the soil in terms of nutrition and soil microflora (Kania and Klapec 2012).

Karthigeyan and Alagesan (2011) reported that millipede compost derived from various organic substances showed significant enhancement in the physiochemical and biochemical constituents. Ramanathan and Alagesan (2012) explored through laboratory experiments and found out that the compost processed by the millipede, *A. magna*, is superior to vermicompost in enriching the nutrients and enhancing the plant growth parameters. Millipede compost derived from various organic substances showed significant enhancement in the biochemical constituents (Kania and Klapec 2012).

7.7.5 As Bioindicators

Several species of millipedes can serve as bioindicators of climatic conditions and improve the structure and content of organic matter and nutrient elements of soil (Loranger et al. 2007; Seeber et al. 2008). But the first study with diplopods, as possible bioindicators, was conducted by Hopkin et al. (1985) involving the assimilation of metals by the species *Glomeris marginata*, and they found out a higher uptake of copper, zinc and cadmium by the animals collected in soils contaminated when compared to those animals collected in non-contaminated environments. Kazuyo and Junsei (2005) explained the strategies of saprophagous and metal accumulation phytophagous soil invertebrates by comparing adults of the Isopoda, Diplopoda and Gastropoda exposed to lead-, cadmium- and zinc-contaminated food and soil for 3 weeks. Kazuyo and Junsei (2005) studied the distribution of elements in the millipedes *Oxidus gracilis* and its relation to environmental habitats. But Triebskorn et al. (1991) exposed several invertebrates such as isopods and diplopods to different toxic substances and used the ultrastructural analysis to demonstrate the applicability of using such animals in biomonitoring.

References

Adis J (1997) Terrstrial invertebrates: survival strategies, group spectrum, dominance and activity patterns. In: Junk W (ed) The central Amazon floodplain. Ecology of a pulsing system – ecological studies, vol 126. Springer, Berlin, pp 299–317

Alagesan P, Ganga G (1989) Millipedes as potential pests of tuber crops. Indian J Exp Biol 27:177–179

Alagesan P, Muthukrishnan J (2005) Bio energetics of the household millipede pest, *Xenobolus carnifex* (Fabricius, 1775), (Diplopoda: Spirobolidae). Peckiana 4:3–14

Alagesan P, Muthukrishnan J (2009) In: Roble SM, Mitchell JC (eds) Population dynamics of the millipede, *Xenobolus carnifex* (Fabricius) (Diplopoda: Spirobolida) in an unique habitat, vol 16. Virginia Museum of Natural History Special Publication, Martinsville, pp 243–250

Almond JE (1985) Embryology and phylogeny in annelids and arthropods. Pergaman Press, Oxford, pp 127–167

Anderson JM, Bignell DE (1980) Bacteria in the food, gut contents and faeces of the litter-feeding millipede, *Glomeris marginata* (Villers). Soil Biol Biochem 12:251–254

Anderson JM, Leonard MA, Ineson P (1985) Faunal biomass: a key component of a general model of nitrogen mineralization. Soil Biol Biochem 17(5):735–737

Ashwini KM, Sridhar KR (2006) Breakdown of plantation residues by pill millipedes (*Arthrosphaera magna*). Curr Sci 90(7):954–959

Bano K (1992) The role of the millipedes, *Jonespeltis splendidus* (Verhoeff) in an ecosystem. Ver Innsbruck 10:327–331

Bano K (1998) Distinguishing characteristics and systematics of the Indian Paradoxosomatid millipedes. Proc Zool Soc 51(1):86–91

Bano K, Bagyaraj DJ, Krishnamoorthy RV (1976) Feeding activity of the millipede, *Jonespeltis splendidus* (Verhoff) and soil humification. Proc Indian Acad Sci 83(1):1–11

Bercovitz K, Warburg MR (1985) Developmental patterns in two populations of the millipede Archispirostreptus syriacus (De Saussure) in Israel (Diplopoda). Bijdragen tot de Dierkunde 55(1):37–46

Blower JG (1985) Millipedes: keys and notes for the identification of the species. Linnean Society of London/Estaurine and Brakisch-water Science Association, London, p 242

Carl J (1941) Diplopoden aus Sud- Indien und Ceylon 2. Teil Nematophora and Juliformia. Rev Suisse Zool 48(22):569–714

Carrel JE, Eisner T (1984) Defensive secretions of the millipede, Glomeris marginata. Proc Natl Acad Sci U S A 84:806–810

Costaneto M (2007) The perception of Diplopoda (Arthropods, Myriapods) by the inhabitants of the country of Pedra Branca, Santa Teresinha, Bahia, Brazil and Costa. Acta Biol Colomb 12:123–134

Crawford CS (1992) Millipedes as model detritovores, Berichte des naturwissenschaftlichmedizinischen. Ver Innsbruck 10:277–288

Crawford CS, Bercovitz K, Warburg MR (1987) Regional environments, life-history patterns and habitat use of spirostreptid millipedes in arid regions. Zool J Linnean Soc Lond 89:63–88

Dangerfield JM (1993) Ingestion of mineral soil/litter mixtures and faecal pellet production in the Southern millipede, *Alloporus uncinatus* (Attems). Pedobiologia 37:159–166

Dangerfield JM, Milner AE (1996) Millipede faecal pellet production on selected natural and managed habitats of Southern Africa: implications for litter dynamics. Biotropica 28:113–120

Dangerfield JM, Telford SR (1991) Seasonal activity pattern of julid millipedes in Zimbabwe. J Trop Ecol 7:281–285

Ebregt E, Struik PC, Odongo B, Abidin PE (2005) Pest damage in sweet potato, groundnut and maize in North-Eastern Uganda by millipedes (Diplopoda). NJAS Wagening J Life Sci 53:49–69

Ebregt E, Struik PC, Abidin PE, Odongo B (2007) Feeding activity of the East African millipede, *Omopyge sudanica* (Kraus) on different crop products in laboratory experiments. NJAS Wagening J Life Sci 54(3):313–323

Eisner T, Davis A (1967) Mongoose throw and smashing millipedes. Science 155:577–579

Enghoff H (1987) Thalassisobates littoralis (Silvestri)–an amphiatlantic millipede (Diplopoda, Julida, Nemasomatidae). Entomologists Mon Mag 123:205–206

Enghoff H, Sutcharit C, Panha S (2007) The shocking pink dragon millipede, *Desmoxytes purpurosea*, a colourful new species from Thailand (Diplopoda: Polydesmida: Paradoxosomatidae). Zootaxa 1563:31–36

Fanenbruck M, Harzch S, Wolfgang-Wagele J (2004) The brain of the remipedia (Crustacea) and an alternative hypothesis on their phylogenic relationship. Proc Natl Acad Sci U S A 101:3868–3873

Golovatch SI, Kime D (2009) Millipede (diplopoda) distributions: a review. Soil Org 81(3):565–597

Hanger-Holler S, Schoen A, Erker W, Marden J, Rupprecht, Decker H, Burmester T (2004) A respiratory haemocyanin from an insect. Proc Natl Acad Sci U S A 101:871–874

Hannibal JT (1986) The enduring myriapods, millipedes and centipedes have been around for millions of years and their fossils are found worldwide. Earth Sci 39(1):21–23

Harzsch S (2006) Neurophylogeny: architecture of the nervous system and a fresh view on arthropod phylogeny. Comp Biol 146:162–194

Hattenschwiler S, Gasser P (2005) Soil animals alter plant litter diversity effects on decomposition. Proc Natl Acad Sci USA 102:1519–1524

Hoffman RL (1969) Classification of the Diplopoda. Museum d' Histoire Naturelle, Geneve, p 78

Hoffman RL (1990) Diplopoda. In: Dindal DL (ed) Social biology guide. Wiley Interscience, New York, pp 1349–1860

Hoffman RL (1999) Checklist of the millipedes of North and Middle America. Virginia Mus Nat Hist Spec Publ 8:584

Hopkin SP, Read HJ (1992) The biology of millipedes. Oxford University Press, Oxford, p 223

Hopkin SP, Watson K, Martin MH, Mould ML (1985) The assimilation of heavy metals by Lithobius variegatus and Glomeris marginata (Chilopoda; Diplopoda). Bijdragen tot de dierkunde 55(1):88–94

Ito F (1998) Colony composition and specialized predation on millipedes in the enigmatic ponerine ant genus *Probolomyrmex* (Hymenoptera, Formicidae). Insect Soc 45:79–83

Johns PM (1967) A note on the introduced millipedes of New Zealand. N Z Entomol 3:60–62

Kaneko N, Ito MT (2004) Biodiversity and ecosystem function of soil animals. Japan J Ecol 54:201–207

Kania G, Klapec T (2012) Seasonal activity of millipede (Diplopods) – their economic and medical significance. Ann Agric Environ Med 19(4):646–650

Karthigeyan M, Alagesan P (2011) Millipede composting: a novel method for organic waste recycling. Recent Res Sci Technol 3:62–67

Kazuyo N, Junsei T (2005) Distribution of elements in the millipede, Oxidus gracilis C. L Koch (polydesmida: paradoxosomatidae) and the relation to environmental habitats. Biometals 18(6):651–658

Kheirallah A (1990) Fragmentation of leaf litter by a natural population of the millipede *Julus scandinavius* (Latzel, 1984). Biol Fertil Soils 10:202–206

Kozlovskaya LS (1976) The role of soil invertebrates in the transformation of organic matter of swampy soils. Nauka Publication, Leningrad, p 211

Krivollutzky DA, Pokarzhevsky AD (1977) The role of soil animals in nutrients cycling in forest and steppe. Ecol Bull 25:253–260

Lavelle P, Spain AV (2001) Biomass alteration of Earthworms in the organic wastes. Soil Ecol 34:141–15

Lawrence RF (1984) The centipedes and millipedes of Southern Africa: a guide. Balkema, Cape Town, p 233

Lee P (2006) Atlas of the millipedes (Diplopoda) of Britain and Ireland. Pensoft, Sofia, Moscow, p 216

Loranger G, Imbert D, Bernhard- Reversat F, Ponge JF, Lavellae P (2007) Soil fauna abundance and diversity in a secondary semi-evergreen forest in Guadeloupe (Lesser Antilles): influence of soil type and dominant tree species. Biol Fertil Soils 44:269–276

Loranger G, Imbert D, Bernhard-Reversat F, Ponge JF, Levelle P (2008) Litter N-content, soil millipede abundance, species richness and feeding preference in a semi-evergreen dry forest of Guadeloupe (Lesser Antilles). Biol Fertil Soils 45:93–98

Mahsberg D (1997) *Pelmatojulus tigrinus*, a key detritivore of a tropical gallery forest (Diplopoda, Spirobolida: Pachybolidae). Entomol Scand 51:269–280

Manton SM (1977) The arthropoda. Habits, functional morphology and evolution. Clarendon, Oxford, p 527

Manton SM (1985) The evolution of arthropod locomotory mechanisms. Part 6. Habits and evolution of the lysiopetaloidae (Diplopods), some principles of leg design in Diplopods and chilopods and limb structure in Diplopods. J Linn Soc Lond (Zool) 43:489–556

Maraun M, Scheu S (1996) Changes in microbial biomas, respiration and nutrient status of beech (*Fagus sylvatica*) leaf litter processed by millipedes (*Glomeris marginata*). Oecologia 107:131–140

Martens MM, Alphei J, Schaefer M, Scheu S (2002) Millipedes and Earthworms increase the decomposition rate of 15N labelled winter rape in and arable field. Isot Environ Health Stud 37(1):43–51

McBrayer JF, Reichle DE (1971) Tropic structure and feeding rates of forest soil invertebrate populations. Oikos 22:381–388

Milkov FN (1997) Nature zones of the USSR. Mysl Publication, Moscow, p 295

Minelli A, Golovatch SI (2001) Myriapods. In: Levin SA (ed) Encyclopedia of biodiversity. Academic, San Diego, pp 291–303

Morgan CR, Schindler SC, Mitchell MJ (1989) Towards organic farming with millipede Arthrosphaera magna. Biol Fertil Soils 7:239–246

Nijima K, Shinohara K (1988) Outbreaks of the *parafontaria laminate* group (Diplopoda: Xystodesmidae). J Ecol 38(3):257–268

O'Neill RV, Reichle DE (1970) Urban infestation by the millipedes, *Oxidus gracilis* Koch. J Tenn Acad Sci 45:114–115

Okai N, Fujimaki R, Saito Y, Kaneko N (2008) Influence of feeding of larvae train millipede *Parafontaria laminate* on acceleration of nitrogen mineralization. Proceeding of the 31th annual meeting of the Japanese society of soil zoology, p 16

Parle JN (1963) Microorganisms in the intestines of earthworms. J Gen Microbiol 31:1–11

Ramanathan B, Alagesan P (2012) Evaluation of vermicompost versus Millicompost. Curr Sci 103(2):140–143

Regier JC, Wilson HM, Shultz JW (2005) Phylogenetic analysis of Myriapoda using three nuclear protein-coding genes. Mol Phylogenet Evol 34:147–158

Rossi W, Balazuc J (1977) Laboulbeniales parasites de Myriapodes. Rev Mycol 41:525–535

Schäfer M (1990) The soil fauna on a beech forest on limestone: trophic structure and energy budget. Oecologia 82:128–136

Scheu S, Wolters V (1991) Influence of fragmentation and bioturbation on the decomposition of ^{14}C-labelled beech leaf litter. Soil Biol Biochem 23:1029–1034

Seeber J, Seeber H, Langel R, Scheu S, Meyer E (2008) The effect of macro-invertebrates and plant litter of different quality on the release of N from litter to plant on alpine pastureland. Biol Fertil Soils 43:333–340

Setala H, Huhta V (1990) Evaluation of the soil fauna impact on decomposition in a simulated coniferous forest soil. Biol Fertil Soils 10:163–169

Shear WA, Kukalova-Peck J (1990) The ecology of Paleozoic terrestrial arthropods: the fossil evidence. Can J Zool 68:1827–1834

Shear WA, Krejca, LK (2007) Revalidation of the millipede genus *Amplaria* Chamberlin 1941(Diploda, Chordeumatida, Striariidae), and description of two new species from caves in Sequoia and Kings Canyon National Parks, California, Zootaxa 1532: 23–39

Shelley RM (2007) Taxonomy of extant diplopoda (millipeds) in the modern era: perspectives for future advancements and observations on the global diplopod community (Arthropoda: Diplopoda). Zootaxa 1668:343–362

Sierwald P, Bond JE (2007) Current status of the myriapod class Diplopod (millipedes): taxonomic diversity and phylogeny. Annu Rev Entomol 52:401–420

Smit A, Van Aarde RJ (2001) The influence of millipedes on selected soil elements: a microcosm study on three species occurring on coastal sand dunes. Funct Ecol 15:51–59

Sriganova BR (1980) Nutrition of soil saprophages. Nauka Publication, Mascow, p 244

Stoev P, Lapava–Gjonova A (2005) Myriapods from ant nest in Bulgaria (Chilopoda, Diplopoda). Peckiana 5:131–142

Swift MJ, Heald OW, Anderson JM (1979) Decomposition in terrestrial ecosystem. Blackwell Scientific Publication, Oxford, p 372

Tarjovorsky K, Vilemin G, Toutain F (1991) Microstructural and ultrastructural changes of the oak leaf litter consumed by millipede *Glomeris hexasticha* (Diplopods). Rev Ecol Biol Sci 128:287–302

Taylor EC, Crawford CS (1982) Microbial gut symbionts and desert detritovores. Sci Rev Arid Zone Res 1:37–52

Telford MJ, Thomas RH (1995) Demise of the atelocerata? Nature 376:123–124

Tilman D, Wedin K, Knops J (1996) Productivity and sustainability influenced by biodiversity on grassland ecosystems. Nature 379:718–720

Triebskorn R, Köhler HR, Zanh T, Vogt G, Ludwig M, Rumpf S, Kratzmann M, Alberti G, Storch V (1991) Invertebrate cells as targets for hazardous substances. Z Angew Zool 78:277–287

Vandel A (1964) Biospéologie. La Biologie des Animaux Cavernicoles. Gauthier-Villard, Paris, p 622

Verhoeff KW (1936) Ueber Einige Indische Chilognathen Gesemmelt Von Herrn S. Jones, Madras. Rec Indian Mus 38(2):103–123

Warren MW, Zou X (2002) Soil macrofuana and litter nutrients in three tropical tree plantations on a disturbed site in Puerto Rico. Ecol Manag 170:161–171

Wilson HM, Anderson LI (2004) Morphology and taxonomy of Paleozoic millipedes (Diplopoda: Chilognatha: Archipolypoda) from Scotland. J Paleontol 78(1):169–184

Zulka KP (1996) Submersion tolerance of some diplopod species. Mem Mus Nat Hist Nat 169:477–481

Spiders: Diversity, Distribution, and Conservation

8

Ganesh Vankhede, Priyanka Hadole, and Akshay Kumar Chakravarthy

Abstract

Spiders are invertebrate animals with two body segments, four pairs of legs and without wings and mouth parts. They exclusively feed on insect pests and protect the cultivated crops. They are at the top of the lower food web in any ecosystem. Spiders maintain ecosystem balance. Nearly 45,000 spider species are known worldwide. Spider silk is one of the best biomaterial on the earth having antimicrobial properties, high tensile strength and is biocompatible with humans. All spiders produce silk, a solid protein strand expelled from spinnerets most normally found at tip of the abdomen. Silk is used to trap insects in webs, by many species, and they hunt freely. It is also used in climbing, form smooth outer covering walls for burrows, and build egg sacs, wrap prey, and temporarily hold sperm, among other applications. An abnormal fear of spiders (arachnophobia) is one of the most common phobias, and spiders are often looked at as something to be eliminated. Spiders have an awesome scope of variety in way of life, albeit all are savage. Arachnids give a substantial number of free biological system administrations.

Keywords

Predatory • Ecosystem balance • Spider silk • Threats

G. Vankhede (✉) • P. Hadole
Department of Zoology, Sant Gadge Baba Amravati University,
Amravati 444602, Maharashtra, India
e-mail: vganeshan2001@rediffmail.com; priya.hadole@gmail.com

A.K. Chakravarthy
Division of Entomology and Nematology, Indian Institute of Horticultural Research (IIHR),
Hessarghatta Lake Post, Bengaluru 560089, Karnataka, India

A.K. Chakravarthy, S. Sridhara (eds.), *Arthropod Diversity and Conservation in the Tropics and Sub-tropics*, DOI 10.1007/978-981-10-1518-2_8

8.1 Introduction

Spiders are arthropods having four pairs of legs, eight eyes, and the capacity to secrete silk immediately after birth. They are cosmopolitan. Spiders secrete venom, not toxic to humans but show cytotoxicity. Many spiders can prepare the web but not all. Though spiders have eight eyes, there are some spiders with six or four or two eyes only. Spiders living in caves have no eyes. The sperm of the male spider shows 9 ± 2 arrangement of microtubules like those in prokaryotes, and hence according to one theory, it is thought that the spiders have come from other planets. The primary distinctions among insects and arachnids are the pedicel between the cephalothorax and the gut and the nearness of spinnerets (Marina and Kleber 2008). All spiders aside from families Uloboridae and Holarchaeidae can infuse venom to secure themselves or to murder and condense prey. Only up to 200 species, however, have bites that pose health problems to humans (Diaz 2004). Several specie's nibbles might be difficult; however it will not create enduring health concerns. Be that as it may, in satisfying their own particular individual reason for survival, upkeep, and multiplication, spiders likewise give a profitable part in biological systems and for people. Environmentally, they offer essential control of insects and other prey populaces. while numerous insects are connected to fatalities or have risky venoms, spiders do not inject venom unless squeezed, or generally debilitated. In any case, they add to the marvel and assorted qualities of nature, and some are even eaten routinely as nourishment/or as delicacy. They are also common animals in popular culture, folklore, and symbolism.

8.2 Evolution

Trigonotarbids, spider-like creatures, were among the most known area arthropods. Like spiders, they breathed through book lungs (the respiratory organ normal for four-paired legs, disconnected developmentally to mammalian lungs), and strolled on four pairs of legs with two extra legs adjusted to use around their mouth. In any case, they were not true spiders, not in any case tribal to them, but rather spoke to autonomous off shoots of the Arachnida. There are confirmations that these arthropods advanced in states of sustenance hardship. Spiders can live without nourishing for long stretch. Arachnids additionally eat a huge amount of prey when they are accessible having the anatomical plausibility to extend their abdomen (Marina and Kleber 2008).

True spiders developed around 400 million years prior and were among the primary species to live ashore. They are recognized by abdominal division and silk-creating spinnerets. The Pedipalpi (counting whip scorpions) are accepted to constitute the sister gathering to the Araneae (Coddington and Levi 1990).

The greater part of the early portioned fossil arachnids had a place with Mesothelae, a gathering of primitive spiders with the spinnerets set underneath the middle of the abdominal area, instead of toward the end as in modern species

(Opisthothelae). They were likely ground-abiding predators of other primitive arthropods. Silk has been utilized basically as a defensive covering for the eggs, a coating for a retreat gap, and later may be for basic ground sheet web and trap entry way development.

As plant and insect life advanced, so did the spiders' utilization of silk. Spiders with spinnerets toward the end of the midriff (Mygalomorphae and Araneomorphae) seemed more than 250 million years back, apparently advancing the improvement of more intricate sheet and labyrinth networks for prey catch both on ground and foliage, and the advancement of the security dragline. (Networks can be utilized as "security draglines" with the reason for tying the spiders; if an insect is brushed off a tree limb, the dragline can help it to come back to its position.)

By the Jurassic period, the refined ethereal networks of the sphere-weaving arachnids had been officially created to exploit the quickly expanding gatherings of spiders. A cobweb safeguarded in golden, thought to be 110 million years of age, shows confirmation of an immaculate circle web. It is trusted that cement catch strings, rather than cribellate strings (more primitive strings with less aggregate stickiness, and subsequently viability of getting prey, than cement catch strings), advanced around 135 million years prior (Opell 1997). The capacity to weave sphere networks is thought to have been "lost," and some times even redeveloped or advanced independently, in various types of spiders since its appearance.

Spiders can be arranged by procedures they use for entrapping. Spiders which seemed earlier in the developmental scale nourish by sitting tight in a tunnel for sustenance to tag along the prey before getting it. Trailing them were arachnids which effectively meandered searching for sustenance and got it by trap or pursuing it down. The advancement of flying insects produced an impetus for spiders to advance better approaches for getting nourishment which could not be pursued earlier; thus most airborne web weavers emerged.

- Most arachnids feast upon insects and different arthropods. Yet some have bigger prey. Case in point, the shrieking spider or bird-eating spider has been archived taking and bolstering on frogs and little birds.
- Spiders cannot eat strong sustenance – they suck out their prey's body liquids and gentler tissues utilizing strong stomach muscles.
- A spider intermittently sheds its skin (exoskeleton) so that it can grow (called shedding).
- Spider silk is used to make networks or wrap up the gotten prey and is made generally of proteins and amino acids. There are no less than seven distinctive sorts of spider silk made for definite uses (e.g., web development, stay lines, wrapping prey, and covering homes), and most arachnids have no less than three sorts of spinnerets. These augmentations toward the end of their mid-region radiate a fluid underweight which turns into the strings, some of which might be exceptionally sticky.

Some tropical rainforest species spend their entire lives in the tree canopy. If a storm, or a miss step, or an encounter with a predator causes these creatures to lose their footing and plummet to the forest floor; they could suffer fatal injuries or quickly get eaten by ground-dwelling animals. At least one type of spider has evolved a heroic solution to this potentially deadly problem. Should they teeter from their canopy perch, species of the *Selenops* genus assume a superman-like pose – aiming head first with two front legs extended and navigating their falls. The maneuver is the first evidence that some spiders can control their aerial descents.

8.3 Ecological Significance

While spiders are generalist predators, in reality their diverse techniques for prey catch frequently decide the sort of prey taken. In this manner, web-building spiders seldom catch caterpillars. Interestingly, crab spiders trap prey in blossoms to catch honey bees, butterflies, and other insects. Guilds are prey-capture methods that tend to take certain types of prey by groups of families. A couple of arachnids are better represented with considerable authority in their prey catch. Spiders of the genus *Dysdera* catch and eat sow bugs, pill bugs, and scarabs, while privateer insects eat just different arachnids. Bolas spiders in the family Araneidae use sex pheromone analogs to catch only the males of certain moth species.

In spite of their by and large expansive prey ranges, spiders form the most critical connection in the control of insects. Consistently on a normal glade, they eat up more than 10 g/m^2 of insects and other arthropods.

Maintain Ecosystem Balance Spiders are polyphagous and eat an assortment of accessible prey. They go after adult insects as well as feast upon their eggs and hatchlings. They help in keeping up the biological system parity. Spiders are great companions of ranchers as they control a wide range of vermin on the yield. Numerous studies have shown that spiders can altogether decrease prey densities. Spiders in a maize yield can discourage populaces of leafhoppers (Cicadellidae), thrips (Thysanoptera), and aphids (Aphididae). Both web-weaving and chasing spiders can control populaces of phytophagous Homoptera, Coleoptera, and Diptera in agroecosystems. Spiders have additionally ended up being viable predators of herbivorous bugs in apple plantations, including the beetle, *Anthonomus pomorum* Linnaeus, and Lepidoptera larvae in the family Tortricidae. Wolf spiders additionally can lessen densities of sucking herbivores (Delphacidae and Cicadellidae) in rice paddies.

Recent patterns in agriculture involving lessened pesticide use and environmental supportability have led to expanded enthusiasm for arachnids as potential biocontrol agents. The sphere-web weavers, Araneidae and Tetragnathidae sustain upon Homoptera, for example, leafhoppers, Diptera, and Orthoptera, particularly grasshoppers. The sheet-web weavers, for example, Linyphiidae, Dictynidae, and Theridiidae, catch Diptera, Hemiptera, and Homoptera (particularly aphids and leafhoppers) and also beetles in the family Curculionidae. The channel web weavers (Agelenidae, Atypidae, Ctenizidae, and Eresidae) go after Orthoptera, Coleoptera,

Fig. 8.1a Select species of spiders from South India

and Lepidoptera. Chasing spiders (Lycosidae, Oxyopidae, Thomisidae, and Salticidae) every now and again catch Orthoptera, Homoptera, Hemiptera, Lepidoptera, Thysanoptera, Diptera, and some Coleoptera and Hymenoptera.

While arachnids are generalist predators, in reality their distinctive techniques for prey catch frequently decide the sort of prey taken. In this manner web-building arachnids once in a while catch caterpillars, and crab spiders that snare prey in blossoms catch more honey bees, butterflies, and some flies than different insects. Gatherings of families that tend to take certain sorts of prey as a result of their prey catch techniques are frequently called societies. A couple of arachnids have more practical experience in their prey catching capacities (Figs. 8.1a and 8.1b). Spiders of the sort *Dysdera* catch and eat sow bugs, pill bugs, and beetles, while pirate

Fig. 8.1b Select species of spiders from South India (**e–h**, Photos of Rajendra Prasad BS)

spiders eat just different arachnids. Bolas spiders in the family Araneidae use sex pheromone analogs to catch the males of certain moth species.

In spite of their by and large expansive prey ranges, spiders are a standout among the most critical connections in the control of the populaces of insects. Consistently on a normal knoll, they eat up more than 10 g/m^2 of insects and different arthropods.

Water Percolation Different tarantulas and Geolycosa (Lycosidae) spiders make tunnels in the dirt amid the midyear for rearing reason; and when the young ones are scattered, these tunnels are cleared by grown-ups, and from thereon in the stormy season, water gets permeated through these tunnels. As per one evaluation, there are more than 300 gaps made by Geolycosa per acre of land. This unquestionably helps the permeation of water in the dirt, expanding the water-holding limit of the soil.

The vast majority of the spiders in nature eat mosquitoes and thus shield us from malaria and other mosquito-borne ailments. Some pisaurid and tetragnathid spiders eat mosquito hatchlings. The mud wasps, numerous pollinators, reptiles and some birds nourish their young ones with spiders as spiders are rich in basic proteins.

8.4 Behavior

Spiders are air-breathing arthropods that have four pairs of legs and chelicerae with teeth that infuse venom (Figs. 8.1a and 8.1b). They are the biggest order of arachnids and rank seventh in all-out species-assorted qualities among different gatherings of life-forms, which is reflected in their extensive differences in behavior (Sebastin and Peter 2009).

Numerous spiders, yet not all, form networks. Different spiders utilize a wide assortment of techniques to catch prey. There are a few perceived sorts of networks which are as given beneath:

- Spiral orb webs, associated primarily with the family Araneidae
- Tangle webs or cobwebs, associated with the family Theridiidae
- Funnel webs
- Tubular webs, which run up the bases of trees or along the ground
- Sheet webs

The net-throwing spider weaves a little net which it appends to its front legs. It then sneaks in sit tight for potential prey and when such prey arrives, rushes forward to wrap its casualty in the net and nibbles and incapacitates it. Subsequently, spiders consume less vitality by prey than primitive seekers which maintain a strategic distance from the the prey thus vitality loss of weaving a huge orb web is comparitively less.

Bolas spiders are uncommon sphere-weaver spiders that do not turn the networks. Rather, they chase by utilizing a sticky "catch blob" of silk on the end of a line, known as a "bolas." By swinging the bolas at flying male moths or moth flies adjacent, the spider may catch its prey rather like an angler catching a fish on a snare. In light of this, they are likewise called calculating or angling spider. The prey is baited to the spider by the creation of up to three pheromone analogs (Yeargan and Quate 1997).

Jumping spiders, wolf spiders, and numerous different sorts of spiders chase unreservedly. Some of these have improved visual perception. They are for the most part vigorous and lithe. Some are astute seekers jumping upon prey as they discover it or notwithstanding pursuing it over short separations. Some will sit tight to pass prey in or close to the mouth of a tunnel. *Dolomedes* spiders chase by holding up at the edge of a pool or stream. They clutch the shore with their back legs while whatever is left of the body lies on the water, with legs extended. When they distinguish the swells from prey, they keep running over the surface to curb it utilizing their principal legs, which are tipped with little hooks; like different spiders they then infuse venom with their empty jaws to execute and process the prey. They predominantly eat insects; however, some bigger species can get little fish (Barbour 1921). Females of the water spider, *Argyroneta aquatica* manufacture submerged "plunging ringer" networks which they load with air and use for processing prey, shedding, mating, and raising offspring. They live totally inside the ringers, shooting out to catch prey creatures that touch the chime or the strings that grapple it (Schütz and Taborsky 2003). A few arachnids chase different spiders utilizing duplicity; the bouncing spider *Portia* imitates the development of caught insect prey on the networks of different spider. This draws in the proprietor of the web whereupon *Portia* jumps and overpowers the proprietor.

Trap door arachnids build tunnels with a stopper like trap door made of soil, vegetation, and silk. The trap door is hard to see when it is shut in light of the fact that the plant and soil materials successfully disguise it. The spider identifies the prey by vibrations and, when it approaches enough, jumps out of its tunnel to make the catch. The *Kaira* arachnid utilizes a pheromone to pull in moths and gets the insects with a bushel shaped from its legs (Levi 1993).

Spiders perform savagery under a variety of circumstances. The most generally known case of savage consumption in spiders is when females tear up males some times amid or after copulation. For instance, the male Australian redback spider (*Latrodectus hasselti*) is killed by the female after he embeds his second palpus in the female's genital opening; in more than 60 % of matings, the female then eats the male. Be that as it may, the hypothesis of the "conciliatory male" may have gotten to be more prominent than reality. Some trust that this type of barbarianism just happens in outstanding cases (Foelix 1982, 1996; Roberts 1995).

Male water spiders (*Argyroneta aquatica*) demonstrate an inclination for mating with bigger females, while ripping apart females smaller than themselves (Schütz and Taborsky 2005). Young of the *Stegodyphus lineatus* eat their mother. Females of *Segestria florentina* some of the time bite the dust while guarding her eggs, and the hatched spiders later eat her. A few spiders, for example, *Pholcus phalangioides*, will go after their own kind when nourishment is rare (Johnson and Andrew 2005).

Death feigning can be used in reproductive behavior of spiders. In the nursery web spider, the male often fakes death to escape from getting eaten by females after mating (Hansen et al. 2008). Spiders may reproduce abiogenetically or sexually. A few species are popular for the way the females eat the male after copulation.

Spiders show differing levels of sociality. Though most arachnids are single and even forceful toward different individuals from their own species, a few species in few families demonstrate a propensity to live in gatherings, regularly alluded to as states. These can form moderately durable conglomerations. Some of these conglomerations contain upwards of 50,000 individuals as in *Anelosimus eximius*. The level of sociality regularly fluctuates between species (interspecies); however can shift within a species (interspecies) too. A large number of these social arachnids show helpful brood care, utilize the same home (web), and have some measure of generational cover (Agnarsson and Coddington 2006). A few species, for example,

Anelosimus eximius, show regenerative division of work. A few researchers consider these species as eusocial because non reproductives or sterile individuals can be viewed as laborers.

Albeit all arthropods use muscles joined within the exoskeleton to flex their appendages. Spiders and some different gatherings still utilize water powered weight to expand them. Arachnids can create weights up to eight times their resting level to amplify legs (Parry and Brown 1959), and jumping spiders all of a sudden expand the circulatory strain in the third or fourth pair of legs (Ruppert et al. 2004).

Expanding is utilized for the mechanical kiting spiders use to scatter through the air. An arachnid or spiderling in the wake of birth will move as high as possible. The spider then stands on raised legs with its midriff pointed upward. After that, it begins discharging a few silk strings from its guts into the air, which naturally frame a triangular-formed parachute. The arachnid can then give itself a chance to be diverted by updrafts of winds.

A noxious spider, once restricted to Africa, is gradually spreading far and wide. Also, it has discovered an approach to Malaysian shores. The brown widow arachnid (*Latrodectus geometricus*) is currently affirmed to exist in Penang, Selangor, and Johor, after Universiti Malaya (UM) researchers concentrated on the physical components and DNA bar coding of the spiders. These records were made in 2011 and 2012 (Santhosh et al. 2015).

The brown widow spider is among the 30 species of the genus *Latrodectus*. They earned the name "widow" on the grounds that the female as far as anyone knows eats the male in the wake of mating. Notwithstanding, this conduct has been just convincingly archived for one species, the red-backed spider (*Latrodectus hasselti*), that is found in Australia, New Zealand, and Japan. The name widow spider in a flash strikes dread as their venom has endowed them with this reputation. Be that as it may, while nibbles from the black widow (*Latrodectus mactans*, found in North America) can be lethal, they have prompted passings in under 1 % of cases. A death from infusion of venom (envenomation) from the brown spider is less known. A case was accounted for in Madagascar in 1991.

The brown widow spider is very versatile and has discovered corners in 58 nations. They likely "hitchhiked" in transportation compartments into South and North America, the Middle East, Central Asia, Indonesia, the Philippines, Singapore, Malaysia, Japan, and Australia. They have yet to be accounted for Europe, except Turkey.

8.5 Species Richness

The spear heading commitment to the scientific classification of Indian spiders is that of European arachnologist Stoliczka (1869). Review of available literature reveals that Blackwall (1867), Karsch (1873), Simon (1887), Thorell (1895), and Pocock (1900a, b) were the pioneer workers on Indian spiders. They discovered numerous species from India. Tikader and Biswas (1981) considered 15 families, 47 genera, and 99 species from Calcutta and encompassing zones with representations and portrayals. Contributions of B. H. Patel, U. A. Gajbe, G. C. Bhattacharya, B. D. Basu, A. T. Biswas, R. Roy, K. Biswas, B. Biswas, V. Biswas, D. Raychaudhuri, G. L. Sadana, and M. Siliwal are noteworthy. They have reported numerous new arachnid species from India.

Arachnids contain one of the largest orders of animals. They are predators and devour countless organisms, but do not harm plants. They live in extraordinary natural surroundings and live in every one of the ecological situations. The prey-seeking capacity, wide host range, high fertility, and polyphagous nature make them a potential predator in biocontrol applications. The different arachnid groups on agrarian area are essential both regarding pest control and conservation purposes(Tikader and Biswas 1981).

The substantial and predictable connection of altitude crosswise over gatherings of arboreal arachnids can have vital ramifications in the light of environmental change. Although the CIT reported significant differences in species richness at high and low elevation, Hajian-Forooshani et al. (2014) found no differences between the accumulation of species in sites at high or low elevations (Fig.8.2).

Number of spider species has been found increasing from Canada to the United States to Mexico and Central America. That is by and large valid for spiders, and for the most part genuine moving south from Scandinavia to North Africa or from Siberia to the Indo-Pacific domain also. However, it is not as a matter of course valid in converse, in moving from the tropics to south temperate regions. Take, for instance, the family Anapidae, a gathering of modest spiders, normally under two millimeters in length that turn little orb networks, which are typically not exactly a centimeter in width, in wood litter and greenery. In the Nearctic locale, there is only one genus, including standout species, limited to California and Oregon (Platnick and Forster 1989). Three genera are reported in the American tropics, two of which are endemic there, and the three genera together incorporate no less than 35 species (Platnick and Shadab 1978, 1979). Be that as it may, the south fauna, in Chile and Argentina, is not similar to the fauna of North America; it incorporates six endemic genera containing no less than 15 species (Platnick and Forster 1989). The distinctions are much more pronounced in the Asian segment. In Asia itself, there are around four genera and eight species, occcuring from Nepal to Malaysia and Japan, and the tropical Eastern Indo-Pacific anapid fauna incorporates no less than two genera and five species. However, the south calm fauna is by a wide margin the largest: New Caledonia has 3 endemic genera, with 8 species; New Zealand has another 3 endemic genera with 13 species; and Australia has another 10 endemic genera with 37 species (Platnick and Forster 1989). One of those Australian genera is known from peaks in Northern Queensland, and its nine species found on various mountains there, could questionably be viewed as tropical as opposed to south mild,

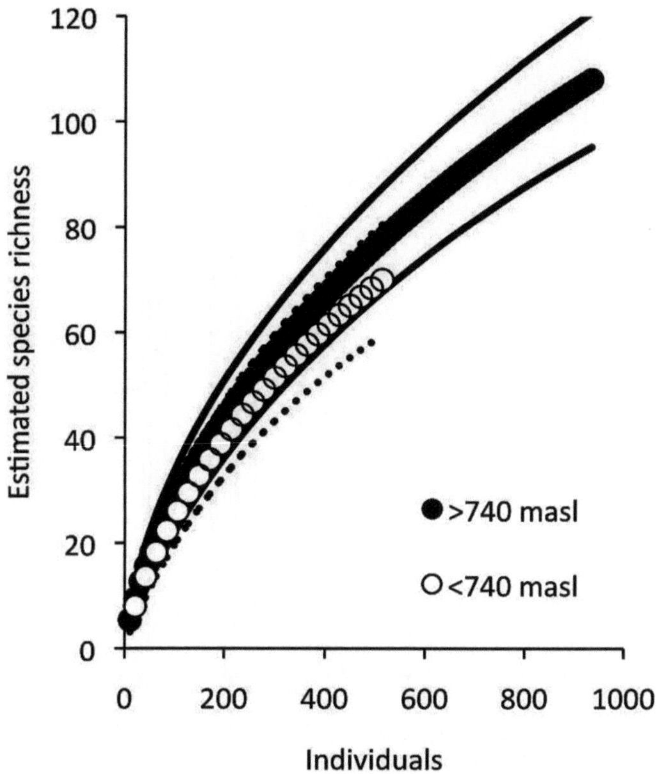

Fig. 8.2 Estimated species accumulation curves for high and low elevations. High-elevation sites (*black*; >740 masl) and low-elevation sites (*white*; <750 masl). The *thin solid lines* and *dotted lines* represent 95 % confidence intervals for high- and low-elevation sites, respectively (Zachary Hajian et al. 2014)

despite the fact that their connections are to other austral as opposed to tropical genera (Platnick 2014).

Tropical forests show high arachnid diversity; however most related studies have inspected a specific utilitarian gathering or layer of the living space and few have evaluated the effects of unsettling influence on tropical Araneae diversity. Orchid I. is 92 km off the southeastern coast of Taiwan, and its timberlands are the northernmost tropical woods in East Asia. In this study, the spiders' differences were contrasted in four sorts of environments and distinctive degrees of human unsettling influences on this island. Natural surroundings sorts analyzed in this study included essential backwoods, developed forests with a little level of unsettling influence, kindling manors with a middle of the road level of aggravation, and meadows produced from the reasonable cutting of woodlands. Two reproduces were utilized for all living space sorts, each containing four 5 × 5 m test plots. Spiders starting from

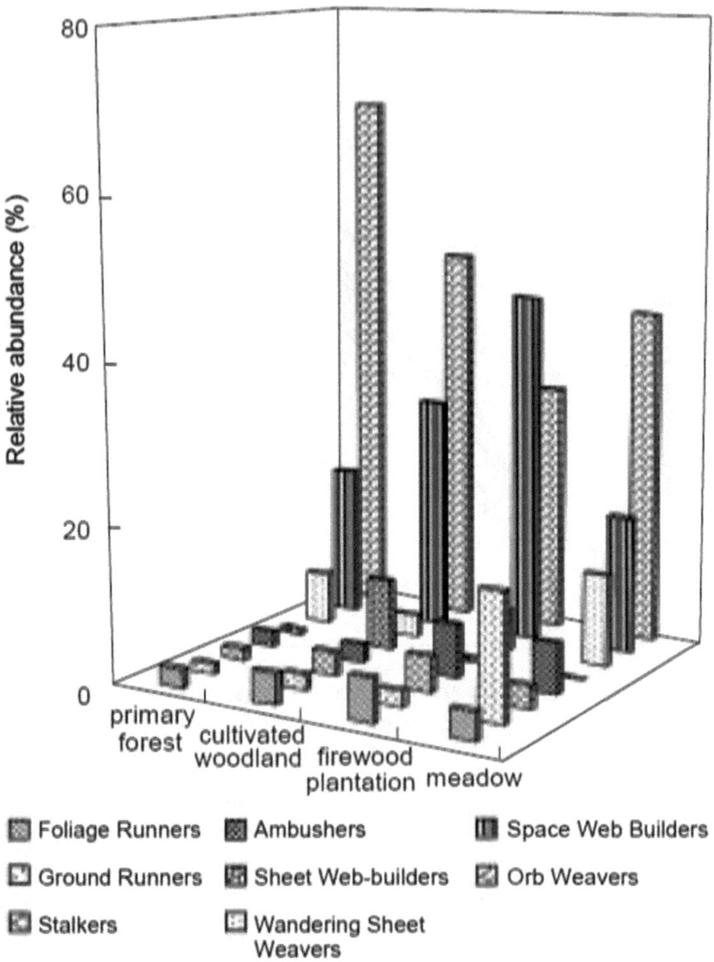

Fig. 8.3 Guild compositions of spiders collected from four different habitats (Data from individual plots were combined Kuan-Chou Chen and I-Min Tso 2004)

the earliest stage, bushes, and shelter were gathered to acquire an extensive representation of differing qualities from all microhabitats in the specimen plots. From the 1718 specimens examples got, 123 species from 19 families were recognized. The plenitude was the most astounding in the primary forest and least in the grassland plots. Plots in the four natural surroundings sorts did not contrast in Margalef species lavishness, Shannon-Weaver capacity, or Simpson index. In any case, plots in the essential woods had fundamentally brought down uniformity because of the high relative abundance of overwhelming orb weaver species. Spiders society organizations of the natural surroundings sorts in Taiwan appear in Fig. 8.3. Consequences of an UPGMA examination utilizing pair-wise Euclidean separations demonstrated that the sample plots could be bunched into four particular gatherings, showing that

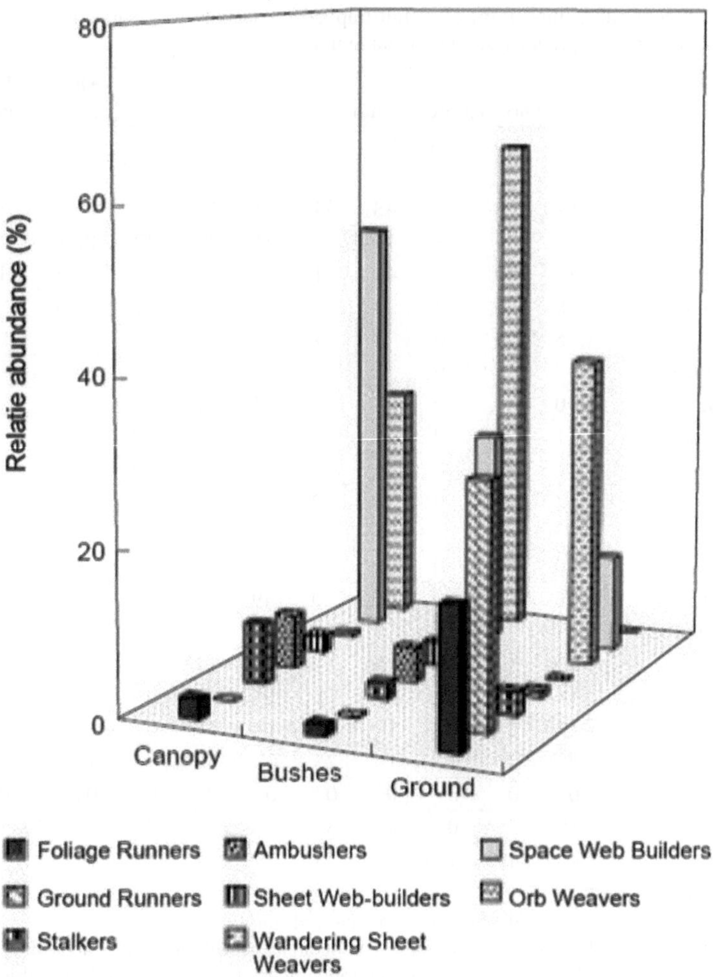

Fig. 8.4 Guild compositions of spiders collected from different layers of the wooded sample plots (Data from the primary forest, cultivated woodland, and firewood plantation plots were combined Kuan-Chou Chen and I-Min Tso 2004)

the synthesis of spiders among living space sorts extensively contrasted. Plots in the essential backwoods, developed forest, and kindling ranch natural surroundings were overwhelmed by space web manufacturers and orb weavers, while those in the meadows contained a much higher extent of meandering sheet weavers and ground runners (Kuan-Chou Chen and I-Min Tso 2004). Grasslands had very little of shrubbery layer and no shade spread, so the anticipating of seekers and ground-level web builders expanded. The arrangements of spider societies of various layers of the living spaces are presented in Fig. 8.4.

In the natural ecosystem (fallow land), the predator/parasitic taxa reported were Formicidae (31 %), Araneae (29.1), Histeridae (13.2 %), and Carabidae (10.1 %). In

Table 8.1 Taxonomic composition of pitfall-trapped predaceous and parasitic arthropods in a continuously cultivated agroecosystem (vegetable field)

Taxon	Total no. captures/month								Mean (X)	% of total captured
	Oct	Nov	Dec	Jan	Feb	Mar	Apr	May		
Insecta										
Formicidae	81	0	29	48	84	11	115	0	46.0	15.2
Histeridae	26	54	8	12	19	25	24	17	23.1	7.6
Carabidae	15	10	2	48	11	91	67	16	32.5	10.7
Hymenoptera[a]	8	6	4	4	3	3	2	1	3.9	1.3
Diptera[b]	6	5	0	3	1	0	4	0	2.4	0.8
Staphylinidae	7	3	0	7	9	1	1	3	3.9	1.3
Anthicidae	0	1	0	1	3	4	5	1	1.9	0.6
Pentatomidae	1	1	0	1	0	0	1	0	0.5	0.1
Syrphidae	1	2	0	1	2	1	0	0	0.9	0.3
Myrmeleontidae	3	0	0	3	2	3	1	0	1.5	0.5
Dermaptera										
Labiduridae	0	3	0	0	1	3	3	0	1.3	0.4
Cicindelidae	0	0	0	0	0	49	0	0	6.1	2.0
Other arthropods										
Araneae	226	391	110	156	140	99	142	20	160.5	53.0
Phalangida	44	8	0	20	11	1	3	0	10.9	3.6
Solifugae	15	2	0	1	1	1	5	0	3.1	1.0
Scorpionida	2	2	0	1	2	0	2	0	1.1	0.4
Chilopoda	0	0	0	0	0	0	2	0	0.3	0.1
Acari	10	5	0	4	3	4	0	0	3.3	1.1
Total no.	445	493	153	310	292	296	377	58	2424	100.0

Source: Faragalla and Taher (1991)
Data for December is for 2 weeks only, due to a natural runoff
[a]Hymenoptera included: Vespidae, Eumenidae, Sphecidae, Mutillidae, and Braconidae
[b]Diptera included: Asilidae, Cecidomyiidae, Sarcophagidae, Tachinidae, and Bombyliidae

the continuously cultivated agroecosystem, these were Araneae (53 %), Formicidae (15 %), Carabidae (10 %), and Histeridae (7.6 %) (Table 8.1). Arachnids in coffee manors are to some degree cryptic as a result of their absence of reaction to agricultural heightening. In Chiapas, Mexico, arboreal spiders were observed to be decidedly affected by shade tree and covering connectedness. Shade-developed espresso offers more suitable territory for arboreal creepy-crawlies because of attributes of shade trees (Zachary Hajian et al. 2014).

True spiders (Araneae) dwarf all different gatherings of arthropods and together with harvestmen (Phalangida) and Solifugae were very much represented in Western Saudi Arabia during 1985–1986 in various developed biological systems. Moreover, five groups of true spiders had been gathered in sufficient numbers that warrant family correlations. Of these, the Lycosidae is the most prevalent and dwarf the other true spider families together with the rest of the arachnids put together. Their most

Table 8.2 Pitfall-trapped spider, harvestman, and sun spider fauna in a ground-dwelling arthropod complex of three ecosystems

Taxa	% of all arachnids captured		
	Natural ecosystem	Vegetable field	Date palm-citrus orchard
Order: Araneae			
Family			
Lycosidae	8.9	36.8	26.4
Zodariidae	0.6	4.2	3.3
Theridiidae	0.4	4.2	3.3
Gnaphosidae	0.9	2.9	3.0
Salticidae	0.1	0.9	0.1
Order: Opiliones or Phalangida (harvestmen or daddy longlegs)	0.5	3.1	2.3
Order: Solifugae (sun spiders or wind scorpions)	0.5	0.9	1.1

Source: Faragalla and Taher (1991)

astounding commonness was in the constantly developed agrobiological system taken after by the date palm-citrus plantation agro-environment and the regular agrobiological system (Table 8.2). It could be expressed from these outcomes that Araneae represent a dominant group in the three agro-environments examined contrasted with predaceous/parasitic hexapod families (Faragalla and Taher 1991).

8.6 Spider Diversity

Spiders comprise one vast gathering of arthropods, with 45,741 spider species placed in 114 families perceived from the World Spider Catalog (2015). Morphology-based IDs of spiders are tedious and risky for a few reasons. It is troublesome or difficult to distinguish adolescents of spiders as after each shed they seem distinctive. Another test for distinguishing proof is striking sexual dimorphism, particularly in some orb weavers (*Nephila*), or the absence of data on indicative characters for one sex. In spite of the fact that there are around 44,957 portrayed spider species, this speaks just a small amount of their aggregate differences. Marina and Kleber (2008) embody information on differences of spiders in temperate ranges approximately 300 N and S from equator including desert and savanna contain higher quantities of arachnids than tropical regions.

Nearly 1800 spider species from 443 genera are known from India. Among the 62 families, 5 families represented more than 100 species each, and they are Salticidae (74 genera and 212 species), Thomisidae (40 genera and 178 species), Araneidae (28 genera and 167 species), Gnaphosidae (30 genera and 148 species), and Lycosidae (19 genera and 134 species) (Kuan-Chou Chen and I-Min Tso 2004).

Out of 1724 spider species recorded from the Indian sub-mainland, 102 species are from mygalomorphs and 1622 species are from araneomorphs. Seventy species are solely recorded from the Karakorum zone. As indicated by an IUCN report eight mygalomorph species have been considered as imperiled/fundamentally jeopardized/close debilitated. Territory misfortune and debasement alongside worldwide pet exchange are significant dangers for these spiders, and thus the species are under threat in the Indian subcontinent:

1. *Poecilotheria formosa* Pocock (1899) – endangered
2. *Poecilotheria hanumavilasumica* – critically endangered
3. *Poecilotheria metallica* Pocock (1899) – critically endangered
4. *Poecilotheria miranda* Pocock (1900a, b) – endangered
5. *Poecilotheria rufilata* Pocock (1899) – endangered
6. *Poecilotheria striata* Pocock (1895) – vulnerable
7. *Haploclastus kayi* Gravely (1915) – endangered
8. *Thrigmopoeus truculentus* Pocock (1899) – near threatened (IUCN 2014)

8.7 Endemic Spiders

The first information on the endemic spider fauna of India came from the female specimen of *Oxyopes indicus* (Walckenaer 1805). Thereafter, the work of Sundevall (1833), Walckenaer (1841), Blackwall (1864, 1867), Simon (1864, 1885, 1887, 1888, 1889, 1892, 1893, 1895, 1897, 1900, 1901, 1902, 1905, 1906), Frauenfeld (1867), Thorell (1887, 1891), Stoliczka (1869), O. P.-Cambridge (1870, 1874, 1877, 1885, 1890, 1892), Butler (1873), Wood Mason (1877), Keyserling (1890), Walsh (1891), Peckham and Peckham (1895), Pocock (1895, 1899, 1900), Leardi (1901, 1902), Strand (1912), Hirst (1909), Narayan (1915), Gravely (1915, 1921, 1924, 1931, 1935), Chamberlin (1917), Hogg (1922), Hingston (1927), Sherriffs (1927), Mukerjee (1930), Reimoser (1934), Caporiacco (1934, 1935), Bhattacharya (1935a, b), Roewer (1942), and Fage (1946) initiated the spider research during British India. These workers reported 408 spider species from India which are endemic to India. Later on, spider research was geared up in India from 1962 by the work of Tikader (1960–1991), Gajbe (1991, 2008), Biswas and Roy (2008), and Patel (1973–2005) who reported more than 500 spider species endemic to India. Siliwal and co-workers (2005–2015) reported 21 mygalomorphs endemic to India. Notwithstanding the Indian specialists, expansive number of arachnologists outside India has additionally reported spiders endemic to present day India. Prominent among them are Prószyński, Platnick, Jäger, Peckham and Peckham, Huber, and Baehr (see Siliwal et al. 2015 for details).

Endemism is the environmental condition of a species being restricted to a characterized geographic area, for example, an island, nation or other characterized zone, or living space sort. Endemic spiders in India contain 1238 species from 58 families and 340 genera. The most characteristic families are Thomisidae, Salticidae, and Gnaphosidae containing 11.66 %, 11.50 %, and 10.51 % precinctive/endemic species separately among the aggregate 1238 endemic known spider species in India. This has all the earmarks of being high endemism (Siliwal et al. 2015).

8.8 Threats to Spiders

Populace of arachnids is diminishing gradually because of urbanization, water and environmental changes. To save these spiders, water must be made accessible by diving trenches in streams before the blustery season with the goal that water will be accessible and keep up the moistness. Today the populace and assorted diversity of pisaurids is influenced. From the information available for the previous five years from India, it is unmistakably seen that no new species have been discovered from these families and that their populace is diminishing. The following are threats responsible for the decreasing population of spiders.

Climate Change Because of environmental change, the seasons have become shaky. There can be downpours in summer or even in winter. Most spiders require the right stickiness for survival, and their life cycles are synchronized with the blustery season and winter. Most spiders lay egg sacs amid September/October, i.e., before the end of the stormy season. In the event that there are overwhelming downpours amid winter or if there is worldwide darkening, there is high mortality. In view of unnatural weather changes, there is water shortage, bringing about a less hygroscopic environment, and spiders cannot endure high temperatures. Streams get dry, and there will be no water in downstream stores. These conditions are unfavorable for the survival of spiders.

Grazing Nibbling by herbivores in timberlands prompts pulverization of the environment of arachnids that possess grasses and bushes, for instance, *Argiope (signature spider)*. Amid browsing, a few spiders are eaten up, and much of the time the orb web is annihilated and these spiders need to spend more vitality for setting up another web. Browsing additionally brings about slackening of soil on the slants, along these lines prompting soil expulsion amid the stormy season as spillover. This influences the development of bushes (living space debasement), influencing the spiders populace. Eating additionally brings about the spread of *Lantana camara* (a weed), which keeps down the development of bushes and grasses, influencing the spider's natural surroundings seriously. In India most woodland living spaces are confronting the issue of this weed smothering the development of endemic bushes and grasses. To preserve arachnids, touching by residential herbivores must be kept away from in backwoods to secure spider living spaces.

Deforestation/Habitat Loss Deforestation annihilates numerous normal territories, influencing composition of animal and plant communities. At the point when timberlands are chopped down and common progression is permitted to happen, an alternate, widely varied vegetation colonizes the territory, influencing the spider diversity and populace.

Besides, soil is lost when the organic litter of the timberland floor is washed away. There are a variety of weeds that become established rapidly in clear-cuttings in the wake of logging and/or blazing, and the vegetation is modified within a brief time frame. The weeds make new territories for a wide range of arthropods. Be that as it may, spiders require appropriate spots to make their networks. They require some investment to change in accordance with the new conditions.

An established instance of habitat misfortune is *Thrigmopoeus truculentus* in the Western Ghats. This spider is classified as "closely threatened" as it does not meet the "confined dissemination criteria." The natural surroundings of *Poecilotheria hanumavilasumica* (*Rameshwaram parachute spider*) are demolished on account of different reasons, and because of the restricted region of dissemination of this spider, its presence is in risk. It is announced as critically endangered.

Forest Fire Woods fires profoundly affect all arthropods in a forest environment. Not just does the smoldering kill a great deal of arthropods straight forwardly, it additionally modifies the ground vegetation and the natural soil spread, which thus alters the conditions for ground-living animals. Workers examined the effect of the wood fire on the soil pH and the number of spiders in the season. In a dry deciduous forest, the plant litter begins rotting after the primary rains and structures, an appropriate environment for soil arthropods and microscopic organisms, which finishes the rotting of the litter. Arachnids eat this arthropod populace, controlling them. The excreta of the spiders fall on the soil, influencing its pH. Subsequently in a sustenance web, bugs assume a pivotal part as predators and are at the highest point of this transitory biological community. At that point in the next months, grasses and bushes establish, giving a decent natural surroundings to different spiders. Lower differing qualities and a smaller populace of spiders were found in Melghat in flame influenced compartments.

Scarcity of Water Spiders from the families Pisauridae and Tetragnathidae like to live along streams and water bodies. On the off chance that water is not accessible amid their life cycle stages, from the laying of egg sacs to advancement up to eight or nine shedding stages, they cannot survive.

Pesticides Agriculturists use pesticides to shield their harvests from nuisances. They splash pesticides to kill the pests, and at the same time, nontarget invertebrates, for example, spiders, are likewise killed. Really, spiders being summed up feeders can feast upon an assortment of nuisances and secure the harvests; however, ranchers use pesticides that execute the arachnids as well. A few uses of pesticides for every season can decimate insect groups. A few pesticides are additionally held in the webs of spiders and can be adverse to those arachnids that ingest their webs day by day. Pests return and assault the product; however arachnids do not return so effortlessly to the fields. This makes an unevenness, and the agriculturists are at a misfortune. It takes more time for the populace densities of spiders to develop after the use of spider sprays contrasted and plant hoppers and leafhoppers since arachnids have a more drawn-out era interim.

To save spiders, the utilization of pesticides must be banned/minimized, and agriculturists ought to be prepared about this. Indeed, even spiders can be utilized as biocontrol operators. Yes, it is conceivable if arachnids are raised by infestation of products. Luckily, the life cycle of farming money crops coincides with the life cycle of spiders, i.e., June to December. Farmers can be prepared for rearing spiders that have high fertility.

Agricultural Practices In India, agricultural practices are changing, and agriculturists have quit utilizing natural manures. In one test, banana fields in which natural compost was utilized had a significant high spider populace (780 *Thelacantha* per section of land + 1240 *Cyclosa* per section of land + 2000 lycosids + other spiders) contrasted with an agro field in which inorganic manures were utilized. The family Lycosidae made up just 5 % of the group in customary fields where inorganic pesticides were utilized; however, they made up 35 % in organic fields. Spider densities were additionally observed to be expanded in banana and orange fields where straw mulch was utilized as a ground spread to avoid water dissipation. Smoldering litter and product remains decimate ground arachnids. To moderate ground-tunneling spiders in agro-environments, agriculturists ought to furrow their fields before March so that these spiders are not demolished as their tunneling exercises start amid the principal week of April. They develop and mature before the onset of the blustery season. Soil unsettling influenced by furrowing annihilates overwintering locales and can kill any spider that effectively displays in the soil. The development of homestead gear through a harvest field harms spider-catching networks and may pulverize web connection locales. Thus, the spider density and diversity are higher in organic fields than in traditional ones. Spiders have an extensive variety of prey species, get critical quantities of their prey, and utilize different scrounging procedures. Crop differing qualities additionally prompt an accessibility of option prey, which may expand the spider-assorted qualities and also lessen the domain size of arachnids, prompting a steady populace of spiders at high densities. Protection of predators like spiders in the field can be proficient by lessening physical unsettling influences to the living space. Consequently, to moderate and improve spider populaces, rural frameworks ought to be controlled in routes useful to the requirements of arachnids.

Use of Pesticides to Kill Mosquito Larvae For mosquito control, Indians use mosquito anti-agents in their homes. This has seriously influenced the number of inhabitants in house spiders, for example, *Pholcus*, salticids, and gnaphosids. Spiders in houses are useful in mosquito control. A room of size 10 × 10 ft must have five or six pholcids on the off chance that every one of the mosquitoes is to be nourished on. Essentially, the larval stages of mosquitoes that possess water bodies can likewise be controlled by spiders, for example, tetragnathids, pisaurids, and lycosids, living along trench, streams, and waterways.

However, we splash pesticides or phenyls to kill the larval stages. Spiders can feed upon grown-up mosquitoes and also on the larval phases of a wide range of mosquitoes. Thus, spiders can be monitored by ceasing the utilization of mosquito killing agents and pesticide sprays.

Urbanization is nothing but living space fragmentation, pulverization, and/or natural surroundings change. Generally agribusiness land or timberland area is utilized for urbanization. This prompts low diversities and higher plenitudes of confined species. Especially, spiders from the families Pisauridae, Tetragnathidae, and Clubionidae lose their territories. Spiders from the families Salticidae, Uloboridae, Therediidae, Pholcidae, Oonopidae, and Araneidae overwhelm urban territories. To moderate spiders from the families Pisauridae, Tetragnathidae, and Clubionidae, which have lost their environments, advancement of kitchen patio nurseries, foundation of parks in cities and towns with moving water in drains, and manor of trees can be taken up. The floor space index (FSI) of developments ought to be expanded, with structures becoming vertically as opposed to on a level plane.

Development of Road Network India being a developing nation, advancement of the street system in India is unavoidable at present. Tree felling is done to augment existing streets. This devastates the living spaces of spiders that possess tree trunks, e.g., spiders from the family Hersiliidae, a few spider salticid genera, for example, *Marengo* and *Myrmarachne* and social arachnids from the family Eresidae. The rundown is long. Indeed, even in woodlands, after the blustery season, utility streets are repaired by cutting grasses developing on old streets. This practice ought to be delayed till the end of December so that the spiders utilizing grass living spaces grow totally and are scattered broadly. Something else, after a delayed time of wrong practices, these species will get to be jeopardized. At present the vast majority of the spiders are named Data Deficient and exploration is required. For preservation of these arachnids, it is proposed here that, before felling of trees for building up the street system, trees be planted first at the fundamental separation. Once these develop, the old trees can be cut for new streets.

8.9 Trade

Species, for example, tarantulas, are gathered and killed on a substantial scale. A portion of the animal groups, for example, dark widows, are gathered for venom in India, and thus their populace is diminishing. Tarantulas, the unpleasant and crawly creepy spiders, are in incredible interest the world over. Also, there is anxiety in the Western Ghats, where the species is lessening due to a sudden forcefulness in the unlawful exchange of these eight-legged arachnids. The poaching of spiders goes unnoticed, and there is no development all things considered against the trade. The International Union for Conservation of Nature has listed *Poecilotheria regalis, Poecilotheria striata, Poecilotheria hanumavilasumica, Poecilotheria miranda, and Poecilotheria metallica* in the Red Data Book as threatened with extinction. *Thrigmopoeus truculentus* is categorized as "near threatened" because it does not meet the restricted distribution criteria. Of the 90 tarantula species found in India, there is a tremendous interest for 62 for decorative purposes. These mygalomorphs ought to be brought into the planned gathering under the Wildlife Protection Act, so they are given security and hence can be moderated.

The essential driver of the decay of spider-assorted qualities is not immediate human over abuse but rather territory decimation. At present a large portion of the spider species in India is information insufficient. The timberland, watering system, and farming divisions ought to work on the whole to take care of the issue of protection of spider by averting fire, sufficiently giving water to solid environments, and taking after eco-accommodating agro and backwoods rehearses. Agriculturists specifically should quit utilizing fertilizers and pesticides and go for organic cultivating, utilizing integrated pest management. Scientists can chip away at spider scientific classification to recognize more spider species from India.

8.10 Spider Silk

Spider silk is a case of a multifunctional fiber that assumes a key part in a spider's life and serves as an incredible model for the adaptable assembling of the up-and-coming era of forte filaments. The discharges of the arachnid are a shelter for people, for instance, silk. Arachnid silk has high rigidity, four to five times more than that of iron. Subsequently, the recombinant silk delivered on substantial scale is utilized to produce fabric from which impenetrable coats are fabricated. On the off chance that spider silk bars of 5 mm measurement are utilized as a part of building development rather than iron poles, seismic tremor confirmation houses can be built as silk is versatile in nature. This is the nature's blessing which is of extraordinary use for the armed force of any nation. Parachute ropes can likewise be fabricated from arachnid silk. German researchers have made guitar strings from arachnid silk. The silk protein atoms can redirect the radio frequencies originating from rockets and consequently can be utilized to mislead the rockets of foe. Along these lines, spider silk proteins are helpful in guard at a huge scale.

Breast Implant Coatings Preclinical studies have demonstrated that silicone bosom inserts are better acknowledged by the body when covered with a nonimmunogenic spider silk film. In particular, capsular fibrosis and aggravation are altogether lessened.

Durable Skin Barrier Solid skin barrier is being produced to give insurance to the skin. Spider silk can viably rehydrate the skin and manufacture a defensive silk film to restore the skin's obstruction capacity.

Wound Care and for Wound Healing Because of the biocompatibility and biodegradability properties of the filaments, spider silk has turned into an alluring material for analysts in regard to biomedical applications.

Antibiotic Surgical strings and gauzes can be set up from the spider silk. This silk is biocompatible with people. The greater part of the birds inside line their homes with arachnid silk in light of the fact that the spider silk is smooth and anti-infection.

In the event that youthful ones are harmed with grass-sharp edges or their pointed tips, they do not get bacterial contamination as they are in contact with antibacterial silk.

Tissue Culture Spider silk covering offers an expanded hydrophilic, somewhat adversely charged surface, which energizes authoritative. Additionally, the customized RGD (Arg-Gly-Asp) arrangement enhances cell connection and empowers particular integrin authoritative. This totally straightforward covering permits exact optical estimation. The tripeptide Arg-Gly-Asp (RGD) is the grouping inside a protein that intervenes cell connection and has been found in spider silk proteins which go about as receptors for cell bond atoms.

Tissue Engineering Arachnid silk gives a reasonable 3D permeable framework that impersonates the in vivo extracellular network and empowers cells to connect, relocate, and develop. Arachnid silk protein is better than its RGD (Arg-Gly-Asp) areas upgrading cell. The mechanical properties of this delicate material make it particularly adjusted for tissue designing.

Reflects UV Rays Arachnid silk covering the egg sacs of spiders reflects UV beams and hence secures the sensitive eggs. With this property, the spider silk is currently used to produce UV-reflecting fabrics.

Drug Delivery Thus, spider silk is unmistakably a case of a multifunctional fiber that assumes a key part in spider's life and serves as a magnificent model for the adaptable assembling of the up-and-coming era of forte filaments. Because of the biocompatibility and biodegradability and also the mechanical properties of the filaments, arachnid silk has turned into an appealing material for scientists in regard to biomedical applications. Drug spider venom is utilized to make pharmaceuticals against malignancy.

Pesticides from Spider Venom Presently, pesticides are likewise made from spider venom. Venom from the world's most noxious arachnids may spare the bumble bees, giving a biopesticide that kills pests yet saves the valuable pollinators. Honey bee populaces, both wild and hostage, are in decrease all through the world. Certain pesticides used to ensure yields can scramble the cerebrum circuits of bumble bees, influencing memory and route abilities they have to discover for sustenance, setting whole hives under risk. Amazingly, the spider venom is not venomous for humans; just slight unfavorably susceptible reactions are seen after envenomation.

8.11 Conservation

"If spiders disappeared, we would face famine." This is how Norman Patnick talks about spider conservation. Arachnids are essential controllers of insects. Without arachnids, the majority of our yields would be devoured by pests. "Arachnids are vital in natural cultivating, which depends on common foes." Venom of spiders contains several distinct chemicals, some of which might be therapeutically dynamic. Natural surroundings misfortune and discontinuity are the worst dangers to spiders. Keeping up environmental heterogeneity and minimizing insect sprays and different chemicals are the approaches to moderate spiders. Spiders are small to the point that they may frequently be neglected in protection arranging. On one acre of land of forest alone, spiders may devour more than 80 lb of insects, every year. The pest populations in developed biological communities will explode without spiders. So they should be conserved.

References

Agnarsson A, Coddington M (2006) Sociality in Theridiid spiders: repeated origins of an evolutionary dead end. Evolution 60(11):2342–2351

Barbour T (1921) Spiders feeding on small cyprinodonts. Psyche 28:131–132

Bhattacharya GC (1935a) A new spider of Bengal, mimicking *Oecophylla smaragdina* (Fabr.). J Bombay Nat Hist Soc 37:960–962

Bhattacharya GC (1935b) A new species of gregarious spider mimicking *Camponotus compressus*. Sci Cult 1:159–160

Biswas B, Roy R (2008) Description of six new species of spiders of the genera *Lathys* (Family: Dictynidae), *Marpissa* (Family: Salticidae), *Misumenoides* (Family: Thomisidae), *Agroeca* (Family: Clubionidae), *Gnaphosa* (Family: Gnaphosidae) and *Flanona* (Family: Lycosidae) from India. Rec Zool Surv India 108:43–57

Blackwall J (1864) Descriptions of seven new species of East Indian spiders received from the Rev. O. P. Cambridge. Ann Mag Nat Hist 14(3):36–45

Blackwall J (1867) Description of seven new species of East Indian spiders received from the Rev. O. P. Cambridge. Ann Mag Nat Hist 14(3):36–45

Butler AG (1873) A monographic list of the species of *Gasteracantha* or crab-spiders, with Calcutta 11:257–287

Cambridge OP (1870) Descriptions and sketches of two new species of Araneida, with characters of a new genus. J Linn Soc Lond Zool 10:398–405

Cambridge OP (1874) On some new species of *Drassides*. Proc Zool Soc London 1874:370–419

Cambridge OP (1877) On some new species of Araneidea, with characters of two new genera and some remarks on the families Podophthalmides and Dinopides. Proc Zool Soc London 1877:557–578

Cambridge OP (1885) Araneida. In: Scientific results of the second Yarkand mission. Calcutta, pp 1–115

Cambridge OP (1890) On some new species and two new genera of Araneida. Proc Zool Soc London 1890:620–629

Cambridge OP (1892) On a new spider from Calcutta. Ann Mag Nat Hist 10(6):417–419

Caporiacco L (1934) Aracnidi dell'Himalaia e del Karakoram raccolti dalla Missione Italiana al Karakoram (1929- VII). Mem Soc Entomol Ital Genova 13:113–160

Caporiacco L (1935) Aracnidi dell'Himalaia e del Karakoram, raccolti dalla Missione italiana al Karakoram (1929-VII). Mem Soc Entomologica Ital Genova 13:161–263

Chamberlin RV (1917) New spiders of the family Aviculariidae. Bull Mus Comp Zool Harv Coll 61:25–75

Chen K-C, Tso I-M (2004) Spider diversity on Orchid Island, Taiwan: a comparison between habitats receiving different degrees of human disturbance. Zool Stud 43(3):598–611

Coddington JA, Levi HW (1990) Systematics and evolution of spiders (Araneae). Annu Rev Ecol Syst 22:565–592

Diaz JH (2004) The global epidemiology, syndromic classification, management, and prevention of spider bites. Am J Trop Med Hyg 71(2):239–250
Fage L (1946) Araignées cavernicoles de l'Inde. Bull Mus Natl Hist Nat Paris 18(2):382–388
Faragalla AA, Taher MO (1991) Predaceous and: parasitic, ground-dwelling arthropods in different ecosystems in Western Saudi Arabia. KAU Sci 3:57–64
Foelix RF (1982) Biology of spiders. Harvard University Press, Cambridge, pp 6(2):306
Foelix RF (1996) Biology of spiders, 2nd edn. Oxford University Press, New York. ISBN 9780195095944
Frauenfeld GR (1867) Zoologische Miscellen. XI. Verh K-Kg Zool-Botanischen Ges Wien 17:425–502
Gajbe UA (1991) Description of male spider *Uloborus danolius Tikader* (Araneae: Uloboridae). Rec Zool Surv India 89:307–308
Gajbe UA (2008) A new species of *Dieta spider* (Araneae: Philodromidae) from Jabalpur, Madhya Pradesh, India. Rec Zool Surv India 108:59–61
Gravely FH (1915) Notes on Indian mygalomorph Spiders. Rec Indian Mus Calcutta 11:257–287
Gravely FH (1921) Some Indian spiders of the subfamily Tetragnathinae. Rec Indian Mus Calcutta 22:423–459
Gravely FH (1924) Some Indian spiders of the family Lycosidae. Rec Indian Mus Calcutta 26:587–613
Gravely FH (1931) Some Indian spiders of the families Ctenidae, Sparassidae, Selenopidae and Clubionidae. Rec Indian Mus Calcutta 33:211–282
Gravely FH (1935) Notes on Indian mygalomorph spiders. II. Rec Indian Mus Calcutta 37:69–84
Hajian-Forooshani Z, Gonthier DJ, Marín L, Iverson AL, Perfecto I (2014) Changes in species diversity of arboreal spiders in Mexican coffee agroecosystems: untangling the web of local and landscape influences driving diversity. Peer J 2:e623. https://doi.org/10.7717/peerj.623
Hansen LS, Gonzales SF, Toft S, Bilde T (2008) Thanatosis as an adaptive male mating strategy in the nuptial gift–giving spider Pisaura mirabilis. Behav Ecol 19:546–551
Hingston RWG (1927) Protective devices in spiders' snares, with a description of seven new species of orb-weaving spiders. Proc Zool Soc London 1927:259–293
Hirst AS (1909) On some new or little-known mygalomorph spiders from the Oriental Region and Australasia. Rec Indian Mus Calcutta 3(4:30):383–390
Hogg HR (1922) Some spiders from South Annam. Proc Zool Soc London 1922:285–312
IUCN (2014) The ICUN red list of threatened species. http://www.iucnredlist.org/details
Johnson and Andrew (2005) Precopulatory sexual cannibalism in fishing spiders (*Dolomedes triton*): a role for behavioral syndromes. Behav Ecol Sociobiol 58(4):390–396
Karsch E (1873) Verzeichniss Westfalischer Spinnen (Araneiden). Verh Naturh Ver Preuss Rheinl Westfalen 10:113–160
Keyserling E (1890) Die Arachniden Australiens. Nürnberg 2:233–274
Leardi Z (1901) Aracnidi di Almora. Atti Soc Ital Sci Nat Museo Civico Storia Nat Milano 40:85–94
Leardi Z (1902) Aracnidi di Mahé e Kandy. Atti Soc Ital Sci Nat Museo Civico Storia Nat Milano 40:345–373
Levi (1993) The orb-weaver genus Kaira (Araneae: Araneidae). J Arachnol 21:209–225
Marina FM, Kleber DC (2008) Diversity of tropical spiders-ground-dwelling species of Brazillian Savannas. Trop Biol Conserv Manag 11:21
Mukerjee D (1930) Description of a new ant-mimicking spider, *Synemosyna transversa* sp. nov. J Bombay Nat Hist Soc 34:200–201
Narayan K (1915) Notes on ant-like spiders of the family Attidae in the collection of the Indian Museum. Rec Indian Mus Calcutta 11:393–406
Opell BD (1997) The material cost and stickiness of capture threads and the evolution of orb-weaving spiders. Biol J Linn Soc 62:443–458
Patel BH (1973) Some interesting theridiid spiders (Araneae: Theridiidae) from Gujarat, India. Bull Br Arachnol Soc 2:149–152

Peckham GW, Peckham EG (1895) Spiders of the *Homalattus* group of the family Attidae. Occas Pap Nat Hist Soc Wis 2:159–183
Platnick (2014) The world spider catalog, version 15. American Museum of Natural History, online at http://research.amnh.org/entomology/spiders/catalog/index.html. doi:10.5531/db.iz.0001
Platnick NI, Forster RR (1989) A revision of the temperate South American and Australasian spiders of the family Anapidae (Araneae, Araneoidea). Bull Am Mus Nat Hist 190:1–139
Platnick NI, Shadab MU (1978) A review of the spider genus *Anapis* (Araneae:Anapidae), with a dual cladistic analysis. Am Mus Novit 2663:1–23
Platnick NI, Shadab MU (1979) A review of the spider genera *Anapisona* and *Pseudanapis* (Araneae, Anapidae). Am Mus Novit 2672:1–20
Pocock RI (1895) On a new and natural grouping of some of the Oriental genera of Mygalomorphae, with descriptions of new genera and species. Ann Mag Nat Hist 15(6):165–184
Pocock RI (1899) Diagnoses of some new Indian Arachnida. J Bombay Nat Hist Soc 12:744–753
Pocock RI (1900a) The fauna of British India, including Ceylon and Burma. Arachnida. Taylor and Francis, London, pp 1–279
Pocock RI (1900b) Great Indian spiders. The genus *Poecilotheria*: its habits, history and species. J Bombay Nat Hist Soc 13:121–133
Reimoser E (1934) Araneae aus Süd-Indien. Rev Suisse Zool 41:465–511
Roberts MJ (1995) Spiders of Britain and Northern Europe. Collins, London. ISBN 13: 9780002199810. ISBN 10: 0002199815
Roewer CF (1942) Katalog der Araneae von 1758 bis 1940. Bremen 1:1–1040
Ruppert EE, Fox RS, Barnes RD (2004) Invertebrate zoology, 7th edn. Brooks/Cole, Belmont. ISBN 0-03-025982-7
Santhosh KN, Pavana D, Thippeswamy NB (2015) Indian scorpions of Karantaka: maintenance in captivity, venom extraction and toxicity studies. J Venom Anim Toxins Incl Trop Dis 21(1):1–7
Sebastian PA, Peter KV (2009) Spiders of India. Universities Press/Orient Blackswan, Hyderabad. ISBN 978-81-7371-641-6
Schütz D, Taborsky M (2003) Adaptations to an aquatic life may be responsible for the reversed sexual size dimorphism in the water spider, Argyroneta aquatica. Evol Ecol Res 5:105–117
Schütz D, Taborsky M (2005) Mate choice and sexual conflict in the size dimorphic water spider Argyroneta aquatica (Araneae: Argyronetidae). J Arachnol 33(3):767–775
Sherriffs WR (1927) South Indian arachnology. Part II. Ann Mag Nat Hist 19(9):533–542
Siliwal M, Kananbala A, Bhubaneshwari M, Raven R (2015) Natural history and two new species of the trapdoor spider genus *Conothele* Thorell 1878 (Araneae: Ctenizidae) from India. J Arachnol 43(1):34–39
Simon E (1864) Histoire naturelle des araignées (aranéides). Paris, pp 1–540
Simon E (1885) Matériaux pour servir à la faune arachnologiques de l'Asie méridionale. I. Arachnides recueillis à Wagra-Karoor près Gundacul, district de Bellary par M. M. Chaper. II. Arachnides recueillis à Ramnad, district de Madura par M. l'abbé Fabre. Bull Soc Zool Fr 10:1–39
Simon E (1887) Etude sur les Arachnides de l 'Asie meridionale faisant partie des collection de l ' Indian Museum (Calcutta). J Asiat Soc Beng 56:101–117
Simon E (1888) Etudes sur le arachnides de l'Asie méridionale faisant partie des collections de l'Indian Museum (Calcutta). II. Arachnides recueillis aux îles Andaman par M. R. D
Simon E (1889) Arachnides de l'Himalaya, recueillis par MM. Oldham et Wood-Mason, et faisant partie des collections de l'Indian Museum. Première partie. J Asiat Soc Bengal 58:334–344
Simon E (1892) Histoire naturelle des araignées. Paris 1:1–256
Simon E (1893) Histoire naturelle das araignées. Paris 1:257–488
Simon E (1895) Histoire naturelle des araignées. Paris 1:761–1084

Simon E (1897) Histoire naturelle des araignées. Paris 2:1–192
Simon E (1900) Descriptions d'arachnides nouveaux de la famille des Attidae. Ann Soc Entomol Belg 44:381–407
Simon E (1901) Histoire naturelle des araignées. Paris 2:381–668
Simon E (1902) Description d'arachnides nouveaux de la famille des Salticidae (Attidae) (suite). Ann Soc Entomologique Belg 46(24–56):363–406
Simon E (1905) Arachnides (1re partie). In: Voyage de M. Maurice Maindron dans l'Inde méridionale. 7e Mémoire. Ann Soc Entomologique France 74:160–180
Simon E (1906) Etude sur les araignées de la section des cribellates. Ann Soc Entomologique Belg 50:284–308
Stoliczka F (1869) Contribution towards the knowledge of Indian Arachnoidea. J Asiat Soc Bengal 38:201–251
Strand E (1912) Über einige Spinnen aus Travancore in Indien. Arch Naturgesch 78(A8):144–148
Sundevall CJ (1833) Conspectus arachnidum. Londini Gothorum, pp 1–39
Thorell T (1887) Viaggio di L. Fea in Birmania e regioni vicine. II. Primo saggio sui ragni birmani. Ann Museo Civico Storia Nat Genova 25:5–417
Thorell T (1891) Spindlar från Nikobarerna och andra delar af södra Asien. Bihang Till Kongliga Svenska Vetenskaps-Akademiens Handlingar 24(2):1–149
Thorell T (1895) Descriptive catalogue of the spiders of Burma. British Museum, London, pp 1–406
Tikader BK (1960) On some new species of spiders (Arachnida) of the family Thomisidae from India. J Bombay Nat Hist Soc 57:173–183
Tikader BK, Biswas B (1981) Spider fauna of Calcutta and vicinity. Rec Zool Surv India Occ Paper No 30:1–148
Walckenaer CA (1805) Tableau des aranéides ou caractères essentiels des tribus, genres, familles et races que renferme le genre Aranea de Linné, avec la désignation des espèces comprises dans chacune de ces divisions. Paris, pp 88
Walckenaer CA (1841) Histoire naturelle des Insects. Aptères. Paris 2:1–549
Walsh JHT (1891) A new trap-door spider from Orissa. J Asiat Soc Bengal 59:269–270
Wood-Mason J (1877) Note on *Mygale stridulans*. Trans Entomol Soc Lond 1877:281–282
World Spider Catalog (2015) World Spider Catalog. Natural History Museum Bern, online at http://wsc.nmbe.ch, version 15.5. Assessed 8 Sept 2015
Yeargan and Quate (1997) Adult male bolas spiders retain juvenile hunting tactics. Oecologia 112:572–576
Zachary Hajian-Forooshani, Gonthier DJ, Marín L, Iverson AL, Perfecto I (2014) Changes in species diversity of arboreal spiders in Mexican coffee agroecosystems: untangling the web of local and landscape influences driving diversity. Peer J 2:e623. https://doi.org/10.7717/peerj.623

Fluctuation of Brown Mite, *Bryobia rubrioculus* (Scheuten) (Acari: Tetranychidae) Population in the Cherry Orchards of Hamedan, Western Iran

9

Mohammad Khanjani

Abstract

Brown mite, *Bryobia rubrioculus* (Scheuten) is the most important pest of cherry and plum trees in West Iran. Mites caused considerable damage in select orchards in West Iran. So its biology was studied during 2010 and 2011. This pest overwintered as egg on fruit tree shoots. Larvae of the first generation appeared early in April at 10.58 °C. This pest completed 4–5 generations/year. Mite population peaked in September. This database can be used to develop ecofriendly management practices.

Keywords

Brown mite • Hamedan • Iran • Fruit trees

9.1 Introduction

Fruit orchards and their products play an important role in the economy of Iran. One can find different mites on the fruit trees especially cherry, black cherry and plum, particularly *Bryobia rubrioculus* which is the most important one. The population of this phytophagous mite in some of the Hamedan province is considerable, and it should be controlled; otherwise optimum harvesting of product would not be possible (Khanjani and Haddad 2009). The insurgency of this pest is visible in orchards

M. Khanjani (✉)
Department of Plant Protection, College of Agriculture, Bu-Ali Sina University, Hamedan, Iran
e-mail: Khanjani@basu.ac.ir

A.K. Chakravarthy, S. Sridhara (eds.), *Arthropod Diversity and Conservation in the Tropics and Sub-tropics*, DOI 10.1007/978-981-10-1518-2_9

sprayed in previous years against other pests such as aphids and plum moth. Surely, this insurgency occurred because of destroying natural enemies. For performing successful control, sufficient time is needed, so the investigation of pest biology in this area became necessary.

Bryobia rubrioculus, the mite on the fruit tree species, has long been confused with *B. praetiosa* Koch, the grass species. *B. rubrioculus* is also known in literatures as *B. arborea* (Morgan and Anderson 1957). The brown mite is an economic pest of orchard trees such as cherry, plum, black cherry, apple, pear, peach and almond.

Brown mite is distributed in Europe (Austria, Belgium, Bulgaria, France, Germany, West Greece, Hungary, Italy, the Netherlands, Poland, Portugal, Romania, Spain, Switzerland) (Momen 1987; Carmona 1992), North America (USA; Canada – Nova Scotia, Ontario, British Columbia) (Herbert 1962), Africa (Mozambique, South Africa, Zimbabwe), Australia, New Zealand, Asia (Afghanistan, China, India, Iraq, Iran, Japan, Lebanon, Turkey) (Charles 1998), and South America (Argentina, Brazil, Chile, Uruguay) (Osakabe et al. 2000). This species has been reported in Iran by Khalil Manesh in 1975 (Khanjani 2004; Khanjani and Hadad-Iraninejad 2009).

9.2 Materials and Methods

9.2.1 Biology and Seasonal Fluctuation

For inspection of biological cycle of brown mite in the orchards, a sampling was done between 2010 and 2011 in two different areas: one mountain zone, Heydareye Ghazi-Khan, and foot of a mountain, Heydareye and Abbas-Abad in Hamedan. After selecting a sampling zone, 20 trees infected by brown mite were marked, and sampling was done twice per week. Sampling units were branches with 30–40 cm length and 8–12 thickness. In this inspection 12 branches (40 cm) from four sides and three levels of a tree were cut and placed in the paper bag with data such as the zone of sampling and the date written on it, and then they were delivered to the Bu-Ali Sina University's Acarology Laboratory. For separation of active and inactive stages, infected branches were plunged in hypochlorite sodium 2 %, and then they were placed in different sifts according to their size; then the content of each one was washed and transferred to Petri dishes which then were observed under stereomicroscope. For fixing the exact time of winter egg hatching and commencement of pest's activity in the new year, the infected branches were cut and sent to the lab, and after observing under stereomicroscope, the exact time of hatching were defined.

9.2.2 Biology in Laboratory

Mites were reared in the lab (25 ± 2 °C, 65 ± 5 %, light period 16:8) with a two-layer tissue in them. A leaf was placed in the Petri dish on a wet tissue. For biological details, larvae were released in the Petri dishes (0.2 m) on a receptacle leaf disc. The life cycle of mites was monitored by changing the leaf disc once in every 3 days. Different trees were used, but the endurance of black cherry leaf was layered so this leaf was used for the rest of the examination. It should be considered that selection of host's type during training plays an important role because two hosts (primitive and lab type) should be proportional.

9.3 Results and Discussion

This species normally occurs on cherry, black cherry and plum trees and is much smaller than other members of the *B . praetiosa*. Not only is body length, width and leg measurements considerably smaller, but the distance between the *c1* setae is only about half or slightly more than that of other species. A rather unusual feature of many specimens is the wide spacing of the duplex setae on tarsus IV. *B. rubrioculus* has propodosomal projection over the rostrum. The peritremes are astomosing distally. The solenidion and proximal tactile setae on tarsus III are approximate and subequal in length; these setae are reversed in position on tarsus IV, are subequal in length and are separated. The life cycle of brown mite includes egg, larvae, protonymph, deutonymph and female adult.

Egg Eggs are small, globular and dark bright red, deposited on leaf undersides. Brown mite eggs hatch in early spring. The eggs hatch in as little as 9.68 ± 0.66 days, depending on temperature, and the newly hatched mites (called larvae) immediately begin to feed.

Larvae Length of body is 243 ± 17 μ; greatest width of body is 214 ± 21 μ. Anterior propodosomal setae are short, slender and 11–13 μ long. Propodosoma coarsely granulates with a few wrinkles. Rest of the body on the dorsal side of the body surface granulates and wrinkles. Newly hatched mites are red but turn green when mites start feeding. The development time of larvae in 60 repeats was 4.12 ± 0.17 days in the laboratory.

Protonymph Length of body is 377 ± 41 μ; greatest width of body is 311 ± 44 μ. Propodosomal lobes are present. Anterior propodosomal setae are short and 8–11 μ long. The remainder of the dorsal setae spatulate, 19–25 μ long. The development time of protonymph in 34 repeats was 4.18 ± 0.21 days in laboratory.

Deutonymph Length of body is 430 ± 57 μ; greatest width of body is 347 ± 38 μ. Propodosomal lobes are strongly developed. Anterior propodosomal setae are 14–19 μ long. The remainder of the dorsal body spatulates, 19–25 μ long. The development time of protonymph in 30 repeats was 4.62 ± 0.13 days in laboratory.

Adult Female Length of body is 652 ± 77 μ; greatest width of body is 491 ± 45 μ. Propodosomal lobes are with a basal width of 115 ± 14 μ; height of outer lobes is 40 ± 8 μ; and height of median lobes is 45 ± 14 μ. The dorsal body surface granulates and wrinkles. Adult mites vary in colour from brownish green to reddish brown. The legs are orange with the front legs approximately twice the length of other legs.

This mite has three resting stages. They are nymphochrysalis (larvae), deutochrysalis (protonymph) and teliochrysalis (deutonymph). The development time of these resting stages in 56, 28 and 21 repeats is as follows: 4.74 ± 0.19, 3.88 ± 0.17 and 5.00 ± 0.22 days in the laboratory. Also, the period of preoviposition, oviposition and postoviposition in 22, 32 and 17 repeats is as follows: 9.18 ± 0.65, 12.56 ± 0.50 and 45.32 ± 0.81 days.

Damage Brown mites feed on the upper surface of the leaf by piercing cells and sucking the contents out. This feeding on young leaves shows up as whitish-grey spots resulting in a stipple appearance when the leaves grow. Attack on newly emerged leaves can result in discolouration of leaves when they fail to grow.

Biology Brown mites feed only during the cooler parts of the days and migrate off the leaves during midday. They are not active during hotter periods such as summer. This mite has two types of eggs: summer eggs and winter eggs. Winter eggs are on the 1-year branches but summer eggs are on the leaves, although they are similar in morphological character (Herbert 1962; Meyer 1987). Brown mite in all generation leaves summer eggs on the leaves, but on the bark branch, spores, especially junction part of spores with branches, they leave two types of eggs. In winter the mite undergoes aestivation as an egg on the 2nd or 3rd year branches about to sprout; conjunction of main and offshoots in plum, cherry and black cherry trees; or on *Chionaspis asiatica* (Herbert 1962; Keshavarze-Jamshidian 2004; Khanjani and Haddad 2009; Eghbalian 2007). Host type is effective on quality and number of generations of this mite. When it is on fruit trees, the period of living for one generation will be shorter, while on stone fruit, this period is longer. Of course, temperature has particular effect on this period; the lower the temperature, the sooner will be the growth. The maximum effect was during larval period, protonymph and deutonymph before and after hatching. Mite's strain is effective on the rate of growth especially hatching. Morgan and Anderson (1957) reported that 90 % eggs of brown mite on apple tree at 23 °C are hatched but Herbert (1962) reported 26 %. But in the current study, (25 °C) more than 90 % eggs hatched. At 25 °C, it takes shorter time for hatching and more eggs would be hatched. At 25 °C also, when the mites were offered leaves of apple, maximum number of eggs hatched. Leaves of stone trees are not parallel in all zones of Hamedan, for example, in mountainous and cold condition like Heydareye Ghazi-Khan, they take 20–23 days more than in plain areas. Accordingly eggs also hatch 20 days later in Abbas-Abad orchards. This delay caused changes in mite's generation period in different zones. Generally, this mite has five generations in plain areas and four generations in mountainous areas in Hamedan. Khanjani and Haddad (2009) also announced four generations in a year on plum and black cherry in Hamedan.

Morgan and Anderson (1957) observed at least four generations on apple in British Columbia. According to their report, this mite produces three generations in apple orchards of Nova Scotia but the third generation is not perfect. They recorded seasonal fluctuation of brown mite between 1960 and 1961. All the growth stages of mite completed their duration 1 or 2 weeks earlier in 1960 than in 1961. The reason may be the increase in average temperature by 5 °C in 1960.

Winter eggs begin to hatch when average temperature during the last 2 weeks became 8.13 ± 2.54 °C, and a peak for hatching was on the 20th of April (Fig. 9.1). At this time, average temperature was 14.9 °C and optimum threshold was 15 °C, and pest's larvae hatch in next 10 days after 30th of March. The maximum population occurred in the second 10 days of September. The peak in population of first active generation occurred at the beginning of April. The period before oviposition took a long time about 1–10 days for which high temperature can be the reason, and then oviposition occurred in 12–14 days gradually. In Hamedan, eggs hatched after 10 days and resulted in second-generation larva damaging fruit trees. A peak in population of second-generation larvae was in the first 10 days of May, third generation second 10 days of June, fourth generation in July and the fifth generation at the end of September or at the beginning of October. The fluctuations of the stages protonymph, deutonymph and adult are as depicted in Figs. 9.2, 9.3 and 9.4.

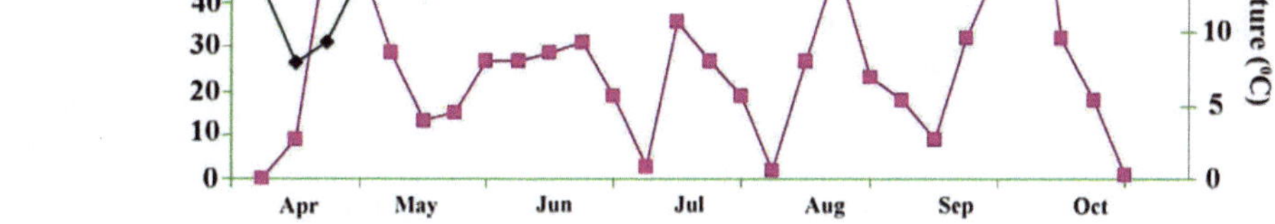

Fig. 9.1 Seasonal fluctuation of *Bryobia rubrioculus* (larva) in Hamedan cherry orchards (2010)

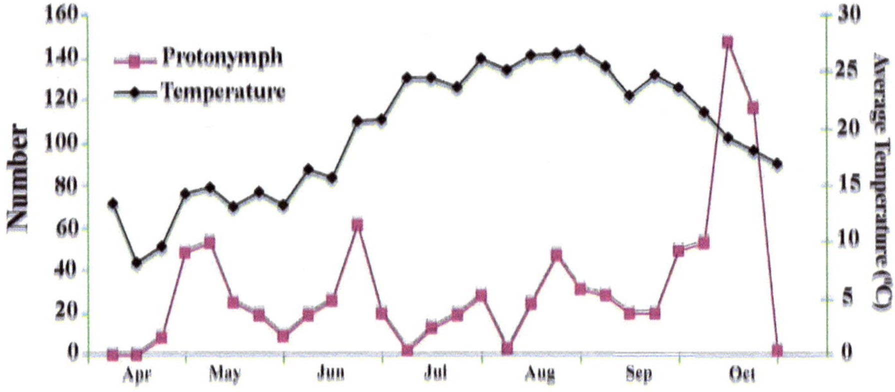

Fig. 9.2 Seasonal fluctuations of *B. rubrioculus* (protonymph) in Hamedan cherry orchards

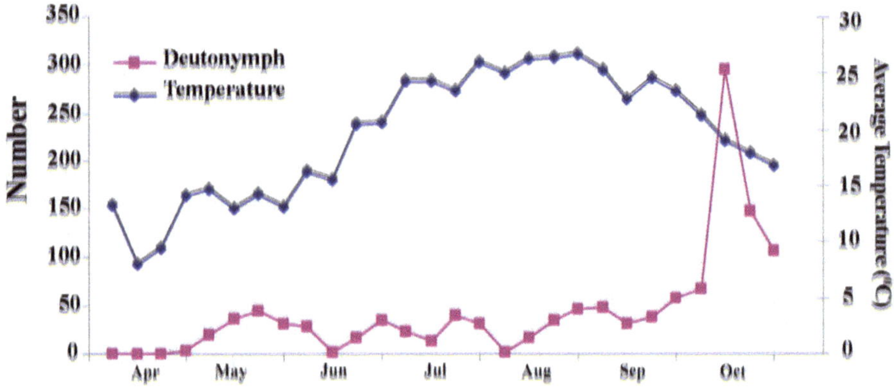

Fig. 9.3 Seasonal fluctuations of *Bryobia rubrioculus* (deutonymph) in Hamedan cherry orchards

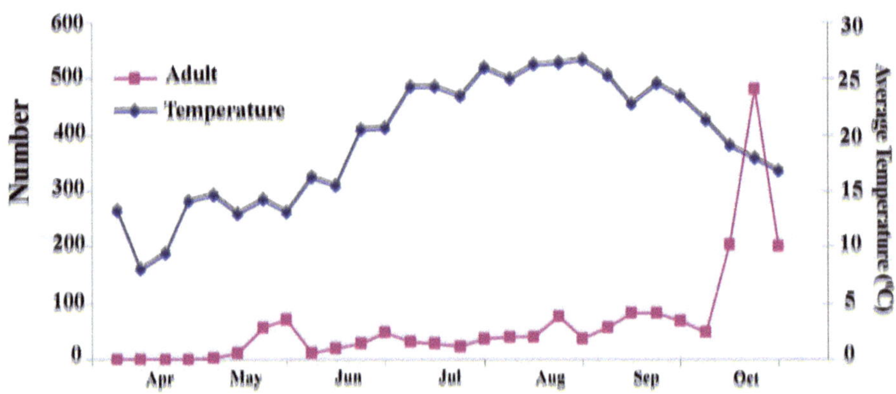

Fig. 9.4 Seasonal fluctuation of *Bryobia rubrioculus* (adult) in Hamedan cherry orchards

References

Carmona MM (1992) Phytophagous and predatory mites of Madeira-II. Baletin de Sanidcid Vegetal Plagas 18(2):469–482

Charles JG (1998) The settlement of fruit crop arthropod pests and their natural enemies in New Zealand: a historical guide to the future. Biocont News Inform 19(2):47–58

Eghbalian AH (2007) Study on biology and natural enemies of brown mite, *Bryobia rubrioculus* scheuten (Acari: Tetranychidae) in plum, sweet cherry and black cherry orchards in Hamedan. MSc thesis, Urmia University, pp 96

Herbert HJ (1962) Life history and habits of the brown mile, *Bryobia arborea* M. and A. (Acarina: Tetranychidae) on apples in Nova Scotia. Can Entomol 94:934–941

Keshavarze-Jamshidian M (2004) Study on biology of *Bryobia rubrioculus* (Scheuten) (Acari: Tetranychidae) and identification of their natural enemies in Karaj. MSc thesis, Esfahan University, pp 71

Khanjani M (2004) Study on mites of fruit trees and their natural enemies in west of Iran. Final report of Research Project, Bu-Ali Sina University Persian, pp 44

Khanjani M, Haddad Irani-Nejad K (2009) Injurious mites of agricultural crops in Iran, 2nd edn. Bu-Ali Sina University Persian, pp 526

Meyer MKPS (1987) African tetranychidae (Acari: Prostigmata) with references to the world genera. Entomol Mem Deep Agric Tech Serv Repu S Afr 69:1–175

Momen FM (1987) The mite fauna of an unsprayed apple orchard in Ireland. Zeit Ange Zoo 74(4):417–431

Morgan CVG, Anderson NH (1957) *Bryobia arborea* n. sp. and morphological characters distinguishing it from *B. praetiosa* Koch (Acarina: Tetranychidae). Can Entomol 89:486–490

Osakabe M, Ehara S, Adhikari S (2000) Damage to young leaves of pear trees by *Bryobia rubrioculus* (Scheute) (Acari: Tetranychidae) in Nepal. J Acarol Soc Jpn 9(1):15–22

Coconut Water as a Promising Culture Media for *Hirsutella thompsonii* Fisher, a Pathogen of Coconut Mite

10

Chandrika Mohan, M.K. Radhika, A. Josephrajkumar, and P. Rajan

Abstract

Coconut eriophyid mite *Aceria guerreronis* Keifer has emerged as a major pest of coconut in the country since 1998. Among the various biocontrol agents, *Hirsutella thompsonii* is a promising fungal pathogen used in field biosuppression of the pest. Influence of various synthetic and laboratory-derived growth media, viz., Potato Dextrose Agar (PDA), Sabouraud Dextrose Agar (SDA), Glucose Yeast Extract Agar (GYA), and Carrot Agar (CA), was compared with a locally available inexpensive substrate, coconut water on growth, and sporulation of *H. thompsonii*. Fungal growth characters, viz., radial growth, conidial production, and micromorphometry of the fungal mycelium and spore, were recorded. Results indicated comparable growth in coconut water media (1.91 cm/20 days) to that of PDA (1.94 cm/20 days). In solid media, laboratory-derived coconut water agar showed significantly higher spore count (12.9×10^4spores/cm^3) followed by laboratory-derived PDA (11.38×10^4spores/cm^3). Regarding conidial production in broth culture, maximum spore production of *H. thompsonii* was observed in Sabouraud Dextrose broth (51.2×10^4spores/cm^3). Coconut water showed comparable spore count (18.2×10^4spores/cm^3) to that of Potato Dextrose broth (19.6×10^4spores/cm^3) and Glucose Yeast Extract broth (18.6×10^4spores/cm^3). There was no significant difference in micromorphometric characters among the media tested. The results suggest the potential of using coconut water for mass production of *H. thompsonii* in the laboratory.

C. Mohan (✉) • M.K. Radhika • A. Josephrajkumar • P. Rajan
Entomology Section, ICAR-Central Plantation Crops Research Institute,
Regional Station, Kayamkulam, Post. Krishnapuram,
Alappuzha 690 533, Kerala, India
e-mail: cmcpcri@gmail.com

A.K. Chakravarthy, S. Sridhara (eds.), *Arthropod Diversity and Conservation in the Tropics and Sub-tropics*, DOI 10.1007/978-981-10-1518-2_10

Keywords
Coconut mite • Coconut water • Fungal growth media • *Hirsutella thompsonii*

10.1 Introduction

The acaropathogenic fungus, *Hirsutella thompsonii* Fisher, is a facultative pathogen of many phytophagous mites especially those belonging to family Eriophyidae. The geographic distribution of the fungus is widespread (Brady 1979). The isolation of the fungal pathogen *H. thompsonii* from citrus rust mite, *Phyllocoptruta oleivora* (Ashmead), and its successful cultivation on an artificial medium (McCoy and Kanavel 1969) introduced the possibility of large-scale production of the fungus. Promising results are reported on the efficiency of *H. thompsonii* to control citrus red mite in the USA, Surinam, Israel, and China (McCoy 1996; Brussel 1975). Biocontrol of the coconut mite, *Aceria guerreronis*, using *H. thompsonii* is being investigated in various countries where eriophyid mite is a major problem of coconut palm (Hall et al. 1980; Espinosa Becerril and Carrillo-Sanchez 1986; Lampedro and Luis 1989; Moore and Howard 1996; Cabrera 2002). *H. thompsonii* has been isolated from samples of coconut mites from India (Beevi et al. 1999; Nair et al. 2005), Sri Lanka (CFC/DFID/APCC/FAO 2004), and from samples of *Colomerus novahebridensis* from the New Hebrides, New Guinea, and Sri Lanka (Hall et al. 1980). Mass production of the fungus has to be standardized for assuring a steady supply for area-wide field delivery in mite management. Nutrients in the culture media affect the growth and sporulation of fungus. Faster growth and better germination are important factors for a successful infection. The use of cheaper locally available substrates reduces the cost of production of the fungus. Hence in this paper, the potential of growing *H. thompsonii* in coconut water, a locally available cheap substrate, is investigated and compared with other standard fungal growth media, viz., Sabouraud Dextrose, Glucose Yeast Extract, Potato Dextrose, and Carrot Agar.

10.2 Materials and Methods

10.2.1 Source of Fungal Culture

Culture of *H. thompsonii* Fisher (IMI 394108) maintained in the biocontrol laboratory of ICAR-Central Plantation Crops Research Institute, Regional Station, Kayamkulam (9°8′ NL and 76°30′EL), Alappuzha District, Kerala, was used in the study. The mother culture was subcultured in the laboratory at 28 ± 2 °C and 80 ± 5 % RH in PDA slants so as to obtain adequate culture for the experiment.

10.2.2 Culture Media

Three commercially available media, viz., Potato Dextrose Agar (PDA), Sabouraud Dextrose Agar (SDA), Glucose Yeast Extract Agar (GYA), and laboratory-derived fresh media, viz., PDA and Carrot Agar (CA), were compared with coconut water for growth and sporulation of the fungus in solid media.

10.2.3 Preparation of Solid Media

Commercially available powder forms of the fungal growth media (HiMedia), viz., Potato Dextrose Agar (39.0 g of PDA powder/1000 ml of distilled water), Sabouraud Dextrose Agar (65.0 g/1000 ml distilled water), and Glucose Yeast Extract Agar (28.4 g/1000 ml distilled water), were used for medium preparation. The powder form of respective media was dissolved in 1000 ml distilled water by heating.

Laboratory-derived media, viz., PDA and Carrot Agar, were prepared by following standard protocols (Aneja 2004) using infusion from fresh potatoes and carrot. pH of the media was maintained at 6 ± 0.2. Laboratory-derived coconut water agar was prepared by dissolving agar by boiling in 500 ml coconut water collected from mature coconuts and made up to 1000 ml by addition of fresh coconut water (pH 5.6).

Influence of Media on Radial Growth Twenty-five milliliter of the respective media were poured into presterilized Petri dishes (90 mm) and allowed to solidify. Following this, the Petri plates were inoculated with 5 mm diameter mycelia mat from active-growing region of 2-week-old fungus culture maintained on PDA. The plates were then sealed with Parafilm and incubated at 28 ± 2 °C and 80 ± 5 % RH. Ten replicates were maintained for each treatment. Radial growth was observed as mean of two perpendicular diameters and was expressed in centimeters. Observations were recorded from fourth day onward on alternate days for 20 days. Data was statistically analyzed using ANOVA and means separated using least significant difference.

Comparison of Media for Conidial Production Twenty-day-old Petri dish culture of *H. thompsonii* was used for observation on conidial production. Culture disk of 1 cm diameter was taken from the medium with a sterilized cork borer and introduced into a sterile test tube containing 5 ml of sterile distilled water. A drop of Tween 80 was added and stirred well with a glass rod to dispense the spores. Spore count was taken under microscope with the help of a hemocytometer. Data was analyzed statistically using ANOVA.

Liquid Broth Media for Spore Count and Mycelia Mat Yield Liquid broth media, viz., Potato Dextrose broth, Sabouraud Dextrose broth, and Glucose Yeast Extract broth, were prepared by individually dissolving 24 g (Potato Dextrose), 30 g (Sabouraud Dextrose) broth powder (HiMedia) in 1000 ml distilled water by boiling.

For preparing Glucose Yeast Extract broth, 20 g yeast extract and 20 g dextrose were added to 1000 ml distilled water. Coconut water (1000 ml) collected from mature coconuts was used as media for comparison of growth and sporulation. Each medium (100 ml) was transferred to a 250 ml conical flask plugged with cotton and sterilized at 121 °C for 20 min at 15 lbs pressure. After cooling these flasks were inoculated with 5 mm diameter mycelia mat. Ten replicates were maintained for each treatment. Flasks were incubated at 28 °C on a shaker at 150 rpm for 2 days and then transferred to incubator maintained at 28 ± 2 °C. After 20 days of incubation, conidial production was counted. For this 1 ml Teepol was added to each flask. Flasks were shaken well and spores were counted under microscope using a hemocytometer. Mycelia growth was filtered from the liquid medium on a pre-weighed filter paper (Whatman No. 1) and dried to constant weight at 60 °C. Data was analyzed statistically for comparing treatments following analysis of variance.

Micromorphology of ***H. thompsonii*** *H. thompsonii* was grown in respective media using slide culture technique (Aneja 2004). The slides after 10 days of growth were observed under compound microscope (Leica DMLB). Morphological characters of *H. thompsonii*, viz., size of conidia, diameter of mycelium, and distance between two phialides, were measured using pre-calibrated compound microscope. The observations were recorded for ten replicates for each growth medium.

10.3 Results

Among the various solid media tested, synthetic PDA (HiMedia) supported maximum growth of the fungus at 20 days of inoculation with a radial growth of 1.94 cm. Coconut water in solid medium with agar is on par with synthetic PDA with radial growth of 1.91 cm (Table 10.1). Synthetic SDA medium showed the least radial growth of 1.54 cm compared to other media tested.

Regarding conidial production, coconut water agar showed the highest spore count of $12.9 \times 10^4/cm^3$ followed by laboratory-derived PDA with a spore count of

Table 10.1 Radial growth of *H. thompsonii* in various growth media at 20 days of incubation

Media	Radial growth (cm)
PDA synthetic	1.94
SDA synthetic	1.54
GYA synthetic	1.73
PDA lab derived	1.65
Carrot Agar lab derived	1.79
Coconut water agar lab derived	1.91
CD ($p = 0.05$)	0.24

Table 10.2 Spore count of *H. thompsonii* in various solid growth media

Media	Spore count (spores/cm^3)
PDA synthetic	7.36×10^4
SDA synthetic	10.2×10^4
GYA synthetic	6.54×10^4
PDA lab derived	11.38×10^4
Carrot Agar lab derived	10.48×10^4
Coconut water agar lab derived	12.9×10^4
CD ($p = 0.05$)	2.15

Table 10.3 Spore count and yield of mycelia mat of *H. thompsonii* in various liquid media

Media	Spore count (spores/cm^3)	Mean dry weight (g/100 ml) of *H. thompsonii* in 20 days of incubation
Glucose Yeast Extract broth	18.667×10^4	1.067
Coconut water	18.217×10^4	1.017
Potato Dextrose broth	19.617×10^4	1.045
Sabouraud Dextrose broth	51.183×10^4	1.306
CD ($p = 0.05$)	1.87	NS

11.38×10^4/cm^3 which were on par. There was no significant difference in spore count of *H. thompsonii* grown in SDA and Carrot Agar (Table 10.2).

In liquid broth media, among the four liquid media tested, Sabouraud broth showed highest spore count of 51.18×10^4 spores/cm^3. Coconut water media showed comparable spore count to that of Potato Dextrose broth and Glucose Yeast Extract broth (Table 10.3). Yield of dry mycelia mat (per 100 ml of broth) did not differ significantly among the various broth culture media tested (1.017–1.306 g/100 ml) (Table 10.3).

There was no significant difference in various micromorphology parameters with respect to the media in which the fungus was grown. Spore of *H. thompsonii* was round to spherical in shape, and size varied from 2.652 μ to 3.138 μ grown in various media (Fig. 10.1). Width of mycelia varied from 2.144 μ in lab-derived PDA to 2.365 μ in coconut water. There was no significant difference among the media used with regard to hyphal width (Fig. 10.2). Length of hyphal internodes also did not differ significantly. It varied from 41.71 μ (SDA) to 43.37 μ (GYA) (Fig. 10.3).

10.4 Discussion

In the present study, *H. thompsonii* grew successfully in all the media tested, and there was significant difference in fungal growth among the media. Growth of *H. thompsonii* in coconut water as solid media and liquid media was comparable with other standard growth media. The observation on radial growth of *H. thompsonii* in the present study also agrees with McCoy and Kanavel (1969) who reported the

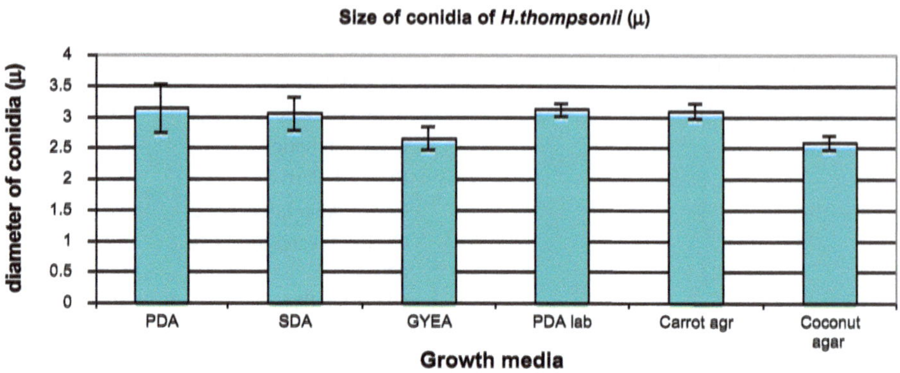

Fig. 10.1 Size of conidia of *H. thompsonii* grown in various media

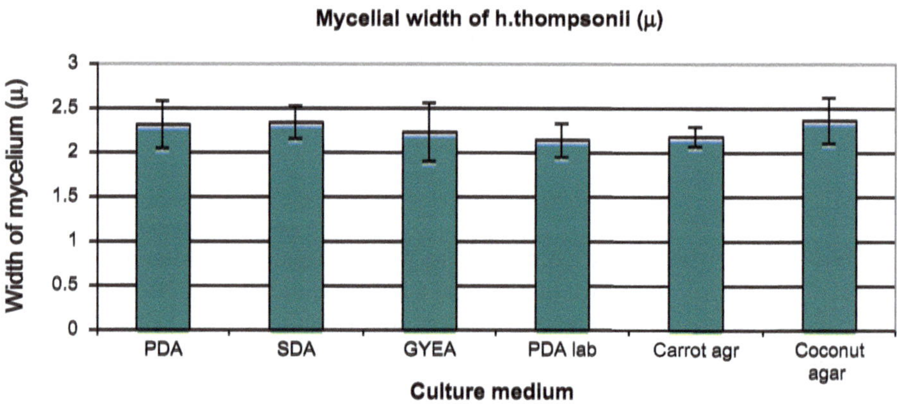

Fig. 10.2 Mycelia width of *H. thompsonii* (mean±SD) grown in various media

highest growth rate of *H. thompsonii* (0.95 mm/day) in PDA followed by v-8 juice agar and modified soil fungus (MSF) agar medium.

Various workers reported that *H. thompsonii* could be cultured on agar media including Potato Dextrose, modified soil fungus medium, Sabouraud Dextrose (McCoy and Kanavel,1969), potato carrot (Kenneth et al. 1979), Sabouraud maltose-peptone grapefruit, and egg yolk (Brussel 1975). MacLeod (1960) found that higher concentration of yeast in a liquid medium inhibited growth of *Hirsutella gigantea.* Growth parameters of *H. thompsonii* on media with different concentrations of carbon and nitrogen sources were reported by McCoy et al. (1978) and Vey et al. (1983).

Aghajanzadah et al. (2006) reported higher mycelia growth of *H. thompsonii* on the synthetic medium (Sabouraud Maltose Broth) than on media prepared using cheap crude and unprocessed materials, viz., Bengal gram, horse gram, red gram,

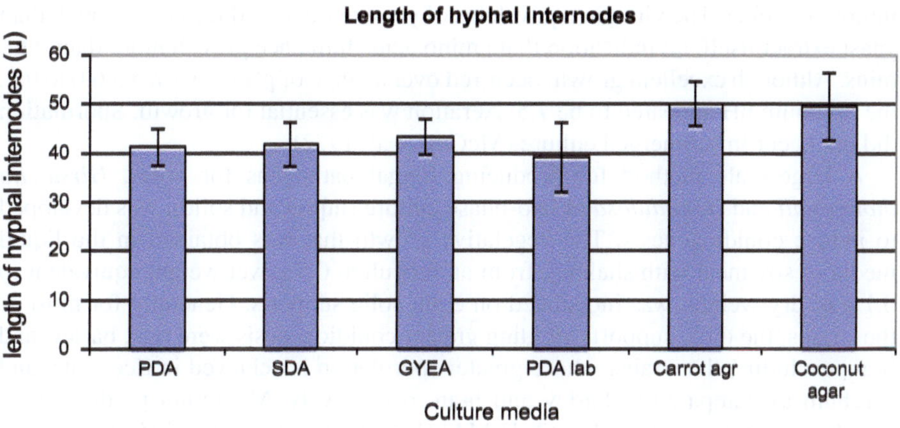

Fig. 10.3 Length of hyphal internodes (mean±SD) of *H. thompsonii* grown in various growth media

maize, barley, sorghum, rice, and wheat, as sources of carbon and nitrogen in different combinations.

Fungal growth media generally are a source of carbon, nitrogen, and vitamins. In general, natural media are based on naturally available materials such as carrot, potatoes, etc., and synthetic or defined media contain precise amounts of carbon, vitamins, and minerals. Both potatoes and coconut water are found to be excellent sources enriched with all common types of carbohydrates, proteins, amino acids, vitamins, and minerals (Aykroyd 1963). In addition coconut water is rich in sugars and minerals along with minor amounts of fat and nitrogenous substances. It also contains protein, calcium, phosphorus, magnesium, iron, and copper in trace amounts (CDB 2012). This unique composition of coconut water with adequate sugar, nitrogen, and minerals could be supporting fungal growth by the easy availability of common metabolites in accessible forms. Previously, mass production of the entomopathogen *Metarhizium anisopliae* infecting coconut rhinoceros beetle using coconut water was reported by Dangar et al. (1991). Coconut water is used as a growth supplement in plant tissue culture/micro-propagation owing to its chemical composition of sugars, vitamins, amino acids, and phytohormones (Jean et al. 2009).

Comparative nutritional studies showed that dextrose and sucrose at an optimum concentration of 5 mg/ml and 10 mg/ml, respectively, were the most effective sources of carbon (produced the greatest increases in weight) among sugars tested for use in the large-scale production of the fungal pathogen *Hirsutella thompsonii* in submerged culture. Inorganic nitrogen in a basal dextrose-salts medium was unsuitable, but a combination of yeast extract and peptone at an optimum concentration of 5 mg/ml and 0.5 mg/ml, respectively, was the most effective source of organic nitrogen. Higher concentrations of yeast extract appeared to have an inhibitory effect. The vitamin equivalents of yeast extract produced less growth than yeast extract itself, an indication that amino acids influence growth more than vitamins. Although excellent growth occurred over a range of pH values from 6.0 to 9.2, the optimum pH appeared to be 7.5. Aeration was essential for growth. Sporulation did not occur in submerged culture (McCoy et al. 1972).

A large-scale method for producing fungal pathogens for mites, *Hirsutella thompsonii* and *H. nodulosa* in two-phase culture (liquid and solid), was developed to induce conidiogenesis. The vegetative growth that was obtained in the liquid media of soy meal with shaking, from an inoculum 0.5 g wet weight equivalent to 0.1 g of dry weight, was inoculated on eight solid supports. Generally for most of the strains, the three supports yielding greater conidiogenesis were rice, barley, and bran, excluding the strains whose greater sporulation is achieved in rice, oats, and sorghum as compared to barley and bran, respectively. Maximum production of conidia was obtained with HtM2, HtM4481, and HtC59 strains of *H. thompsonii*, which reached on solid support 334.75, 269.68, and 137.12×10^7 conidia, respectively (McCoy et al. 1978).The virulence of *Hirsutella thompsonii* (Fisher) to *Brevipalpus phoenicis* (Geijskes) was evaluated in laboratory grown on complete and solid culture media (MC-S), complete and liquid culture media (MC-L), rice (APC), and powdered rice (APC-SM) by Rossi-Zalaf et al. (2008). Adults were confined to arenas prepared with citrus leaves in acrylic dishes containing water agar. Conidial suspensions were prepared at different concentrations (3.2×10^5 to 1×10^7spores/ml) and applied on mites to establish the table curve response on the fourth day. In the field, adults were maintained in arenas prepared with fruits which were placed in plants. In this test, four treatments were tried: *H. thompsonii* cultured on rice (APC) at two concentrations (20 kg/ha and 10 kg/ha), *H. thompsonii* produced by liquid fermentation(MC-L) (5 L/ha), and control (sterile water). Adult survival, number of eggs, and nymphs per fruit were observed 10 and 20 days after the fungus application. The lowest LC_{25}value calculated was from pathogen produced in MC-S (1.9×10^5conidia/ml). The LC_{25}values calculated to APC and APC-SM did not differ statistically. The LC_{25}values to MC-L and MC-S were 1.9×10^6infective cells/ml and 2.2×10^5conidia/ml. In the field, concentration and time to death differed between treatments and control. The applications resulted in reduction of adult survival and number of eggs.

In the present study, coconut water supported higher growth and better sporulation of *H. thompsonii* as evidenced by radial growth, spore production, and yield of dry mycelia mat. The micromorphometry of the hyphae and conidia produced was also on par with measurement for these characters in other standard media. Hence the present study indicates that coconut water is a potential media for laboratory multiplication of *H. thompsonii*.

References

Aghajanzadah S, Mallik B, Chandrasekhar SC (2006) Mass culturing of *Hirsutella thompsonii* Fisher, a fungal pathogen of eriophyid mites. Pak J Biol Sci 9:876–879

Aneja KR (2004) Experiments in microbiology, plant pathology, tissue culture and mushroom production technology, 3rd edn. New Age International Publishers, New Delhi, p 568

Aykroyd WR (1963) The nutritive value of Indian foods and planning of satisfactory diets, revised 6th edn. Gopalan C, Balasubramanian SC (eds). Indian Council of Agricultural Research, New Delhi, p 255

Beevi SP, Beena S, Lyla KR, Varma AS, Mathew MP, Nadarajan L (1999) *Hirsutella thompsonii* var. *synnematosa* Samson, McCoy & O'Donnell on coconut mite *Aceria* (*Eriophyes*) *guerreronis* (Keifer): a new report from India. J Trop Agric 37:91–93

Brady RLK (1979) Report No 608. Commonwealth Agricultural Bureaux, The Cambrian News Ltd, Great Britain, p 2

Brussel Van EW (1975) Interrelations between citrus rust mite, *Hirsutella thompsonii* and grey spot on citrus in Surinam. Landbouwproefst Surinam Agric St Bull 98:43

Cabrera RI (2002) Biological control of the coconut mite Aceria guerreronis (Acari: Eriophyidae) with the fungus Hirsutella thompsonii and its possible integration with other control methods. In: Fernando LCP, de Moraes GJ, Wickramanada IR (eds) International workshop on coconut mite (*Aceria guerreronis*). Proceedings. Coconut Research Institute, Lunuwila, pp 89–103

CDB (2012) Health benefits of tender coconut water – analysis of mature and tender coconut water: http://coconutboard.nic.in

Dangar TK, Geetha L, Jayapal SP, Pillai GB (1991) Mass production of the entomopathogen *Metarhizium anisopliae* in coconut water waste form copra making industry. J Plant Crops 19(1):54–69

Espinosa Becerril A, Carrillo-Sanchez JL (1986) El hongo *Hirsutella thompsonii* Fisher en el control del eriófido del cocotero *Eriophyes guerreronis* (Keifer). Agric Tech 12:319–323

Hall RA, Hussey NW, Mariau D (1980) Results of a survey of biological control agents of the coconut mite *Eriophyes guerreronis*. Oléagineux 35:395–398

Jean WH, Young Liya Ge, Yan Fei Ng, Swee Ngin Tan (2009) The chemical composition and biological properties of coconut (*Cocos nucifera* L.) water. Molecules 14:5144–5164

Kenneth R, Muttah TJ, Gerson U (1979) *Hirsutella thompsonii*, a fungal pathogen of mites two. Biology of the fungus in vitro. Ann Appl Biol 91:21–28

Lampedro L, Luis RJ (1989) Selection of *Hirsutella thompsonii* Fisher strains to fight coconut mite *Eriophyes guerreronis* Keifer. 1. Pathogenecity experiments. Rev Mex Micol 5:225–231

MacLeod DM (1960) Nutritional studies of the genus *Hirsutella*. III. Acid-hydrolysed casein and amino acid combinations as sources of nitrogen. J Insect Pathol 2:139–146

McCoy CW (1996) Pathogens of eriophyid mites. In: Lindquist EE, Sabelis MW, Bruin J (eds) Eriophyid mites, their biology, natural enemies and control. Elsevier, Amsterdam, pp 481–490

McCoy CW, Kanavel RF (1969) Isolation of *Hirsutella thompsonii* from citrus rust mite *Phyllocoptruta oleivora* and its cultivation on various synthetic media. J Invertebr Pathol 14:386–390

McCoy CM, Hill AJ, Kanavel RF (1972) A liquid medium for the large-scale production of *Hirsutella thompsonii* in submerged culture. J Invertebr 19(3):370–374

McCoy CW, Hill AJ, Kanavel RF (1978) Large-scale production of the fungal pathogen *Hirsutella thompsonii* in submerged culture and its formulation for application in field. Entomophaga 20(3):229–240

Moore D, Howard FW (1996) Coconuts. In: Lindquist EE, Sabelis MW, Bruin J (eds) Eriophyoid mites: their biology, natural enemies and control. Elsevier, Amsterdam, pp 561–570

Nair CPR, Rajan P, Chandrika M (2005) Coconut eriophyid mite *Aceria guerreronis* Keifer – an overview. Indian J Plant Protect 33(1):1–10

Rossi-Zalaf LS, Sergio BA, Solange VA (2008) Effect of culture media on virulence of Hirsutella thompsonii (Fischer) (Deuteromycetes) to control Brevipalpus phoenicis (Geijskes) (Acari: Tenuipalpidae). Neotrop Entomol 37(3):312–320

Vey A, Quiot JM, Mazet I, McCoy CW (1983) Toxicity and pathology of crude broth filtrate produced by *Hirsutella thompsonii* var. *thompsonii* in shake culture. J Invertebr Pathol 61:131–137

Cladoceran Diversity, Distribution and Ecological Significance

11

M. Karuthapandi and D.V. Rao

Abstract

Cladocerans invariably constitute a dominant component of limnological and zooplankton systems. Cladocerans have a key role in aquatic food chain and contribute to secondary production in aquatic ecosystems. Substantially cladocerans represent a very old group of Palaeozoic origin and currently 620 species are reported globally. In India so far nearly 130 species are recorded, of which the highest taxon was documented from Northeastern India. Cladocerans have been distributed in all the biogeographic regions with more endemic species in the Australasia. *Indialona ganapati* is the only endemic taxa, restricted to central India. Cladocerans are an important group for biomonitoring, and the group has been used as indicator as well as test organisms for estimation of pesticide toxicity levels and other environmental pollutants. They are also one of the major natural food sources for the fishes, especially fry and fingerlings.

Keywords

Cladoceran • Species diversity • Importance • Limnology

11.1 Introduction

Cladocerans are also called as water fleas and are minute crustaceans commonly varying from 0.2 to 6 mm. Most cladocerans are filter-feeders, straining out minute organic particles like bacteria, detritus and algae. They are present in all types of water bodies and it inhabits pelagic, littoral and benthic zones. They are a vital

M. Karuthapandi (✉) • D.V. Rao
Freshwater Biology Regional Centre, Zoological Survey of India,
Hyderabad 500048 Andhra Pradesh, India
e-mail: kpandi83@gmail.com

A.K. Chakravarthy, S. Sridhara (eds.), *Arthropod Diversity and Conservation in the Tropics and Sub-tropics*, DOI 10.1007/978-981-10-1518-2_11

constituent component of zooplankton community and particularly freshwater monophyletic group of the micro-crustacean zooplankton. They are major food source of many freshwater organisms like fish and other crustaceans and play an important part in the food chain, trophic levels and energy transfer in the ecosystem. They have been used as indicator as well as test organisms for estimation of toxicity levels of pesticides and other environmental pollution.

They are minute, and the trunk and appendages of most cladoceran are enclosed in bivalve carapace. The eye and ocellus are usually present. Some cladocerans develop conspicuous head and tail spines, helmet or 'neck-teeth'. Interestingly the body is not segmented and bears a folded carapace. Sexual reproduction results in dormant eggs. These eggs can be carried along by wind to far-off places and hatch under favourable conditions allowing many species to have cosmopolitan distribution. Except for a few species, the life cycle is by asexual reproduction, with occasional periods of sexual reproduction, and this phenomenon is known as cyclical parthenogenesis.

11.2 Diversity

Globally, 620 species of cladocerans are reported from biogeographic regions, viz. Palaearctic (245), Nearctic (189), Neotropical (186), Afrotropical (134), Oriental (107), Australasian (158), Pacific Oceanic Islands (33) and Antarctic (12). Among the global cladoceran faunal diversity, order Anomopoda contains 537 species, of which 43 % constitutes Chydoridae and 19.5 % Daphniidae (Table 11.1). The families Chydoridae, Daphniidae, Ilyocryptidae and Sididae have been studied more than other families. The largest number of documented species is from Europe, North America, Australia and South America and the smallest number from Africa and Southern Asia (Forro et al. 2008). A high diversity of cladocerans has been detected in the littoral zone of stagnant waters, as well as in freshwater aquatic ecosystems.

About 137 species have been recorded so far from India (Chatterjee et al. 2013), out of which nearly a hundred species have been well studied by Michael and Sharma (1988). Systematic investigations on Indian freshwater Cladocera was first conducted by Baird (1859), followed by several workers, viz. Biswas (1964a, b, 1965, 1971), Das and Akhtar (1970), Patil (1976), Nasar (1977) and Qadri and Yousuf (1977). Cladoceran of Rajasthan was studied by Biswas (1971), Nayar (1971) and Venkataraman (1990); Tamil Nadu by Michael (1973), Venkataraman and Krishnaswamy (1984) and Venkataraman (1983); Little Andaman by Venkataraman (1991); West Bengal by Venkataraman (1998); Nilgiri Biosphere Reserve by Raghunathan and Rane (2001); Damodar River by Venkataraman (2003); Melghat Tiger Reserve by Rane (2005); Kashmir by Siraj et al. (2006) and Sharma and Chandrakiran (2011); Northeast India by Sharma and Sharma (2009a, 2010, 2012); Maharashtra by Padhye and Kotov (2010); and Madhya Pradesh including Chhattisgarh by Rane (2011). Fernando (1980a, b) recorded 61 species of Cladocera from the Indian subcontinent with detailed knowledge on the absence of large Cladocera. Sharma and Michael (1987) identified 87 species, and later Michael

Table 11.1 Cladoceran diversity in the world, oriental and Indian regions

	India	Oriental	World
Order Anomopoda	114	89	537
Family Daphniidae	27	17	121
Family Moinidae	8	3	29
Family Dumontiidae	0	0	1
Family Ilyocryptidae	3	5	28
Family Bosminidae	5	4	14
Family Acantholeberidae	0	0	1
Family Ophryoxidae	0	0	3
Family Macrothricidae	8	12	60
Family Neothricidae	0	0	3
Family Eurycercidae	1	1	8
Family Chydoridae	62	48	269
Order Ctenopoda	16	15	50
Family Holopediidae	1	0	3
Family Sididae	15	15	47
Order Haptopoda	1	1	1
Family Leptodoridae	1	0	1
Order Onychopoda	5	1	32
Family Polyphemidae	1	1	2
Family Podonidae	4	0	17
Family Cercopagidae*	0	0	13
Total	**137**	**107**	**620**

*Indicates Invasive species not considered

and Sharma (1988) documented 93 species from India, of which over 60 species are from tropical and subtropical region and 15–20 species from higher elevated lakes and latitudes in the north.

Raghunathan (1989) recorded 106 species and Sharma (1991b) listed 109 species which was confirmed later by Murugan et al. (1998). Among the 109 species of Cladocera in India, three species were synonymised by Orlova (1998). Forty-nine species of Cladocera were reported from Bihar; among these 29 species are new records to Bihar and two species are new records to India (Sharma 2001). Detailed studies of 190 species belonging to 49 genera and 10 families of cladoceran have been recorded from India, of which 18 species are endemic to India (Raghunathan and Kumar 2002). According to the annotated checklist of Indian Cladocera by Chatterjee et al. (2013), only 130 species are valid from India. But many records remain doubtful because Indian literature abounds with checklists riddled with identification errors. In India, the cladoceran fauna from the northeast is the best documented (Sharma and Sharma 1990, 2007, 2008, 2009a, b, 2011), with more than 60 species identified. Cladoceran species richness and abundance often may be influenced by the abiotic and biotic factors. The richest cladoceran faunal diversity was recorded in Deepor Beel, Assam, India (Sharma and Sharma 2009b).

11.3 Distribution

The species richness of cladocerans as a whole is projected to four times more than what is known to science in the world (Fig. 11.1). This is projected from molecular studies. Detailed studies, combining morphological analyses and molecular tools, are especially promising for delineating species boundaries in groups with relatively uniform morphology, fewer qualitative characters and widespread phenotypic plasticity (Forro et al. 2008). Among other zoogeographical regions, Australasia is rich in endemics, represented by one family, one subfamily, one tribe, 11 genera and more than 83 species, while known endemics in Oriental and Neotropical regions are of a lower rank or fewer (one tribe, one genus and more than 21 species and three genera and more than 98 species, respectively). The Afrotropical region, though poorly studied, seems to be especially deprived of known higher-level endemic cladoceran taxa, being represented by a single endemic genus and more than 24 endemic species (Forro et al. 2008).

The equatorial region has few *Daphnia* species (Fig. 11.2), all belonging to the subgenus *Ctenodaphnia*, while the more northern parts have more species of *Daphnia*. The limnetic Cladocera lacks the carnivorous Polyphemidae and Leptodoridae at lower latitudes (equatorial). The common limnetic species of the equatorial region are eurytopic and extend throughout the subcontinent (Fernando and Kanduru 1984). *Leydigia acanthocercoides* inhabits aquatic weeds in polluted ponds (Alam and Khan 1998). Venkataraman (2003) reported that *Ceriodaphnia cornuta*, *Moina micrura*, *Macrothrix spinosa* and *Chydorus barroisi* can be used as indicators of pollution. Daphnids and chydorids are contributing to higher diversity in the water bodies of India (Sharma and Michael 1987). *Indialona ganapati*, the endemic taxa, is restricted to central India. Nearly, 60–65 cladoceran species are found in tropical and subtropical regions in India (Fernando and Kanduru 1984; Sharma and Michael 1987). The highest cladoceran species (58) were recorded in

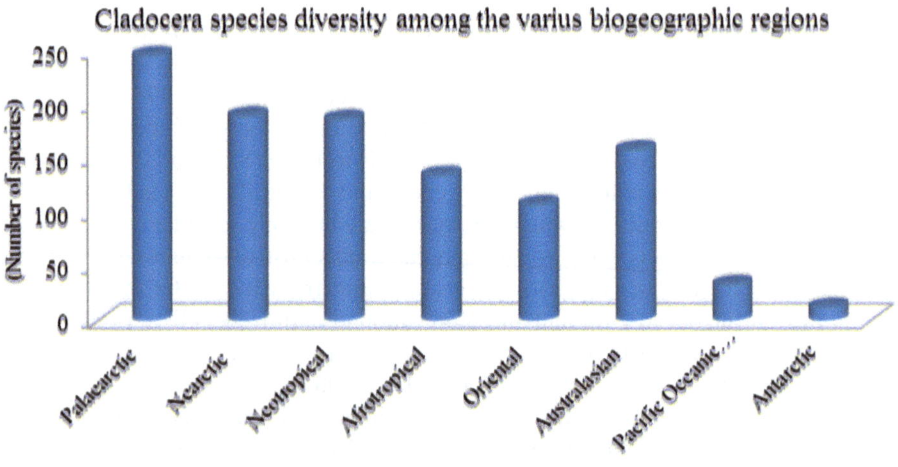

Fig. 11.1 Distribution of Cladocera species from various biogeographic regions

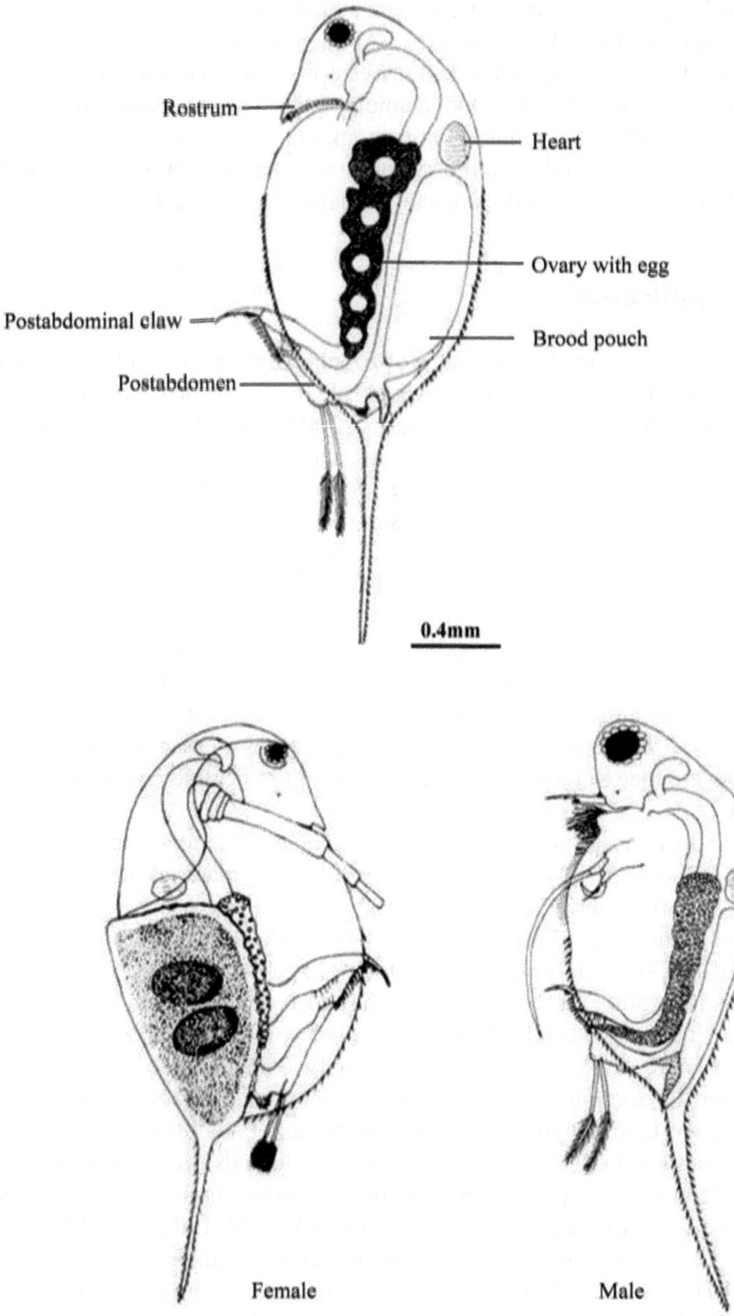

Fig. 11.2 Morphological features of Cladocera *Daphnia* sp.

and around Deepor Beel, a Ramsar Site and a biodiversity hot spot of global importance in India (Sharma and Sharma 2013). About 51 Cladocera species belong to six families that have been recorded from Western Ghats and surrounding areas of Maharashtra and Goa (Padhye and Dumont 2015). Sharma and Sharma (2014) reported that one should not highlight on comparing Cladocerans with poorly documented habitats/patches because the sampling may be incomplete or species inventors may be lacking. Species documentation depends on expertise also.

11.4 Significance

Cladocerans are often the target groups of zooplankton studies; only limited reports are available on their ecology, diversity and role in aquatic productivity in freshwater environs in India (Sharma and Sharma 2009a). Cladocerans are one of the important groups for biomonitoring studies and are an important part of trophic cascades of the aquatic system and highly responsive to pollutants; it can even react to very low concentration of contaminants (Ferdous and Muktadir 2009; Sharma and Chandrakiran 2011). Cladocera have been used as indicators as well as test organisms for estimation of toxicity levels of pesticides and other environmental pollutants (Frear and Boyd 1967; Muirhead-Thompson 1971; Canton and Adema 1978). According to Forro et al. (2008), cladoceran has immense direct and indirect impact in economic front on fish food and phytoplankton. Besides, a high diversity of the Cladocera can be found in the littoral zone of stagnant and ephemeral aquatic bodies. These habitats are often adversely affected by human activities and especially in the absence of temporary water may result in decline of diversity and local extinction.

Large cladocerans, especially *Daphnia*, feed on a wide variety of phytoplanktons and other suspended matters, such as decayed plant material and clay particles. They may greatly reduce phytoplankton abundance. There are several genera of carnivorous cladocerans. It occurs normally from spring to early summer, reaching densities of 10–30 animals per litre in ponds, lakes and reservoirs. In a special case, a high density of 500 cladocerans per litre has been reported from a waste stabilisation pond (Stemberger et al. 2001). Large cladocerans and calanoid copepods in general are more sensitive to pesticide toxicity than microzooplankton, such as *Bosmina* and *Ceriodaphnia*. Cladocerans are an abundant group of animals in the invertebrate world linked to macrophytes (Rocha and Por 1988). Cladocerans are an important component of most freshwater lakes and ponds. In addition to serving an important food item for planktivorous fishes and invertebrates, they are grazers on smaller organisms like algae and detritus (Pennak 1978; Balayla and Moss 2004). They play a crucial role in the recycling of nutrients in aquatic ecosystems (Hudson et al. 1999; Urabe et al. 2002).Because of their intermediate tropic position, these minute animals have a vital part in the energy through aquatic food web as well as in regulating the transfer of contaminants and pollutants to higher trophic levels (Hall et al. 1997) in aquatic ecosystems. The integral role of cladocerans in

Table 11.2 Monthly densities of cladocerans (No/m^3) in Guntur pond during October 2006 to March 2007 (Gulam Mohideen et al. 2008)

S. no.	Species	October 2006	November 2006	December 2006	January 2007	February 2007	March 2007	Mean ± SD
1	*Daphnia magna*	18	22	24	18	16	20	20 ± 2.83
2	*Daphnia similis*	18	22	20	16	18	14	18 ± 2.38
3	*Ceriodaphnia cornuta*	46	48	52	45	47	44	47 ± 2.83
4	*Ceriodaphnia reticulate*	36	32	35	38	33	36	35 ± 2.19
5	*Moina micrura*	40	45	43	36	34	30	38 ± 5.69
6	*Macrothrix goeldi*	16	14	11	15	12	10	13 ± 2.37
7	*Macrothrix hirusticornis*	20	18	14	22	18	16	18 ± 2.83
8	*Chydorus ciliatus*	12	10	16	14	10	12	12 ± 2.34
9	*Leygidia acanthocercoides*	24	22	28	30	22	24	25 ± 3.29
10	*Alona pulchella*	18	16	20	14	16	12	16 ± 2.83

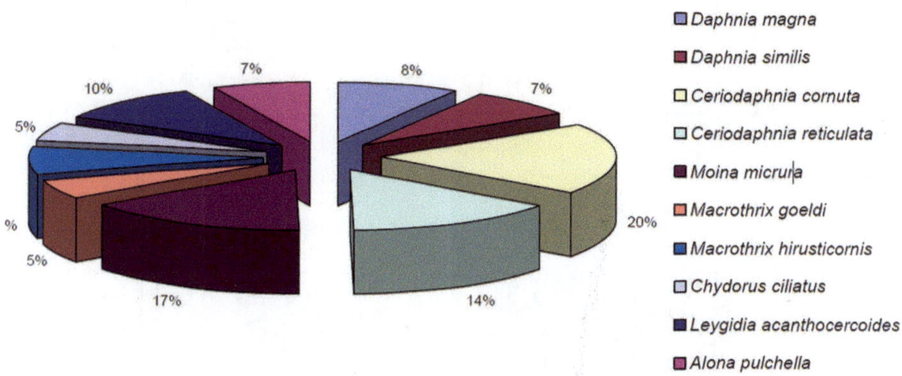

Fig. 11.3 Percentage composition of cladocerans in Guntur pond, Tiruchirappalli (Gulam Mohideen et al. 2008)

freshwater ecosystem make them useful ecological and palaeolimnological indicators (Stemberger et al. 2001; Jeppesen et al. 2011; Korhola and Rautio 2001).

The role of cladocerans as live food is gaining importance due to the scarcity of *Artemia* cyst. The cladoceran neonates of different sizes can be used as live food to feed economically important animals such as shrimps and fishes. So it is necessary to identify suitable species for mass culture (Table 11.2 and Figs. 11.3, 11.4a, 11.4b, and 11.4c; Gulam Mohideen et al. 2008). Mustaq (1990) found that certain cladoceran species breed well in polluted waters and can serve as biological indicators of water pollution. Sharma and Sharma (2008, 2009b) conducted studies on these fish

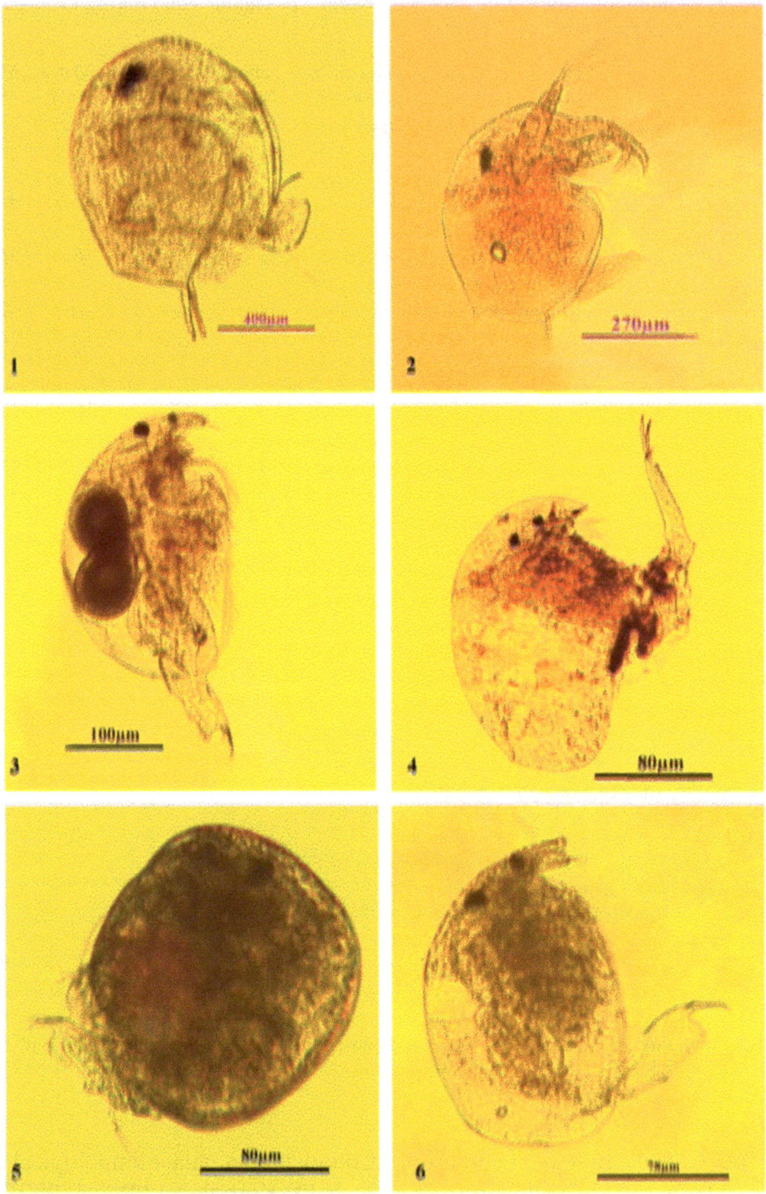

1. *Bosmina (Bosmina) longirostris* (O. F Muller, 1776)
2. *Bosminopsis deitersi* Richard, 1895 3. *Alona* sp. 4. *Camptocercus* sp.
5. *Chdorus sphericus* (O. F Muller, 1776)
6. *Coronatella rectangula rectangula* (Sars, 1862)

Fig. 11.4a The microscopic views of the cladocerans identified during the study

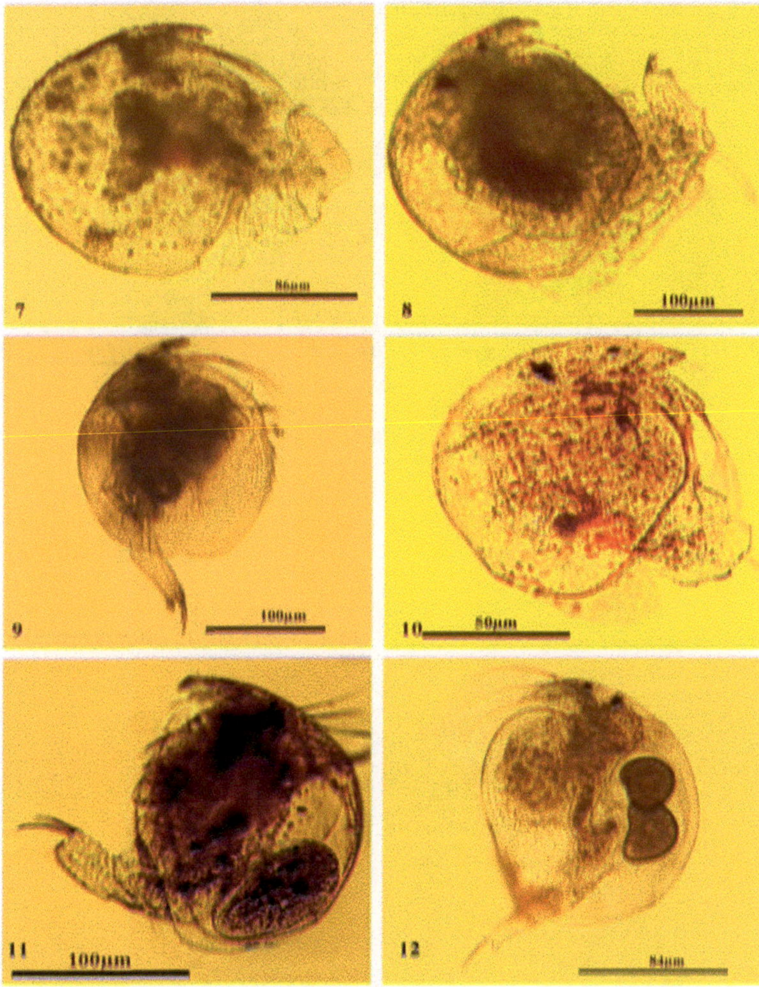

7. *Dunhevedia crassa* King, 1853 8. *Ephemeroporus barroisi* (Richard, 1894)
9. *Euryalona orientalis* (Daday, 1898) 10. *Indialona ganapati* Petkovski, 1966
11. *Karualona karua* (King, 1853) 12. *Kurzia longirostris* (Daday, 1898)

Fig. 11.4b The microscopic views of the cladocerans identified during the study

food organisms in aquatic biotopes of conservation importance in Northeastern India. They also (Sharma and Sharma 2011) carried out studies on alpha diversity of cladoceran in Meghalaya, where biodiversity is poorly documented. Ariane et al. (2015) worked on community of Cladocera in a tropical hyper-neurotrophic structure in Brazil. They reported low species richness of the Cladocera community at eutrophic reservoir with high values of total phosphorus concentration in Danish lakes. The dense cyanobacterial blooms recorded continuously at Garcas Reservoir led to the replacement of large cladocerans like *Daphnia gessneri* for small cladocerans like *Bosmina huaroensis*. Such replacements have also been recorded in a

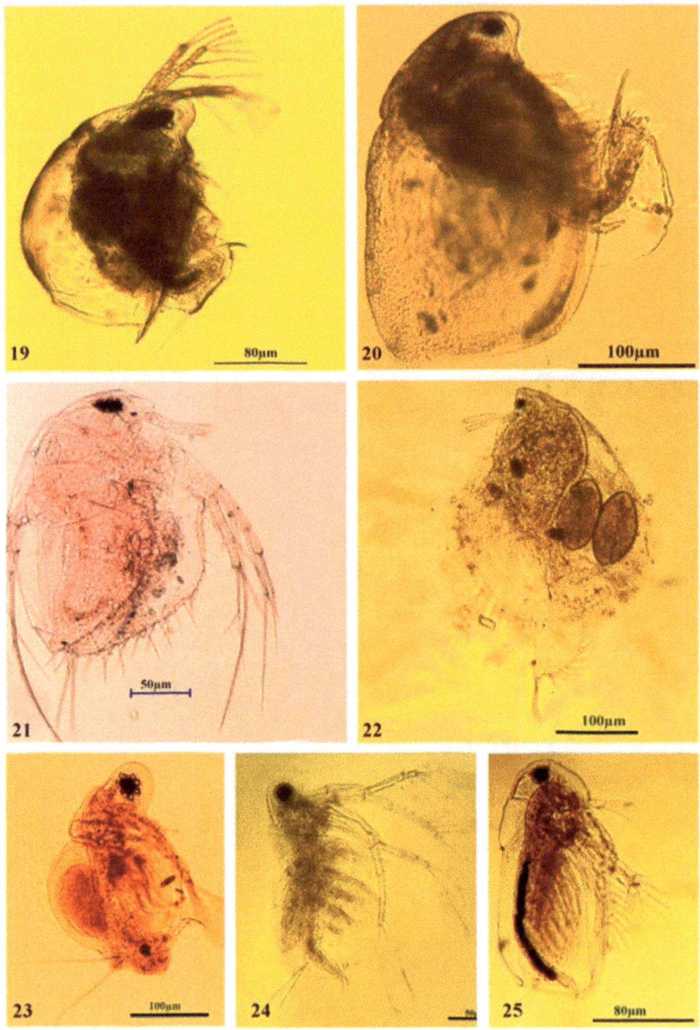

19. *Scapholeberis kingii* Sars, 1888 **20.** *Simocephalus* sp. **21.** *Macrothrix spinosa* King, 1853
22. *Ilyocryptus spinifer* Herrick, 1882 **23** *Moina* sp. **24.** *Diaphanosoma excisum* Sars, 1885
25. *Latonopsis australis* Sars, 1888.

Fig. 11.4c The microscopic views of the cladocerans identified during the study

hyper-neurotrophic water body in Argentina (Echaniz et al. 2008). In eutrophic lakes, large cladocerans (e.g. *Daphnia*) disappear because they have a large carapace opening which allows entry of cyanobacterial filaments in the filtering chamber, thus favouring action of toxins present in cyanobacteria. In smaller cladocerans, the carapace opening is too small for filaments and colonies of cyanobacteria to enter, and toxins cannot harm the Cladocera. Small cladocerans like *Bosmina* have a distinct mode of feeding. *Bosmina* can also devour bacteria and detritus (Demott and Kertoot 1982). Studies on dynamics of Cladocera and other organisms contribute to understanding of zooplankton community as an index to assess aquatic ecosystems in tropical and subtropical areas.

Knowledge on cladoceran diversity is still at infant stage. So few detailed studies on Cladocera are available. Site-specific studies have revealed that Cladocera have ecological significance in any aquatic ecosystem. Cladocera can be utilised to monitor the health of aquatic ecosystems. A nationwide programme can be established to detect changes and trends in the aquatic Cladocera that will result in making conservation of water bodies more effective. Cladocera are an important prey for many of the aquatic organisms. So conservation of Cladocera will result in the conservation of aquatic life at large leading to the conservation of water bodies.

Acknowledgements The authors are very grateful to Dr. A. K. Chakravarthy, Head, Division of Entomology and Nematology, Indian Institute of Horticultural Research, Bangalore, for motivating us to write this chapter. Also we are thankful to the Director of the Zoological Survey of India, Kolkata, for his constant support and encouragements.

References

Alam A, Khan AA (1998) On the first recorded of a cladoceran *Leydigia acanthocercoides* (Fischer 1854) (Chydoridae) from Aligarh, Uttar Pradesh. J Bomb Nat Hist Soc 95:143–144

Ariane CD, Sendalz S, Moraes MAB, Mercante CTJ (2015) Dynamics of cladocera community in a Tropical Hypereutrophic Environment Sao Paulo, Brazil. J Water Resour Protect 7:379–398

Baird W (1859) Description of the two new species of Entomostracous Crustacea from India. Proc Zool Soc London 213–234

Balayla DJ, Moss B (2004) Relative importance of grazing on algae by plant-associated and open-water microcrustacea (Cladocera). Arch Hydrobiol 161:199–224

Biswas S (1964a) Five species of Daphniidae from Simla Hills in India with a new record of *Alona costata* Sars from Kameng Division, NEFA. J Zool Soc India 16(1&2):92–98

Biswas S (1964b) A new species of genus *Latona* (Sididae) Straus, 1860 from Rajasthan, India. Proc Zool Soc Calcutta 17:149–152

Biswas S (1965) A new species of genus *Chydorus* Leach, 1843 from Rajasthan, India. Crustaceana 11(2):113–114

Biswas S (1971) Fauna of Rajasthan, India. Part 11(Crustacea: Cladocera). Rec Zool Surv India 63(1–4):95–141

Canton JH, Adema DMM (1978) Reproducibility of short-term and reproduction toxicity experiments with *Daphnia magna* and comparison of the sensitivity of *Daphnia magna* with *Daphnia pulex* and *Daphnia cucullata* in short-term experiments. Hydrobiologia 59:135–140

Chatterjee T, Kotov AA, Damme KV, Chandrasekhar SVA, Padhye S (2013) An annotated checklist of the Cladocera (Crustacea: Branchiopoda) from India. Zootaxa 3667:1–89

Das SM, Akhtar S (1970) A report on freshwater Cladocera from Dal Lake. Kashmir Sci 7(1–2):133–137

Demott WR, Kertoot WC (1982) Competition among cladocera: nature of the interaction between Bosmina and Daphnia. Ecology 63:1949–1966

Echaniz SA, Vignatti, Bunino P (2008) El zooplancton de un lago somero hipereutrófico de la región central de Argentina. Cambios después de una década. Biota Neotrop 8(4):63–71

Ferdous Z, Muktadir AKM (2009) A review: potentiality of zooplankton as bioindicator. Am J App Sci 6(10):1815–1819

Fernando CH (1980a) The fresh water zooplankton of South-East Asia. Trop Ecol Dev 845–853

Fernando CH (1980a) The freshwater zooplankton of Sri Lanka with a discussion of tropical freshwater zooplankton composition. Int Revue Qes Hydrobiol 65(i):85–125
Fernando CH, Kanduru A (1984) Some remarks on the latitudinal distribution of Cladocera on the Indian subcontinent. Hydrobiologia 113:69–76
Forro L, Korovchinsky NM, Kotov AA, Petrusek A (2008) Global diversity of Cladocerans (Cladocera; Crustacea) in freshwater. Hydrobiologia 595:177–184
Frear DEH, Boyd JE (1967) Use of *Daphnia magna* for the microbioassay of pesticides. 1. Development of standardised techniques for rearing Daphnia and preparation of dosage- mortality curves for pesticides. J Econo Ento 60(5):1228–1236
Gulam Mohideen BM, Shahul Hameed P, Shajitha C (2008) Studies on the diversity and abundance of cladocerans in Guntur pond, Tamil Nadu. In: Sengupta M, Dalwani R (eds) Proceedings of Taal: The 12th World Lake conference 470–476
Hall BD, Bodaly RA, Fudge RJP, Rudd JWM, Rosenberg DM (1997) Food as the dominant pathway of methyl mercury uptake by fish. Water Air Soil Poll 100:13–24
Hudson JJ, Taylor WD, Schindler DW (1999) Planktonic nutrient regeneration and cycling efficiency in temperate lakes. Nature 400:659–661
Jeppesen E, Noges P, Davidson TA, Haberman J, Noges T, Blank K, Lauridsen T, Sondergaard M, Sayer C, Laugaste R, Johansson LS, Bjerring R, Amsinck SL (2011) Zooplankton as indicators in lakes: a scientific-based plea for including zooplankton in the ecological quality assessment of lakes according to the European Water Framework Directive (WFD). Hydrobiologia 676:279–297
Korhola A, Rautio M (2001) Cladocera and other branchiopod crustaceans. In: Smol JPH, Birks JB, Last WM (eds) Tracking environmental change using lake sediments, vol. 4: Zoological Indicators. Kluwer Academic Publishers, Dordrecht: 5–41
Michael RG (1973) A guide to the study of freshwater organisms. Madurai Univ J Supplement I. pp186
Michael RG, Sharma BK (1988) Indian Cladocera (Crustacea: Brachiopoda: Cladocera), Zoological Survey of India, Calcutta. pp xvii+262
Muirhead-Thompson RC (1971) Pesticides and freshwater fauna. Academic, New York, p 248
Murugan N, Murugavel P, Koderkar MS (1998) Cladocera: The biology, classification, identification and ecology. Indian Association of aquatic Biologists (IAAB), Hyderabad. 1–47. http://wgbis.ces.iisc.ernet.in/energey/water/paper/Tr-115/ref.htm
Mustaq M (1990) A study on cladoceran fauna of river Kaveri, Tiruchirapalli. Mphil thesis, Bharctidasan university, Tamil Nadu pp 65
Nasar SAK (1977) Investigations on the seasonal periodicity of zooplankton in a freshwater pond in Bhagalpur, India. Acta Hydrochem Hydrobiol 5:577–584
Nayar CKG (1971) Cladocera of Rajasthan. Hydrobiologia 37(3–4):509–519
Orlova BMJO (1998) A revision of the cladocera genus *Simocephalus* (Crustacea: Daphniidae). Bull Nat Hist Muse London 64(1):1–62
Padhye SM, Dumont HJ (2015) Species richness of Cladocera (Crustacea: Branchiopoda) in the Western Ghats of Maharashtra and Goa (India), with biogeographical comments. J Limnol 74(1):182–191
Padhye S, Kotov AA (2010) Cladocera (Crustacea: Branchiopoda) in India hot water springs. Invert Zool 7(2):155–158
Patil SG (1976) Plankton ecology of a few water bodies of Nagpur. Ph.D., Thesis submitted to Nagpur University, Nagpur
Pennak RW (1978) Freshwater invertebrates of the United States, 2nd edn. Wiley, New York
Qadri MY, Yousuf AR (1977) Some new records of Crustacea from Kashmir. Curr Sci 46:859–860
Raghunathan MB (1989) Indian cladocera (Crustacea). Indian Rev Life Sci 9:137–152
Raghunathan MB, Kumar SR (2002) Checklist of Indian Cladocera (Crustacea). Zoos' Print J 18(8):1180–1182

Raghunathan MB, Rane PD (2001) Cladocera-Crusatracea. Zool Surv India, Fauna of conservation Area Series No. 11. Fauna of Nilgiri Biosphere Reserve pp, 31–37
Rane PD (2005) Cladocera: Crustacea. Zool Surv India, Fauna of Melghat Tiger Reserve. Concervation Area Series, vol 24. pp 451–500
Rane PD (2011) Cladocera (Arthropoda: Crustacea), ZoolSurv India, Fauna of Madhya Pradesh (including Chhattisgarh). State Fauna Ser 15(3):45–83
Rocha C, Por F (1988) Preliminary comparative data on the fauna of the pleuston in the southern Pantanal, Brazil, with emphasis on the microcrustaceans. Ver Internat Ver Theoret und Ang Limnol 26:2137–2140
Sharma BK (1991) Cladocera in Animal Resources of India (Protozoa to Mammalia). State of the Art Report Zoological Survey of India, Calcutta, pp 205–223
Sharma BK (2001) Contributions to the Cladoceran fauna (Crustacea: Branchiopoda: Cladocera) of Bihar. Rec Zool Surv India 99(1–4):31–43
Sharma KK, Chandrakiran (2011) Comparative analysis of cladoceran communities from three subtropical freshwater ponds of Jammu: patterns, composition and diversity. Bioscan 6(2):233–237
Sharma BK, Michael RG (1987) Review of taxonomic studies on freshwater cladocera from India with remarks on biogeography. Hydrobiologia 145:29–33
Sharma BK, Sharma S (1990) On the taxonomic status of some Cladoceran taxa (Crustacea: Cladocera) from Central India. Revue Hydrobiol Trop 23:105–133
Sharma BK, Sharma S (2007) New records of two interesting chydorid cladocerans (Branchiopoda: Cladocera: Chydoridae) from the floodplain lakes of Assam, India. Zoos Print J 22(8):2799–2801
Sharma S, Sharma BK (2008) Zooplankton diversity in floodplain lakes of Assam. Rec Zool Suv India (Occ. Paper) 290:1–370
Sharma BK, Sharma S (2009a) Biodiversity of freshwater Rotifers (Rotifera: Eurotatoria) of Tamil Nadu. Rec Zool Surv India 109:41–60
Sharma BK, Sharma S (2009b) Microcrustacea (Crustacea: Branchiopoda) of Deepor Beel, Assam, India: richness, abundance and ecology. J Threat Taxa 1(8):411–418
Sharma BK, Sharma S (2010) Notes on new records of monogonont rotifers (Rotifera: Eurotatoria) from floodplain lakes of Assam. Rec Zool Suv India 110(2):49–55
Sharma BK, Sharma S (2011) Fauna diversity of cladocera (Crustacea: Branchiopoda) of Nokrek Biosphere Reserve, Meghalaya, Northeastern India. J Threat Taxa 3(10):2120–2127
Sharma BK, Sharma S (2012) Notes on some rare and interesting Cladocera (Crustacea: Branchiopoda: Anomopoda: Chydoridae) from Deepor Beel, Assam, India. J Threat Tax 4(1):2304–2309
Sharma S, Sharma BK (2013) Faunal diversity of aquatic invertebrates of Deepor Beel (a Ramsar site), Assam, Northeast India. Wetland Ecosystem Series, vol 17. Zoological Survey of India, Kolkata, pp 1–226
Sharma BK, Sharma S (2014) Faunal diversity of Cladocera (Crustacea: Branchiopoda) in wetlands of Majuli (the largest river island), Assam, northeast India. Opusc Zool Budapest 45(1):1–12
Siraj S, Yousuf AR, Bhat FA, Parveen M (2006) Cladoceran community in Dal lake, Kashmir. Proc Nat Acad Sci India 76(B) IV:343–350
Stemberger RS, Larsen DP, Kincaid TM (2001) Sensitivity of zooplankton for regional lake monitoring. Can J Fish Aqua Sci 58:2222–2232
Urabe J, Kyle M, Makino W, Yoshida T, Andersen T, Elser JJ (2002) Reduced light increases herbivore production due to stoichiometric effects of light: nutrient balance. Ecology 83:619–627
Venkataraman K (1983) Taxonomy and Ecology of Cladocera of Southern Tamilnadu, Ph.D. Thesis Madurai Kamaraj University, Madurai
Venkataraman K (1990) Cladocera of Keoladeo National park, Bharatpur, and its environs. J Bomb Nat Hist 89:17–26

Venkataraman K (1991) The freshwater Cladocera (Crustacea: Branchiopoda). Zool SurIndia, State Fauna Series 3, Fauna of West Bengal, Part 10: 251–284
Venkataraman K (1998) Freshwater Cladocera of Little Andaman. J Andaman Sci Assoc 7(2):43–49
Venkataraman K (2003) Crustacean Zooplankton of Damodar River. Rec Zool Surv India 101(1–2):209–229
Venkataraman K, Krishnaswamy S (1984) On the occurrence of *Leydigia ciliate* Gautheir 1939 (Cladocera: Chydoridae) from Madurai, South India. Curr Sci 53(40):1097–1098

Diversity and Distribution of Sphecoid Wasps in Kerala, India: Bioindicators of Habitat Quality

12

Baaby Job and J.L. Olakkengill

Abstract

Sphecoid wasps, superfamily Apoidea, series Spheciformes with 9706 species in the world come under 318 genera. The life histories of Spheciformes include hunting strategies and solitary to communal nesting to eusociality. They are mainly beneficial and relatively harmless to man. The diversity and distribution of sphecoid wasps at the genus level, with 9 subfamilies and 74 genera, are being reported in India, out of which 8 subfamilies and 35 genera occur in Kerala. A comprehensive analysis of these wasps in Kerala is provided. The food items and habitat patches the wasps prefer are specific so that the changes in the population of wasps can be related to changes in the food and habitat.

Keywords

Apoidea • Eusociality • Sphecoid • Spheciformes

12.1 Introduction

The vast group of Apocrita of Hymenoptera, despite their importance, remain unexplored, especially Spheciformes which has been greatly neglected. The Spheciformes are a highly diverse assemblage of solitary hunting wasps, most of which are brightly colored and fast moving. It comprises 9706 described species under 318 genera (Pulawski 2012) with representations in all biogeographical regions and shows great diversity in morphological and biological characteristics. The oldest known record of sphecoid wasps is from the early Cretaceous, 135 million years ago. The sphecoid wasps can be distinguished from the other hymenopteran groups

B. Job (✉) • J.L. Olakkengill
Department of Zoology, St. Thomas College, Thrissur, Kerala 680 001, India
e-mail: baabyjobc@yahoo.com

A.K. Chakravarthy, S. Sridhara (eds.), *Arthropod Diversity and Conservation in the Tropics and Sub-tropics*, DOI 10.1007/978-981-10-1518-2_12

by the presence of a pronotum with a lateral lobe separated from the tegula, the presence of a cleaning pecten on the inner side of hind basitarsus, and simple unbranched setae (Roche and Gadallah 1999). Adult sphecids feed on a variety of food from nectar and honeydew to spiders and insects across several orders. Prey paralysis and provisioning of nest is a common feature. Nests are constructed in soil, wood, plant stems, twigs, and crevices or holes in rocks, stones, walls, etc. Mating strategies include prenuptial flights, territorial defenses, and courtship activities (Ashmead 1894; Bohart and Menke 1976; Capinera 2003).

12.1.1 Economic Importance

Spheciformes are natural control agents of pests such as aphids, biting flies, cutworms, grasshoppers, and leafhoppers and are used as biocontrol agents in the cultivated fields and act as predators, pollinators, parasites, and parasitoids (Borror and Delong 1988). They are valuable bioindicators responding to environmental disturbances and environmental variations and reflect the diversity patterns of other taxa. They are regarded as nuisance by the people who cannot abide by the presence of wasps, together with the fear of their overrated stinging power, and two species have been reported to be pests in apiaries (Capinera 2003).

12.1.2 Sphecoid Wasps in India and Kerala

Though most species of Spheciformes have been collected and described from central and northern parts of India, the sphecoid fauna of the country is still imperfectly known. The pioneer studies of Indian Spheciformes were done by Fabricius (1781). Bingham (1897) recorded as many as 168 species of sphecoid wasps as occurring in the Indian subcontinent. Bohart and Menke (1976) published on the world fauna. In Kerala, Sudheendra kumar and Narendran (1989, 1985) and Madhavikutty (2004) worked on the sphecoid wasps. Some authors treat Spheciformes as a single family Sphecidae (Bohart and Menke 1976; Gauld and Bolton 1988) or as a series of nine families (Finnamore et al. 1993), giving family status to subfamilies and their tribes being treated as subfamilies. This paper uses the classification set up by Bohart and Menke (1976).

In determining the fauna of a country, faunistical studies on small regions are important because individual habitats and the microclimate in a small region play an important role on the distribution of insects (Gulmez and Tuzum 2005). The present paper aims to provide data toward complete knowledge of the Indian fauna by analyzing the diversity and distribution of these wasps in Kerala.

12.2 Materials and Methods

The insects were collected from urban and semi-urban areas of Kerala from 2010 to 2012. The insects were caught using sweep net and were killed with 10 % acetone analytical grade as agent. Field studies were carried out from 9:00 am to 17:00 pm. The specimens were mounted using no. 3 entomological pins and identified according to Bohart and Menke (1976) using Leica MZ6 Stereo Zoom microscope.

12.2.1 Study Area

Kerala is one of the smallest states in India with an area of 38,863 km^2, contributing 1.3 % of total area of India. Kerala is located between north latitude 8°18′ and 12°48′ and east longitude 74°52′ and 77°22′.

(a) *Topography*
The topography consists of eastern highlands comprising Western Ghats, covered with dense forests and high mountains with tea and coffee plantations, midlands along the Central Kerala with paddy fields, pepper and tapioca fields to a flat coastal belt of coconut trees; interconnected with canals and rivers.

(b) *Climate*
Kerala lies in the tropics and is subjected to humid tropical wet climate. It receives an average rainfall of 3107 mm. Average daily temperature is around 36.7 °C with minimum 19.8 °C. The state enjoys four climatic seasons—summers (from March to June), southwest monsoon (from June to August), northeast monsoon (from September to October) and the winter season with a chilly climate from December to February.

12.3 Results

Out of the nine subfamilies and 74 genera reported in India (Bohart and Menke 1976), Kerala has recorded eight subfamilies and 35 genera (Sudheendra kumar and Narendran 1989; Suresh et al. 1999) (Table 12.1). A total of 175 specimens belonging to 15 genera were collected from the state and listed in Table 12.2 and Fig. 12.1. The genus *Chalybion* Dahlbom showed the widest range of distribution suggesting the ability of this group to exist in varied environmental conditions, provided with their food sources.

12.4 Discussion

The collected specimens were represented in four subfamilies—Ampulicinae, Sphecinae, Larrinae, and Nyssoninae—and 15 genera. Subfamily Sphecinae were the most represented with six genera and showed almost cosmopolitan distribution.

Table 12.1 Diversity of sphecoid wasps in Kerala, India

Sl. no.	Genus	India	Kerala
Subfamily Ampulicinae (Fig. 12.2)			
1	*Ampulex* Jurine	P	P
2	*Dolichurus* Latreille	P	P
3	*Trirogma* Westwood	P	P
Subfamily Sphecinae (Fig. 12.3)			
4	*Sceliphron* Klug	P	P
5	*Chlorion* Latreille	P	P
6	*Chalybion* Dahlbom	P	P
7	*Sphex* Linnaeus	P	P
8	*Isodontia* Patton	P	P
9	*Prionyx* Vander Linden	P	P
10	*Podalonia* Fernald	P	P
11	*Ammophila* W. Kirby	P	P
12	*Parapsammophila* Taschenberg	P	P
Subfamily Pemphredoninae (Fig. 12.4)			
13	*Mimumesa* Malloch	P	A
14	*Psen* Latreille	P	A
15	*Psenulus* Kohl	P	P
16	*Diodondus* Curtis	P	A
17	*Pemphredon* Latreille	P	A
18	*Polemistus* Saussure	P	P
19	*Stigmus* Panzer	P	A
20	*Carinostigmus* Tsuneki	P	P
21	*Spilomena* Shuckard	P	A
22	*Ammoplanellus* Gussakovskij	P	P
Subfamily Astatinae (Fig. 12.5)			
23	*Astata* Latreille	P	P
24	*Dryudella* Spinola	P	A
25	*Dinetus* Panzer	P	A
Subfamily Laphyragoginae			
26	*Laphyrogogus* Kohl	P	A
Subfamily Larrinae (Fig. 12.6)			
27	*Larra* Fabricius	P	P
28	*Liris* Fabricius	P	P
29	*Paraliris* Kohl	P	A
30	*Dicranorhina* Shuckard	P	A
31	*Gastrosericus* Spinola	P	P
32	*Tachytes* Panzer	P	P
33	*Tachysphex* Kohl	P	P
34	*Parapiagetia* Kohl	P	A
35	*Holotachysphex* de Beaumont	P	A
36	*Prosopigastra* A. Costa	P	P
37	*Palarus* Latreille	P	A

(continued)

Table 12.1 (continued)

Sl. no.	Genus	India	Kerala
38	Lyroda Say	P	P
39	*Paranysson* Guérin-Méneville	P	A
40	*Solierella* Spinola	P	P
41	*Miscophus* Jurine	P	A
42	*Nitela* Latreille	P	A
43	*Pison* Jurine	P	P
44	*Trypoxylon* Latreille	P	P
Subfamily Crabroninae (Fig. 12.7)			
45	*Oxybelus* Latreille	P	P
46	*Encopognathus* Kohl	P	A
47	*Entomognathus* Dahlbom	P	A
48	*Lindenius* Lepeletier and Brulle	P	A
49	*Rhopalum* Stephens	P	A
50	*Isorhopalum* Leclercq	P	A
51	*Crossocerus* Lepeletier and Brulle	P	A
52	*Crabro* Fabricius	P	A
53	*Piyuma* Pate	P	P
54	*Vechita* Pate	P	A
55	*Hingstoniola* Turner and Waterson	P	A
56	*Dasyproctus* Lepeletier and Brulle	P	P
57	*Ectemnius* Dahlbom	P	A
58	*Lestica* Billberg	P	A
Subfamily Nyssoninae (Fig. 12.9)			
59	*Alysson* Panzer	P	A
60	*Nursera* Cameron	P	A
61	*Nysson* Latreille	P	A
62	*Synnevrus* A. Costa	P	A
63	*Brachystegus* A. Costa	P	A
64	*Dienoplus* W. Pax	P	A
65	*Gorytes* Latreille	P	A
66	*Lestiphorus* Lepeletier	P	A
67	*Ammatonius* A. Costa	P	P
68	*Hoplisoides* Gribodo	P	A
69	*Stizus* Latreille	P	P
70	*Stizoides* Guerin-Meneville	P	A
71	*Bembecinus* A. Costa	P	P
72	*Bembix* Fabricius	P	P
Subfamily Philanthinae (Fig. 12.8)			
73	*Philanthus* Fabricius	P	A
74	*Cerceris* Latreille	P	P

P Present, *A* Absent

Table 12.2 Genera of sphecoid wasps collected from Kerala, India, 2010–2012

Sl. no.	Genus	Kasargod	Kannur	Wayanad	Kozhikode	Malappuram	Palakkad	Thrissur	Ernakulam	Kottayam	Idukki	Pathanamthitta	Kollam	Trivandrum
1	*Ampulex* Jurine	X						X	X					
2	*Dolichurus* Latreille								X					
3	*Trirogma* Westwood													
4	*Sceliphron* Klug	X	X	X	X	X	X	X	X	X	X	X	X	X
5	*Chlorion* Latreille			X										
6	*Chalybion* Dahlbom	X	X	X	X	X	X	X	X	X	X	X	X	X
7	*Sphex* Linnaeus			X				X	X	X		X	X	X
8	*Isodontia* Patton								X					
9	*Ammophila* W. Kirby	X	X	X		X	X	X	X		X			
10	*Larra* Fabricius					X		X						
11	*Liris* Fabricius	X		X					X	X	X			X
12	*Tachysphex* Kohl													
13	*Pison* Jurine							X						
14	*Trypoxylon* Latreille	X				X								
15	*Bembix* Fabricius					X		X						

Fig. 12.1 Distribution patterns of sphecoid wasps (Based on Table 12.2)

Most of the collected specimens showed marked variations at the times of their collections. All of the specimens were collected in the morning, the insect's activity increasing with temperature, but collections were less after around 1:00 pm and increased toward around 4–5:00 pm in the evening, displaying their diurnal nature. The wasps also showed variations in the areas collected. While most of the Sphecinae and Ampulicinae were collected from open areas, *Liris* Fabricius and *Larra* Fabricius were collected from shady areas with patches of light, while *Bembix* Fabricius was collected from sandy areas. *Chalybion* Dahlbom and *Sceliphron* Klug were collected mainly from areas with loamy soil. Subfamily Laphyragoginae, although reported in most other parts of India, has not been reported yet in Kerala (Table 12.1).

Ampulex spp (http://ispeakforthefleas.blogspot.in)

Dolichurus spp (http://www.padil.gov.au)

Fig. 12.2 Sphecoid wasps: subfamily Ampulicinae

12.4.1 Subfamily Ampulicinae

These wasps are called as "cockroach wasps" because of their prey choice; one cockroach prey is provided in each cell as food for the larvae. These wasps represented in six genera, out of which three, *Ampulex* Jurine, *Dolichurus* Latreille, and *Trirogma* Westwood are reported from Kerala (Fig. 12.2). *Ampulex* Jurine is readily identified in the field by metallic blue or black and red integument. *Trirogma* Westwood is medium to large sized wasps, with white mandibles in males. *Dolichurus* Latreille are small black wasps, often with red terminal abdominal segments. They run or skip over leaf litter, occasionally disappearing beneath the fallen leaves to search prey. Among the three genera collected, *Ampulex* Jurine was the most common, being collected from three Kerala districts: Ernakulam, Thrissur, and Kasargod.

12.4.2 Subfamily Sphecinae

These are called 'thread-waisted wasps' on account of their cylindrical petiole. They are cosmopolitan wasps, with large size and bright colors. They are common inhabitants of fields, forests, and even garage lots. The genera *Chalybion* Dahlbom and *Sceliphron* Klug were the most common specimens collected followed by *Sphex* Linnaeus and *Ammophila* W. Kirby. *Isodontia* Patton was collected from the Koothattukulam area in Ernakulam, Kerala (Figs. 12.3, 12.4, and 12.5). These display a variety of nesting habits from fossorial to nesting in preexisting cavities in twigs and woods to social behavior. Preys include spiders, cockroaches, crickets, grasshoppers, etc.

Chlorion Latreille: Length ranging from 16 to 37 mm, metallic green or blue wasps. They can be identified from their well-developed foretarsal rake, spiracular groove, and second submarginal cell receiving the first recurrent vein. Prey species are crickets belonging to Gryllidae.

Sceliphron Klug: Cosmopolitan wasps with two subgenus—*Scelipheron* and *Proscelipheron*. These are medium-sized wasps with lengths ranging from 12 to 32 mm and black body with yellow markings and are closely associated with human habitations. They are called "mud dauber wasps" with reference to their habit of

Sceliphron spp. (http://www.pestnet.org) *Sphex* Linnaeus (http://www.waspweb.org)

Ammophila spp. (http://napamosquito.org) *Prionyx spp.*(http://www.pbase.com)

Fig. 12.3 Sphecoid wasps: subfamily Sphecinae

Stigmus spp (http://www.bwars.com) *Carinostigmus* spp (http://fukker666.blog32.fc2.com)

Fig. 12.4 Sphecoid wasps: subfamily Pemphredoninae

building nests with mud collected from a moist spot. They can be identified from other groups of the subfamily by expanded third maxillary palpal segment. In subgenus *Scelipheron*, the hypostomal carina ends near the mandible socket, while this is evanescent about halfway to the socket in *Proscelipheron*. Each mud nest consists of six to seven cells of mud, each cell provisioned with spiders.

Astata spp. (http://www.natureconservationimaging.com) *Dryudella* spp.(http://bugguide.net)

Fig. 12.5 Sphecoid wasps: subfamily Astatinae

Chalybion Dahlbom: These are cosmopolitan metallic blue wasps with lengths ranging from 11 to 32 mm. They are the most common wasps, associated with human dwellings and habitats. They are characterized by the absence of propodeal enclosure and submarginal cell, one receiving both recurrent veins. These wasps mass provision their mud nests with spiders, which is then sealed with a white cementing substance.

Sphex Linnaeus: These are cosmopolitan and moderate to large wasps, with black or golden yellow colors. They are identified with the length of petiole as measured along dorsum less than the combined lengths of hind tarsomeres II–IV and complete spiracular groove on the propodeal side. Members of the genus *Sphex* dig nests in soil with gravel and provision their nests with prey—Acrididae and Locustidae, agricultural pests.

Isodontia Patton: "Grass carrier wasps." These are moderate to large wasps, 11–33 mm, mostly black. These wasps use preexisting cavities like hollow plant stems and crevices between stones for nesting. They are identified by the absence of a complete spiracular groove and the anterior veinlet of the third submarginal cell exceeding the length of posterobasal veinlet. They are nonfossorial; hence, either the females lack a fore tarsal rake or the rake is poorly developed. Preys include Gryllidae and Tettigoniidae.

Ammophila W. Kirby: These wasps range in length from 8 to 37 mm. They are easily diagnosed by partly or all-red gaster, and the legs are commonly partly red, with the legs and thorax covered with appressed silver hairs. They are characterized by the presence of a long petiole, appearing two segmented; apex of sternum I does not reach the base of II, with a long intervening space connected by ligaments. *Ammophila* are generally solitary nesters, feeding on lepidopteran larvae.

12.4.3 Subfamily Larrinae

These are the largest subfamily with 18 genera reported in India, often dark-colored wasps, sessile or petiolate. Most species are fossorial and are commonly known as "digger wasps." The most abundant in the collection was *Liris* Fabricius, followed by *Larra* Fabricius and *Trypoxylon* Latreille. *Pison* Jurine was collected from

Chelakkottukara area of Thrissur district, Kerala, and *Tachysphex* Kohl from Ramapuram in Kottayam district, Kerala (Figs. 12.6, 12.7, and 12.8). Prey species range from spiders, Orthoptera, Hemiptera, Dipterans, and hymenopterous adults to lepidopterous larva. *Larra* Fabricius: These are small to large wasps, 6–25 mm long, black wasps with a shiny gaster. Hind-ocellar scars are very small and obscure. They are identified by the last tarsomere evenly arcuate, propodeal side punctate, and pronotal collar flat or arcuate in front view. The subgenus *Crotolarra* is distinguished from the subgenus *Larra* by the absence of spine rows on the outer face of

Fig. 12.6 Sphecoid wasps: subfamily Larrinae

Fig. 12.7 Sphecoid wasps: subfamily Crabroninae

***Philanthus* Fabricius (http://www.biolib.cz)**

Fig. 12.8 Sphecoid wasps: subfamily Philanthinae

foretibia. *Larra* females don't construct their own nests and prey paralysis is temporary, the prey being mole crickets (Gryllotalpidae).

Liris Fabricius: These are dull black wasps with a fine appressed vestiture on the abdomen. They range in length from 5 to 30 mm. They have a hind tibia with a sharp polished carina, propodeum converging posteriorly from above, and reduced hind ocellus and dentate mandibles. Prey species include crickets from the family Gryllidae.

Tachysphex Kohl: These are 4–18-mm-long black wasps with the abdomen and legs often partly or all red. Diagnostic features include inner orbits converging above, hind-ocellar scars oval or oblong or accent marked, and long axes forming an angle of 80°–130°. They have a pair of polished prominences just above the antennal sockets. Prey consists of Orthoptera and nest is constructed in open sandy or vegetated areas.

Pison Jurine: Black wasps with two or three submarginal cells in the fore wing, entire mandible, emarginate eyes, and sessile abdomen. Prey species include a variety of spiders.

Trypoxylon Latreille: They are distinguished by single submarginal cell, emarginate eyes, and slender clavate abdomen. Most of these exist in preexisting cavities and are called "keyhole wasps," with prey mainly being spiders.

12.4.4 Subfamily Nyssoninae

The second largest family is called "sand wasps" as the majority of these are found in sandy habitats. The only genus collected was *Bembix* Fabricius; these were collected from sandy mounds in parks from Nilambur area of Malappuram district and

Nysson spp.(http://www.arthropodafotos.de) *Gorytes* spp.(http://www.bwars.com)

Fig. 12.9 Sphecoid wasps: subfamily Nyssoninae

Thumburmuzhi in Thrissur, Kerala (Fig. 12.9). They are identified by totally reduced anterior ocelli and reduced palpal segments and long labrum. They nest gregariously and prey consists of Diptera. The digger wasps focus predominantly on habitats influenced by anthropogenic disturbance, to allow burrowing activity and subsequent nesting (Srba and Heneberg 2012).

The abiotic factors driving the presence and nesting of solitary hymenopterans include light intensity, moisture, and soil (Schrimer et al. 2008; Hranitz et al. 2009; Murray et al. 2009), and the biotic variables include floral abundance (Banaszak 1996), availability of pollen and nectar sources (Petanidou and Votou 1990), changes in the availability of the preferred prey (O'Neill 2001; Polidori et al. 2007), or these factors combined (Potts et al. 2005). In addition to specialization at different sizes or types of prey, variability in the abiotic factors of the microhabitat conditions like temperature, moisture, soil type, vegetation cover, and ground inclination was found both necessary and sufficient to allow the parallel presence of viable populations of several closely related sphecoid wasps (Srba and Heneberg 2012).

12.5 Conclusion

The occurrence of Spheciformes in a habitat is conditioned by moisture, but other factors like the soil exposure also seem to be the determinant, relating to nesting requirements. Soil type and prey abundance are important for the settlement of these wasps (Gayubo et al. 2000). The collections of these wasps are a tedious process, yet they have to be catalogued because of economic importance, especially in being bioindicators of habitat quality and ecological disturbance (Gayubo et al. 2005). The richness of wasp species has been shown to correlate with landscape complexity and habitat diversity (Steffan Dewanter 2005).

The creations of protected areas were one of the first measures taken for the protection of biodiversity and are still the most widely used approach (Vieira et al. 2011). Aculeate hymenopterans, especially Spheciformes and Apiformes, gather exceptional characteristics as bioindicators, are a part of several functional niches (predators, kleptoparasites, and pollinators), have economic importance (pollinators and pest management), and are proven to be good bioindicators (predicting the diversity of other groups of animals and for all the species of a given area (Klein et al. 2002). In England, the Invertebrate Species-Habitat Information System [ISIS], a classification system for conservation based on invertebrate communities and their relation with habitats, is already being developed (Webb and Lott 2006).

Most of our conservation strategies are centered around hotspots, but little known niches like these also have tremendous value in biodiversity conservation. Hence the establishment of systems to evaluate the effectiveness of management of these areas is crucial to validate their importance for conservation and guide the managers towards conservation goals. Our current knowledge of these wasps is still imperfect, and a proper understanding of their habitat requirements is a key prerequisite to allow efficient conservation of individual species, before they are lost.

Acknowledgments The first author wishes to acknowledge Council of Scientiific and Industrial Research, New Delhi for the "MANF-09" financial assistance. The authors wish to acknowledge the late Dr. T. C. Narendran for continued support, encouragement, and guidance. The authors are also grateful to Dr. Karmaly, St. Xavier's college, Aluva, Kerala.

References

Ashmead WH (1894) The habits of Aculeate Hymenoptera-III. Psyche 7:59–66

Banaszak J (1996) Ecological basis of conservation of wild bees. In: Matheson JH (ed) The conservation of bees. Academic, New York, pp 55–62

Bingham CT (1897) The Fauna of British India Hymenoptera, Vol I. Taylor and Francis, UK

Bohart RM, Menke AS (1976) Sphecid wasps of the world a generic revision. University of California Press, Berkeley

Borror DJ, DeLong DM (1988) Introdução ao estudo dos insectos. Editora Edgard Blücher Ltd, Brasil, pp 417–486

Capinera JL (2003) Encyclopedia of Entomology, 2nd edn. Springer, London

Dewander S (2005) Landscape context affects trap nesting bees, wasps and their natural enemies. Ecol Entom 27:631–637

Fabricius JC (1781) Mantissa insectorum I. Hamb.et kilonii, pp 552

Finnamore AT, Charles D, Michener (1993) Super family Apoidea. In: Goulet H, Huber JT (eds) Hymenoptera of the world–an identification guide to families. Research Branch, Agriculture, Canada, pp 279–346

Gauld ID, Bolton B (1988) The hymenoptera of British Museum (Natural History). Oxford University Press, Oxford

Gayubo SF, González JA, Torres F (2000) Estudio de una comunidad de esfécidosen la zona natural de "Las Arribes del Duero" (Salamanca, Oeste español) (Hymenoptera, Sphecidae). Fragm Entomol 32:181–209

Gayubo SF, González JA, Asís JD, Tormos J (2005) Conservation of European environments: the Spheciformes wasps as biodiversity indicators (Hymenoptera: Apoidea: Ampulicidae, Sphecidae and Crabronidae). J Nat Hist 39(29):2705–2714

Gulmez Y, Tuzum A (2005) Spheciformes (Hymenoptera: Apoidea) from Ankara province, Subfamilies Sphecinae, Pemphredoninae and Astatinae. J Entomol Res Soc 7(1):41–57

Hranitz JM, Barthell JF, Thorp RW, Overall LM, Griffith JL (2009) Nest site selection influences mortality and stress responses in developmental stages of *Megachile apicalis*. Spinola (Hymenoptera: Megachilidae). Environ Entomol 38:484–492

Klein AM, Steffen Dewenter I, Buchori D, Tscharntke T (2002) Effects of land-use intensity in tropical agroforestry systems on coffee flower-visiting and trap-nesting bees and wasps. Conserv Biol 16(4):1003–1014

Madhavikutty M (2004) Investigation in the systematics of Sphecid wasps (Hymenoptera, Apoidea, Spheciformes) of Kerala at alpha level. Ph.D thesis, University of Calicut India

Murray TE, Kuhalmann M, Potts SG (2009) Conservation ecology of bees: population species and communities. Apidologie 40:211–236

O'Neill KM (2001) Solitary wasps. Behaviour and natural history. Comstock Publishing Associates, Ithaca

Petanidou T, Vokou D (1990) Pollination and Pollen energetic in Mediterranean ecosystems. Am J Bot 77:986–992

Polidori C, Boesi R, Pesarini C, Papadia C, Bevacqua S, Federic M, Andriettic F (2007) Temporal relationship between the prey spectrum and population structure of Weevil hunting wasp *Cerceris arenaria* (Hymenoptera: Crabronidae). Zool Study 46:83–91

Potts SG, Valliamy B, Roberts S, O'Toole C, Dafni A, Ne'eman G, Willmer P (2005) Role of nesting resources in organizing diverse be communities in a Mediterranean landscape. Ecol Entomol 30:78–85

Pulawski WJ (2012) Catalogue of Sphecidae at California Academy. www.research.catacademy.org

Roche GC, Gadallah NS (1999) The Sphecid wasps of Egypt (Hymenoptera: Sphecidae): Introduction and generic key. Egyption J Biol 1:104–117

Schrimer S, Sengonca C, Blaesar P (2008) Influence of some abiotic factors on some biological and ecological characteristics of the aphid *Aphelinus asychis* (Hymenoptera: Aphelinidae) Parasitizing *Aphis gossypii* (Sternorrhyncha: Aphididae). Eur J Entomol 105:121–129

Srba M, Heneberg P (2012) Nesting habitat segregation between closely related terricolous Sphecid species (Hymenoptera: Spheciformes): key role of soil physical characteristic. J Insect Cons 16(4):557–570

Sudheendra kumar VV, Narendran TC (1985) Alpha taxonomy of three new species of Sphecidae (Hymenoptera) from the Malabar region (India). J Entomol Res 9(1):50–53

Sudheendrakumar VV, Narendran TC (1989) Sphecoid wasps of Kerala, India. Sphecos 18:11–12

Suresh PV, Sudheendrakumar VV, Binoy CF, Mathew G, Narendran TC (1999) The macro hymenopteran fauna of Parambikulam Wildlife Sanctuary. Zoo's Print 14(4):1–2

Vieira C, Oliveire NG, Gayubo SF (2011) On the use of Apiformes and Spheciformes (Insecta: Hymenoptera) populations as management tool. Biodivers Conserv 20(12):519–530

Webb JR, Lott DA (2006) The development of ISIS: a habitat-based invertebrate assemblage classification system for assessing conservation interest in England. J Insect Cons 10:179–188

Diversity and Distribution of Chalcid Wasps in Kerala: Key Biological Control Agents in Cultivated Ecosystems

13

N. Preethi and P. Lakshmi Devi Menon

Abstract

Chalcidoidea is one of the most abundant and speciose group of highly diverse parasitic Hymenoptera. The economic importance of the superfamily Chalcidoidea lies in their worldwide use as biological control agents. They are parasitic on economically important pests of Lepidoptera and Coleoptera and some of them are keystone species. Chalcididae is a family consisting about 90 genera and 1500 species in the world. In the present study, the diversity and distribution of family Chalcididae in Kerala, India, has been reviewed. The specimens were mainly collected from Kerala using sweep net and Malaise trap. A total of 216 species, 30 genera, and 5 subfamilies (Chalcidinae, Dirhininae, Epitraninae, Haltichellinae, Smicromorphinae, unplaced) have been recorded from India. Of this, 83 species belonging to 21 genera were reported from Kerala. Among the 21 genera, 6 are exclusively recorded from Kerala. The subfamily Haltichellinae has the largest number of recorded species (8 genera and 39 species). This is followed by Subfamily Chalcidinae with 2 genera and 24 species. *Brachymeria* is the predominant genus reported with 23 species from Kerala. *Antrocephalus* is the next with 15 species. A distribution map of chalcidids has also been provided.

Keywords

Chalcidoidea • Chalcididae • Distribution • Diversity • Kerala

N. Preethi (✉)
Department of Zoology, NSS College, Manjeri, Malappuram 676122, Kerala, India
e-mail: preethinambrayil@yahoo.co.in

P.L.D. Menon
Department of Zoology, S.H. College, Thevara, Cochin, Kerala, India
e-mail: lakshmi82ani@gmail.com

A.K. Chakravarthy, S. Sridhara (eds.), *Arthropod Diversity and Conservation in the Tropics and Sub-tropics*, DOI 10.1007/978-981-10-1518-2_13

13.1 Introduction

Chalcidids are wasps that are cosmopolitan and abundant in the tropics and sub tropics, especially in South America. The family currently includes 87 genera and 1464 species placed in 5 subfamilies as follows: Chalcidinae (25 genera/767 species), Dirhininae (3 genera/65 species), Epitraninae (1 genus/64 species), Haltichellinae (55 genera/560 species), Smicromorphinae (1 genus/6 species), and unplaced (2 genera/2 species). The chalcid wasps are important and have diverse lifestyles. These are important and interesting, but also difficult parasitic wasps to study taxonomically (Bouček 1951). The group consists of tiny parasitic wasps which are brightly colored with a gorgeous metallic sheath. Despite the chalcids' diminutive size, these wasps cause death of the eggs and larvae of some of the most destructive insect pests, including flies, beetles, aphids, leafhoppers, scale insects, whiteflies, and plant hoppers. In addition to their devouring other insects in the embryonic or larval forms, certain chalcid species display a number of unusual, and therefore interesting, habits. For example, some chalcids are well known for *polyembryony*. Another chalcid phenomenon – *hyperparasitism* – can work against the gardener and occurs when a chalcid uses another parasitic insect for its host, thus negating the beneficial effect of the victim parasite.

13.2 Morphology and Biology

Several species of chalcidids are larval endoparasitoids. All are solitary. This group of wasps include few ectoparasitic and hyperparasitic species also. These species have a wide host range. They are principal parasitoids of Lepidoptera, although Coleoptera and Diptera can also be parasitized. Hyperparasitic species prey on tachinid and braconid parasitoids. Chalcididae is economically important because several of its species parasitize agricultural insect pests.

Gibson (1993) recorded that in Chalcididae, the body will be without a metallic luster (except for *Notaspidium*). These wasps are either black or brown yellow or red. The head and mesosoma are heavily sclerotized, usually coarsely punctate at least in part. The head dorsally bears a projection (horn) between the scrobal depression and each eye. The gena is carinate or ridged. Antennae are inserted below or above the level of ventral margin. The flagellum is shorter than the length of the eye with five to seven segments. The first segment is ringlike or plain. The pronotum has a rectangular collar. The mesopleuron has a shallow femoral depression over most of its height. These wasps are winged, with the forewing not folded lengthwise and with venation sometimes reduced. The postmarginal vein in the wing is absent or long. A tarsus bears five tarsomeres. The metacoxa is long and in cross section subcircular to large and subtriangular. The metafemur is large, compressed, and ventrally serrated over atleast apical one-third. The metatibia is usually distinctly curved, either truncate with 2 apical spurs or obliquely pointed with 1 spur or none. The metasoma bears a transverse or long, slender petiole which is rarely inserted into the propodeum. Postpetiolar terga are not jointed. The ovipositor sheath extends a bit (Gibson 1993).

This group of wasps are cosmopolitan in distribution with circa 195 nominal genera and 1875 nominal species. These wasps could be distinguished morphologically rapidly. It is hypothesized that Chalcididae could be paraphyletic relative to Leucospidae (Gibson 1993). The hind legs are modified with multiple functions, at least in females. Females of select species retain the prey between the dentated femur and curved tibia during oviposition, while others stand upright on the hind legs so that the front and middle legs are free for movement. Females use the hind legs in back-to-back fighting, and in *Lasiochalcidia igiliensis* females use hind legs to hold mandibles of ant lion larvae (Neuroptera: Myrmeleontidae) while laying eggs into the membrane of the exposed throat.

Bouček (1988) has extensively studied these wasps from Australia and documented keys to genera from other regions. Delvare and Bouček (1992) have reviewed and updated the genera of the family and the species of the Chalcidini, a New World species.

"Many species of this family are primary parasites (occasionally hyperparasites) of moths and butterflies. Other insects such as flies, beetles, wasps and bugs are also attacked. Even the ants and ant-lions fall prey to some of these chalcidids."

13.3 Historical Review

Investigations on chalcidids date back to 200 years when Linnaeus discovered and identified a few species, viz., *Sphex sispes* (*Chalcis sispes*), *Vespa minuta* (*Brachymeria minuta*), etc. Soon Linnaeus was followed by Fabricius who was the first to coin the term *Chalcis*. Later, the names of Walker, Westwood, Dalla Torre, Dalman, Spinola, Haliday, Saunders, Motschulsky, Foerster, Cresson, Klug, and Kirby stand out among the early workers who studied this group. Unfortunately, the types of selected species described by these workers no longer exist (Bouček 1951). Since the monumental work on the classification of Chalcidoidea by Ashmead (1904), knowledge of the family has been greatly enhanced by the studies of Cameron, Crawford, Schmitz, Waterston, Girault, Gahan, Ruschka, and Masi. Recently, the information on the family has been greatly increased by the studies of Bouček (world fauna), Steffan (Palaearctic and Ethiopian fauna), Burks (Nearctic fauna), Erdős (Hungarian fauna), Habu (Japanese fauna), Nikol'skaya (Russian fauna), Mani and his students, and Narendran (Indian fauna). In the recent years, documentation on the Indian Chalcididae has been made by Gahan (1930, 1942), Waterston (1922), Masi (1927, 1929a), Mani (1935, 1936, 1938), Joseph et al. (1973), Narendran (1976, 1977, 1984), Bouček (1982), Bouček and Narendran (1981), and Husain and Agarwal (1981–1982).

13.4 Classification

Narendran (1989) from India has extensively studied and recorded observations on the family Chalcididae. This family is divided into the following six subfamilies (Narendran 1984) and tribes: Brachymeriinae (Brachymeriini, Cratocentriini,

Phasganophorini), Haltichellinae (Haltichellini, Hybothoracini), Chalcidinae, Epitraninae, Dirhininae (Dirhinini, Aplorhinini), and Smicromorphinae.

These insects are found near flowers (esp. Asteraceae, Umbelliferae, Fabaceae) generally during warm times of the year. They are shade lovers. Commonly known species are parasitoids of a wide range of insect pests, and hyperparasitoids are also common in this family via Braconidae and Ichneumonidae (Steffan 1959; Habu 1960; Bouček 1988; Delvare 1995). Investigations have been carried out in few geographical regions. The species composition and distribution of Chalcididae in the Palaearctic region have been documented by Bouček (1952, 1956), Steffan (1962, 1976), Masi (1929a, b), Nikol'skaya (1952, 1960), and Habu (1966).

13.5 Materials and Methods

Kerala, a costal state lying between Lakshadweep Sea and the Western Ghats, is a narrow, fertile strip of land on the southwest coast of India. The landscape of Kerala is a beautiful stretch of land with warm beaches on the east, rolling mountains and deep valleys in the interiors.

13.5.1 Geography of Kerala

Kerala is located between latitude 8° 18′ north and 12° 48′ north and longitude 74° 52′ east and 7° 22′ east. The green land covers 1.18 % of India. It is on the tropical Malabar Coast of southwestern India. Tamil Nadu is in the east and Karnataka north of Kerala. Due to its landscape and physical features, it has three distinct regions – hills and valleys, midland plains, and coastal region. Forests cover 27 % of Kerala. Some of the forests are so dense that their flora and fauna in patches such as Silent Valley are yet to be documented, completely assessed and recorded. Medicinal herbs, abundant in these forests, are used in Ayurveda. Kerala has 14 districts. Based on geographical, historical, and cultural similarities, the districts are generally grouped into three groups:

North Kerala: Kasaragod, Kannur, Wayanad, Kozhikode, and Malappuram
Central Kerala: Palakkad, Thrissur, Ernakulam, and Idukki
South Kerala: Thiruvananthapuram, Kollam, Alappuzha, Pathanamthitta, and Kottayam

Nearly half of Kerala's population is engaged in agriculture, horticulture, plantation crops, and forestry. The chief crops include rice, coconut, tea, coffee, rubber, cashews, and spices – including pepper, cardamom, vanilla, cinnamon, and nutmeg. The state has mineral resources like limonite, silica, quartz, kaolin, bauxite, sillimanite, zircon, and rutile. Agriculture-based industries in Kerala include traditional manufacturing such items such as coir, handlooms, and handicrafts, small-scale industries, and some cottage industries.

Climate Kerala state lies closer to the equator. Yet the weather in Kerala is pleasant and salubrious throughout the year. This is because of the land's nearness to the sea and the presence of the fortlike Western Ghats on the east. Kerala has been a dry land because of the dry winds blowing from the north, but for the Western Ghats which prevent this wind from entering the land, Kerala receives copious rains (average 3000 mm a year) each year. The temperature in Kerala varies from 28° to 32 °C (82° to 90 °F) on the plains but drops to about 20 °C (68 °F) in the highlands. The highland landscape attracts tourists, who enjoy a cool and pleasant climate throughout the year. Owing to its location and geographical features, the climatic condition in Kerala is diverse. It can be divided into four seasons – winter, summer, southwest monsoon, and northeast monsoon. The topography consists of a hot and wet coastal plain gradually rising in elevation to the high hills and mountains of the Western Ghats. Kerala's climate is wet and maritime tropical, heavily influenced by the seasonal, heavy monsoon rains.

13.5.2 Collection and Curation

The materials were collected from different districts of Kerala. The main collection method involved sweeping. The specimens were collected mainly in the early morning and late evening. As chalcids are shade lovers, these specimens were abundant during this time. The rainy season was not favorable in collecting the specimens. After collection they were sorted and stored in 70 % alcohol-containing vials, which were labeled and refrigerated. The specimens were identified using Leica microscope after they were mounted dry onto cards and labeled. Later they were dried and preserved in wooden insect boxes. Naphthalene balls were added to prevent insect attacks.

13.5.3 Sampling methods

(a) *Sweep Net* Sweeping using insect net is probably the most rewarding way of collecting chalcids because many specimens can be collected rapidly. If material is collected in alcohol, then a good day's collection will provide enough material for study.

The sweep-net method described here is based on the design first drawn up 50 years ago by Dr. Z. Bouček, a well-known taxonomist on chalcidoids. The insect net has been adapted so that it can be dismantled easily for travel. The best frame is prepared out of aluminum. The triangular head increases the surface area in contact with the ground when sweeping grasslands or dense vegetation.

(a) *Malaise traps*
A malaise trap is a large, tent-like, three-dimensional structure used for attracting and trapping insects in flight. The trap is terylene netting and it can be of any color. Insects fly into the tent wall and are funneled into a collecting vessel attached to the highest point. The trap was invented by René Malaise in 1937.

(b) *Yellow pan or Moericke trap*
This trap is an excellent tool for collecting chalcids, particularly mymarids and encyrtids and other groups of insects. Species that are rarely swept or collected in Malaise traps can often be collected using this trap. The principle on which this trap works is that many insects associated with herbaceous vegetation are attracted to yellow color. This trap is made up of a shallow tray, about 60–75 mm deep and with 300–400 cm^2. The tray is painted with yellow on the inside. The tray can be placed on the ground in a suitable habitat such as grassland, scrub, a forest trail, or clearing. It is filled with water (plus a drop or two of detergent to break the surface tension), saturated salt solution, or a 50/50 mix of ethylene glycol and water. If only water is poured, the pan then must be emptied once a day; otherwise, specimens will become partly macerated, particularly in hot weather conditions. Pans having salt solution (5–10 %) may be left a couple of days, and that containing ethylene glycol/water mix can run for a week without changing the solution.

13.5.4 Survey of Literature

The literature pertaining to the genera of Chalcididae were collected and reviewed. The review represents a collection of much of the accessible literature that provides information on the taxonomy, diversity, distribution, and abundance of chalcids. Chalcid-related websites and publications as well as a number of natural history and biodiversity websites have been consulted. Online electronic survey was conducted to provide contemporary data for assessing the diversity, distribution, and abundance of chalcids in India, with specific reference to Kerala. The data from these accounts had been used to plot the distribution map.

13.6 Results

Species Richness Of the 40 genera described from the Oriental region, the most abundant genus is *Brachymeria* with 115 species. From Kerala, 83 species were recorded that belong to 21 genera distributed in 5 subfamilies (Fig. 13.1).
The subfamily with the greatest number of genera and species is Haltichellinae, with 8 genera (35 % of the total) and 39 species (47 % of the total). The Chalcidinae is represented by 2 genera (9 %) and 24 species (29 %). The genus with greatest number of species was *Brachymeria* with 23 species, followed by *Antrocephalus* with 15, *Hockeria* with 9, *Dirhinus* with 6, *Kriechbaumerella* and *Psilochalcis* with 5 each, *Epitranus* with 4, and *Oxycoryphe* and *Haltichella* with 2 each, while the remaining genera were represented with only 1 species. Approximately 61 % of the genera were represented by only one species.

The different species recorded under the chalcid genera from Kerala are listed below:

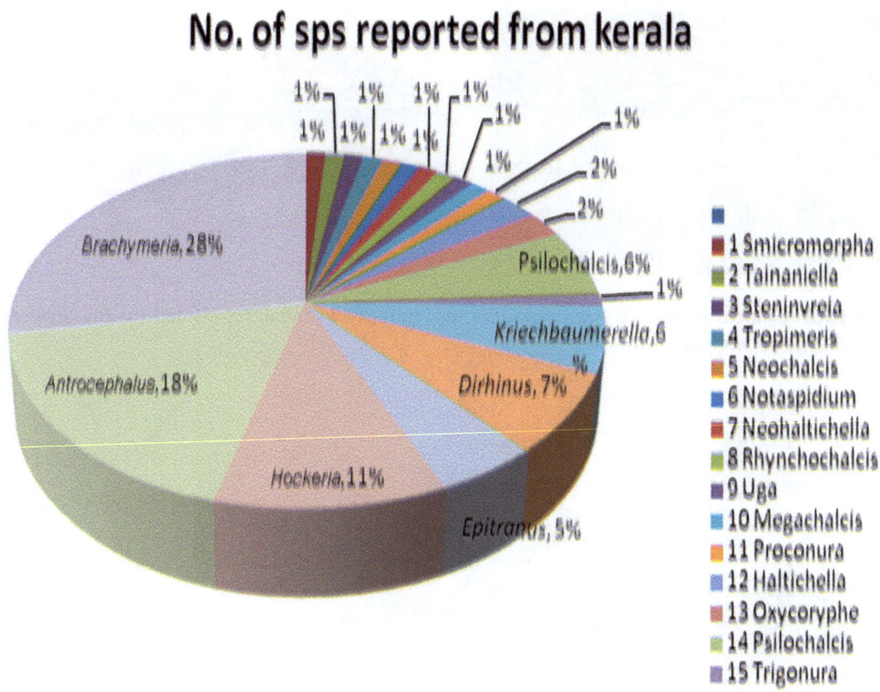

Fig. 13.1 Species richness of Chalcididae in Kerala

1. *Antrocephalus* spp. (Fig. 13.2 a)
   A. abui Narendran, *A. brevidentata* Roy and Farooqi, *A. brevigaster* Masi, *A. cariniceps* (Cameron), *A. decipiens* (Masi), *A. dividens* (Walker), *A. fascicornis* (Walker), *A. hakonensis* (Ashmead), *A. lugubris* (Masi), *A. maculipennis* Cameron, *A. masii* Özdikmen, *A. mitys* (Walker), *A. narendrani* Sureshan, *A. peechiensis* Narendran, *A. phaeospilus* Waterston
2. *Brachymeria* spp. (Fig. 13.2 b)
   B. apicicornis (Cameron); *B. atteviae* Joseph, Narendran, and Joy; *B. aurea* (Girault); *B. banksi* (Ashmead); *B. carinata* Joseph, Narendran, and Joy; *B. coxodentata* Joseph, Narendran, and Joy; *B. croceogastralis* Joseph, Narendran, and Joy; *B. excarinata* Gahan; *B. femorata* (Panzer); *B. hime* Habu; *B. inermis* (Fonscolombe); *B. jambolana* Gahan; *B. lasus* (Walker); *B. lugubris* (Walker); *B. margaroniae* Joseph, Narendran, and Joy; *B. marmonti* (Girault); *B. megaspila* (Cameron); *B. menoni* Joseph, Narendran, and Joy; *B. nephantidis* Gahan; *B. nosatoi* Habu; *B. scutellocarinata* Joseph, Narendran, and Joy; *B. taiwana* (Matsumura); *B. thracis* (Crawford)
3. *Dirhinus* spp. (Fig. 13.3 a)
   D. alticornis (Masi), *D. auratus* Ashmead, *D. bakeri* (Crawford), *D. banksi* Rohwer, *D. claviger* Bouček and Narendran, *D. madagascariensis* (Masi)

Fig. 13.2 Chalcid wasps: (a) *Antrocephalus* spp. (http://www.nbaii.res.in) and (b) *Brachymeria* spp. (http://www.nbaii.res.in)

Fig. 13.3 Chalcid wasps: (a) *Dirhinus* spp. (http://blog.insectmuseum.org) and (b) *Haltichella* spp. (http://bugguide.net)

4. *Epitranus* spp.
   E. elongatulus (Motschulsky), *E. erythrogaster* Cameron, *E. nigriceps* Bouček, *E. parvidens* (Strand)
5. *Haltichella* spp. (Fig. 13.3b)
   H. clavicornis (Ashmead), *H. macrocera* Waterston
6. *Hockeria* spp. (Fig. 13.4a)
   H. anupama Narendran, *H. atra* Masi, *H. carinata* Narendran, *H. fronta* Narendran, *H. hayati* Narendran, *H. manii* Narendran, *H. menoni* (Narendran), *H. tarsata* (Dalla Torre), *H. tristis* (Strand)
7. *Kriechbaumerella* spp.
   K. destructor (Waterston), *K. kraussi* Narendran, *K. ornatipennis* (Cameron), *K. rufimanus* (Walker), *K. titusi* Narendran
8. *Megachalcis malabarica* Narendran
9. *Neochalcis breviceps* (Masi)
10. *Neohaltichella thresiae* Narendran
11. *Notaspidium grisselli* Narendran

Fig. 13.4 Chalcid wasps: (a) *Hockeria* spp. (http://bugguide.net) and (b) *Proconura* spp. (http://ponent.atspace.org)

Fig. 13.5 Chalcid wasps: (a) *Psilochalcis* spp. (http://bugguide.net) and (b) *Steninvreia* spp. (http://www.waspweb.org)

12. *Oxycoryphe* spp.
    O. komui Narendran, *O. thresiae* Narendran
13. *Proconura caryobori* (Hanna) (Fig. 13.4b)
14. *Psilochalcis* spp. (Fig. 13.5 a)
    P. adhara (Narendran), *P. anupama* (Narendran), *P. carinigena* (Cameron), *P. ghanii* (Habu), *P. keralensis* Narendran
15. *Rhynchochalcis thresiae* Narendran
16. *Smicromorpha keralensis* Narendran
17. *Steninvreia noyesi* Narendran
18. *Tainaniella malabarica* Narendran
19. *Trigonura steffani* Narendran
20. *Tropimeris monodon* Bouček
21. *Uga menoni* Kerrich

13.7 Diversity in Kerala and Other States

The genera *Dirhinus*, *Megachalcis*, *Neochalcis*, and *Tropimeris* were reported equally from Kerala and other states. Nine genera, viz., *Conura*, *Indoinvreia*, *Cratocentrus*, *Chalcis*, *Neoirichohalticella*, *Stenochalcis*, *Hayatiella*, and *Bucekia*, are not reported from Kerala yet (Table 13.1, Fig. 13.6).

The number of species of *Antrocephalus* reported was three times higher in Kerala than other states, while the number of species reported of *Brachymeria*, *Hockeria*, *Epitranus*, *Trigonura*, and *Proconura* was less than half of that reported from other states.

Table 13.1 Diversity of genera of Chalcididae in Kerala and other Indian states

Sl. no.	Name of genus	No. of species reported from Kerala	No. of species reported from outside Kerala
1	*Bucekia*	0 %	1 %
2	*Hayatiella*	0 %	1 %
3	*Neoirichohalticella*	0 %	1 %
4	*Stenochalcis*	0 %	1 %
5	*Chalcis*	0 %	1 %
6	*Cratocentrus*	0 %	1 %
7	*Indoinvreia*	0 %	2 %
8	*Smicromorpha*	1 %	0 %
9	*Tainaniella*	1 %	0 %
10	*Steninvreia*	1 %	0 %
11	*Tropimeris*	1 %	1 %
12	*Neochalcis*	1 %	1 %
13	*Notaspidium*	1 %	0 %
14	*Conura*	0 %	2 %
15	*Neohaltichella*	1 %	2 %
16	*Rhynchochalcis*	1 %	0 %
17	*Uga*	1 %	0 %
18	*Megachalcis*	1 %	1 %
19	*Lasiochalcidia*	0 %	2 %
20	*Proconura*	1 %	3 %
21	*Haltichella*	2 %	3 %
22	*Oxycoryphe*	2 %	1 %
23	*Psilochalcis*	6 %	5 %
24	*Trigonura*	1 %	4 %
25	*Kriechbaumerella*	6 %	5 %
26	*Dirhinus*	7 %	5 %
27	*Epitranus*	5 %	8 %
28	*Hockeria*	11 %	11 %
29	*Antrocephalus*	18 %	4 %
30	*Brachymeria*	28 %	38 %

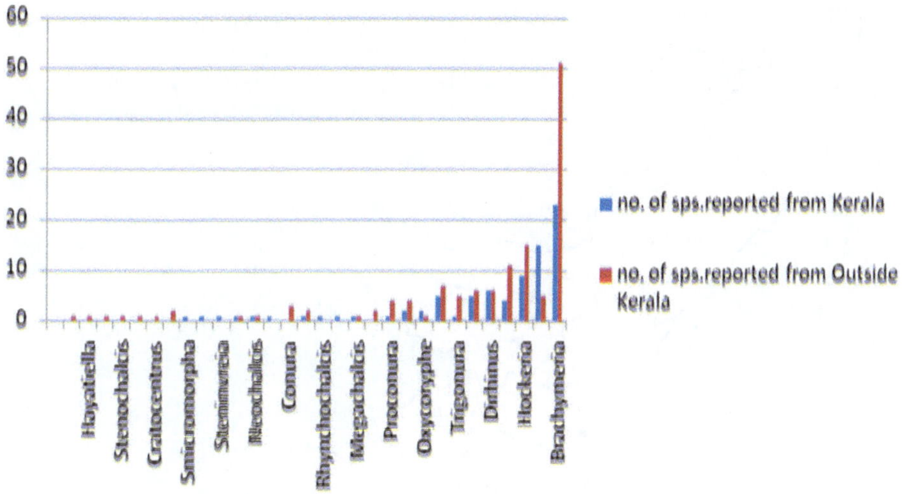

Fig. 13.6 Diversity of genera of Chalcididae in Kerala and the rest of Indian states

13.8 Distribution

The genera *Antrocephalus*, *Brachymeria*, *Proconura*, *Psilochalcis*, *Tropimeris*, *Haltichella*, and *Epitranus* were recorded from all the 14 districts of Kerala (Fig. 13.7). The districts of Kasaragod, Kannur, Ernakulam, Kottayam, Alappuzha, Pathanamthitta, Kollam, and Thiruvananthapuram had reports of only the above mentioned seven genera. The species of *Kriechbaumerella* were reported from Kozhikode, Malappuram, Palakkad, Thrissur, and Idukki, while *Neochalcis* were from Kozhikode, Malappuram, Thrissur, Palakkad, and Idukki. While the genus *Dirhinus* was reported from Wayanad, Kozhikode, and Malappuram, the genus *Hockeria* was reported from Kozhikode, Malappuram, Thrissur, and Idukki. The genus *Rhynchochalcis* was reported from Wayanad and Thrissur. Nine genera were reported from one district only. The genera *Uga*, *Megachalcis*, *Steninvreia*, *Oxycoryphe*, *Smicromorpha*, *Lasiochalcidia*, *and Tainaniella* were from Malappuram; *Indoinvreia* was from Kozhikode and *Chalcis* sp. from Palakkad.

13.9 Discussion

The diversity of chalcids in Kerala reflects the favorable climatic and topographic conditions for the sustenance of host plants (for phytophagous forms) and host animals (for parasitic or hyperparasitic forms). The comparison may also reflect the limitations of the sampling technique. The difference in the richness between different regions may also be due to the variations in the regions or the extent of studies carried out in those genera or regions. The distribution of chalcid genera in the different districts of Kerala shows that certain genera like *Antrocephalus*, *Brachymeria*,

Fig. 13.7 Distribution of genera of Chalcididae in districts of Kerala

Proconura, *Psilochalcis*, *Tropimeris*, *Haltichella*, and *Epitranus* are distributed almost uniformly throughout the state. Some of the genera like *Dirhinus*, *Uga*, and *Megachalcis* are reported only from Malabar regions.

Acknowledgments We thank our mentor Dr. T. C. Narendran for guidance. Preethi, N. expresses her gratitude to the Principal and the Registrar of Sainik School for their support. Lakshmi Devi Menon is grateful to Dr. P. A. Sebastian, S.H. College Thevara, and Fr. Johson Pallackapillil, Principal of S. H. College, for support and encouragement.

References

Ashmead WH (1904) Classification of the chalcid flies of the superfamily Chalcidoidea, with descriptions of new species in the Carnegie Museum, collected in South America by Herbert H. Smith. Mem Carnegie Mus 1(4):251

Bouček Z (1951) The first revision of the European species of the family Chalcididae (Hymenoptera). Acta Entomol Mus Natl Prague 27:1–29

Bouček Z (1952) *Steffanisa rubrocincta* n.g., n.sp., and notes of some other species (Chalcididae, Hym.). Acta Entomol Musei Natl Pragae 27:201–205

Bouček Z (1956) A contribution to the knowledge of the Chalcididae, Leucospidae and Eucharitidae (Hymenoptera, Chalcidoidea) of the Near East. Bull Res Counc Isr 5B:227–259

Bouček Z (1982) Oriental chalcid wasps of the genus *Epitranus*. J Nat Hist 16:577–622

Bouček Z (1988) Australian Chalcidoidea (Hymenoptera) a biosystematic revision of genera of fourteen families, with a reclassification of species. CAB International, Wallingford, pp 717–719

Bouček Z (1992) The New World genera of Chalcididae. Mem Amer Ent Inst 53:49–117, 443–446

Bouček Z, Narendran TC (1981) Indian chalcid wasps (Hymenoptera) of the genus *Dirhinus* parasitic on synanthropic and other Diptera. Syst Entomol 6:229–251

Delvare G (1995) 11.3 Chalcididae. In: Hanson P, Gauld ID (eds) The Hymenoptera of Costa Rica. Oxford University Press, Oxford, pp 289–298

Delvare G, Bouček Z (1992) On the New World Chalcididae (Hymenoptera). Mem Am Entomol Inst 53:1–466

en.wikipedia.org/wiki/ Chalcididae. Last referred on 28 Dec 2012

en.wikipedia.org/wiki/ Malaise_trap. Last referred on 28 Dec 2012

Gahan AB (1930) Synonymical and descriptive notes on Parasitic Hymenoptera. Proc U S Natl Mus 77(8):1–12

Gahan AB (1942) Descriptions of five new species of Chalcidoidea, with notes on a few described species (Hymenoptera). Proc US Natl Mus 92:41–51

Gibson GAP (1993) Superfamilies Mymarommatoidea and Chalcidoidea. In: Goulet H, Huber JT (eds) Hymenoptera of the world: an identification guide to families. Canada Communications Group, Ottawa, pp 668

Habu A (1960) A revision of the Chalcididae (Hymenoptera) of Japan with description of sixteen new species. Bulletin of National Institute of Agricultural Sciences. Tokyo (C) 11:131–363

Habu A (1966) Descriptions of some *Brachymeria* species of Japan (Hymenoptera: Chalcididae). Kontyu 34(1):22–28

Husain T, Agarwal MM (1981) A new species of the genus *Tropimeris* Steffan (Hymenoptera: Chalcididae) from India. J Entomol Res 5(1):36–38

Husain T, Agarwal MM (1982a) Taxonomic studies on Indian Epitraninae (Hymenoptera: Chalcididae). Orient Insects 15(4):413–432

Husain T, Agarwal MM (1982b) Taxonomic studies on Haltichellinae of India (Hymenoptera: Chalcididae) PT.I. Haltichellini. Orient Insects 16:333

Joseph KJ, Narendran TC, Joy PJ (1973) Oriental Brachymeria. A monograph on the Oriental species of *Brachymeria* (Hymenoptera: Chalcididae). University of Calicut, Zoology Monograph Know India: Geography of Kerala

Malaise R (1937) A new insect-trap. Entomol Tidskr, Stockholm 58:148–160

Mani MS (1935) New Indian Chalcidoidea (Parasitic Hymenoptera). Rec Indian Mus 37:241–258

Mani MS (1936) Chalcids (Parasitic Hymenoptera) From India. Rec Indian Mus 38:125–129

Mani MS (1938) Catalogue of Indian insects. Part 23 – Chalcidoidea. pp. ii+174pp Delhi New combination for *Chalcis elongatula* Motschulsky, pp 55

Masi L (1927) Diagnosi di una nuova species di *Belaspidia* (Hymen. Chalcididae). Boll Soc Entomol Ital 59:82–83

Masi L (1929a) Sopra un nuovo genere di Haltichellini e sulle diverse forme attribute al genere *Euchalcis* Duf. (Hymen. Chalcididae). Mem Soc Entomol Ital 6:208–222

Masi L (1929b) Contributo alla conoscenza dei calcididi orientali della sottofamiglia Chalcidinae. Boll Lab Entomol RIstituto Super Agrar Bologna 2:155–188

Narendran TC (1976) Note on two little known species of *Antrocephalus* Kirby (Hymenoptera: Chalcididae) from India. Entomon 1(2):185–188

Narendran TC (1977) The systematic position of the genus *Tainania* (Hym.: Chalcididae). Entomophaga 22(3):295–297

Narendran TC (1984) Key to Indian genera of the family Chalcididae (Hymenoptera: Chalcidoidea). Entomophaga 29(4):431–438

Narendran TC (1989) Oriental Chalcididae (Hymenoptera: Chalcidoidea). Zoological Monograph. Department of Zoology, University of Calicut, Kerala

Nikol'skaya MN (1952) Chalcids of the fauna of the USSR (Chalcidoidea). Opredeliteli Faune SSSR, Zool Inst Akade Nauk SSSR, Moscow Leningrad 44:575

Nikol'skaya MN (1960) Chalcididae and Leucospidae in central Asia (Hymenoptera, Chalcidoidea). Trudy Zool Inst Akad Nauk SSSR, Leningrad 27:220–246

Steffan JR (1959) Revision de la tribu des Cratocentrini (Hymen. Chalcididae). Acta Entomol Mus Natl Pragae 33:317

Steffan JR (1962) Chalcidoides de l'Institut National d'Entomologique de Rome. Fragm Entomol 4(3):25

Steffan JR (1976) Les *Euchalcidia* Masi du basin méditerranéen (Hym. Chalcididae). Bull Soc Entomol France 81(1/2):52–63

Waterston J (1922) On Chalcidoidea (Mainly bred at Dehra Dun, UP, from pests of Sal, Toon, Chir and Sundri). Indian For Rec 9:23

www.newkerala.com/states-of-india/kerala.php. Last referred on 28 Dec 2012

www.prokerala.com/kerala/climate.htm. Last referred on 28 Dec 2012

www.sel.barc.usda.gov/hym/chalcids/chalcidd.html. Last referred on 28 Dec 2012

Diversity of Economically Important Indian Microgastrinae (Hymenoptera: Braconidae) with New Records from India

14

Ankita Gupta

Abstract

Microgastrinae (Hymenoptera: Braconidae) is a hyperdiverse subfamily of koinobiont parasitic wasps which are obligate endoparasitoid of larval Lepidoptera. Indian Microgastrinae encompasses 21 genera and 231 species including introduced species for various biological control programmes. Of these, 191 species are currently recognised as valid, and 40 species are questionable due to the lack of type specimens or erroneous identification. Amongst the four species that are introduced in India as parasitoids for control of their respective pest, only one has been recovered, whilst the rest have not been established or their present status is not known. Indian microgastrine genera include *Apanteles*, *Buluka*, *Cotesia*, *Distatrix*, *Dolichogenidea*, *Diolcogaster*, *Fornicia*, *Parapanteles*, *Hypomicrogaster*, *Microgaster*, *Microplitis*, *Neoclarkinella*, *Parapanteles*, *Parenion*, *Philoplitis*, *Pholetesor*, *Protapanteles*, *Promicrogaster*, *Protomicroplitis*, *Snellenius* and *Wilkinsonellus*. Nearly 42 species under 15 genera have been documented from all across India during 2010–2012. Also 35 species of hosts from 14 families of Lepidoptera are also recorded in association with parasitic wasp species. *Apanteles* followed by *Cotesia* is recorded as the most diverse and maximum encountered genus in India. New records of microgastrine wasps from host families Crambidae, Arctiidae and Hesperiidae are recorded: *Apanteles mamitus* Nixon, a solitary endoparasitoid, parasitic on larvae of *Spoladea recurvalis* (Fabricius) (Lepidoptera: Crambidae) on the host plant *Amaranthus* sp., is recorded for the first time from India and cashew leaf roller, *Pleuroptya balteata* (Fabricius) (Lepidoptera: Crambidae) on *Anacardium occidentale* L., is recorded as a new host for *Apanteles javensis* Rohwer. *Dolichogenidea kunhi* Gupta and Kalesh is recorded from a larva of *Borbo*

A. Gupta (✉)
Insect Systematics Division, National Bureau of Agricultural Insect Resources, No. 2491, H. A. Farm Post, Hebbal, Bangalore 560 024, Karnataka, India
e-mail: drankitagupta7@gmail.com

A.K. Chakravarthy, S. Sridhara (eds.), *Arthropod Diversity and Conservation in the Tropics and Sub-tropics*, DOI 10.1007/978-981-10-1518-2_14

cinnara (Wallace) (Lepidoptera: Hesperiidae) from Maharashtra and Tamil Nadu. A new distribution record for *Neoclarkinella vitellinipes* (You and Zhou) is obtained from southern India. Distinct host-range extension is noticed for species *A. phycodis*, *A. javensis* and *D. kunhi*. The present study provides a comprehensive list of microgastrine species with host-parasitoid association, information on diagnostic characters of species with new records, species distribution, taxonomic relationships and GenBank accession numbers of all molecularly characterised species studied during 2010–2012.

Keywords
Endoparasitoids • Diversity • Koinobiont • Microgastrinae

14.1 Introduction

Microgastrine wasps (Hymenoptera: Braconidae) are primary parasitoids of Lepidoptera. Some of them parasitise major economic pests and have been largely used for biocontrol programmes. Microgastrinae form a diverse subfamily within Braconidae, with about 2000 species widespread throughout the regions of the world, except Antarctica (Yu et al. 2012). Beyond this consensus, the phylogenic subdivision of Microgastrinae is still debated, though several attempts were proposed to stabilise it (cf. Whitfield et al. 2002 for a review). As a large part of the world fauna is still poorly known, the actual richness of Microgastrinae is currently expected to range between 5000 and 10,000 species.

Within the Braconidae, the subfamily Microgastrinae, with an estimated global diversity of 5000–10,000, is one of the most important groups of parasitoids in terms of species richness and economic importance. This fact has caused the Microgastrinae to be one of the most studied parasitic wasps by DNA barcoding. More than 100 species of Microgastrinae have been used in the biological control of Lepidoptera pests, and this total is likely to rise.

Microgastrinae is the most important group of koinobiont parasitoids of Lepidoptera. These tiny wasps attack caterpillars and develop in large broods as ‘gregarious’ species or develop one per host as ‘solitary’ species depending on the size of caterpillars, but there are few exceptions. The subfamily Microgastrinae is universally recognised as a monophyletic taxon, characterised by its wing venation and number of flagellomeres invariably fixed to 16. However, the latter’s strong apomorphic feature was nevertheless reconsidered with the discovery of the unusual genus *Kiwigaster* (Fernández-Triana et al. 2011), with the males and females having different numbers of antennal flagellomeres (females 17, males 18).

Recently Rodriguez et al. (2012) extrapolated a new range of estimates for the global species richness of microgastrine wasps. The new estimates range from minimum 17,000 to maximum 46,000+ for the species. Looking at the enormous species richness, economic importance in biological control programmes, lack of taxonomic expertise, lack of availability of type specimens and erroneous identification, the Indian Microgastrinae remains a taxonomically challenging group. Some of these parasitic wasps attack major economic pests and have been largely used for biocontrol (Rao et al. 1971; Sankaran 1974).

Surveys were undertaken by the National Bureau of Agricultural Insect Resouces (NBAIR), Bangalore, India, during 2010–2012 to various parts of India to document the diversity of parasitic wasps. Gupta and Kalesh (2012) documented eight species of parasitic wasps bred from various life stages of five species of lepidopteran hosts belonging to the family Hesperiidae, viz. *Thoressa evershedi* (Evans), *Pelopidas mathias* (Fabricius), *Udaspes folus* (Cramer), *Borbo cinnara* (Wallace) and *Caltoris* sp., inhabiting Western Ghats, Kerala, India. Also, *Glyptapanteles hypermnestrae* bred from parasitised larvae of *Elymnias hypermnestra* (Linnaeus) (Lepidoptera: Nymphalidae) was reported from Maharashtra, India (Gupta and Pereira 2012).

Looking at the vast diversity of these tiny parasitic wasps and their interesting host-parasitoid relationships, the present study is undertaken to provide a consolidated review with the documentation of new records from India during 2010–2012. Also steps were initiated to molecularly characterise the microgastrine fauna.

14.2 Materials and Methods

Routine exploration surveys were undertaken by the author and the staff of Insect Biosystematics Division of the National Bureau of Agricultural Insect Resouces (NBAIR), Bangalore, India, during 2010–2012. The collection localities covered were nine states: Jammu and Kashmir, Uttar Pradesh, Uttarakhand, Maharashtra, Karnataka, Andhra Pradesh, Kerala, Tamil Nadu and Andaman and Nicobar Islands. The wasp specimens were collected through laboratory rearing of field-collected parasitised hosts and also through field sampling (mass collecting) using sweep nets, yellow pan traps and Malaise traps. The Malaise traps and yellow pan traps were installed/set for the entire duration of stay during the exploration visit to distant places and were daily monitored. The larval rearing was conducted on the natural host plant from which the respective larvae were collected. The specimens received for identification from all across India are also included in the studies.

The morphological characterisation was based on thorough taxonomic studies of specimens collected and identified by the author. Wasp specimens were processed using hexamethyldisilazane and card mounted and identified consulting relevant taxonomic literature. The wasp images were taken using Leica M205-A stereozoom microscope with Leica DC 420 inbuilt camera using AutoMontage software (version 3.8). The images were further processed using Adobe Photoshop.

Molecular characterisation was done with DNA extraction and PCR (polymerase chain reaction) amplification. DNA was extracted using DNeasy Blood and Tissue Kit (QIAGEN) and stored at −80 °C until used following the manufacturer's protocols. Amplification, sequencing and sequence analysis closely followed standard methods. PCR amplification of the CO1 gene was done by using the universal primer. PCR was performed with a total reaction mixture of 50 μl consisting of 10x

Taq Buffer (complete buffer with $MgCl_2$), 10 mMdNTP mix (Genei), universal primers HCO1-2198 (20 pm/µl) and LCO1-1490 (20 pm/µl) (Folmer et al. 1994), template DNA (50 ng/µl), Taq DNA polymerase (Genei) (1U/µl) and sterile water. The DNA extracted was amplified under the following conditions: initial denaturation at 94 °C for 5 min, followed by 30 cycles of denaturation (94 °C for 1 min), annealing (45 °C for 1 min), extension (72 °C for 1 min) and a final extension step at 72 °C for 10 min. To detect any possible contamination, a negative control was kept in the PCR amplification. The PCR amplification was performed in a thermal cycler (Bio-Rad, C1000 Thermal Cycler). The amplified genes by PCR were visualised on 1.8 % gel with a low-range ladder (Fermentas MassRuler 1000 bp). The PCR products from which the partial sequences of CO1 region were obtained were sequenced. The PCR products sequenced were edited by BioEdit, aligned using BLAST 2 and verified by BLASTN for species homology. Nucleotide sequences were submitted to GenBank. The sequences were aligned for studying the similarity between the species using MegAlign, DNASTAR Inc. (Lin and Wood 2002).

14.3 Results and Discussion

The present study is based on taxonomic examination/identification of approximately 2500 specimens of microgastrine wasps collected/reared by select methods in comparison with examination of type specimens studied at the National Collections, New Delhi. All specimens of the present study are housed in the repository of the NBAIR, Bangalore, India.

Nearly 42 species under 15 genera are documented from all across India. The 15 genera collected include *Apanteles*, *Buluka*, *Cotesia*, *Distatrix*, *Dolichogenidea*, *Diolcogaster*, *Fornicia*, *Glyptapanteles*, *Microgaster*, *Microplitis*, *Neoclarkinella*, *Parapanteles*, *Pholetesor*, *Protapanteles* and *Snellenius*.

Fifteen (15) lepidopteran host families are documented as hosts for Microgastrinae wasps: Arctiidae, Crambidae, Gracillariidae, Hesperiidae, Limacodidae, Lycaenidae, Lymantriidae, Papilionidae, Plutellidae, Pieridae, Pyralidae, Sphingidae, Noctuidae, Nymphalidae and Oecophoridae, the majority being associated with moths. More than 60 species of host caterpillars, majority belonging to Lepidoptera, were reared out, of which 35 species could be identified to the genus/species level. Gregarious nature was noticed for 12 reared species: *A. javensis*, *A. taragamae*, *A. machaeralis*, *A. phycodis*, *A. stantoni*, *C. flavipes*, *C. ruficrus*, *Distatrix papilionis*, *G. hypermnestrae* and for three unidentified species of the genus *Glyptapanteles*, *Diolcogaster* and *Parapanteles*.

During this study period, two new species of parasitic wasps were discovered, *Dolichogenidea kunhi* (Gupta and Kalesh 2012) and *Glyptapanteles hypermnestrae* (Gupta and Pereira 2012), already published by the author. A complete list of host-parasitoid association for the species of Microgastrinae wasps documented during 2010–2012 is provided in Table 14.1. The status of the four introduced species in India for various biological control programmes is given in Table 14.2.

Table 14.1 Host-parasitoid association documented for the Microgastrinae wasps (2010–2012)

S. no.	Species	Host	Host family	Host plant	Distribution record
1.	*Apanteles bisulcatus* Cameron	*Glyphodes* (= *Diaphania*) *pulverulentalis* Hampson	Crambidae	*Morus alba*	Karnataka
2.	*Apanteles folia* Nixon	*Arhopala amantes* (Hewitson)	Lycaenidae	*Lagerstroemia speciosa* L.	Karnataka
3.	*Apanteles galleriae* Wilkinson	*Galleria mellonella* L.	Pyralidae	–	Jammu and Kashmir, Karnataka
4.	*Apanteles hyposidrae* Wilkinson	*Acria* sp.	Oecophoridae	*Elaeis guineensis* Jacq.	Andhra Pradesh
5.	*Apanteles javensis* Rohwer	*Pleuroptya balteata* (Fabricius)	Crambidae	*Anacardium occidentale* L.	New-host record Karnataka
	Apanteles javensis Rohwer	*Pelopidas mathias* (Fabricius)	Hesperiidae	*Brachiaria mutica* (Forsk.) Stapf	Kerala
6.	*Apanteles mamitus* Nixon	*Spoladea recurvalis* (Fabricius) (=*Hymenia recurvalis* (Fabricius))	Crambidae	*Amaranthus* sp.	First distribution record from India (Karnataka)
7.	*Apanteles machaeralis* Wilkinson	*Eutectona machaeralis* (*Walker*)	Crambidae	*Tectona grandis* L.	Karnataka
8.	*Apanteles opacus* (Ashmead)	Sweep net		Weed	Karnataka
9.	*Apanteles phycodis* Viereck	*Argina syringa* Cramer	Arctiidae	–	Karnataka and Tamil Nadu
10.	*Apanteles stantoni* (Ashmead)	*Parotis marginata* (Hampson)	Crambidae	*Gardenia jasminoides* J. Ellis	Karnataka
11.	*Apanteles tachardiae* Cameron			–	Orissa
12.	*Apanteles taragamae* Viereck	*Diaphania indica* Saunders	Crambidae	*Cajanus cajan* (L.) Millsp.	Cosmopolitan

(continued)

Table 14.1 (continued)

S. no.	Species	Host	Host family	Host plant	Distribution record
13.	*Buluka* sp.	Sweep net	–	–	Karnataka
14.	*Cotesia erionotae* (Wilkinson)	*Udaspes folus* (Cramer)	Hesperiidae	*Zingiber zerumbet* (L.) Smith	Karnataka
15.	*Cotesia flavipes* Cameron	*Chilo partellus* (Swinhoe) on maize	Crambidae	*Zea mays* L.	Cosmopolitan
16.	*Cotesia glomerata* (Linnaeus	*Pericallia ricini* Fabricius	Arctiidae	–	Karnataka
17.	*Cotesia marginiventris*	*Plutella xylostella*, *Pieris brassicae*	Plutellidae Pieridae	*Brassica oleracea* L.	Southern India
18.	*Cotesia ruficrus* (Haliday)	*Spodoptera* sp.	Noctuidae	Weed	Karnataka
19.	*Cotesia vestalis* (=*plutellae*) (Haliday)	*Plutella xylostella* (L.)	Plutellidae	*Brassica oleracea* L.	Cosmopolitan
20.	*Diolcogaster* sp1.	Malaise trap	–	–	Andaman Islands
21.	*Diolcogaster* sp2.	Sweep net	–	*Terminalia* sp.	Tamil Nadu
22.	*Dolichogenidea* sp1.	*Melanephia mosara* Swinhoe	Noctuidae	*Ipomoea pes-caprae*	Wandoor (Andaman Islands)
23.	*Dolichogenidea* sp2.	*Caloptilia acidula* Myerick (Aonla leaf roller)	Gracillariidae	*Phyllanthus emblica* (=*Emblica officinalis*) L.	Karnataka
24.	*Dolichogenidea kunhi* Gupta and Kalesh	*Thoressa evershedi* (Evans)	Hesperiidae	*Ochlandra* sp.	Kerala
	Dolichogenidea kunhi Gupta and Kalesh	*Borbo cinnara* (Wallace)	Hesperiidae	Weed	New distribution record: Tamil Nadu and Maharashtra
25.	*Distatrix papilionis* (Viereck)	*Papilio demoleus* L.	Papilionidae	*Citrus aurantifolia* (Christm.) Swingle	Karnataka, Assam, Andhra Pradesh and Tamil Nadu
26.	*Fornicia ceylonica* Wilkinson	*Spatulifimbria castaneiceps* Hampson	Limacodidae	*Ricinus communis* L.	Karnataka

(continued)

Table 14.1 (continued)

S. no.	Species	Host	Host family	Host plant	Distribution record
27.	*Glyptapanteles* sp1	*Ocinara varians* (Walker)	Sphingidae	*Ficus religiosa* L.	Karnataka
28.	*Glyptapanteles* sp2	Sweep net	–	*Artemisia vulgaris* L.	Karnataka
29.	*Glyptapanteles* sp3	Sweep net	–	–	Karnataka
30.	*Glyptapanteles hypermnestrae* Gupta and Pereira	*Elymnias hypermnestra* (Linnaeus)	Nymphalidae	*Cocos nucifera* L.	Maharashtra
31.	*Microplitis maculipennis* (Szepligeti)	*Acanthodelta janata* L.	Noctuidae	*Ricinus communis* L.	Karnataka, Andhra Pradesh, Orissa, Maharashtra and Tamil Nadu.
32.	*Neoclarkinella vitellinipes* (You and Zhou)	Yellow pan trap	–	–	Karnataka
33.	*Microgaster* sp.	Sweep net	–	Weed	Kashmir
34.	*Microplitis prodeniae* Rao and Kurian	Sweep net	–	–	Karnataka
35.	*Microplitis demolitor* Wilkinson	*Helicoverpa armigera* (Hubner)	Noctuidae	–	Karnataka
36.	*Microplitis indicus* Marsh	*Helicoverpa* sp.	Noctuidae	–	Himachal Pradesh
37.	*Microplitis manila* Ashmead	*Spodoptera* sp.	Noctuidae	–	Jammu and Kashmir and Uttar Pradesh
38.	*Microplitis prodeniae* Rao and Kurian	*Spodoptera litura* (Fabricius)	Noctuidae	*Amaranthus* sp.	Karnataka
39.	*Microplitis spodoptera* Rao and Kurian	*Spodoptera mauritia* (Boisduval)	Noctuidae	*Trigonella foenum-graecum* L.	Karnataka and Bihar
40.	*Parapanteles* sp.	Sweep net	–	–	Maharashtra
41.	*Protapanteles obliquae* (Wilkinson)	*Diaphania* (=*Glyphodes*) *pyloalis* Walker	Crambidae	–	Karnataka
42.	*Protapanteles colemani* (Viereck)	*Orgyia postica* Walker	Lymantriidae	*Cajanus cajan* (L.) Millsp.	Karnataka

Source: author's (Ankita Gupta) unpublished research data

Table 14.2 Status of four introduced wasp species in India for biological control programmes

S. no.	Species	Remarks	Status
1.	*Apanteles malevolus* Wilkinson	Introduced in 1937–1938 from Myanmar (Burma) for the biological control of *Eutectona machaeralis* (Walker) on teak	Released in the field, present status not known
2.	*Apanteles scutellaris* Muesebeck	Introduced in India in 1965 from the USA for the target pest *Phthorimaea operculella* (Zeller) on potato	Present status not established in India
3.	*Apanteles subandinus* Blanchard	Introduced in India in 1944–1945 from South America for the target pest *Phthorimaea operculella* (Zeller) on potato and was recovered later	Recovered. Also reported from Maharashtra
4.	*Cotesia sesamiae* (Cameron 1906) junior synonym of *Cotesia ruficrus* (Haliday 1834) [nomen nudum]	Introduced from Uganda in 1967 for the control of *Chilo partellus* (Swinhoe), sugarcane borers, *C. infuscatellus* (Snellen), *C. sacchariphagus indicus* (Kapur), *C. auricilius* (Dudgeon), *C. tumidicostalis* (Hampson), *Scirpophaga excerptalis* (Walker) and *Acigona steniellus* (Hampson) for the crops sorghum, maize, sugarcane	Inundative releases were made in Maharashtra and Punjab. The status shows it is not established

Source: Rao et al. (1971), Sankaran (1974)

In this paper new records of larval parasitoids from host caterpillars belonging to the families Crambidae, Arctiidae and Hesperiidae are provided with illustrations. Marked host-range extension is observed for *Apanteles phycodis* Viereck and *A. javensis* Rohwer. The species, *A. mamitus* Nixon bred from larvae of *Spoladea recurvalis* (Fabricius) (Lepidoptera: Crambidae) on the host plant *Amaranthus* sp., is recorded for the first time from India; the cashew leaf roller, *Pleuroptya balteata* (Fabricius) (Lepidoptera: Crambidae), and *Argina syringa* Cramer (Lepidoptera: Arctiidae) are recorded as new hosts for *A. javensis* and *A. phycodis*, respectively. Also *Cotesia glomerata* (Linnaeus), a solitary larval endoparasitoid, is bred from larvae of *Pericallia ricini* Fabricius (Lepidoptera: Arctiidae) on the host plant *Ricinus communis* L., and a new distribution record for *Neoclarkinella vitellinipes* from Karnataka, India, is given.

(1) **Microgastrine wasps reared from host family Crambidae and Arctiidae include**: *Apanteles mamitus* Nixon, *Apanteles javensis* Rohwer, *Apanteles phycodis* Viereck and *Cotesia glomerata* (Linnaeus)

***Apanteles mamitus* Nixon 1965 (Plate** 14.1 **– Figs.** 14.1, 14.2, 14.3, 14.4 and 14.5)

This is the first ever record of this species from India.

Plate 14.1 *Apanteles mamitus* Nixon

Fig. 14.1 Adult female wasp, dorsal view

Fig. 14.2 Propodeum and part of metasoma; dorsal view

Fig. 14.3 Metasoma, dorsal view

Fig. 14.4 Host caterpillar, *Spoladea recurvalis* (Fab.)

Brief diagnosis: Vertex punctuate; first tergum black, markedly rugose and considerably narrower at apex; tergites (2+3) reddish yellow; hind tibia 1.2× longer than ovipositor sheath.

Material examined: Two females, August 2012, Bangalore, Karnataka, India, solitary endoparasitoid, ex. caterpillar of *Spoladea recurvalis* (Fabricius) (Lepidoptera: Crambidae) on the host plant *Amaranthus* sp. coll. Ankita. All specimens deposited in NBAIR, Bangalore, India (Code NBAII/Bra/Mic/Apan/mam/0812).

Distribution: India (new record), China, the Philippines and Vietnam.

Fig. 14.5 Host, adult of *S. recurvalis*

Apanteles javensis **Rohwer 1919 (Plate** 14.2 – **Figs.** 14.6, 14.7 and 14.8)

Apanteles javensis Rohwer is recorded for the first time from the host family Crambidae. It is associated with the family Hesperiidae from larvae of *Pelopidas mathias* (Fabricius) (Gupta and Kalesh 2012), *Parnara conjuncta* Herrich-Schaffer, *Polytremis pellucid* (Murray) and from the family Nymphalidae from larvae of *Mycalesis perseus* (Fabricius) (Yu et al. 2012).

Brief diagnosis. Female. Black; antennae, tegulae, hind legs large, midlegs except apex of femora, front legs at base of femora, basal sternites and first latero-tergite deep brown red; rest of the portion of legs red testaceous. Basal part of hind tibiae lighter than apical part. Stigma pallid with margins brown. Propodeum with areola well marked in the apical half; areola shining. Ovipositor sheath equal in length to basal joint of the hind tarsus. First tergite of metasoma narrow apically and broad basally; punctuate along the marginal third in apical half (except at apex); second tergite completely smooth and shining.

Material examined: Five females, 21.xii.2011, Bangalore, Karnataka, India, gregarious larval endoparasitoid, ex. caterpillar of cashew leaf roller, *Pleuroptya balteata* (Fabricius) (Lepidoptera: Crambidae) on *Anacardium occidentale* L., coll. Ankita. All specimens deposited in the NBAIR, Bangalore, India (Code NBAII/Bra/Mic/Apan/jav/211211).

Distribution: India, China, Indonesia, Japan, Sri Lanka and Thailand.

Apanteles phycodis **Viereck 1913 (Plate** 14.3 – **Figs.** 14.9, 14.10, 14.11, 14.12 and 14.13)

Apanteles phycodis Viereck, prevalent only in India, is recorded as a gregarious parasitoid on larva of *Phycodes radiata* Ochsenheimer (Lepidoptera: Brachodidae) on *Ficus* sp. Thus *Argina syringa* (Lepidoptera: Arctiidae) is a new host record as well as marked host extension of *A. phycodis* parasitic wasp into the family Arctiidae.

6

7

8

Plate 14.2 *Apanteles javensis* Rohwer

Brief Diagnosis: Follows the detailed description in Wilkinson (1928). Legs mostly red testaceous except apical one third of hind tibia red black; coxae black. Tegulae black; stigma white with its margins brown. First tergite of metasoma nearly twice as long as broad; longitudinally striate. Second tergite slightly sculptured. Ovipositor sheaths longer than hind tarsi.

Fig. 14.6 Adult female wasp, dorsal view

Fig. 14.7 Host caterpillar, *Pleuroptya balteata* (Fab.)

Fig. 14.8 Host, adult of *P. balteata*

Plate 14.3 *Apanteles phycodis* Viereck

Fig. 14.9 Adult female wasp, dorsal view

Fig. 14.10 Mesosoma and metasoma, dorsal view

Fig. 14.11 Adult female wasp, lateral view

Fig. 14.12 Cocoons of *A. phycodis*

Fig. 14.13 Host, adult of caterpillar of *Argina syringa* Cramer

Material Examined: Six females mounted on card, 30.iii.12, Periyakulam, Tamil Nadu, India, coll. Rajeshwari, ex. caterpillar of *Argina syringa* (Lepidoptera: Arctiidae). All specimens deposited in NBAIR, Bangalore, India (Code NBAII/Bra/Mic/Apan/phy/30312).

Distribution: India, Tamil Nadu (new distribution record) and Karnataka.

Neoclarkinella vitellinipes **(You and Zhou** 1990) **(Plate** 14.4 – **Figs.** 14.14, 14.15, 14.16 and 14.17)

The genus *Neoclarkinella*, an elusive of old world tropics, is documented from India, Thailand, Malaysia, the Philippines and China. It is recognised by the presence of a complete transverse carina in the propodeum along with a complete median longitudinal carina (shown in the Figs. 14.15 and 14.16). The first metasomal mediotergite has a large, rounded depression centrally in the anterior half (usually of a different, paler colouration compared to the rest of the tergite), and the first mediotergite itself strongly narrows towards the posterior margin.

Brief diagnosis of ***N. vitellinipes***: T1 4x as long as apex, strongly narrowed at apex; distinctive 'U'-shaped basal depressed area of T1 and lateral sides yellow; T1 punctate baring apical patch and basal depression; T2 subtriangular; propodeum with mid-longitudinal carina and basal carinae; the presence of large triangular scutellar lunules.

Distribution: India (Karnataka, new distribution record), Thailand, Malaysia, the Philippines and China.

Material Examined: Two females, five males, 8.viii.12, IWST, Malleswaram, Karnataka, India, yellow pan trap, coll. Ankita. All specimens deposited in NBAIR, Bangalore, India (NBAII/Bra/Mic/Neo/vite/8812).

Dolichogenidea kunhi **Gupta and Kalesh**

Brief Diagnosis: Body and head colour black; scape blackish. Hind tibia slightly longer than ovipositor sheath. Tergite (2+3) distal to the basal area smooth and shiny without any coarse sculpture. Ovipositor sheaths thickened in apical one third with apical attenuation. First discoidal cell not wider than high. First metasomal tergal plate longer than wide with coarse sculpture at basal half and longitudinal carinae at the apical half; parallel sided; slightly wider apically; bearing a median longitudinal depression. Second tergum wider than long; shorter than third tergum. Tergite (2+3) distal to the basal area smooth and shiny without any coarse sculpture; considerably longer than the basal area itself. Ovipositor sheaths distinctly projecting beyond apex of gaster. Hind tibia slightly longer than ovipositor sheath. Ovipositor sheaths long and hairy throughout; gently decurved. Third anterior of mesopleuron coarsely puncto-reticulate and setose, posteriorly largely smooth and nitid. Metapleuron generally smooth except for very shallow punctures posteriorly, anterior pit deep. Hind coxae laterally and dorsally shagreened with shallow punctures.

Plate 14.4 *Neoclarkinella vitellinipes* (You and Zhou)

Fig. 14.14 Adult female wasp, dorsal view

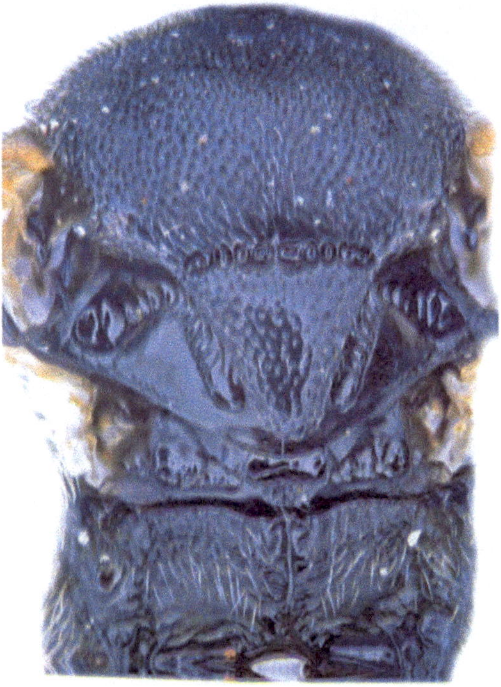

Fig. 14.15 Mesosoma

Fig. 14.16 Propodeum and part of metasoma, dorsal view

Fig. 14.17 Adult female wasp, lateral view

Material examined: Six females, five males, 2012, Chennai, Tamil Nadu, Mumbai, Maharashtra, ex. larva of *Borbo cinnara* (Wallace) (Lepidoptera: Hesperiidae), coll. NBAII. All specimens are deposited in NBAII, Bangalore, India (Code NBAII/Bra/Mic/Doli/kunh/2012).

Distribution: New distribution record from Maharashtra and Tamil Nadu. First discovered from Western Ghats, Kerala.

Cotesia glomerata **(Linnaeus 1758) (Plate** 14.5 – **Figs.** 14.18, 14.19, 14.20, 14.21, 14.22 and 14.23)

Brief diagnosis: Propodeum with a weak median longitudinal carina, no transverse basal carinae; second tergite with lateral sulci stronger at base than apex.

Material examined: Eight females, 3.viii.2012, UAS, Gangenahalli, Karnataka, India, solitary endoparasitoid, ex. *Pericallia ricini* Fabricius (Lepidoptera: Arctiidae) on the host plant *Ricinus communis* L., coll. Ankita. All specimens deposited in NBAIR, Bangalore, India (Code NBAII/Bra/Mic/Cot/glom/0812).

Distribution: Cosmopolitan India: Karnataka, Assam, West Bengal, Uttar Pradesh, Punjab, Himachal Pradesh, Bihar, Uttarakhand, Andaman and Nicobar Islands, Tamil Nadu, Manipur and Haryana.

(2) Molecular characterization of Indian Microgastrinae

Molecular characterisation of Microgastrinae specimens based on marker genes COI and ITS2 was done and DNA barcodes were generated for the six species. The GenBank accession no. of the six species is indicated in Table 14.3. All the details of the six species molecularly characterised are in the public domain of BOLD systems with their respective DNA barcodes (http://www.barcodinglife.com).

The paper includes a comprehensive list of species of parasitic wasps and their host relationships for 2010–2012. It also deals with the advancement of comparative knowledge of parasitoid-wasp biology and ecology in combination with molecular and morphological approaches. Nearly 42 species under 15 genera are documented from all across India. Fifteen (15) lepidopteran host families are documented as hosts for Microgastrinae wasps. More than 60 species of host caterpillars, majority belonging to Lepidoptera, were reared out, of which 35 species could be identified to genus/species level. Gregarious nature was noticed for 12 reared species. Two species of parasitic wasps, *A. javensis* Rohwer and *A. phycodis* Viereck, which have shown host-range extension, are illustrated. New distribution records are given for *A. mamitus* Nixon and *Neoclarkinella vitellinipes* (You and Zhou 1990).

The Microgastrinae species that are successful in biological control/natural control of some major pests throughout India are *Apanteles taragamae* Viereck for the control of *Pammene critica* Meyrick earlier known as *Grapholita* (*Cydia*) *critica*, a major pest of (Linnaeus) *Opisina arenosella* Walker, which is a major pest of coconut in southern India; *Cotesia flavipes* Cameron for *Chilo partellus* Swinhoe on maize and sorghum, *C. auricilius* Dudgeon and *C. infuscatellus* Snellen, *C. tumidicostalis* Hampson on sugarcane, *Sesamia inferens* (Walker) on wheat and rice, *Syllepte derogate* Fabricius on cotton and *Spilarctia obliqua* Walker on many crops.

Plate 14.5 *Cotesia glomerata* (Linnaeus)

Fig. 14.18 Adult female wasp, dorsal view

Fig. 14.19 Mesosoma with propodeum

Fig. 14.20 Metasoma

Fig. 14.21 Cocoon of *C. glomerata*

Fig. 14.22 Parasitised caterpillar of *Pericallia ricini* Fab

Fig. 14.23 Adult of *P. ricini*

Table 14.3 GenBank accession no. of six Indian species of subfamily Microgastrinae (Hymenoptera: Braconidae)

S. no.	Species of Microgastrinae (Hymenoptera: Braconidae)	GenBank accession no.
1.	*Apanteles galleriae* Wilkinson	JN790942
2.	*Apanteles hyposidrae* Wilkinson	JQ308797
3.	*Apanteles machaeralis* Wilkinson	JQ844449
4.	*Apanteles mohandasi* Sumodan & Narendran	JX083405
5.	*Apanteles taragamae* Viereck	JX083404
6.	*Fornicia ceylonica* Wilkinson	JN613568

Source: Molecular Entomology Division, NBAIR, Bangalore, India. All deposited in BOLD systems (http://www.barcodinglife.com)

Cotesia glomerata (Linnaeus) for *Pieris brassicae* (Linnaeus) and *P. rapae* (Linnaeus) on cabbage and *Cnaphalocrocis medinalis* (Guenee) on rice; *Cotesia ruficrus* (Haliday) for *Pelopidas mathias* (Fabricius) on rice, *P. brassicae* (Linnaeus) and *P. rapae* (Linnaeus) on cabbage, *Chilo* spp., *Pectinophora gossypiella* (Saunders) on cotton, *Plutella xylostella* (Linnaeus) on cabbage and *H. armigera* (Hubner) on many crops; and *Cotesia vestalis* (Haliday) for *P. xylostella* (Linnaeus) on cabbage and *Protapanteles obliquae* (Wilkinson) for *Spilarctia obliqua* Walker on mulberry and many other crops.

All these Microgastrinae species are cosmopolitan in habitat; however, *Cotesia kazak* and *Fornicia ceylonica*, solitary parasitoids, were found parasitising larvae of *Helicoverpa armigera* (Hubner) on many crops and *Spatulifimbria castaneiceps* Hampson on castor, respectively, and were confined to the southern parts of India. The most commonly encountered species from southern India is *Distatrix papilionis*, a gregarious parasitoid from the larvae of *Papilio demoleus* Linnaeus on lime, and was recorded from eastern India also. It is usually noticed in the months of August–September in the southern part of India.

During the studies, *Apanteles stantoni* was frequently documented from larvae of *Parotis marginata* (Hampson) on the host plant *Gardenia jasminoides*. Recently *Apanteles folia* was found parasitising lycaenid larvae *Arhopala amantes* Hewitson on *Lagerstroemia speciosa* Linnaeus and *Glyptapanteles hypermnestrae* from larvae of *Elymnias hypermnestra* (Linnaeus) on coconut. It is also noticed that *Microplitis maculipennis*, which is associated with a common pest of castor, *Acanthodelta janata* (Linnaeus), and *Apanteles machaeralis* from teak pest, *Eutectona machaeralis* (Walker) on *Tectona grandis*, are well distributed throughout India.

Thus rearing attempt showing host-parasitoid interactions was initiated in the year 2010 which is still continuing. Many species are still awaiting identification. The major constraints felt during the course of the study are 'missing types' in museums; scanty, confusing literature dealing with species described based on 'single male' specimens; and tiny size of these wasps. Hence to explore the richness of parasitic wasp diversity, future studies should aim at more exploratory surveys, host rearing and deep uninterrupted taxonomic studies.

Acknowledgements The author is sincerely thankful to Director, NBAIR, Bangalore and the Indian Council of Agricultural Research for providing necessary funds.

References

Cameron P (1906) On the Hymenopterous parasites of the mealie stalk borer (Sesamia fusca, Hampson). Trans South Afr Philos Soc 16:334–336

Fernández-Triana JL, Ward DF, Whitfield JB (2011) *Kiwigaster* gen. nov. (Hymenoptera: Braconidae) from New Zealand: the first Microgastrinae with sexual dimorphism in number of antennal segments. Zootaxa 20(2932):24–32

Folmer O, Black M, Hoeh W, Lutz R, Vrijenhoek R (1994) DNA primers for amplification of mitochondrial cytochrome c oxidase subunit I from diverse metazoan invertebrates. Mol Mar Biol Biotechnol 3(5):294–299

Gupta A, Kalesh S (2012) Reared parasitic wasps attacking hesperiids from Western Ghats (Kerala, India) with description of a new species of *Dolichogenidea* (Hymenoptera: Braconidae) as a larval parasitoid of *Thoressa evershedi* (Evans) (Lepidoptera: Hesperiidae). Zootaxa 3413:29–43

Gupta A, Pereira B (2012) A new species of *Glyptapanteles* (Hymenoptera: Braconidae: Microgastrinae), a larval parasitoid of *Elymnias hypermnestra* (Linnaeus) (Lepidoptera: Nymphalidae), along with some new host records of parasitoids from Peninsular India. Zootaxa 3227:54–63

Lin CP, Wood TK (2002) Molecular Phylogeny of the North American *Enchenopa binotata* (Homoptera: Membracidae) Species Complex. Ann Entomol Soc Am 95(2):162–171

Rao VP, Ghani AM, Sankaran T, Mathur KC (1971) A review of the biological control of insects and other pests in Southeast Asia and the Pacific region. Commonwealth Institute of Biological Control Technical Communication, (6) Slough, U.K.: CAB

Rodriguez JJ, Fernández-Triana J, Smith MA, Janzen DH, Hallwachs W, Erwin TL, Whitfield JB (2012) Extrapolations from field studies and known faunas converge on dramatically increased estimates of global microgastrine parasitoid wasp species richness (Hymenoptera: Braconidae). Insect Conservation and Diversity, 6:530–536

Sankaran T (1974) Natural enemies introduced in recent years for biological control of agricultural pests in India. Indian J Agric Sci 44:425–433

Whitfield JB, Mardulyn P, Austin AD, Downton M (2002) Phylogenetic relationships among microgastrine braconid wasp genera based on data from the 16S, COI and 28S genes and morphology. Syst Entomol 27:337–359

Wilkinson DS (1928) A revision of the Indo-Australian species of the genus *Apanteles* Forster (Hymenoptera: Braconidae). Bull Entomol Res 19:79–146

You LS, Zhou ZH (1990) [Two new species of *Apanteles* Foerster (Hymenoptera: Braconidae: Microgastrinae) from South China.] (in Chinese with English summary). Entomotaxonomia 12(2):151–155

Yu D, van Achterberg C, Horstmann K (2012) Taxapad 2012, Ichneumonidae 2011. Database on flash drive. Available from: http://www.taxapad.com, Ottawa, Ontario, Canada

A Study on the Diversity and Distribution of Genus *Camponotus* Mayr (Hymenoptera: Formicidae) in Kerala: Ecologically Significant Agents in Ecosystem Functioning

15

Presty John and K.A. Karmaly

Abstract

Ants are diverse, abundant, important, widespread and the most successful of all insect groups in tropical and temperate ecosystems. The genus *Camponotus* belongs to the subfamily Formicinae with 1087 extant species in the world. In India the genus *Camponotus* is represented by 59 species (without subspecies, total 62) (Bharti, Halteres 3:79–87, 2011). The present study was carried out in Kerala to study the diversity and distribution of *Camponotus*. Field survey and sampling of specimens were carried out. Of all the sites surveyed, *Camponotus compressus* (Fabricius), *Camponotus parius* Emery, *Camponotus mitis* (Smith) and *Camponotus sericeus* (Fabricius) were widespread. Calicut district recorded the maximum diversity containing 18 species. Review on distribution and diversity of *Camponotus* Mayr in India is also provided.

Keywords

Camponotus • Distribution • Diversity • Kerala

P. John (✉)
Department of Zoology, S.H College, Thevara, Cochin, Kerala 682 013, India

Current address: Kerala Forest Research Institute, Peechi, Kerala 680 653, India
e-mail: prestyjohn@gmail.com

K.A. Karmaly
St. Xaviers College for Women, Aluva, Kerala 683 101, India
e-mail: dr.karmaly@gmail.com

A.K. Chakravarthy, S. Sridhara (eds.), *Arthropod Diversity and Conservation in the Tropics and Sub-tropics*, DOI 10.1007/978-981-10-1518-2_15

15.1 Introduction

Ants are amongst the most ecologically and numerically dominant organisms in almost every terrestrial habitat throughout the world (Stork 1987, 1991) and are a common, widespread and well-known group, possibly the most successful of all the insect groups. Ants can be found in any type of habitat from the Arctic Circle to the equator (Brian 1978), although they are absent in Iceland, Greenland and Antarctica (Hölldobler and Wilson 1990), and some islands lack native ants (Wilson and Taylor 1967), and they outnumber in individuals than most other terrestrial animals. The number of species declines with increasing latitudes, altitudes and aridity (Kusnezov 1957; Fowler and Claver 1991; Farji Brener and Ruggiero 1994; Samson et al. 1997).

The diversity of ants depends on many factors, both biotic and abiotic, such as elevation (Samson et al. 1997), vegetation type (Wilson 1958; Bestelmeyer and Wiens 2001), predation (Kaspari 1996b), temperature (Bestelmeyer 2000), humidity (Kaspari 1996a), sampling methods (Watanasit 2003; Noon-Anant et al. 2005; Watanasit et al. 2007) and habitat preferences. Ants utilise various substrates for nesting, including dead wood; in colder climates ants prefer to nest under because they have low-specific heat when dry and thus heat up faster than the surrounding soil rocks (Hölldobler and Wilson 1994).

Ants are important not only because of their diversity (Alonso et al. 2000), they are also known to be ecologically significant invertebrates in many ecosystems (Hölldobler and Wilson 1990). They positively affect physical and chemical soil properties, turning forest soil, plant and animal distribution, dispersing seeds and helping with decomposition and nutritional recycling (Maryati 1997; Hölldobler and Wilson 1990; Douwes et al. 2012; Clark and Blom 1991). A reduction in ant biodiversity has profound effect on sequences in ecosystem functioning because of the important role of ants as ecosystem engineers (Folgarait 1998). Ants also serve as food for birds (e.g. woodpeckers) and for many vertebrates (e.g. anteater) (Torgersen and Bull 1995). Species of carpenter ants (*Camponotus* spp.) attain direct pest status because they cause indirect damage to plants by tending aphids and scale insects. Some species cause structural damage to buildings and invade homes. It is also found that in Brazilian tropical savannas, this genus is actively engaged in several food webs (Oliveira and Brandão 1991; Del-Claro and Oliveira 2000).

Globally, there are about 12,571 extant ant species (Bolton 2012). All ant species fall into the family Formicidae, superfamily Vespoidea of the order Hymenoptera in the class Insecta. Carpenter ants belong to the genus *Camponotus* (Hymenoptera: Formicidae), belonging to the subfamily Formicinae. Formicinae is the second largest subfamily in Formicidae with *Camponotus* as the most diverse genus (Wilson 1976). This genus was first described by Gustav Mayr in 1862, which derives their common name from the sandpaper-like finish of the galleries of nests which are built in wood. The genus *Camponotus* occurs in essentially all terrestrial habitats where ants are found. They construct nests in wooden materials such as partially broken-down tree trunks or stumps, living trees and hard-dried wood (Hansen and

Klotz 2005). This habit of nest construction leaves them in skirmish with humans when the ants move into buildings (Akre and Hansen 1990; Fowler 1990) to utilise a wide array of nesting materials. Currently there are 1,092 species, 477 subspecies and 32 fossil species in the world (Bolton 2012).

15.1.1 Relevance

The objective of this study was to find out the diversity and distribution of ants of the genus *Camponotus* in different habitats such as forest, grassland and human of the 14 districts of Kerala. On the contrary, the number of ant species recognised as threatened is higher than in other insects (IUCN 2008). We have a poor understanding of the global biodiversity and distribution of ants. This study provides an account of the distribution and diversity of the *Camponotus* species in the state of Kerala, India, and a brief review on the diversity of this genus in India. An attempt is also made to provide a detailed checklist on the distribution of the *Camponotus* species in Kerala (district-wise), graphs showing the diversity patterns of the subgenera of *Camponotus* and the diversity of *Camponotus* in South, North and Central Kerala.

15.1.2 *Camponotus* Mayr in India

The present review covers India. Forel (1910, 1912a, b, 1914), Wheeler (1909, 1921, 1927a, b, 1929, 1930–1931), Viehmeyer (1922), Santschi (1928), Donisthorpe (1933), Stitz (1935) and Karmaly and Narendran (2006) compiled the detailed information regarding *Camponotus* Mayr in Kerala as well as India, which includes a checklist and key to the species, detailed description of the *Camponotus* species, etc. Ramesh et al. (2010) reported four species of *Camponotu* s from Kalpakkam, viz. *C. compressus*, *C. sericeus*, *C. variegatus variegatus* and *Camponotus sp1.* Aravind Chavhan and Pawar (2011) reported *C. barbatus taylori*, *C. wasmanni* and *C. parius* from Amaravati, Maharashtra. A list of species of ants present in India under the genus *Camponotus* Mayr was prepared by Bharti (2011) and is represented by 59 (including subspecies) species.

15.2 Materials and Methods

15.2.1 Study Area

The study was conducted in Kerala (latitude 88°17′30″ N and 12°47′40″ N and longitude 74°27′47″ E and 77°37′12″ E) covered with land on all the three sides. Kerala is engulfed with the Arabian Sea to the west, the Indian Ocean to its south and the Western Ghats to the east. Kerala is tropical with a total area of 38,863 km^2. The state is situated in the south western tip of the Indian Peninsula. It has a fertile coastal strip and is about 550 km long and not wider than 100 km. Kerala has been identified as one of the world's 25 biodiversity hotspots. It has 14 districts, each of which has its own features.

Though Kerala has a large area under forest, it is quite rich in vegetation. It has diverse types of soils such as sandy, loamy, red, black, ferruginous and peat. The major crops of the state are pepper, cassava, paddy, cashew and coconut, and the cash crops are coffee and tea, vanilla, spices, cashew nut and nutmeg. Plantation crops like rubber also dominate the agricultural produce of Kerala. Kerala's climate remains wet most of the time as it is heavily influenced by seasonal rains brought by the monsoons. This coastal state has hot and humid climate during April–May and pleasant, cold climate in December–January. Summer extends from April to June. Summer is followed by the south west monsoon that starts pouring in June and continues till September. With the arrival of winter, there is certain drop in the temperature and cold wind prevails at that time. Winter in Kerala lasts from November to January or February. The maximum temperature throughout the year is around 36.7 °C and the minimum, 19.8 °C.

15.2.2 Sampling

Ants were collected for 2 years (2010–2011) from 14 districts of Kerala (i.e. Alappuzha, Ernakulam, Idukki, Kannur, Kasaragod, Kollam, Kottayam, Kozhikode, Malappuram, Palakkad, Pathanamthitta, Trivandrum, Thrissur and Wayanad). Collections were made the whole year. Species were collected by all-out search method, brush method, sweep net and pitfall traps (Fig. 15.1). All collected ants were brought back to the Department of Zoology, St. Xavier's College for Women, Aluva, and were separated from invertebrate material. Collected samples were mounted on rectangular cards and pinned with Asta insect pins of size 38 × 0.53 mm of 3 (Newey Goodman Ltd., UK). Remaining samples were preserved in 70 % alcohol for identification and for further studies.

15.2.3 Preservation and Identification

Ants up to the genus level were identified using stereoscopic binocular microscope (Gentner 40×), and species identification was carried out under Leica MZ6 stereo zoom microscope (Germany made). Genus-level identification was carried out using the keys of Bolton (1994, 1995, 2003), and species-level identification was done using the keys of Bingham(1903) and Bolton (1994) and the monograph of Indian Ants of the genus *Camponotus* by Karmaly and Narendran (2006).

15.3 Results

A total of 29 species of *Camponotus* were identified including the reported species from Kerala. The identified species of *Camponotus* falls into ten subgenus, viz. *Myrmoturba*, *Myrmoscericus*, *Taneamyrmex*, *Dinomyrmex*, *Colobopsis*,

Sweep net

***Camponotus sericeus* (Fabricius)**

***Camponotus compressus* (Fabricius)**

Pitfall trap

***Camponotus iruidus* Forel**

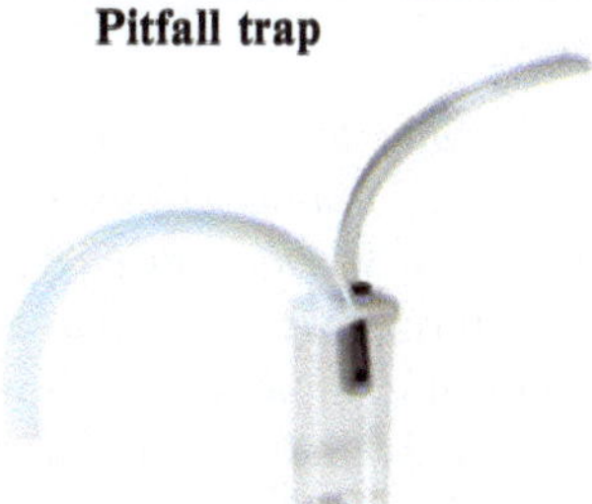

Aspirator

***Camponotus parius* Emery**

Fig. 15.1 Different types of traps and species of *Camponotus* (Species images courtesy: Google)

Table 15.1 List of *Camponotus* species at subgeneric level

No	Subgenus	Species of *Camponotus* Mayr
1	*Colobopsis*	(a) *Camponotus badius* (Smith)
		(b) *Camponotus phragmaticola* Donisthorpe
		(c) *Camponotus strictus* (Jerdon)
2	*Dinomyrmex*	(a) *Camponotus angusticollis* (Jerdon)
		(b) *Camponotus ashokai* Karmaly & Narendran
		(c) *Camponotus carin* Emery
3	*Myrmentoma*	(a) Camponotus *sericeus* (Fabr.)
4	*Myrmoscericus*	(a) *Camponotus parius* Emery
		(b) *Camponotus rufoglacus* (Jerdon)
		(c) *Camponotus dolendus* Forel
5	*Myrmotarsus*	(a) *Camponotus misturus fornaronis* Forel
6	*Myrmotemnus*	(a) *Camponotus binghamii* Forel
		(b) *Camponotus reticulatus* Roger
		(c) *Camponotus varians* Roger
7	*Myrmothrix*	(a) *Camponotus nicobarensis* Mayr
8	*Myrmoturba*	(a) *Camponotus barbatus* Roger
		(b) *Camponotus barbatus taylori* Forel
		(c) *Camponotus invidus* Forel
		(d) *Camponotus irritans* (Smith)
		(e) *Camponotus lamarckii* Forel
		(f) *Camponotus mitis* (Smith)
		(g) *Camponotus siemsseni* Forel
9	*Orthanotomyrmex*	(a) *Camponotus mendax* Forel
10	*Taneamyrmex*	(a) *Camponotus compressus* (Fabr.)
		(b) *Camponotus variegatus variegatus* (Smith)
		(c) *Camponotus variegatus infuscus* Forel
		(d) *Camponotus variegatus somnificus* Forel

Myrmotemnus, *Orthanotomyrmex*, *Myrmothrix*, *Myrmentoma* and *Myrmotarsus*. The subgenus *Myrmoturba* contains seven species; five under *Taneamyrmex*; three in *Dinomyrmex*, *Colobopsis* and *Myrmotemnus*; two species in *Myrmoscericus*; and single species comes under *Orthanotomyrmex*, *Myrmothrix*, *Myrmentoma* and *Myrmotarsus*. The details are listed in Table 15.1. Graphical comparison is shown in Fig. 15.2. Two species, viz. *Camponotus keralensis* Karmaly and Narendran and *C. ashokai* Karmaly and Narendran, which were reported by Karmaly and Narendran 2006 were not discussed in this study. *C. compressus*, *C. parius*, *C. mitis* and *C. sericeus* exhibited the highest distribution. Calicut district contained the highest species diversity containing 18 species. The wide range of distribution and diversity suggested that this may be due to the capability of these groups to exist in varying ecological conditions. The northern part of Kerala contains 23 species, South Kerala has 18 species and Central Kerala is inhabited by 15 species. A list of diversity is provided in Table 15.2. Graphical representation based of *Camponotus* species in area-wise pattern (South, Central and North Kerala) is represented in Fig 15.3.

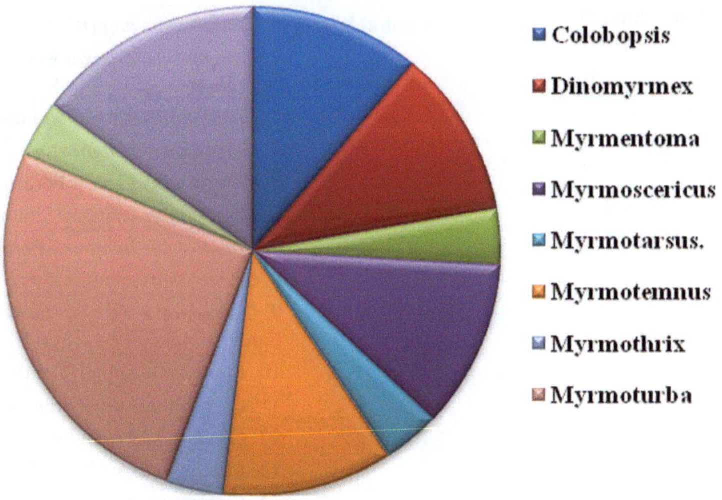

Fig. 15.2 The diversity of subgenera of *Camponotus* Mayr

Table 15.2 Distribution of species of *Camponotus* Mayr in different zones of Kerala

South Kerala	1. *Camponotus badius* (Smith)
	2. *Camponotus barbatus* Roger
	3. *Camponotus barbatus taylori* Forel
	4. *Camponotus binghamii* Forel
	5. *Camponotus compressus* (Fabr.)
	6. *Camponotus invidus* Forel
	7. *Camponotus irritans* (Smith)
	8. *Camponotus keralensis* Karmaly & Narendran
	9. *Camponotus lamarckii* Forel
	10. *Camponotus mendax* Forel
	11. *Camponotus misturus fornaronis* Forel
	12. *Camponotus mitis* (Smith)
	13. *Camponotus nicobarensis* Mayr
	14. *Camponotus phragmaticola* Donisthorpe
	15. *Camponotus reticulatus* Roger
	16. *Camponotus rufoglacus* (Jerdon)
	17. *Camponotus sericeus* (Fabr.)
	18. *Camponotus siemsseni* Forel

(continued)

Table 15.2 (continued)

Central Kerala	1. *Camponotus angusticollis* (Jerdon)
	2. *Camponotus ashokai* Karmaly & Narendran
	3. *Camponotus binghamii* Forel
	4. *Camponotus compressus* (Fabr.)
	5. *Camponotus invidus* Forel
	6. *Camponotus irritans* (Smith)
	7. *Camponotus lamarckii* Forel
	8. *Camponotus mendax* Forel
	9. *Camponotus misturus fornaronis* Forel
	10. *Camponotus mitis* (Smith)
	11. *Camponotus nicobarensis* Mayr
	12. *Camponotus rufoglacus* (Jerdon)
	13. *Camponotus sericeus* (Fabr.)
	14. *Camponotus siemsseni* Forel
	15. *Camponotus variegatus variegatus* (Smith)
North Kerala	1. *Camponotus angusticollis* (Jerdon)
	2. *Camponotus barbatus taylori* Forel
	3. *Camponotus binghamii* Forel
	4. *Camponotus carin* Emery
	5. *Camponotus compressus* (Fabr.)
	6. *Camponotus dolendus* Forel
	7. *Camponotus invidus* Forel
	8. *Camponotus irritans* (Smith)
	9. *Camponotus lamarckii* Forel
	10. *Camponotus mendax* Forel
	11. *Camponotus misturus fornaronis* Forel
	12. *Camponotus mitis* (Smith)
	13. *Camponotus nicobarensis* Mayr
	14. *Camponotus phragmaticola* Donisthorpe
	15. *Camponotus reticulatus* Roger
	16. *Camponotus rufoglacus* (Jerdon)
	17. *Camponotus sericeus* (Fabr.)
	18. *Camponotus strictus* (Jerdon)
	19. *Camponotus timidus* (Jerdon)
	20. *Camponotus varians* Roger
	21. *Camponotus variegatus variegatus* (Smith)
	22. *Camponotus variegatus infuscus* Forel
	23. *Camponotus variegatus somnificus* Forel

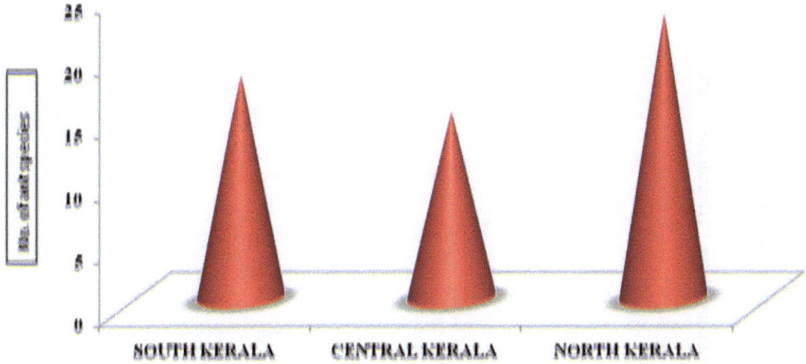

Fig. 15.3 Ant species diversity in the three zones of Kerala

15.4 Discussion

In Kerala, ants occupy a variety of habitats such as leaf litter, trees, soil and dead logs and some species prefer human-modified habitats. *Camponotus* is also the most populous genus in India, with 62 species including subspecies. The earliest available record from the region is probably the data provided by Karmaly and Narendran (2006).

In this study, the distribution pattern of all the 29 species was also noted (Table 15.3 and Fig. 15.4). Of these, two species were not collected during the study period. In the present study, a total of 30 species belonging to ten subgenera were reported. The subgenus *Myrmoturba* was the most dominant genus, then comes *Taneamyrmex*. Whilst *Dinomyrmex*, *Colobopsis*, *Myrmotemnus* and *Myrmoscericus* showed less species diversity and *Orthanotomyrmex*, *Myrmothrix*, *Myrmentoma* and *Myrmotarsus* showed extremely low species diversity, *C. compressus*, *C. parius*, *C. mitis* and *C. sericeus* exhibited the highest distribution. This may be due to the wide niche utilisation of the species habitat. Amongst these, the subgenus *Myrmoturba* has the highest share with seven species. Amongst the different zones of Kerala, i.e. north, south and central, the species diversity is more in North Kerala. The species *C. ashokai* Karmaly and Narendran is endemic to Palakkad region (Central Kerala) and *C. badius* and *C. phragmaticola* to Trivandrum (South Kerala). *C. velox* (Jerdon 1951) is also recorded from Kerala, but the locality is unknown, whilst *C. tharso* is also reported from Kerala, but Bharti (2011) reported that its presence in India couldn't be verified.

Camponotus compressus is amongst the most diverse species reported from the study and is commonly known as Indian black ant or colloquially as "Kulpe", an ant that is notorious for its painful bite in the coastal regions of India which rarely results in anaphylaxis (Sadananda et al. 2012).

Although the genus *Camponotus* has been called carpenter ants, some species did not nest inside wood. For example, *C. sericeus* can build a nest on the ground (Modi and Linsenmair 2003).

Table 15.3 District-wise distribution and diversity of species of the genus *Camponotus* Mayr (Hymenoptera: Formicidae: Formicinae) in Kerala

No	Name of species	Alappuzha	Ernakulam	Idukki	Kannur	Kasargod	Kollam	Kottayam	Kozhikode	Malappuram	Palakkad	Pathanamthitta	Trivandrum	Thrissur	Wayanad
1	*Camponotus angusticollis* (Jerdon)		*						*	*	*			*	
2	*Camponotus ashokai* Karmaly & Narendran										*				
3	*Camponotus badius* (Smith)												*		
4	*Camponotus barbatus* Roger						*						*		
5	*Camponotus barbatus taylori* Forel						*		*				*		
6	*Camponotus binghamii* Forel		*	*		*	*		*		*				*
7	*Camponotus carin* Emery								*	*					
8	*Camponotus compressus* (Fabr.)		*	*			*		*		*		*	*	*
9	*Camponotus dolendus* Forel								*	*					
10	*Camponotus invidus* Forel		*			*			*				*	*	

11	*Camponotus irritans* (Smith)		*			*	*		*	*			*		
12	*Camponotus keralensis* Karmaly & Narendran							*							
13	*Camponotus lamarckii* Forel							*	*	*				*	*
14	*Camponotus mendax* Forel	*	*			*									*
15	*Camponotus misturus fornaronis* Forel		*	*					*	*				*	*
16	*Camponotus mitis* (Smith)		*	*			*		*	*	*	*	*	*	*
17	*Camponotus nicobarensis* Mayr								*	*					
18	*Camponotus parius* Emery		*	*	*		*	*	*	*		*	*	*	*
19	*Camponotus phragmaticola* Donisthorpe												*		
20	*Camponotus reticulatus* Roger	*							*	*	*				
21	*Camponotus rufoglacus* (Jerdon)	*	*	*		*	*		*	*			*		

(continued)

Table 15.3 (continued)

No	Name of species	Alappuzha	Ernakulam	Idukki	Kannur	Kasargod	Kollam	Kottayam	Kozhikode	Malappuram	Palakkad	Pathanamthitta	Trivandrum	Thrissur	Wayanad
22	*Camponotus sericeus* (Fabr.)	*	*	*	*	*	*	*	*	*				*	*
23	*Camponotus siemsseni* Forel											*	*		
24	*Camponotus strictus* (Jerdon)									*					
25	*Camponotus timidus* (Jerdon)								*	*					
26	*Camponotus varians* Roger									*					
27	*Camponotus variegatus variegatus* (Smith)		*						*	*				*	
28	*Camponotus variegatus infuscus* Forel		*						*	*				*	
29	*Camponotus variegatus somnificus* Forel									*					

*Indicates the presence of species

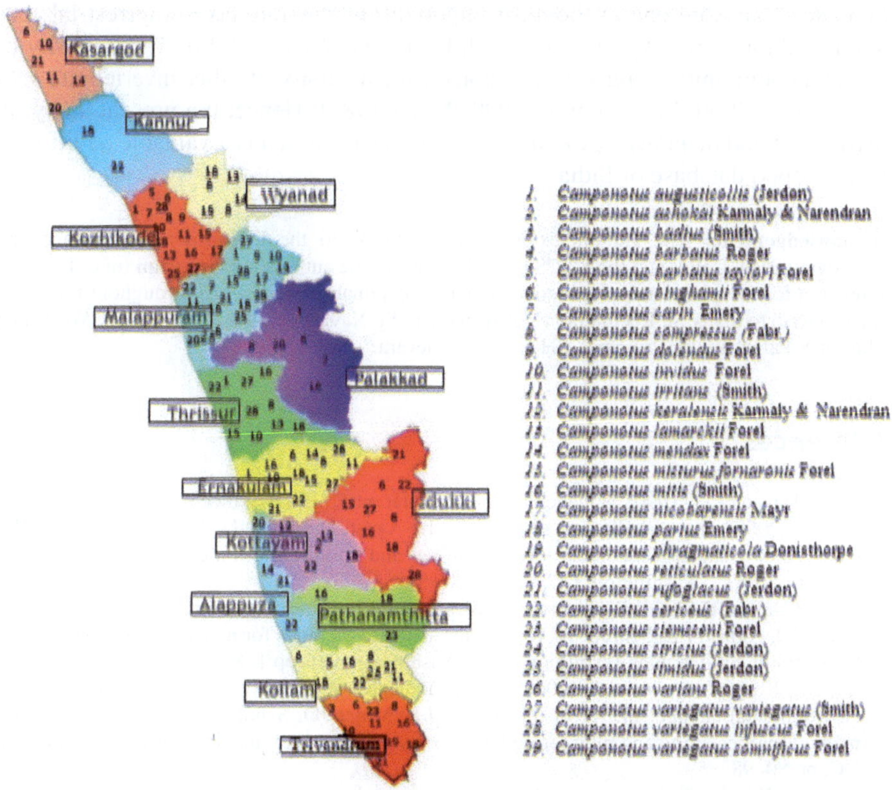

Fig. 15.4 Distribution of *Camponotus* species in Kerala

Ant-foraging activity increased with temperature if ants were sampled on a winter day, or decreased with temperature, if it was recorded on a summer day, irrespective of whether the community was in the sown pasture or natural grassland. This is a common pattern for ants. Usually when it rains, the air and soil temperature decreases, but it causes higher humidity. The rain can destroy the ants, which build their nests in the soil and on the ground, especially dwelling ants. Hölldobler and Wilson (1990) showed that rainfall (precipitation) could disturb the feeding behaviour of ants. They will not come out from their nest to forage after the rain has ceased. From observation of days of heavy rain, there were only a few numbers of ants in the collecting samples. For the *Camponotus*, an increase of individual numbers correlated with a higher temperature of both air and soil (Watanasit et al. 2007).

Ants outnumber all other terrestrial organisms and occur in virtually all types of habitats (Wheeler 1910), which is particularly conspicuous in the tropical region (Fittkau and Klinger 1973). Nowadays, this genus is also one of the most engaged in several food webs at cerrados (Oliveira and Brandão 1991; Del-Claro and Oliveira 2000). *Camponotus* Mayr is distributed throughout the state, but climatic conditions play an important role in collecting the species. In biomass and ecological dominance, ants are one of the most important invertebrate taxa in terrestrial ecosystems (Alonso and Agosti 2000; Hölldobler and Wilson 1990). The diversity of ants in a community is a good indicator for the diversity of other invertebrate species (Alonso 2000; Lawton et al. 1998; Majer 1983). Hence, the present study on distribution and diversity of *Camponotus* species in Kerala is a valuable addition to the arthropod database of India.

Acknowledgements The authors extend their gratitude to the Department of Science and Technology, New Delhi, and the KSCSTE, Trivandrum. The authors are also thankful to Dr. T. C. Narendran for his moral support, encouragement and valuable suggestions throughout this work and also wish to thank Sr. Reethamma V A, Principal, St. Xavier's College for Women, Aluva, and Johnson X Palackapillil, Principal, SH College, Thevara.

References

Akre RD, Hansen LD (1990) Management of carpenter ants. Appl Myrmecol 693–700

Alonso LE (2000) Ants as indicators of diversity. In: Agosti D, Majer JD, Alonso LE, Schultz TR (eds) Ants: standard methods for measuring and monitoring biodiversity. Smithsonian Institution Press, Washington, DC, pp 80–88

Alonso LE, Agosti D (2000) Biodiversity studies, monitoring, and ants: an overview. In: Agosti D, Majer JD, Alonso LE, Schultz TR (eds) Ants: standard methods for measuring and monitoring biodiversity. Smithsonian Institution Press, Washington, DC, pp 1–8

Alonso LE, Kaspari M, Agosti D (2000) Ants as indicator of diversity and using ants to monitor environmental change. In: Agosti D, Alonso LE, Majer JD, Schultz TR (eds) Ant: standard method for measuring and monitoring biodiversity. Smithsonian Institution Press, Washington, DC, pp 80–98

Bestelmeyer BT (2000) The trade-off between thermal tolerance and behavioural dominance in a subtropical South American ant community. J Anim Ecol 69:998–1009

Bestelmeyer BT, Wiens JA (2001) Local and regional scale responses of ant diversity to a semiarid biome transition. Ecography 2:381–392

Bharti H (2011) List of Indian ants (Hymenoptera: Formicidae). Halteres 3:79–87

Bingham CT (1903) The fauna of British India including Ceylon and Burma. Hymenoptera. Ants and Cuckoo Wasps. Taylor and Francis, London

Bolton B (1994) Identification guide to the ant genera of the world. Harvard University Press, London, p 222

Bolton B (1995) A new general catalogue of the ants of the world. Harvard University Press, Cambridge, MA, p 504

Bolton B (2003) Synopsis and classification of Formicidae. Mem Am Entomol Inst 71:1–370

Bolton B (2012) An online catalogue of the ants of the world. Ant Cat. http://www.antcat.org/

Brian MV (1978) Production ecology of ants and termites, vol 13, IBP. Cambridge University Press, Cambridge

Chavhan A, Pawar SS (2011) Distribution and diversity of ant species (Hymenoptera: Formicidae) in and around Amravati City of Maharashtra, India. World J Zool 6(4):395–400

Clark WH, Blom PE (1991) Observations of ants (Hymenoptera: Formicidae: Myrmicinae, Formicinae, Dolichoderinae) utilizing carrion. South-West Nat 36:140–142

Del-Claro K, Oliveira PS (2000) Conditional outcomes in a neotropical ant-homoptera mutualistic association. Oecologia 124:156–165

Donisthorpe H (1933) Descriptions of new species of Formicidae and a synonomical note. Ann Mag Hist 10(11):194–198

Douwes P, Johan A, Björn C, Urban W (2012) National nyckeln till Sveriges flora og fauna. Steklar: myror – getingar, Hymenoptera: Formicidae – Vespidae Art Databanken. SLU, Uppsala, pp 1–382

Farji-Brener AG, Ruggiero A (1994) Leaf-cutting ants (Atta and Acromyrmex) inhabiting Argentina: patterns in species richness and geographical range sizes. J Biogeogr 21:391–399

Fittkau EJ, Klinger H (1973) On biomass and trophic structure of the central Amazonian rain forest ecosystem. Biotropica 5:2–14

Folgarait PJ (1998) Ant biodiversity and its relationship to ecosystem functioning: a review. Biodivers Conserv 7:1221–1244

Forel AH (1910) Glanures myrmeologiques. Ann Soc Entomol Belg 54:6–32

Forel AH (1912a) Sauter's Formosa- Ausbeute Formicidae. Entomol Mitt Berlin 1(3):67–81

Forel AH (1912b) Einiqe neue und interassante Ameisen aus Sumatra. Zool Jahr Suppl 15:62–63

Forel AH (1914) Le genre *Camponotus* Mayr et les genres voisins. Rev Suisse Zool 22:257–276

Fowler HG (1990) Carpenter ants (*Camponotus* spp.): pest status and human perception. Appl Myrmecol 525–532

Fowler HG, Claver (1991) Leaf-cutter ant assemblies: effects of latitude, vegetation, and behavior. In: Huxley CR, Cutler DF (eds) Ant-plant interactions. Oxford University Press, New York, pp 51–59

Hansen LD, Klotz JH (2005) Carpenter ants of the United States and Canada, pp 1–204

Hölldobler B, Wilson EO (1990) The ants. Springer, Berlin, pp 1–732

Hölldobler B, Wilson EO (1994) Journey to the ants. The Belknap Press of the Harvard University Press, Cambridge, MA, p 228

IUCN (2008) Red list of threatened species. http://www.iucnredlist.org. Retrieved on 13 March

Jerdon (1951) *Camponotus (Myrmepomis) velox.* Chapman and Capeo, Monogr. Inst Sci Tech Manila (Check List Ants Asia), 1: 236

Karmaly KA, Narendran TC (2006) Indian ants: genus Camponotus: Teresian. Teresian Carmel Publications, Kerala, p 165

Kaspari M (1996a) Litter ant patchiness at 1 m^2 scale; disturbance dynamics in three Neotropical forests. Oecologia 107:265–273

Kaspari M (1996b) Testing resource-based models of patchiness in four Neotropical litter ant assemblages. Oikos 76:443–454

Kusnezov N (1957) Number of species of ants in faunae of different latitudes. Evolution 11:298–299

Lawton JH, Bifnel LDE, Bolton B, Blowmers GF, Eggleton P, Hammond PM, Hodda M, Holt RD, Larsen TB, Mawdsley NA, Stork NE, Srivastava DS, Watt AD (1998) Biodiversity inventories, indicator taxa and effects of habitat modification in tropical forest. Nature 391:72–76

Majer JD (1983) Ants: bio-indicators of mine site rehabilitation, land-use, and land conservation. Environ Manag 7:375–385

Maryati M (1997) Ants: an indicator for the tropical rain forest. Manual for international plot course on environmental evaluation using insects as indicators of biodiversity: ant ecology, taxonomy collecting methods and identification 17 March–7 April 1997. Tropical Biology and Conservation Unit, Universiti Malaysia Sabah and International Institute of Entomology, Kota Kinabalu

Mayr G (1862) Myrmecologische studien. Verh K-Kg Zool-Bot Ges Wein 12:649–776

Modi K, Linsenmair KE (2003) Finding its place in a competitive ant community: leaf fidelity of *Camponotus sericeus*. Insect Soc 50:191–198

Noon-anant N, Watanasit S, Wiwatwitaya D (2005) Species diversity and abundance of ants in lowland tropical rain forest of Bala Forest, Narathiwat Province, Southern Peninsular Thailand. Nat Hist Bull Siam Soc 53:203–213

Oliveira PS, Brandão CRF (1991) The ant community associated with extrafloral nectaries in the Brazilian cerrados. In: Cutler DF, Huxley CR (eds) Ant–plant interactions. Oxford University Press, Oxford, pp 198–212

Ramesh T, Jahir HusainK, Satpathy KK, Selvanayagam M, Prasad MVR (2010) Diversity, Distribution and Species Composition of Ants. Fauna at Department of Atomic Energy (DAE) Campus Kalpakkam, South India. World J Zool 5(1):56–65

Sadananda Naik B, Prabhu KM, Jyothi CS (2012) Anaphylaxis due to Indian Black Ant Bite [*Camponotus compressus*]. J Indian Soc Toxicol 8:43–45

Samson DA, Rickart EA, Gonzales PC (1997) Ant diversity and abundance along an elevational gradient in the Philippines. Biotropica 29:349–363

Santschi F (1928) Nouvelles fourmis de Chine et du Turkestan russe. Soc Entomol Belg Bull Ann 68:31–46

Stitz H (1935) Schwedisch-Chineische Wissenschaftliche Expedition nach den nordwstlichen provinzen Chinas. 3 Formicidae. Ark Zool 27A(1):1–9

Stork NE (1987) Guild structure of arthropods from Bornean rainforest trees. Ecol Entomol 12:69–80

Stork NE (1991) The composition of the arthropod fauna of Bornean lowland rain forest trees. J Trop Ecol 7:161–180

Torgersen TR, Bull EL (1995) Down logs as habitat for forest-dwelling ants–the primary prey of pileated woodpeckers in Northeastern Orego. University press, Cornell

Viehmeyer H (1922) Neue Ameisen. Arch Naturgesch 88:203–220

Watanasit S (2003) Evaluation of sampling techniques for ants in rubber plantations. In: Proceedings of the 2nd ANeT Workshop and Watanasit, S., Sonthichai, S. and Noon-anant, N. Preliminary survey of ants at Tarutao National Park, Southern Thailand. Songklanakarin J Sci and Tech 25:115–122

Watanasit S, Saewai J, Phlapplueng A (2007) Ants of the Klong U-Tapao, Southern Thailand. Asian Myrmecology 1:69–79

Wheeler WM (1909) Ants of Formosa and the Philippines. Am Mus Nat Hist Bull 26:333–345

Wheeler WM (1910) Ants: their structure, development and behavior. Columbia University Press, New York

Wheeler WM (1921) Chinese ants. Bull Mus Com Zool Harv Univ 64:529–547

Wheeler WM (1927a) Few ants from China and Formosa. Am Mus Novit 259:1–4

Wheeler WM (1927b) Chinese ants collected by Professor S F Light and Professor N Gist Gee. Am Mus Novit 255:1–12

Wheeler WM (1929) Some ants from China and Manchuria. Am Mus Novit 361:1–11

Wheeler WM (1930–1931) A list of known Chinese ants. Peking History Bull 5:1–53

Wilson EO (1958) Patchy distribution of ant species in New Guinea rain forests. Northwest Sci 65:26–38

Wilson EO (1976) Which are the most prevalent ant genera? Stud Entomol 19:187–200

Wilson EO, Taylor RW (1967) The ants of Polynesia (Hymenoptera: Formicidae). Pac Insects Mon 14:109

Diversity of Tephritid Flies in Sub-Himalayan Region of West Bengal: Baseline Data for Developing Rational Management Practices

16

Nripendra Laskar, Dipak Kumar Sinha, Tapan Kumar Hath, and Hirak Chatterjee

Abstract

Six species of tephritid fruit flies, viz., *Bactrocera cucurbitae* (Coquillett), *Bactrocera diversa* (Coquillett), *Bactrocera caudata* (Fabricius), *Bactrocera tau* (Walker), *Bactrocera nigrotibialis* (Perkins), and *Dacus longicornis* (Wiedemann), were trapped in cue lure trap in cucurbit fields, and mean trap catches were 45.00 %, 16.67 %, 11.67 %, 9.50 %, 9.50 %, and 4.80 %, respectively. Using methyl eugenol trap, four species, viz., *B. dorsalis* (Hendel), *B. zonata* (Saunders), *B. versicolor* (Bezzi), and *B. correcta* (Bezii), were trapped in guava orchards of Cooch Behar district of West Bengal with a mean trap catch of 55.85 %, 20.14 %, 21.00 %, and 3.01 %, respectively. From infested fallen and harvested fruits of mango from a mango orchard of Malda district of West Bengal, the fruit fly species recovered were *B. dorsalis* (Hendel), *B. zonata* (Saunders), and *B. versicolor* (Bezzi), and their proportions were 52.25 %, 27.34 %, and 20.10 %, respectively. Again, after rearing from infested fallen and harvested guava fruits in Cooch Behar district (West Bengal), five species of tephritid fruit flies, viz., *B. dorsalis* (Hendel), *B. zonata* (Saunders), *B. versicolor* (Bezzi), *B. caudata* (Fabricius), and *B. nigrotibialis* (Perkins), were obtained. Proportionally, the highest population of *B. dorsalis* (Hendel) (37.86 %) was noted. Rearing from infested fallen and harvested citrus in orchards of Darjeeling district (West Bengal), *B. minax* (Enderlein) and *B. dorsalis* (Hendel) were recorded. Altogether nine tephritid flies were detected through cue lure and methyl eugenol trap catching and infested fallen and harvested fruits of mango,

N. Laskar (✉)
Department of Agricultural Entomology, Uttar Banga Krishi Viswavidyalaya, Pundibari, Cooch Behar, West Bengal 736165, India
e-mail: nripendralaskar@yahoo.co.in

D.K. Sinha • T.K. Hath • H. Chatterjee
Palli Siksha Bhavana (Institute of Agriculture) Visva-Bharati, Santiniketan, Bolpur, Birbhum, West Bengal 731235, India

A.K. Chakravarthy, S. Sridhara (eds.), *Arthropod Diversity and Conservation in the Tropics and Sub-tropics*, DOI 10.1007/978-981-10-1518-2_16

guava, and citrus. The dominant species were *B. cucurbitae* (Coquillett), *B. dorsalis* (Hendel), *B. minax* (Enderlein), *B. zonata* (Saunders), and *B. versicolor* (Bezzi). It was also detected that 31.67 %, 36.33 %, and 45.67 % of fruits of mango, guava, and citrus, respectively, were found infested by fruit flies in Northwest Bengal.

Keywords
Cue lure • Fruit flies • Methyl eugenol • Tephritid

16.1 Introduction

Fruit flies are the most important pests of fruits and vegetables. They belong to Tephritidae (subfamily: Dacinae), one of the largest families of Diptera (Drew 1989) containing predominantly medium-sized, pictured-winged, and highly ornamented flies. Distribution of fruit flies is cosmopolitan covering tropical, subtropical, and temperate regions, and they occupy habitats ranging from rain forest to open savannah except in Arctic and Antarctic regions (McPheron and Steck 1996; Norrbom et al. 1998; Agarwal and Sueyoshi 2005; De-Meyer et al. 2010).

The fruit flies of Indian subcontinent have received considerable attention by Bezzi (1916), Hering (1941, 1956), and Kapoor (1971, 1993). Of late, Drew and Hancock (1994) revised the *Bactrocera dorsalis* complex, including several Indian species. Ranganath and Veenakumari (1995) recorded 11 species of *Bactrocera* in the Andaman Islands, some of which have Indian origin, while others occur in Southeast Asia. Tsuruta et al. (1997) published a list of host plants of Dacinae in Sri Lanka, while Tsuruta et al. (1998) provided a pictorial key to Sri Lankan pest species. White and Evenhuis (1999) added two more *Bactrocera* species of the subgenus *Zeugodacus*. Tsuruta and White (2001) recorded 28 species of Dacinae fauna in Sri Lanka, some of which also occur in India.

Most of the flies are highly polyphagous and infest a wide range of hosts causing considerable economic losses. Adult females puncture the soft and tender fruits with ovipositor and lay eggs under the epidermis. The eggs hatch into maggots that feed inside the fruit on pulp, and infested fruit can be identified by brown resinous secretions that ooze out from the punctures made by the female flies for oviposition. Fruits with fruit fly larvae in them decay quickly. It is sometimes possible to cut out the damage for home consumption of the remaining part of the fruit, but infested fruits are generally unsalable and can certainly not be exported. Crop losses vary from a few percent up to 90 % (Dhillon et al. 2005).

Infestation of cucurbits by *D. cucurbitae* generally ranged from 40 to 80 % between July and October (Pruthi 1941), while the damage was estimated to the tune of 3–100 % on little gourd (Patel 1976). According to Mote (1975), yield loss of cucurbitaceous vegetables in India due to this devastating pest may be as high as 40–80 %. The infestation by fruit flies to guava ranges from 20 to 46 % (Hasseb 2007). Van Schoubroeck (1999) observed crop losses of citrus due to fruit fly infestation between 35 and 75 % in mid and high altitude of Bhutan. Economic losses due to olive fruit fly infestation in the Mediterranean countries have been up to 15 % (Mazomenos et al. 2002) and 30 % (Muskatim et al. 2012) of the olive crop. On mango also, the tephritids are serious pests.

Galande et al. (2010) found seven predominant fruit fly species at Pune region, Maharashtra, India, using methyl eugenol and cue lure traps as well as by rearing infested fallen and harvested fruits of guava. Six fruit fly species were reported recently for the first time from Himachal Pradesh, India, by Prabhakar et al. (2012). The authors also noted that out of 47 species of fruit flies reported from Himachal Pradesh, 13 belonged to genera *Bactrocera* and *Dacus* and majority of them are economically important pests of horticultural crops in several countries the world over.

Owing to typical life history, the flies have tremendous potentiality to cause irreparable damage to crops and are difficult to manage. Apart from direct yield losses, the fruit flies cause major economic impact especially through quarantine and regulatory measures. One of the most important management strategies against this dangerous pest is the use of sex attractants. The sex attractants are also species specific. So, for development of a successful eco-friendly management strategy, accurate identification of pest species is essential. It is also helpful in regulating the entry of pest species to a pest-free zone. In the geographical region under consideration, no investigations have so far been conducted on the detection of fruit fly species and damage they cause on different host plants of economic importance. Present studies were undertaken to detect the species composition of predominant tephritid fruit flies under the sub-Himalayan West Bengal.

16.2 Materials and Methods

The male adult fruit flies were collected using sex attractants methyl eugenol and cue lure trap in guava orchard and cucurbit field, respectively. The traps were installed at different northern districts of West Bengal, viz., Cooch Behar, Darjeeling, North Dinajpur, and Malda during summer months of 2011. At each location four traps of both methyl eugenol and cue lure were installed at 2 m from the soil surface at 50 m distance between two traps. The trapped flies were collected weekly, counted, preserved, and identified.

Five kilograms of infested fallen as well as harvested fruits of mango, guava, and citrus was collected from Malda, Cooch Behar, and Darjeeling, respectively. From each location four orchards were selected for collecting the infested fruits. No plant protection measures with regard to insect pest management were undertaken in the sampled fields. The collected fruits were held for rearing in cages in sand and coco pit (1:1) (Galande et al. 2010) to facilitate pupation and adult emergence.

The adult flies collected using sex attractants as well as those from rearing infested fruits were critically examined under stereo-binocular microscope for characteristic morphological features and identified following the appropriate taxonomic keys (Drew and Raghu 2002). Specimens were also sent to Dr. V. V. Ramamurthy, national coordinator, Network Project on Insect Biosystematics, Indian Agricultural Research Instiute, New Delhi, for identification. The adult fruit flies of a particular species from each lot were recorded and preserved following the standard preservation technique.

In calculating the percent fruit infestation of mango, guava, and citrus, the fresh and infested fruits were counted at each harvest (four orchards each of mango, guava, and citrus at Malda, Cooch Behar, and Darjeeling, respectively), and cumulative of total harvest of a season was determined. The percent fruit infested then was obtained as follows:

$$\%\,\text{fruit infested (in number)} = \text{Number of fruits infested} / \text{Total number of fruits observed} \times 100$$

and

$$\%\,\text{fruit infested (in weight)} = \text{Weight (in kg) of fruits infested} / \text{Total weight of fruits observed} \times 100$$

16.3 Results

Six species of male fruit flies were collected using cue lure traps at four locations of sub-Himalayan West Bengal from four districts during summer 2011 (Table 16.1) from cucurbit field. The species were *Bactrocera cucurbitae* (Coquillett), *Bactrocera diversa* (Coquillett), *Bactrocera caudata* (Fabricius), *Bactrocera tau* (Walker), *Bactrocera nigrotibialis* (Perkins), and *Dacus longicornis* (Wiedemann). All species belonged to Tephritidae, subfamily Dacinae, tribe Dacini. Populations of the trapped fruit fly species revealed that *B. cucurbitae* (45.00 %) was the frequently occurring species followed by *B. dorsalis* (16.67 %), *B. caudata* (11.67 %), and *B. tau* (9.50 %). Least frequent was *D. longicornis* (4.80 %) (Fig. 16.1).

Using methyl eugenol trap installed at four different locations of sub-Himalayan West Bengal, four fruit fly species were collected, viz., *B. dorsalis* (Hendel), *B. zonata* (Saunders), *B. versicolor* (Bezzi), and *B. correcta* (Bezzi) (Table 16.2). The highest catch was noted in *B. dorsalis* (55.85 %) followed by *B. versicolor* (21.00 %) and *B. zonata* (20.14 %). The minimum catch was observed in *B. correcta* (3.01 %). Species composition of fruit flies trapped with methyl eugenol is depicted in Fig. 16.2.

Table 16.1 Fruit fly species collected from cucurbit fields using cue lure trap in West Bengal

	Mean percentage of fruit fly species					
Location	*B. cucurbitae*	*B. diversa*	*B. caudata*	*B. tau*	*B. nigrotibialis*	*D. longicornis*
Cooch Behar	46.70	17.80	10.24	9.70	9.30	6.27
Darjeeling	48.20	16.55	17.92	9.00	6.50	1.91
North Dinajpur	44.50	12.60	15.40	11.50	10.78	5.26
Malda	40.60	19.69	14.56	7.80	11.42	5.76
Mean	45.00	16.67	11.67	9.50	9.50	4.80

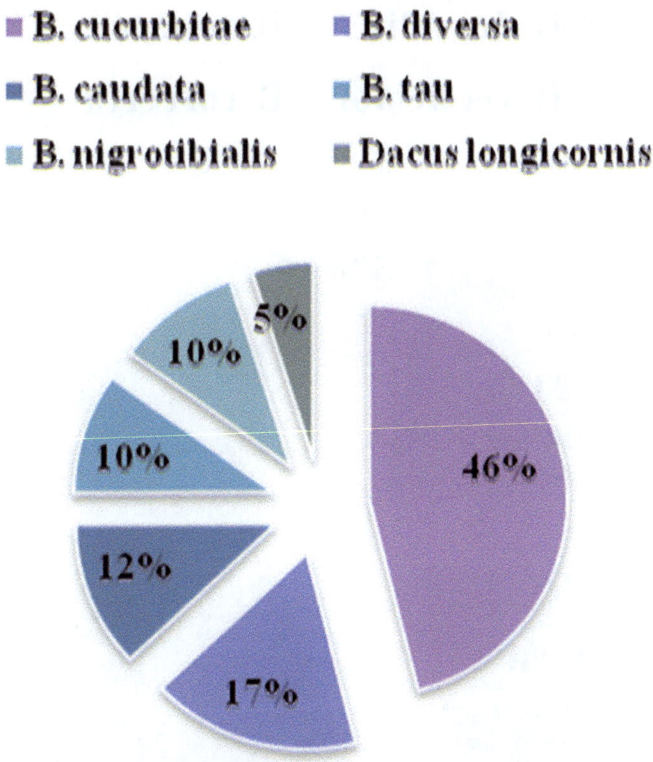

Fig. 16.1 Species composition of fruit flies infesting mango (cue lure trap)

Table 16.2 Fruit fly species collected using methyl eugenol trap in cucurbit fields of West Bengal

Location	Mean percentage of fruit fly species			
	B. dorsalis	*B. zonata*	*B. versicolor*	*B. correcta*
Cooch Behar	58.25	22.30	17.32	1.98
Darjeeling	51.32	20.57	24.06	4.17
North Dinajpur	60.70	16.32	21.48	1.32
Malda	53.13	21.37	21.14	4.57
Mean	55.85	20.14	21.00	3.01

Fruit infestation of mango by fruit flies was determined from four orchards at Malda both on number and weight basis. The percent fruit infestation varied from 28.23 to 45.75 % in number, the average being 37.67 % and 27.40–41.10 % on weight basis averaging 35.20 %. The fruit fly species recovered from rearing infested mango fruits were *B. dorsalis* (Hendel), *B. zonata* (Saunders), and *B. versicolor* (Bezzi). Mean percentages of the fruit fly species were recorded as 52.25 %, 27.34 %, and 20.10 % of *B. dorsalis*, *B. zonata*, and *B. versicolor*, respectively (Table 16.3) (Figs. 16.3 and 16.4).

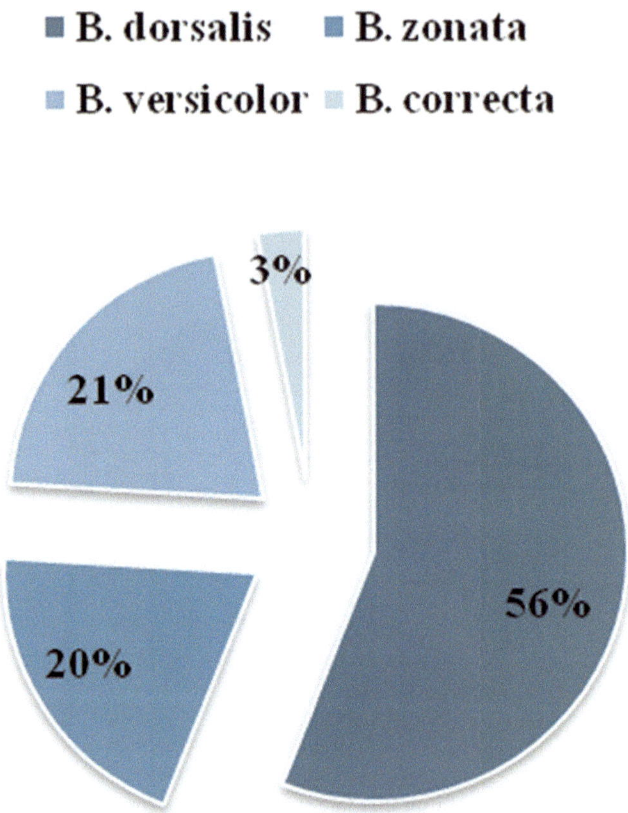

Fig. 16.2 Species composition of fruit flies infesting mango (methyl eugenol trap)

Table 16.3 Fruit infestation and composition of fruit fly recovered from rearing infested mango fruits from Malda

	Percent fruit infestation		Mean percentage of adult fly recovered after rearing		
Location	By no.	By wt.	*B. dorsalis*	*B. zonata*	*B. versicolor*
Orchard-1	40.80	38.26	57.70	25.25	16.32
Orchard-2	28.23	27.40	48.90	30.09	20.57
Orchard-3	35.90	41.10	46.32	27.67	26.30
Orchard-4	45.75	34.20	56.08	26.35	17.21
Mean	37.67	35.20	52.25	27.34	20.10

The extent of guava fruit fly infestation was determined at four different orchards of Cooch Behar district, West Bengal, both on number and weight basis. Observations revealed that 36.33 % and 33.60 % fruits were infested by fruit flies on number and weight basis, respectively. The infestation varied from 28.64 to 41.53 % and 28.30 to 39.50 % in number and weight basis, respectively (Table 16.4; Figs. 16.5 and 16.6).

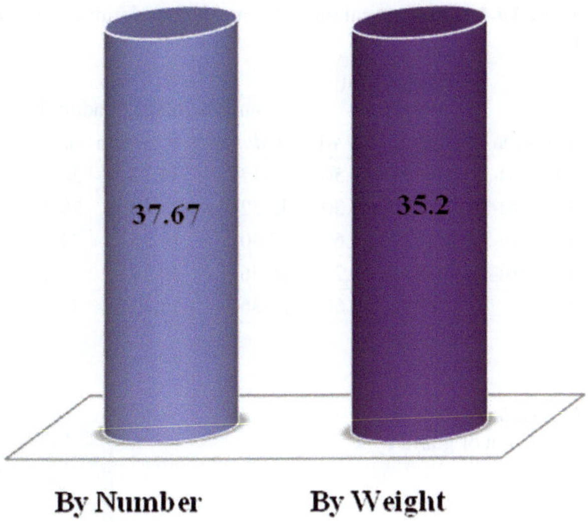

Fig. 16.3 Percent infestation of mango fruits by fruit flies

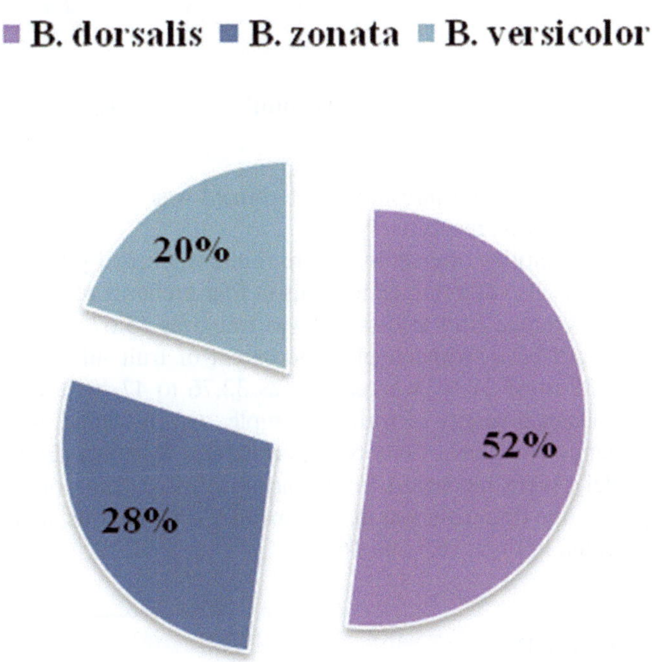

Fig. 16.4 Species composition of fruit flies recovered from infested mango

Table 16.4 Fruit infestation and composition of fruit fly recovered from rearing on guava (Cooch Behar)

	Percent fruit infested		Mean percentage of adult fly recovered after rearing				
Location	By no.	By wt.	*B. dorsalis*	*B. zonata*	*B. versicolor*	*B. caudata*	*B. nigrotibialis*
Orchard-1	39.50	39.50	33.17	34.27	24.10	2.80	5.20
Orchard-2	28.64	28.30	35.32	27.20	29.20	6.90	1.92
Orchard-3	35.67	35.60	42.40	25.56	24.72	4.27	2.75
Orchard-4	41.53	31.27	40.56	30.34	23.38	3.41	2.53
Mean	36.33	33.60	37.86	29.30	25.40	4.34	3.10

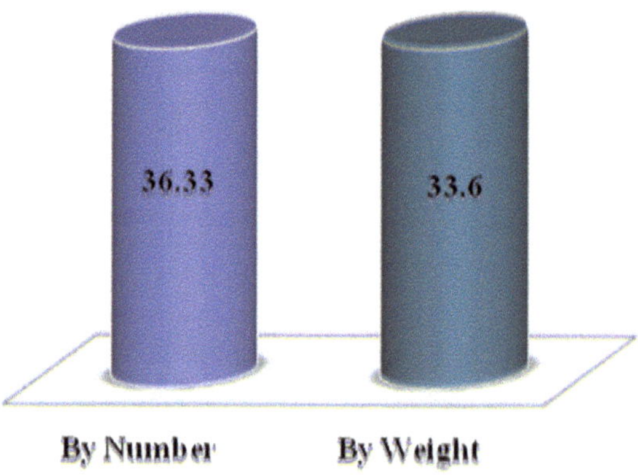

Fig. 16.5 Percent infestation of guava by fruit flies

The mean percentage of fruit infestation determined was 36.33 % and 33.60 % by number and weight basis, respectively.

Considerable quantities of crop loss were found in citrus by fruit flies. Percent fruit infestation was also determined in citrus in four orchards of Darjeeling district of West Bengal. Average fruit infestation was found as 45.67 % and 40.89 % on number and weight basis, respectively. The extent of fruit infestation in number varied from 38.55 to 53.21 %, whereas it was 32.76 to 47.40 % on weight basis (Table 16.5). Species diversity of fruit fly complex was recorded, which was least compared to mango and guava. Only two species, viz., *B. minax* (Enderlein) and *B. dorsalis* (Hendel), were recovered from rearing infested fruits of citrus. Among these two species, the dominant was *B. minax* (60.00 %). Results are also diagrammatically presented in Figs. 16.7 and 16.8.

16.4 Discussions

A good number of tephritid flies have been found to exist in sub-Himalayan region, West Bengal. Six species of adult male fruit flies were trapped using cue lure. The frequently occurring species was *Bactrocera cucurbitae* and the least frequent was

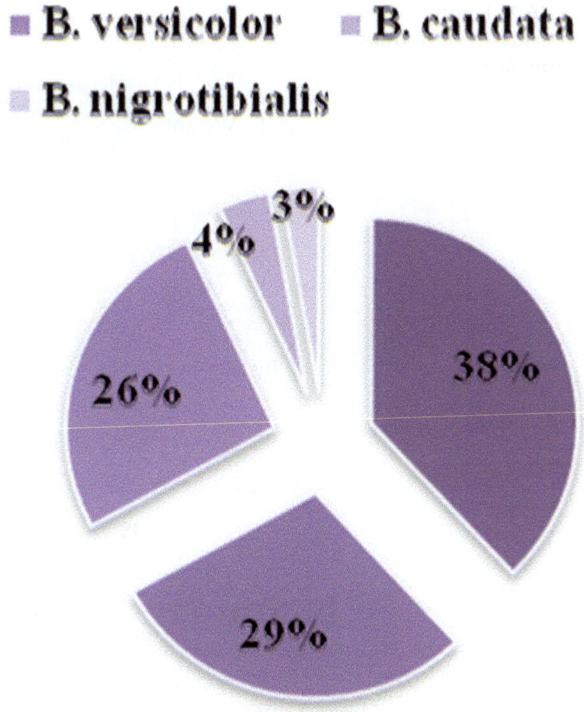

Fig. 16.6 Species composition of fruit flies in infested guava

Table 16.5 Fruit infestation and composition of fruit fly recovered from infested fruits of citrus from Darjeeling

Location	Percent fruit infested		Mean percentage of adult fly recovered after rearing	
	By no.	By wt.	*B. minax*	*B. dorsalis*
Orchard-1	48.30	46.60	59.06	41.12
Orchard-2	42.65	36.85	55.60	43.87
Orchard-3	53.21	32.76	63.72	36.51
Orchard-4	38.55	47.40	61.62	38.50
Mean	45.67	40.89	60.00	40.00

Dacus longicornis. Using methyl eugenol-baited traps, four species of tephritid flies were noted dominated by *B. dorsalis* followed by *B. zonata*. The observations are in corroboration with earlier reports (Galande et al. 2010; Kawashita et al. 2004; Hasyim et al. 2008). However, the authors have not detected *D. longicornis* (Wiedemann) using cue lure as sex attractant. The fruit flies were detected as the major yield-reducing factor for mango, guava, and citrus.

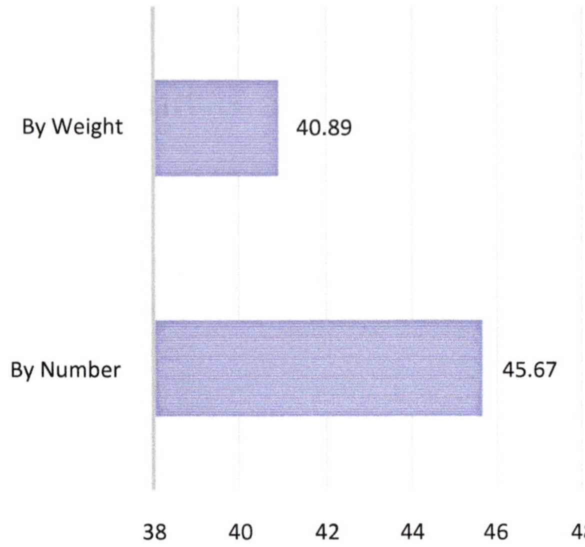

Fig. 16.7 Percent infestation of citrus fruits by fruit flies

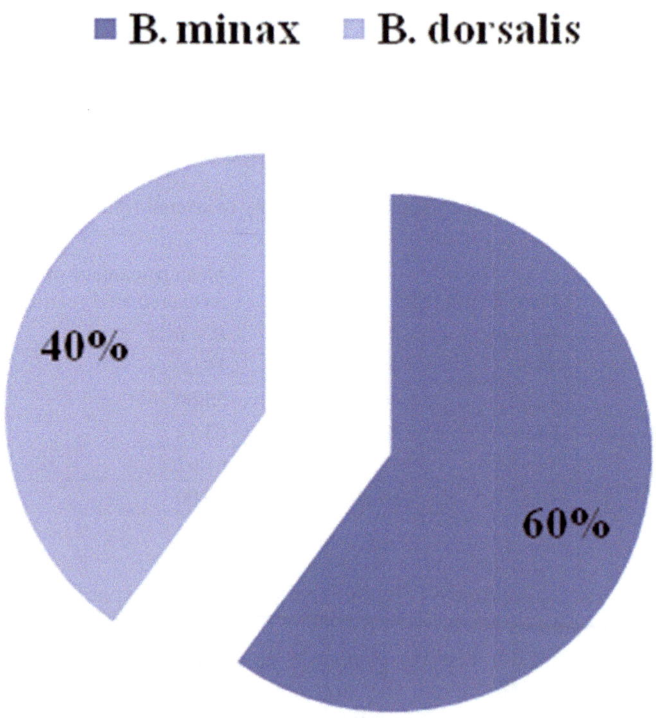

Fig. 16.8 Species composition of fruit flies recovered from infested citrus fruits

Mango The fruit fly species recovered from rearing infested fruits of mango are *B. dorsalis*, *B. zonata*, and *Brugmansia versicolor*, all found trapped in methyl eugenol trap. The trap catch in methyl eugenol and species composition recovered from infested mango fruits showed proportional similarity being the highest percentage of *B. dorsalis* followed by *B. zonata* and *B. versicolor*. However, none of the species were attracted in cue lure-baited trap. The findings are in agreement with Gupta and Bhatia (2000) but contradictory to Khan et al. (2007) and Jalaluddin et al. (1999). Fruit fly species trapped in cue lure and methyl eugenol but not recovered from rearing infested fruits of mango indicated that, those species did not utilize mango as host and certainly infest other hosts grown in the vicinity of mango orchards.

B. zonata has been recorded from most states of India. It is expected that this fly is widely distributed throughout India and Pakistan. In Africa (Mauritius), adventive populations have been recorded. From North America (California), three individuals were trapped (Carey and Dowell 1989), but eradicated (Spaugy 1988). The record from Moluccas (Amboina) is a misidentification (White and Elson-Harris 1994) and the record for Sumatra in this same publication cannot now be traced and is presumed to be a misidentification.

Guava From rearing infested fruits of guava, the fruit fly species recovered were *B. dorsalis*, *B. zonata*, *B. versicolor*, *B. caudata*, and *B. nigrotibialis*. Most of them were also attracted in Methyl eugenol trap except *B. caudata* and *B. nigrotibialis*. However, the percentage recovery of *B. caudata* and *B. nigrotibialis* were meager and also not found to be trapped in methyl eugenol. In cue lure-baited trap, two species of fruit fly were detected. The results are in partial agreement with Galande et al. (2010).

Citrus The fruit fly species recovered from infested citrus fruits were *B. minax* and *B. dorsalis*. The predominant species was *B. minax*. But this species was neither attracted in the sex attractant methyl eugenol nor in cue lure. On the contrary, no species was recovered from rearing infested citrus fruits that was attracted in cue lure trap. It might be that the fly species attracted in cue lure-baited trap exploit other host plants adjacent to citrus orchards. Only methyl eugenol attracted one of the two species, i.e., *B. dorsalis*. This is in agreement with Drew and Raghu (2002).

Fruit flies (Tephritidae: Diptera) (Fig. 16.9) are the important pests of mango, guava, and citrus that cause considerable crop loss in production of mango (37.67 % and 35.20 % on number and weight basis, respectively), guava (36.33 % and 33.60 % on number and weight basis, respectively), and citrus (45.67 % and 40.89 % on number and weight basis, respectively). Altogether ten species of fruit flies were found in sub-Himalayan region. The species were *B. cucurbitae*, *B. dorsalis*, *B. diversa*, *B. zonata*, *B. versicolor*, *B. caudata* and *B. nigrotibialis*, *B. tau*, *Dacus longicornis*, and *B. correcta*. However, from rearing infested fruits of mango, guava, and citrus, only six species were recovered, viz., *B. zonata*, *B. versicolor*, *B. caudata*, *B. nigrotibialis*, and *B. minax*. Thus, altogether 11 species were detected both

Fig. 16.9 Different species of fruit flies in the Srinagar and Budgam districts on different cucurbit crops

by using sex attractants and rearing infested fruits of mango, guava, and citrus. It may also be concluded that for successful management of fruit fly in guava and mango, methyl eugenol can be utilized. But in citrus where the major limiting factor was *B. minax*, methyl eugenol may not be effective.

Studies on the diversity of fruit fly species and developmental biology of the predominant species in the Northwestern Himalaya revealed that *Bactrocera tau* (Walker) was the predominant species followed by *B. scutellaris* (Bezzi) and *B. cucurbitae* (Coquillett) infesting bottle gourd, cucumber, and summer squash. *B. tau*, the predominant species, completed nine generations from February to November under laboratory conditions (Prabhakar et al. 2009).

The study of Bhagat (2014) on fruit fly fauna of Jammu and Kashmir, in northern tip of India, deals with 48 species of fruit flies, belonging to 21 genera under families Drosophilidae (subfamily Drosophilinae) and Tephritidae (subfamily Dacinae, Tephritinae, and Trypetinae), occurring in diverse areas and localities of Jammu and Kashmir Himalayan region. The fruit fly species are responsible for causing great damage to valuable tropical and temperate fruit and vegetable crops of these regions. The crops attacked in Jammu region include ber, citrus, guava, mango, and phalsa. The crops affected in Kashmir region include apple, cherry, pomegranate, walnut, and members of Cucurbitaceae.

16.4.1 Fruit Fly Diversity in India Specially in the Himalayan Orchards and Vegetable Fields

Fruit flies are widely distributed in temperate, tropical, and subtropical regions of the world. Some of the species are polyphagous. There are about 325 species of fruit flies in the Indian sub continent. The major fruit fly pests in India belong to Bactrocera, which are mostly localized in distribution. *B. cucurbitae* (Coquillett), *B. dorsalis* (Hendel), and *B. zonata* (Saunders) are pests on cucurbits and mango, respectively. Among the invasive species, the capsule flies, *Acanthiophilus helianthi* (Rossi) and *Carpomya vesuviana* (Costa), are the cause of concern. Outbreaks of capsule fly are serious on safflower and *C. vesuviana* (ber fly) is serious pest on ber. The useful fruit flies include species that feed on weeds. Among them notable are *Urophora stylata* and *Dacus sorocula*.

References

Agarwal ML, Sueyoshi M (2005) Catalogue of Indian fruit flies (Diptera: Tephritidae). Orient Insects 39:371–433

Bezzi M (1916) On the fruit flies of the genus *Dacus* occurring in India, Burma and Ceylon. Bull Entomol Res 7:99–121

Bhagat RC (2014) Fruitfly fauna (Insecta: Diptera) of Jammu and Kashmir, Himalaya, India: checklist and biodiversity. Int J Food Agri Vet Sci 4(1):18–23

Carey JR, Dowell RV (1989) Exotic fruit pests and California agriculture. Calif Agric 43:38–40

De Meyer M, Robertson MP, Mansell MW (2010) Ecological niche and potential geographic distribution of the invasive fruit fly *Bactrocera invadens* (Diptera: Tephritidae). Bull Entomol Res 100(1):35–48

Dhillon MK, Singh R, Naresh JS, Sharma HC (2005) The melon fruit fly, *Bactrocera cucurbitae*: a review of its biology and management. J Insect Sci 5:40

Drew RAI (1989) The tropical fruit flies (Diptera: Tephritidae: Dacinae) of the Australasian and Oceanian regions. Mem Queensland Mus 26:1–52

Drew RAI, Hancock DL (1994) *Bactrocera dorsalis* complex of fruit flies (Diptera: Tephritidae: Dacinae) in Asia. Bull Entomol Res Suppl 2:1–68

Drew AI, Raghu S (2002) The fruit fly fauna (Diptera: Tephritidae: Dacinae) of the rainforest habitat of the Western Ghats, India. Raffles Bull Zool 50(2):327–352

Galande SM, Ukey NS, Jadhav SS, Shelke SS (2010) Composition of fruit fly species in Pune region of Maharashtra and losses caused to guava crop in the region. Pest Manag Econ Zool 18(1/2):290–294

Gupta D, Bhatia R (2000) Population fluctuation of fruit flies, *Bactrocera spp*. in sub mountainous mango and guava orchards. J Appli Hort 2(1):47–49

Hasseb M (2007) Current status of insect pest problems in guava. Acta Hortic 735:453–467

Hasyim A, Murryati, de-Kogel WJ (2008) Population fluctuation of adult males of the fruit fly, *Bactrocera tau* Walker (Diptera: Tephritidae) in passion fruit orchards in relation to abiotic factors and sanitation. Indones J Agril Sci 9(1):29–33

Hering EM (1941) Entomological results from the Swedish Expedition 1934 to Burma and British India. Diptera, Trypetidae. Arkiv Foer Zool 33B(11):6–7

Hering EM (1956) Trypetidae (Dipt.) von Ceylon. Verh Naturforschenden Gesselschaftin Basel 67:62–74

Jalaluddin SM, Natarajan K, Sadakathulla S, Balasubramaniyam S (1999) Discovery of the guava fruit fly, *Bactrocera correcta* (Bezzi). Enotomon 24(2):195–196

Kapoor VC (1971) Four new species of fruit flies (Tephritidae) from India. Orient Insects 5:477–482

Kapoor VC (1993) Indian fruit flies (Insecta: Diptera: Tephritidae). International Science Publisher, New York

Kawashita T, Rajapakase GBJP, Tsuruta K (2004) Population survey of *Bactrocera* fruit flies by lure trap in Sri Lanka. Res Bull Plant Prot Serv Jpn 40:83–87

Khan MA, Muhammad A, Waseem A (2007) Management of fruit flies (Diptera: Tephritidae) of the most perishable fruits. Entomol Res 35(2):79–84
Mazomenos BE, Pantazi-Mazomenou A, Stefanou D (2002) Attract and kill of the olive fruit fly Bactrocera oleae in Greece as a part of an integrated control system. Use of pheromones and other semiochemicals in integrated production IOBC wprs. Bulletin 25:1–11
McPheron BA, Steck GJ (1996) Fruit fly pests: a world assessment of their biology and management. St Lucie Press, Delray Beach
Mote ON (1975) Control of fruit fly, *Dacus cucurbitae* on bitter gourd and cucumber. Pesticides 9(8):36–37
Muskatim M, Soumya EA, Iraqui HM, Saad KI (2012) Biocontrol potential of a *Pseudomonas aeruginosa* strain against *B. oleae*. Afr J Micro Res 6(26):5472–5478
Norrbom AL, Carrol LE, Thompson FC, White IM, Friedberg A (1998) Systematic database of names. In: Thompson FC (ed) Fruit fly expert identification system and systematic information database, vol 9, Myia. Backhuys Publishers, Leiden, pp 65–299
Patel ST (1976) The bionomics and control measure of Ethiopian fruit fly, *D. ciliates* Loew (Tephritidae: Diptera). M. Sc. (Agri.) thesis submitted to Gujrat Agricultural University, Sardar Krushinagar, Udaipur
Prabhakar CS, Sood P, Mehta PK, Choudhary A (2009) Distribution and developmental biology of fruitflies infesting cucurbits in the north-western Himalaya. J Insect Sci 22(3):300–308
Prabhakar CS, Pankaj S, Mehta PK (2012) Pictorial keys for predominant *Bactrocera* and *Dacus* fruit flies (Diptera: Tephritidae) of north western Himalaya. Arthropods 1(3):101–111
Pruthi HS (1941) Report of the Imperial Entomologist. Sci Rep Indian Agric Res Inst New Delhi, 40:15–16
Ranganath HR, Veenakumari K (1995) Notes on the Dacine fruit flies (Diptera: Tephritidae) on Andaman and Nicobar Islands. Raffles Bull Zool 43:235–238
Spaugy L (1988) Fruit flies. Two more eradication projects over. Citrograph 73:168
Tsuruta K, White IM (2001) Eleven new species of the Genus *Bactrocera* Macquart (Diptera: Tephritidae) from Sri Lanka. Entomol Sci 4:69–87
Tsuruta K, White IM, Bandara HMJ, Rajapakse H, Sundaraperuma SAH, Kahawatta SBMUC, Rajapakse GBJP (1997) A preliminary note on the host plants of fruit flies of the Tribe Dacini (Diptera: Tephritidae) in Sri Lanka. Esakia 37:149–160
Tsuruta K, Bandara HMJ, Rajapakse H, Sundaraperuma SAH, Kahawatta SBMUC, Rajapakse GBJP (1998) Pictorial key to dacine fruit flies associated with economic plants in Sri Lanka. Res Bull Plant Prot Serv Jpn 34:23–35
Van Schoubroeck F (1999) Learning to fight a fly: developing citrus IPM in Bhutan. Ph.D. thesis, Wageningen University and Research Centre, Wageningen, Netherlands, pp 200
White IM, Elson-Harris MM (1994) Fruit flies of economic significance: their identification and bionomics. CAB International, Oxon, p 601
White IM, Evenhuis NL (1999) New species and records of Indo-Australian Dacini (Diptera: Tephritidae). Raffles Bull Zool 47:487–540

Diversity of Soft Scale Insects (Hemiptera, Sternorrhyncha, Coccoidea) in Sri Lanka

17

U.G.A.I. Sirisena, G.W. Watson, K.S. Hemachandra, and H.N.P. Wijayagunasekara

Abstract

The soft scale insects belong to one of 15 families of phytophagous scale insects of economic importance. A field survey of perennial plants in 16 agro-ecological zones of Sri Lanka was conducted in 2011–2012 to identify the soft scale species present and their host ranges. Fourteen species belonging to eight genera were identified. Two introduced species, *Ceroplastes sinensis* Del Guercio and *Pulvinaria urbicola* Cockerell, were recorded from Sri Lanka for the first time. The distribution and host range of the recorded species are documented and their potential threat as crop pests is discussed. Published literature on soft scales is reviewed.

Keywords

Soft scale Insects • *Ceroplastes sinensis* • Crop pest • Sri Lanka

U.G.A.I. Sirisena (✉)
Post Graduate Institute of Agriculture, University of Peradeniya, Peradeniya, Sri Lanka
e-mail: anuraindra@gmail.com

G.W. Watson
California Department of Food & Agriculture, Plant Pest Diagnostic Center, Sacramento, CA, USA
e-mail: gillian.watson@cdfa.ca.gov

K.S. Hemachandra • H.N.P. Wijayagunasekara
Department of Agriculture Biology, Faculty of Agriculture, University of Peradeniya, Peradeniya, Sri Lanka
e-mail: kshema@pdn.ac.lk

A.K. Chakravarthy, S. Sridhara (eds.), *Arthropod Diversity and Conservation in the Tropics and Sub-tropics*, DOI 10.1007/978-981-10-1518-2_17

17.1 Introduction

The soft scale insects feed mainly on perennial (but occasionally annual) plants. Members of this family, about 1100 described species, are distributed in all geographical regions of the world. They feed on almost any live organ of the host plant, including the roots, although most species develop on the leaves or twigs or trunk (Ben-Dov 1997). Coccidae are noxious pests, causing direct damage by depleting the host plant of nutrients and fluids, and by damaging tissues, and indirect damage by excretion of sugary honeydew, which accumulates on plant surfaces. The sticky honeydew deposits often result in the development of black sooty mould, significantly reducing the market value of the crop.

Soft scales are well represented in Sri Lanka, with 71 species in 28 genera (Ben-Dov et al. 2013). Edward Ernest Green, Government Entomologist in Ceylon from 1899 to 1913, pioneered the study of Coccoidea in Sri Lanka. His work was published as *The Coccidae of Ceylon* in five volumes (Green 1896b, 1899, 1904, 1909, 1922), some supplementary notes (Green 1900, 1905) and a final check list of all the Coccoidea species known from Ceylon (Green 1937). Andrew Rutherford followed Green as Government Entomologist (Williams 1999); he described additional scales including three species of Coccidae (Rutherford 1914, 1915). Apparently, no systematic studies have been conducted on Coccidae of Sri Lanka since. A field survey of 16 (out of 46) agro-ecological zones of Sri Lanka was therefore carried out in 2011–2012.

17.2 The Soft Scales

The soft scale insects (Insecta, Hemiptera, Coccoidea, Coccidae), named as *Cochenilles coccines* (in French), *Schildlaus* (in German), *coccini* (in Italian) and *coccidos* (in Spanish), establish a group among the 21 families of scale insects. Whereas scale insects are classified under the superfamily Coccoidea, other taxonomists of the group rank the latter suborder Coccinea. Within them, the Coccidae are among the 18 families of the progressive coccoids (termed as the Neococcoidea). This group is from the Margaroid scale insects; (termed Archeococcoidea) all instars possess only two pairs of thoracic spiracles, and the adult males have only simple unicorneal eyes, while the Margaroids there are abdominal spiracles and compound eyes in the males. The Coccidae are placed among the lecanoid families of Coccoidea, showing great affinity with the Aclerdidae, Kermesidae and Lecanodiaspididae.

Although species of soft scales are noxious plant pests, others are ranked as beneficial insects. Thus the honeydew of several species is foraged by honeybee and of great benefit to apiculture. The voluminous waxy tests of several species are harvested in some countries and processed into commercial products.

The adult female or male (Fig. 17.1) completes its life cycle at the site of settling stage of first nymph or crawler; however, the majority of species retain functional

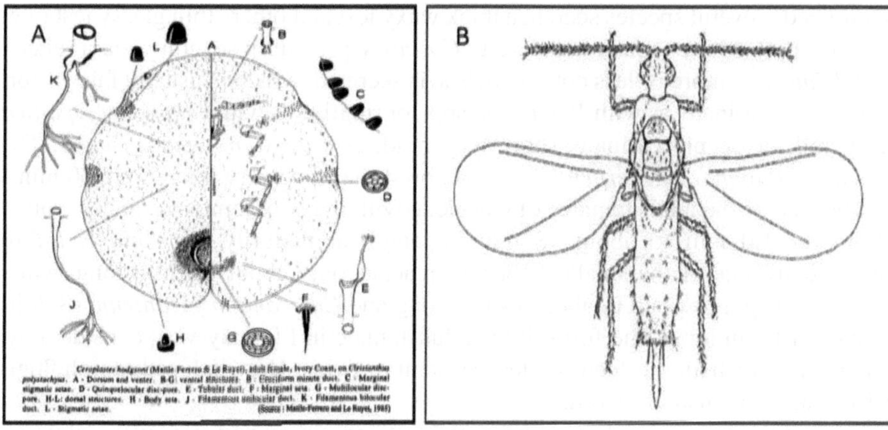

Fig. 17.1 Morphology of (**a**) adult female and (**b**) male coccid

legs in all instars. Consequently, development of the insects takes place by feeding on different organs of the host plant.

The inactive life habit of nymphs and adult females have made Coccidae to face several difficulties. These pressures have been solved in various ways. Namely, (1) the waxy cover overwhelms the pressure of living in a harsh environment; (2) the activity of the host makes them avoid the build-up of honeydew around and above the insect and thus the honeydew droplets are ejected away from the host plant and the coccid; and (3) natural enemies are effectively evaded through a successful symbiosis evolved between species of Coccidae and ants. An additional mechanism is to encapsulate their eggs or larvae of parasitoids.

Sexual and parthenogenetic reproduction recorded in soft scales. Some of the cosmopolitan, widely distributed species, e.g. *Coccus hesperidum* L., *Saissetia oleae* (Olivier) and *Parasaissetia nigra* (Nietner), are known to reproduce parthenogenetically throughout their distribution. Few species of this family contain two chromosome systems. The *Comstockiella* system, in which reproduction is sexual and diploid arrhenotoky, which is a type of facultative parthenogenesis.

Species of Coccidae are currently classified on the basis of external features of the adult female (Steinweden 1929; Borchsenius 1957). The strong distention and thickness of the body at full maturity becomes swollen, strongly convex, heavily sclerotized, hard and brittle in some typical females in the genus *Physokermes* which makes them difficult to study. Young adult females can be studied satisfactorily, especially in univoltine species which are found only for a few days each year. Young and old females however can be most often recognized as Coccidae by the presence of a pair of anal plates.

Unlike males, which have prepupal and pupal instars, adult females emerge directly from the second or third instar nymphs (according to species) and continue to grow slightly or considerably, changing significantly in shape and colour (Gill 1988). Each young adult female may grow from two to eight times the size of its previous instars. Adult females of many species become swollen and heavily sclerotized. Several species secrete a thick waxy test and others thin glassy test (e.g. *Inglisia* spp.). Many remain naked (e.g. *Coccus* spp.), but in several genera related to *Pulvinaria*, a more or less cottony ovisac is secreted, which is usually fairly short and slightly carinate, as with *Pulvinaria* spp., or ringlike as with *Takahashi japonica* Cockerell, or deeply carinate, short and broad, as with *Ceronema africana*. The ovisacs *of Pulvinarisca serpentina* (Balachowsky) can be very long, up to 30 mm.

The size of the adult females of Coccidae with the wax removed ranges from a little over 1.0 mm to about 18 mm in length at maturity. *Vinsonia stellifera* (Westwood), one of the smallest species, reaches only about 1.5 mm long, while *Toumeyella* sp. grows up to about 14 mm long and *Eulecanium giganteum* (Shinji) to about 18 mm long. The form of the adult female in life may vary considerably, depending on whether it feeds on leaves, stem or twigs. Host-induced morphological variation are also observed.

The whole dorsal cuticle may become hard and glossy and variously coloured, either plain, spotted or brightly coloured. The ageing process varies the colour in the population. *Toumeyella liriodendri* (Gmelin), for example, varies from greyish-green to pink-orange or dark brown (Hamon and Williams 1984).

The dorsal cuticle becomes thick and sclerotized where as the ventral cuticle hardly changes after the last moult. The cuticle of swollen, spherical and fully grown females *Saissetia coffeae* (Walker) is about four times thicker after the final moult, and more than ten times thicker than the ventral cuticle of females that produce an ovisac, in which the dorsum remains flat and soft, hence the vernacular name "soft scales". The dorsums remain membranous on females and are entirely enclosed within ovisac, as with species of *Eriopeltis*, while the dorsal cuticle of females not enclosed within an ovisac may become slightly sclerotized, as with *Pulvinaria* spp.

The young female is usually termed as neotonic because its general appearance is similar to the second and third instar female compared with the adult male. The shape of the slide-mounted specimen is narrow to broadly oval, subcircular or sometimes pyriform and dorsoventrally flattened. Adult females have well-developed legs and antennae, except in a few genera where these appendages are reduced or vestigial. Only a few species, such as *Houardia troglodytes* Marchal, are legless. The head, thorax and abdomen are fused. Members of Coccidae are carried with two anal plates located at the anal cleft. The cleft may be very deep, parallel-sided or strongly divergent. The margin of the anal cleft is usually devoid of setae, but few species possess an anal cleft with marginal or submarginal setae which may be slender or spinelike and may be fused along with its whole length. The antennae, eyes and mouthparts indicate the head, while the legs and spiracles mark the thoracic segments.

Most species contain abdominal segmentation visible on the mid-region on the ventral surface and can be detected by the arrangement of the ventral body setae, and the ventral abdominal segments I and II are fused together. The anal opening is on XI abdominal segment. The anal ring is on Xth segment, and the XIth is composed of the anal plates, while the VIIIth segment bears the vulva. The cuticle is sparingly covered with setae and pitted by openings of secretary gland pores, ducts and glandular dorsal tubercles. Both sexes are sometimes found with cuticular glands, whereas the ventral multilocular disc pores associated with the vulva, the dorsal and ventral tubular ducts and the dorsal preopercular pores are restricted to adult females. For many species, the morphology is quite constant. However, host-induced variation of morphological characters occurs on some Coccidae and has been recorded for two worldwide pests, *Coccus hesperidum* (L.) in Rhodesia (Hodgson 1994) and *Parasaissetia nigra* (Nietner) (Ben-Dov 1997).

17.3 Materials and Methods

17.3.1 Collection and Preservation

In field, soft scale insects were collected on pieces of infested plant material and placed in labelled Ziploc polythene bags (15 × 20 cm). Live specimens were not picked off from the host plant individually with forceps because this can damage them. Bagged samples were kept cool and taken to laboratory for sorting and preservation using a dissection microscope. Small pieces of infested plant material were isolated and placed in 80 % alcohol in labelled screw-topped Nalgene vials (3 or 5 ml capacity), to kill and preserve the insects. To prevent body blackening, the labelled vials of freshly killed material were sealed and immediately stood in a water bath at 100°C for 15–20 min, to denature the enzymes and ensure optimal fixation of the body contents. Cooled samples were stored in a refrigerator until they were required for slide preparation. Preserved samples were taken to the Plant Pest Diagnostic Center, California Department of Food and Agriculture, Sacramento, California, USA (CDFA-PPDC), for preparation and identification.

17.3.2 Preparation of Slide Mounts

Permanent slide mounts were prepared using a method modified from that given in Watson and Chandler (2000). Full details of the method used are given in Sirisena et al. (2013). Small- to medium-sized adult females were used in preference to very large specimens. If external wax was visible, soaking in 95 % alcohol dissolved much of the wax or hardened any residue sufficiently to enable any large lumps of external wax to be physically removed, before chemical processing began. To facilitate removal of the body contents, a neat hole was made in the ventral cuticle on one side of the abdomen. Body contents were removed by warming in 10 % KOH and pumping the body with a micro-spatula; the cuticle was rinsed in water and acidified before staining in acid fuchsin. After dehydration with alcohols, dewaxing in 50:50 histoclear phenol and clearing in clove oil, the specimens were slide-mounted in Canada balsam and dried at 40 °C for 3 months. The resultant archival preparations were deposited in the reference insect collections at the Plant Pest Diagnostic Center, California Department of Food and Agriculture, Sacramento, California, USA (CDFA-PPDC), and the Department of Agricultural Biology, Faculty of Agriculture, University of Peradeniya, Sri Lanka.

17.3.3 Authoritative Identification

Slide-mounted specimens were examined using a Zeiss compound microscope with phase contrast illumination and magnifications of ×25–×800. Species were identified using published keys and comparison with reference specimens in the California State Collection of Arthropods (CSCA) at CDFA-PPDC.

17.4 Results and Discussion

Fourteen species of soft scales in eight genera were identified in the survey samples collected from Sri Lanka in 2011 (Table 17.1). *Ceroplastes sinensis* Del Guercio and *Pulvinaria urbicola* Cockerell were recorded from Sri Lanka for the first time (Fig. 17.2). The genera and species collected on the survey are discussed below.

17.4.1 Genus: *Ceroplastes*

Six species of *Ceroplastes* have been recorded from Sri Lanka (Ben-Dov et al. 2013); three of them were collected in the present study (Table 17.1). This genus is distinctive in having the anal plates situated at the apex of a dorsal sclerotized prominence. The species can be very difficult to identify; for taxonomic illustrations of five of them, see Avasthi and Shafee (1986); Gill (1988); Williams and Watson (1990). *C. stellifer* (which also called *Vinsonia stellifera* (Westwood) until recently) is a tropicopolitan and polyphagous species often found in conspicuous numbers on the leaves of economically important trees and shrubs. It can be recognized easily in the field by its colourless, translucent wax cover, shaped like a 7-pointed star.

The following three *Ceroplastes* species have thick, soft, opaque white or pink wax in life. In *C. floridensis*, a tropical and temperate species known from numerous host plants, the slide-mounted adult female possesses submarginal tubular ducts, each with a short bulbous inner ductule. *C. rubens*, a tropicopolitan species, is distinguished by having short legs, each with the tibia and tarsus fused. It attacks many crops but seems to particularly favour *Ixora coccinea* in Sri Lanka.

An adult female in poor condition, collected from a nursery on *Psychotria* sp. in Batalagoda (7° 32′ N, 80° 28′ E, 87 m AMSL), was tentatively identified as *C. sinensis*. This is a temperate species found on numerous host species (Ben-Dov et al. 2013). Our slide-mounted specimen possesses seven-segmented antennae, some fine filamentous ducts near the margin and predominantly trilocular dorsal pores. It may be the first record of *C. sinensis* from Sri Lanka. It is difficult to explain the presence of a temperate species on tropical island, especially at a low

Table 17.1 Distribution and host range of soft scale species in 16 agro-ecological zones in Sri Lanka

Coccidae species	Location(s)	Host range
Ceroplastes floridensis Comstock	Hambantota	*Annona squamosa* L.
Ceroplastes rubens Maskell	Ibbagamuwa	*Ixora coccinea* L.
Ceroplastes sinensis Del Guercio	Ibbagamuwa	*Psychotria* L. sp.
Ceroplastes stellifer (Westwood)	Ibbagamuwa, Hambantota	*Psychotria* L. sp., *Syzygium samarangense* (Blume) Merr. & L.M. Perry
Coccus hesperidum Linnaeus	Batalagoda	*Annona squamosa* L.
Coccus viridis (Green)	Ibbagamuwa, Hambantota	*Psychotria* L. sp., *Aegle marmelos* (L.) Corrêa
Drepanococcus cajani (Maskell)	Peradeniya	Unidentified endemic plant
Parasaissetia nigra (Nietner)	Dambulla, Hambantota	*Annona squamosa* L.
Protococcus acutissimus (Green)	Kiriwaneliya	*Persea americana* Mill.
Pulvinaria polygonata Cockerell	Hambantota	*Citrus aurantifolia* (Christm.) Swingle
Pulvinaria psidii Maskell	Madatugama, Werellagama	*Psidium guajava* L., *Syzygium cumini* (L.) Skeels
Pulvinaria urbicola Cockerell	Mukalanyaya, Hambantota	*Capsicum frutescens* L., *Aegle marmelos* (L.) Corrêa
Saissetia coffeae (Walker)	Ibbagamuwa, Dambulla	*Psychotria* L. sp., *Coccinia grandis* (L.) Voigt
Taiwansaissetia sp., possibly *formicarii* (Green)	Mukalanyaya, Batalagoda, Madatugama	*Coffea arabica* L., *Mangifera indica* L.

elevation; it may have been a single specimen on imported nursery stock. More collections will be needed to confirm whether *C. sinensis* is established on the island or if the species might be *C. rusci* (also not known from Sri Lanka), which is very similar morphologically and has a tropical distribution (Ben-Dov et al. 2013).

17.4.2 Genus: *Coccus*

Of 11 species of *Coccus* previously recorded from Sri Lanka (Ben-Dov et al. 2013), two were found during the survey (Table 17.1). Members of this genus usually remain relatively flat and soft-bodied and lack any ventral submarginal band of ducts.

Fig. 17.2 Major coccid pests; (**a**) *Coccus viridis* (**b**) *Pulvinaria polygonata* (Source:http://www.sel.barc.usda.gov/scalekeys/SoftScales/key/Soft_scales/Media/Html/), (**c**) *Pulvinaria psidii* (Source: Long Beach, Los Angeles County, California, USA, November 1, 2013), (**d**) *Saissetia coffeae* (Source: http://www.azoresbioportal.angra.uac.pt/imagens/Imagenes/Novembro21// Saissetia_coffeae.jpg)

Both species were illustrated by Williams and Watson (1990). *C. hesperidum* is one of the most cosmopolitan and polyphagous of all scale insects; in the slide-mounted female, the tubular ducts by the mesocoxae each have the inner ductule no wider than the outer ductile, and there are no tubular ducts associated with the hind coxae. *C. viridis* is also a polyphagous pest and often causes damage to citrus and coffee; in the slide-mounted female, the tubular ducts by the meso- and metacoxae each have the inner ductule wider than the outer ductule.

17.4.3 Genus: *Drepanococcus*

Drepanococcus is a distinctive genus in which each spiracular fold extends on to the dorsal surface and terminates in a single, long spiracular seta. *D. cajani* is distinguished from other *Drepanococcus* species by the segmentation of the antennae; the antennal segments are almost all shorter than, wide (illustrated by Williams and Watson 1990).

17.4.4 Genus: *Parasaissetia*

Although *Parasaissetia* is superficially similar to *Saissetia* in life, it differs in lacking an H-shaped ridge on the dorsum in the early stages, and the dorsal surface of even the very young adult female has slightly raised, polygonal tessellations. The dorsum of the adult female becomes very heavily sclerotized at maturity, and in slide mounts the dorsal setae are parallel-sided and often capitate (illustrated by Gill 1988). In life, the adult female of *P. nigra* varies considerably in form depending on whether it is on a leaf (when the insect is relatively wide and flat) or a twig (when the insect may be more elongate and highly convex). The colour varies from pale to extremely dark brown or black, depending on age.

17.4.5 Genus: *Prococcus*

This genus contains only one species (Ben-Dov et al. 2013). In life, the adult female of *P. acutissimus* is very long and narrow, bluntly pointed at each end and slightly convex, becoming dark brown at maturity. In slide mounts, the species is easily recognizable by its elongate body and reduced legs, which are shorter than the antennae (illustrated by Williams and Watson 1990). In Sri Lanka, *P. acutissimus* was commonly found on avocado leaves.

17.4.6 Genus: *Pulvinaria*

The genus *Pulvinaria* is large and in need of revision; the species are morphologically complex and difficult to identify. The adult female secretes a white ovisac of sticky wax filaments that is often longer than her own body; she possesses a ventral submarginal band of tubular ducts of two or more types and tubular ducts present in central parts of the venter. Five species of *Pulvinaria* have been reported previously from Sri Lanka (Ben-Dov et al. 2013). In addition, this study found *P. urbicola* Cockerell for the first time, at two widely separated localities. The adult female is green in life and secretes a ribbed white ovisac; Qin and Gullan (1992) illustrated this species. *P. urbicola* seems to prefer *Capsicum* sp. and other species of Solanaceae. *P. polygonata* are widespread throughout southern Asia; although it is apparently polyphagous, in Sri Lanka it preferred feeding on citrus, heavily colonizing the lower leaf surfaces. In life, the immatures and young adult females are yellow but later turn brownish with maturity, also becoming covered with cottony white wax. *P. psidii* is a tropicopolitan species found on a large number of host plants. The live adult female is bright green and secretes a short, white ovisac that is not ribbed. Williams and Watson (1990) illustrated both *P. polygonata* and *P. psidii.*

17.4.7 Genus: *Saissetia*

In the early stages, species of *Saissetia* have a H-shaped ridge on the dorsum; this gradually disappears as the adult female expands and matures, darkening from pale to chestnut brown as she does so. The dorsum of the adult female becomes very heavily sclerotized at maturity, and in slide mounts the dorsal setae are conical with sharp apices. *S. coffeae* is tropicopolitan and polyphagous. It attacks important crop and ornamental plants and is an important pest on coffee in Sri Lanka. The appearance of the adult female in life varies according to age and substrate, from fairly flat and pale when young to hemispherical and rich brown at maturity. Apart from the H-shaped ridge, the young adult female has a smooth dorsal surface, but it sometimes becomes textured with small raised polygons at maturity. In microscope slide mounts, the adult female of *S. coffeae* is easily separated from other species of *Saissetia* by possessing two types of tubular duct in the ventral submarginal band of ducts; the other species possess only one type of tubular duct.

17.4.8 Genus: *Taiwansaissetia*

Green (1896a) described *Lecanium formicarii* from Sri Lanka, associated with ants; this species was subsequently designated as the type species of the genus *Taiwansaissetia* by Tao et al. (1983). This genus often has more than three stigmatic setae in each cleft, and the dorsal setae are setose – sharp, slender and slightly curved (illustrated by Hodgson 1994). In this study, some *Coccus*-like specimens were collected in close association with the ants *Oecophylla smaragdina* Emery and *Anoplolepis gracilipes* (Smith). The specimens have dorsal setae which are small, sharp, slender and slightly curved; at least some stigmatic clefts contain four stigmatic setae; and the membrane under the median margins of the anal plates has a tessellated texture. It is possible that these scales may represent a variant of *T. formicarii*, bearing in mind the high degree of morphological variation in that species discussed by Hodgson (1994), or they may represent an undescribed species. More collections and further study of this species are needed. With the two new country records, the total number of soft scale insects in Sri Lanka is now 29 genera containing 73 species, 34 % of which are endemic.

The diversity of soft scale insects in Sri Lanka is high and many species are pests on important crops. There is evidence that potentially invasive species continue to be accidentally introduced on imported live plant material, so monitoring and curtailing their spread is essential. Coccids in general are ecologically important, forming a link at different trophic levels in the ecosystems.

References

Accessed online on 2 Jan 2013 at http://www.cdfa.ca.gov/plant/PPD/publications/tech_series.html

Avasthi RK, Shafee SA (1986) Species of Ceroplastinae (Homoptera: Coccidae) from India. J Bombay Nat Hist Soc 83:327–338

Ben-Dov Y (1997) Chapter 1.1 Morphology, systematics and phylogeny. 1.1.1 Diagnosis. In: Soft scale insects: their biology, natural enemies and control (vol 7A). Elsevier, Amsterdam, pp 3–4

Ben-Dov Y, Miller DR, Gibson GAP (2013) Scale net. Accessed online on 31 Dec 2012 at http://www.sel.barc.usda.gov/SCALENET/query.htm

Borchsenius NS (1957) Subtribe mealybugs and scales (Coccoidea). Soft scale insects Coccidae, vol IX. (In Russian), Fauna SSSR. Zool Inst Akad Nauk SSSR. N.S 66:1–493

Gill RJ (1988) The scale insects of california, part 1. The soft scales (Homoptera: Coccoidea: Coccidae). Technical Series in Agricultural Systematics and Plant Pathology, Department of Food and Agriculture, California 1(11):132

Green EE (1896a) Catalogue of Coccidae collected in Ceylon. Indian Mus Notes 4:2–10

Green EE (1896b) The Coccidae of Ceylon. Part I. Dulau & Co., London

Green EE (1899) The Coccidae of Ceylon. Part II. Dulau & Co., London

Green EE (1900) Supplementary notes on the Coccidae of Ceylon. J Bombay Nat Hist Soc 13:66–76

Green EE (1904) The Coccidae of Ceylon. Part III. Dulau & Co., London

Green EE (1905) Supplementary notes on the Coccidae of Ceylon. J Bombay Nat Hist Soc 16:340–357

Green EE (1909) The Coccidae of Ceylon. Part IV. Dulau & Co., London

Green EE (1922) The Coccidae of Ceylon. Part V. Dalau and Co., London

Green EE (1937) An annotated list of the Coccidae of Ceylon, with emendations and additions to date. Ceylon J Sci 20:277–341

Hamon AB, Williams ML (1984) Arthropods of Florida and neighboring land areas, vol 11. The soft scale insects of Florida (Homoptera: Coccoidea: Coccidae). Florida Department of Agriculture and Consumer Services, Contribution No. 600, Florida Department of Agriculture, Gainesville, pp 194

Hodgson CJ (1994) The scale insect family Coccidae: an identification manual to genera. CAB International, Wallingford

Qin TK, Gullan PJ (1992) A revision of the Australian pulvinariine soft scales (Insecta: Hemiptera: Coccidae). J Nat Hist 26:103–164

Rutherford A (1914) Some Ceylon Coccidae. Bull Entomol Res 5:259–268

Rutherford A (1915) Some new Ceylon Coccidae. J Bombay Nat Hist Soc 24:111–118

Sirisena UGAI, Watson GW, Hemachandra KS, Wijayagunasekara HNP (2013) Preparation of Hemiptera: Sternorrhyncha for identification. Trop Agric Res J 24(2):39–149

Steinweden JB (1929) Bases for the generic classification of the coccoid family Coccidae. Ann Entomol Soc Am 22:197–245

Tao CCC, Wong CY, Chang YC (1983) Monograph of Coccidae of Taiwan, Republic of China (Homoptera: Coccoidea). J Taiwan Mus 36:57–107

Watson GW, Chandler LR (2000) Identification of mealybugs important in the Caribbean Region, Revised edn. CABI Bioscience, Egham

Williams DJ (1999) E.E. Green's the Coccidae of Ceylon, 1896–1922. Entomol Mon Mag 135:189–191

Williams DJ, Watson GW (1990) The scale insects of the tropical South Pacific region. Part 3: The soft scales (Coccidae) and other families. CAB International Institute of Entomology, London

Insect Conservation: A Synthesis of Management Approaches

18

Ritesh Suresh Inamke, Deepak Panpatte, Noushad Parvez, Leena Pathak, Ankit Patel, Kuldeep Khatri, and Janardan Jani

Abstract

Threats to insect diversity vary from habitat loss and invasive alien species to environmental degradation and pollution. Several of the threats are symbiotic, with the cumulative influence of habitat loss and global climate change being the predominant factors. Recent research on insect conservation has elucidated few fundamental principles for conservation and management. There are six rudimental principles that are interrelated and together provide guidelines for conservation management of insects. They are maintaining reserves, quality landscape heterogeneity, reducing contrast between left out patches and disturbed patches outside reserves, landscapes simulating natural conditions and perturbance, and connecting homogeneous patches over large areas.

Keywords

Conservation • Heterogeneity • Permeating • Synergistic

18.1 Introduction

18.1.1 Historical Background

Insects as a group are dominating earth. The increasing diversity of insects at the family level is witnessed over the last 400 million years. Presently around 600 families are living (Fig. 18.1). While at the species level there is no such considerable steady increase, several species have vanished by the end of Cretaceous period.

R.S. Inamke (✉) • D. Panpatte • N. Parvez • L. Pathak • A. Patel • K. Khatri • J. Jani
Biological Control Research Laboratory, Anand Agricultural University,
Anand 388 110, Gujarat, India
e-mail: riteshinamke100@gmail.com

A.K. Chakravarthy, S. Sridhara (eds.), *Arthropod Diversity and Conservation in the Tropics and Sub-tropics*, DOI 10.1007/978-981-10-1518-2_18

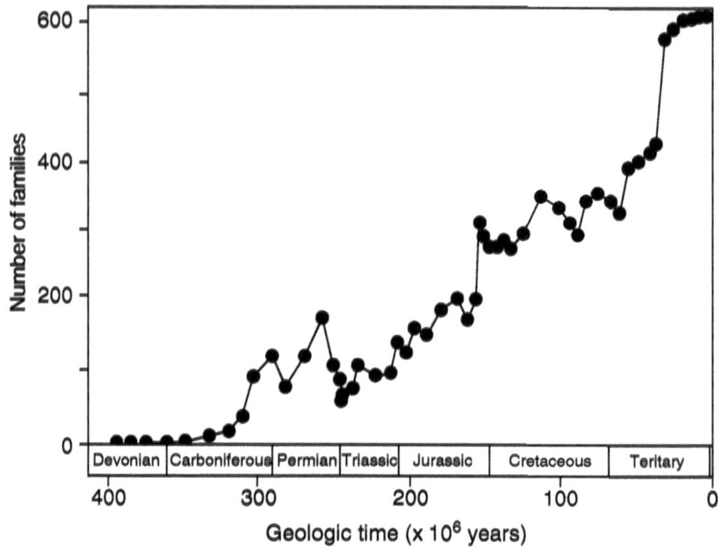

Fig. 18.1 Family-level diversity of fossil insects over the last 400 million years (Samways 2005)

Most of the extinct species were specialists. In the past few hundreds of thousands of years, with the advancement and retreat of glaciers, few insect species got extinct. Population of insects in the Northern Hemisphere have responded to the cold and thaws by moving southward during the glacial regime. Generation of new species is also a result of insects shifting up and down mountains.

18.2 Ecological Importance of Insects

18.2.1 Insects and Ecosystems Processes

Some insects have immense influence on soils, especially in arid areas. The funnel ant *Aphaenogaster longiceps* in Australia can move some 80 % of the surface that is moved by all soil fauna cumulated. Termites are consequential engineers and in West Africa their nests can even cover virtually a tenth of the land surface. Termites additionally influence the ecumenical carbon cycle, and in tropical forests they can engender 1.5 % of CO_2 and 15 % of methane engendered from all sources. The effect of termite activities can be so extensive that they can regulate and structure pant and micro arthropods.

18.2.2 Interaction with Plants

Plants are massive and species poor in comparison with insects which are tiny and species rich. This indicates that several insect species are establishing on a few plants. Some of these insects can be so abundant that they transfer an abundance of times the amount of energy as the vertebrates in the same area. Apart from that grasshoppers can expeditiously cycle nutrients as their frass is so soft and friable compared to that of vertebrates.

18.2.3 Insect Pollinators

It is interesting to note that evolution has driven flowers to be highly diverse and inviolate. The factor for such flower diversity has been the mutualism between the specialist insect and the flower. The current situation is that many flowering plants are threatened because their pollinator insects have disappeared and that select plant species are the "living dead." In addition many crop plants are under threat because populations of wild pollinator insects are dwindling. This is particularly true in the tropics and subtropics.

18.3 Challenges

Of millions of species only about 10 % have scientific denominations. Many insect species taxonomic revisions still are required. Many species, even prevalent ones, are multispecies complexes. Describing all unknown species before they become extinct is the taxonomic challenge. Still, there are liable to be much extinction, even of species that have never and will never be described. Addressing this taxonomic challenge is not a facile task, albeit several approaches are making this possible. These include undertaking full inventories of diminutive but paramount and tractable geographical areas, such as the Seychelles with its high number of endemics, or undertaking an ecumenical assessment of a particular taxonomic group as is being done for dragonflies. These approaches are supplemented with utilizer-amicable keys for nonspecialists engaged in conservation and with the deployment of computer apperception of specimens. Another great challenge for insect conservation is the perception challenge. Even among some general conservation practitioners, insects are often considered consequential or given scant attention.

18.3.1 Challenges Ahead

While many people are very fond of colorful butterflies that visit flowers, a few have a kindred affinity for cockroaches, the word "cockroach" even being utilized in irreverence. The challenge is that humans are historically tuned to optically discerning insects at immensely colossal as conveyors of disease and irritants the least. Yet all organisms have the right to live, and this includes the astronomical majority of insects that neither impact on humans nor are even for a moment in the collective conscience of human kind.

18.3.2 Taxonomic Gaps

It is essential to ameliorate taxonomic erudition on insect groups that would genuinely benefit from conservation. For example, less than one third of Indian insect fauna is documented. An estimated 35 % of Indian geographical area is surveyed for designating insect species, mainly by scientists of the Zoological Survey of India, Kolkata. There is a shortage of manpower in arthropod taxonomy and it is dwindling year by year in India. Secondly, the statuses of insect/arthropod amassments in India are below international standards. Thirdly, there are Draconian laws for impeding taxonomic research on arthropods. The exchange of arthropod specimens, dead or alive, involves elaborate procedures, and there are restrictions on amassment and preservation of arthropod specimens additionally. Conclusively adolescent biologists and students are not magnetized to taxonomic research.

18.4 Threats

18.4.1 Environmental Contamination

An amalgamation of elevating human population and more consumption of resources and energy, as quantified by gross domestic product, incremented by 460 % over the last century, with estimates that there will be a further elevation of 240 % by 2050 has adversely deteriorated the environment to say the least. Among the concerns is that this human pressure will have cascading effects on ecosystems, with loss of plant species leading to loss of insect species. Hawaii has lost five moth species because of plant extinctions. Simulations suggest that loss of just 5–10 % of keystone members of aliment webs can radically alter ecosystem function.

Many effects of environmental contamination are sublethal and not facilely detected. The insecticide deltamethrin can reduce fitness of larval and adult butterflies when applied at only 1/640 of the field dose. Despite the ostensible paramountcy of environmental contamination, little is kenned about its impact on insect species. Species respond differently to particular contaminant and concentration. Furthermore, there can be adverse interactive effects between impacts of contamination and other forms of stress, such as habitat fragmentation. Differences in the replications of species in the same victualing guild are optically discerned on Mayotte island in the Indian Ocean, where some dragonflies are much more susceptible to stream contaminants such as detergent than are others. Some insects are minute affected by pollutants, with some herbivorous insects even benefiting from low calibers of sulfur dioxide and nitrous oxide. In contrast, albeit the larvae of the butterfly *Parnassus apollo* can excrete metals, it cannot abide high calibers on its host plant. Relaxation of cumbersomely hefty metal pollution has enabled it to widen its geographical range once again.

18.4.2 Loss of Natural Habitat

It is estimated that by 2050 another 109 ha of natural ecosystems will be converted to agriculture, with a 2.5-fold increase in nitrogen- and phosphorus-driven eutrophication. These vicissitudes will be synergistic with pollution, habitat fragmentation, impact of invasive alien organisms, and ecumenical warming. These impacts will not affect all species equipollently, with specialists liable to decline the most, albeit some mundane species may additionally decline dramatically, as did the Rocky Mountain locust, *Melanoplus spretus*.

It was so abundant in the Midwest of North America in the tardy 1800s that it caused the wheels of locomotives to slip, yet by 1906 it was extinct. Some species even benefit from incremented edge effects, such as truculent ants at the interface between natural habitat and the agricultural matrix, where they heavily affect soil-dwelling arthropods of the transition zone. Land transformation leads to a mosaic of landscape patches, which is highly isolating for many species. Less mobile species may be tolerant of such isolation, which may be the confined spatial environment in which they evolved.

18.4.3 Invasives

Invasive alien species are a major threat to several indigenous and endemic organisms. Alien plants can replace indigenous ones and overrun ecosystems, even adversely influencing the hydrology. Such effects reduce diversity. Interestingly, the impacts of invasive alien plants are not always negative. Alien plants give shelter when there otherwise might not be, and alien water weeds can provide incremented habitat for insects like dragonflies and other generalists.

Invasive alien vertebrates can have direct and indirect influence on insects. On sub-Antarctic Marion Island, alien mice weigh up to 194 g ha^{-1} of invertebrate biomass, and alien rats have been implicated in local extinction of several insects including the Lord Howe Island stick insect, *Dryococelus australis*. The cane toad, *Bufo marinus*, was introduced in Australia to control certain beetle pests and is now having a major impact on many nontarget native insects, as are mosquito fish *Gambusia* spp., in Hawaii introduced to control mosquitoes have since affected indigenous damselflies *Megalagrion* spp. Invasive arthropods are a potential threat to human health and agricultural, horticultural, and forestry ecosystems. The entry of invasives takes place through mostly exchange of planting material and human carriers in India. There are more than 60 arthropods liable to invade the Indian subcontinent, anytime.

18.4.4 Classical Biological Control

The introduction of foreign biological control agents to control foreign pests has had economic and environmental advantages; inevitably it does carry some risks particularly for nontarget species. Generally adverse impact is likely species or genus specific. The main concern is that once classical biological control agents are introduced and established, they cannot be recalled and are therefore a new and permanent feature of the host landscape, thus violating a sense of place. While the adverse impacts of classical biological control are often difficult to prove, there is nevertheless evidence that some facets of it are detrimental to indigenous species. For instance, the tachinid fly *Compsilura concinnata*, introduced into the United States several times to control various pests, has been responsible in the decline of local saturniid moths. The suppression of alien weeds with insect herbivores has in many cases been successful. But there have also been adverse effects. Indigenous prickly pear Cacti (*Opuntia* spp.) in the United States and Mexico are currently threatened by the cactus moth, *Cactoblastis cactorum*, which is spreading in North America. Even insect pathogens carry risks. The bacterium *Bacillus thuringiensis israelensis*, utilized for suppressing mosquito populations, causes mortality of several aquatic insect larvae. Another form of *B. thuringiensis* used against lepidopteran pests has an impact on indigenous North American moths. Strong taxonomic support is required for surveillance and monitoring of biological control agents. Extensive collaborative arrangements and networking are also required for preventing ill effects of classical biological control activities in the tropics and subtropics, in particular (Table 18.1).

The activity of classical biological control is deliberate, and once control agents have been introduced and established, they cannot be withdrawn and are consequently an incipient and perpetual feature of the host landscape, thus breaching a sense of place. While the adverse impacts of classical biological control are often

Table 18.1 Global number of species of animal groups, plants, and fungi

Class	Described	Estimated
Mammals	4630	5000
Birds	9750	10,000
Reptiles	8000	9000
Amphibians	4950	5500
Fishes	25,000	30,000
Insects	970,000	8,000,000
Arachnids	75,000	75,000
Molluskans	7000	200,000
Crustaceans	40,000	150,000
Nematodes	25,000	440,000
Fungi	72,000	1,500,000
Plants	27,000	32,000
Total	1,750,000	14,000,000

arduous to prove, there is nevertheless evidence that some facets of it are detrimental to indigenous biotas. For example, the tachinid fly *Compsilura concinnata*, introduced in the Coalesced (combined) States many times to control sundry pests, has been responsible for the decline of local saturniid moths. The suppression of alien weeds with insect herbivores has in many cases been prosperous and has had economic and ecological benefits; there have been some side effects. Indigenous prickly pear cacti (*Opuntia* spp.) in the Coalesced States and Mexico are currently threatened by the cactus moth, *Cactoblastis cactorum*, which is spreading in North America. Even insect pathogens carry peril. The bacterium *Bacillus thuringiensis israelensis*, used to control mosquitoes, causes mortality of several aquatic insect larvae. Another form of *B. thuringiensis* used against Lepidopteran pests has an impact on indigenous North American moths. Vigorous taxonomic support is required for surveillance and monitoring of biological control agents. Extensive collaborative arrangements and networking are additionally required for averting side effects of classical biological control programs.

18.5 Insect and Arthropod Decline

Geographical boundary changes and its impact along with anthropogenic effects have caused about critical terminations or decreases of creepy, crawly and relative arthropod populations. The effects are as follows:

Urbanization Urban areas are frequently arranged in areas good for farming. Destinations at the intersections of significant waterways and long coastlines are favored to set up urban areas. In India, urbanization, industrialization, and contamination have had especially solid effect on local arthropods. For example, in Bangalore due to loss of blossoming plants in greenhouses, certain butterfly species have turned out to be less abundant.

The other case example is that of San Francisco, California, USA, which now covers what was once one of the major seaside hill biological systems in Western North America. Three hill butterflies, *Cercyonis sthenele, Glaucopsyche xerces,* and *Icaricia icarioides* which were endemic there, are currently wiped out. Three different butterflies, *Speyeria callippe, Incisalia mossii bayensis*, and *Icaricia icarioides missionensis*, are restricted to rough summits and inland slants in hills now constrained to San Bruno Mountains.

Acid Rain There is scant information. Although not properly documented, the occurrence of acid rain has been circumstantially correlated with the declines and losses of *Rhopalocera* species in North America.

Electric Lights Although no direct evidence can be cited, lights along streets and highways, particularly mercury vapor lamps, have been implicated in the population losses of nocturnal insects, particularly large moths.

Stream Pollution Pollution of streams, concretely through acid mine drainage and siltation, has probably resulted in profound transmutations in aquatic insect communities, albeit some studies have demonstrated rather resilient recuperation from chemical spills and acidification once the polluting source has been abstracted. Certain species of dragonflies may be categorically sensitive to pollution, and dragonflies and gomphids acclimated to main river channels in North America are extinct.

Agricultural Conversion
Conversion of natural habitats for cultivation is one of the most extensive land uses and has resulted in the greatest loss of native and endemic insect populations. However, in temperate altitudes the recent alluvial soils best suited to farming do not conventionally harbor insect communities, abundant in endemics, and relatively little extinction has occurred so far. The population of two sphinx moths, *Euproserpinus euterpe* and *E. weisti*, was greatly reduced by the planting of cereal crops on sandy soils (Opler 1979). It is in the tropics where the most serious losses of endemic insects has occurred due to conversion of natural habitats. The conversion of lowland and mountain forests on the Hawaiian Islands to pineapple and sugarcane plantings as well as to pasture resulted in the loss of several hundred species of native insects.

Exotic Introductions The effects of sundry preludes (whether intentional or not) of exotic animals and plants on native arthropods may be direct or indirect. Introduced plants may out consummate native plants and may indirectly lead to loss of native insects.

Exploitation Excessive collection of insects is a threat to insect populations (together with pesticides), and there are no documented cases of extinctions of even local insect populations due to indiscriminate application of pesticides. Most insect conservationists feel that amassing a sizably voluminous numbers of specimens from declining populations is at least unethical, and as a result several entomological societies have established amassing policies or guidelines. Endeavors to eliminate local populations of a bee and a butterfly by intensive amassing in the course of population studies in fact had the antithesis effect.

Genetic Engineering Genetically modified organisms (GMOs) are being increasingly used in integrated pest management programs. The GMOs can pose risk to select, locally adapted, insects. It has been discussed that adverse effects drastically reduced at the immensely colossal, regional spatial scale. Furthermore, GM plants are not a general answer for pest management, as there are transgenic plants with *B. thuringiensis* and insecticidal toxins resistant to the diamondback moth, *Plutella xylostella*. For insect conservation, the authentic risk of GM crops is what Woiwod has called the "pernicious side": In India exordium of Bt-cotton has resulted in incremented yields without much utilization of pesticides in certain areas. Likewise, scientists in developing countries are developing GM crops to overcome adverse effects of certain cultivated varieties.

Impacts of Climate Changes Climate change can affect trophic interactions, with all components of virtual webs from pathogens and mycorrhizae to predators and parasitoids being affected directly and indirectly. Insect herbivores in elevated carbon dioxide grew more gradually, consumed more plant material, took longer to develop, and suffered higher mortality compared with controls. Competitive interactions are additionally liable to be affected, as visually perceived in *Drosophila* assemblages in which different species were favored by particular temperatures. Nevertheless, some interactions have remained in step with climate change, with the winter moth larvae, *Operophtera brumata*, tracking transmuted budburst and the orange tip butterfly, *Anthocharis cardamines*, keeping pace with victual plant phenology. Climate change also affects geographical distribution and sex ratio.

As insects typically migrate more expeditious than trees, many temperate plant species are liable to have incipient encounters with herbivore insects shifting their geographical range from warmer areas to cooler areas.

However, there is a warning from Kuchlein and Ellis's (1997) study of microlepidoptera in the Netherlands, which suggests little point in monitoring individual species to assess the conservation status of specific ecosystems.

The greatest concern is that climate change will be interactive and synergistic with other adverse factors, leading to multiple impacts on species. Indeed, Travis has called the synergism between climate change and habitat loss a "deadly anthropogenic cocktail" for biodiversity. This is borne out by British butterflies, of which 89 % of the habitat specialists, compared with only 50 % of the mobile generalists, have declined in geographical distribution.

Pesticides Although often cited as a major factor responsible for the loss of insect populations, four decades of wide use of organic pesticides has not resulted in the extinction of any insect, except for the possible loss of some ectoparasites and symbolists due to pesticide residues (Debach 1974). More normally, insecticide usage causes only temporary reductions and changes in the relative abundance of native insect populations. Pesticide use in an island ecosystem should be viewed with great caution. In India, however, there is a renewed interest in botanicals and organic farming practices, and government and nongovernment agencies are trying to reduce the use of synthetic chemicals for crop protection including storage and domestic animal pests and pests in urban areas.

18.6 Insect Conservation Planning at the Regional Scale

Insects are successfull animals, both in terms of numbers and abundance of species. Insect multifariousness at family stratum has been increasing over the last 400 million years, with about 600 families living today. At the species level there has not been such a steady increase, with many species lost at the end of the Cretaceous. Most of extinct species were specialists during the past few hundreds of thousands of years. With the advance and retreat of glaciers, there have been few insect species extinctions. Insect population in the Northern Hemisphere have responded to these chills and thaws by moving southward during the glacial. These trends were not affected by the human fragmented landscapes that are steadily increasing.

During the Pleistocene and early Holocene, mammalian herbivores probably played a significant role in opening up the landscape, as they do today on the African savanna. This vertebrate impact has been highly significant for many insect species because it leads to a myriad of microhabitat cases. Beginning 6000 years ago, this began to vary as humans suddenly, in geological and evolutionary time, altered the landscape. Trees were felled and indigenous game was replaced by domestic livestock. There has been an acceleration of anthropogenic impact on insect populations, with an estimated 1120 species having gone extinct since the 1600. Some approximations are that half a million insect may go defunct in the next 300 years, while some projections suggest that perhaps a quarter of all insect species are under threat of imminent extinction.

18.6.1 Systematic Reserve Selection

At the global level, it has been identified that at least 25 areas are hotspots of world biodiversity. These are likely to be major areas for insect diversity, but this still has to be demonstrated. At the regional scale, insects play a major part in systematic conservation planning, its purpose to identify locations and landscapes that are a priority for conservation action (i.e., prioritizing). There are several ways to combine targeted sites or reserve areas, and the final result must be flexible enough for practical conservation management, including making allowances for climate change. As some sites may be common and others rare or even unique, it is essential to include irreplaceability, which is a concept that embodies the potential contribution of a site to a particular conservation goal, combined with determining the extent to which the options for meaningful conservation are lost if the site is lost.

18.6.2 Conservation Planning

Surrogates of insect species diversity can be useful. Such surrogates may be alternatives or complements, such as higher taxa, species richness, rarity, endemism, threat status, and/or alternative taxa. Other types of surrogates include vegetation types, land systems or classes, and environmental domains. However, none of these surrogates is perfect, and the risk of using them is that important aspects of regional insect diversity may not be given. For instance, although British butterfly family richness may be a good indicator of species richness, rare and threatened species will go unrecorded. When different types of taxa are compared, there may not be concordance, leading to biases depending on which taxa are used.

18.7 Conservation of Endangered Species

In India, Article 48 of the constitution of India specifies that "the state shall endeavor to protect and improve the environment and to safeguard the forests and wild life of the country." Article 51-A states that "It shall be duty of every citizens of India to protect and improve the natural environment including forests, lakes, rivers and wild life and to have compassion for living creatures." The Indian government has established 18 biosphere reserves, along buffer zones that are open to economic uses. Protection for insects became a part of the US government policy with the passage of the Endangered Species Act (ESA) of 1973, a law that extended protection to all endangered plant life and animals, including insects. In 1976, eight species of butterflies were proposed as endangered species. Three of these species, Lange's metalmark (*Apodemia mormo langei*), San Bruno elfin (*Incisalia mossii bayensis*), and mission blue (*Icaricia icarioides missionensis*), were from the Bay Area. Since that time, the number of listed species of insects in the Bay Area has increased to eight (Table 18.2) (Cushman 1993; Powell and Parker 1993; Weiss 1993).

18.8 Insect Decline in Urban Ecosystem

Effects of Urbanization The primary effect of urbanization on insects is through habitat loss. The loss of 43 % of the species of butterflies from San Francisco County probably may be due to habitat loss (Hafernik and Reinhard 1995). In Palo Alto, in

Table 18.2 Endangered insects in Asia

Order	Common name	Scientific name
Lepidoptera	Lange's metalmark	*Apodemia mormo langei*
	San Bruno elfin	*Incisalia mossii bayensis*
	Behrens's silver spot	*Speyeria zerene behrensii*
	Mission blue	*Icaricia icarioides*
		Missionensis
	Callippe silver spot	*Speyeria callippe callippe*
	Myrtle's silver spot	*Speyeria zerene myrtleae*
	Bay checker spot	*Euphydryas editha bayensis*
Coleoptera	Delta green beetle	*Elaphrus viridis*

Cushman (1993), Powell and Parker (1993), and Weiss (1993)

Santa Clara County, Blair and Launer (1997) found that none of the butterflies that occur on the rural cessation of an urban–rural gradient occur in the urbanized business district – again largely the result of habitat loss. Besides the direct loss of habitat, urbanization could affect insects via habitat fragmentation and by transmuting the quality of habitat that remains embedded in the urban matrix (Rickman and Connor 2003; Cappiella 2001).

Habitat loss and fragmentation of a land area are a result of urbanization, yet neither is uniquely caused by the conversion of natural habitats to urban land uses. Habitat loss and fragmentation are innate to the process of land conversion. Changes in habitat quality and area could be manifested as altered host plant quality, soil attributes, microclimate, or enemy attack. Pesticides; air pollution; transmutations in light, nutrient, and moisture regimes; soil compaction; and exotic species could cause such transmutations in habitat quality. Transmutations in habitat quality caused by urbanization may either decline or increase. Some species of herbivorous insects that occur in urban areas persist at lower densities than in natural habitats, while other species are more abundant and inflict more damage to their host plants than in their natural habitats (see Nuckols and Connor 1995; Oksanen et al. 1996; Kozlov 1996; Speight et al. 1998).

Higher densities of herbivorous insects in urban areas may occur for two reasons: because of the urban environmental stresses of plants such as water and temperature imbalances, soil compaction, and air pollution, which makes plants more susceptible to insect attack, or because populations of the natural enemies of herbivorous insects in urban habitats are scarce.

Exotic plants can serve as alternative hosts to support more sizably voluminous populations than in natural habitats (Shapiro 2002). Rickman and Connor (2003) examined the leaf mining moths on *Quercus agrifolia* (coast live oak) in the Bay Area to assess the effects of urbanization apart from those caused by habitat loss and fragmentation. By culling sites where natural habitats had been lost to either urban or agricultural land use, they endeavored to distinguish the effects of urbanization from those caused by agriculture. It appears that the effects of urbanization, in addition to habitat loss and fragmentation, are species specific.

Introduced Species Published literature on the impact of invasives is scanty. The extent of incursions into San Francisco Bay and to physical plant groups gives overall measure of the potential significance of intrusions into the Bay Area on local bugs. More than 200 species that now represent 95 % of biomass in some straight territories have been acquainted with making San Francisco Bay "the most attacked amphibian environment in North America" (Cohen and Carlton 1995). For plant groups, 543 species have been brought into the Bay Area, which sums to 10 % of the verdure of California (Randall et al. 1998). The introduced species generally will have more potential to adapt than the local species.

In studies looking at the effect of urbanization, Pullin (1995), Hardy and Dennis (1999), and Wood and Pullin (2002) found that butterfly abundance in a place depended more on host plant accessibility than on urbanization. On the off chance that butterflies can effectively utilize developed, local host plants, then proper plantings in yards or parks can encourage butterfly species richness. Most biologists agree that if nectar and host plants are made accessible, then butterflies will utilize garden destinations and complete advancement in nurseries. A majority of gardens might be too little and detached to be utilized by butterflies and mostly they may serve as populace sinks instead of sources. Further research into this aspect is required for conserving butterflies.

Insect Conservation in Urban Setup Urban landscapes can hold a few animal varieties from compelling reasons that may increment in recurrence under worldwide environmental change (Shochat et al. 2004; Waite et al. 2007). Further, if urban habitats are improved, urban territories will attain a significant part as wellsprings of species that keep up biodiversity at the urban–rural interface (Carpaneto et al. 2005; Dunn et al. 2006). The possibility that urban communities can serve as accepted habitats for a few animal categories delineates the need to consider urban habitats in preservation science (Cook and Faeth 2006). Where invertebrate conservation is intended, target species or their natural surroundings are conserved through one of the two strategies – land protection and biological site plan. Land preservation shields profitable natural surroundings from advancement though biological site plan. Land protection is frequently arranged over an extensive geological scale keeping in mind the end goal to address issues of territory fracture, regularly as for the life history needs of solitary animal varieties or species bunch (Samways 2005). In countries like Japan and China, the urban scenes are maintained to the point that indigenous and a portion of the uncommon biodiversity components are monitored. Urban scenes likewise give chances to concentrate long haul sways on plant pollinators from biological and monetary stances. The impacts of expanded levels of CO_2, higher temperatures, or changes in the length of day season on pollinating bugs can likewise be examined (Breuste 2004).

Insect Conservation Through Understanding of Ecology Globally, open doors for biological site plan reach out to urban open space, school properties, private advancements, home greenery enclosures, storm water administration ranges, roadside passages, parks, gardens, and, most as of late, green rooftops. Best-known illustrations come in light of risked creepy crawly pollinators. These outlines for the most part take care of host plant provisioning and year-round basic natural surroundings. Developers encourage conservation of wildlife in their own premises. For instance, the Buglife.org is a European community for data that backs bug protection through environment administration. The US National Wildlife Federation has the Backyard Habitat Program (www.nwf.org/lawn/) that offers data on the most proficient method to outline private and schoolyard spaces for biodiversity with a solid accentuation on valuable insects.

Generation of baseline data and exchange of this data to those in charge of area advancement in the developing city is not proficient (McGeoch 2002). Entomologists can be key persons in conveying bug preservation to the manufactured environment if a solid coordinated effort is fashioned with outline experts in charge of urban setups and, critically, for the real configuration of advancement destinations. In making plans for the constructed environment, scene planners change and make living spaces. In each activity, there is open door for protection, stewardship, and reclamation. There is as of now a point of reference for coordinated effort among architects and researchers in backing of preservation area arranging endeavors around the world (Forman 1995; Collinge 1998). Proficient configuration associations that expressly bolster the science–design joint effort happen worldwide and incorporate, among others, the International Federation of Landscape Architects (2003), the American Society of Landscape Architects (2006) that received an unequivocal code of natural morals in 2000, the European Foundation for Landscape Architecture (2008), the Canadian Society of Landscape Architecture, the Australian Institute of Landscape Architects (2008), and so on. Comparable objectives are shared by the Chinese Society of Landscape Architecture (2008) and the Philippine Association of Landscape Architects. In tropical and subtropical countries, such coordinated efforts and society are urgently required.

Studies have shown that there is critical distinction between ecology "in" cities, where habitats and organisms are studied within urban areas, and the ecology "of" cities, where cities are seen as systems that interact in complex ways with the landscapes in which they are embedded (Grimm et al. 2000; Pickett et al. 2001). Ecology "for" cities is a vital step in integration, because it recognizes the importance of public desires, sense of place, and human designs in the management of urban space for biological value (Ryan 2006). This is more than a restatement of the human ecosystem framework (Pickett and Cadenasso 2006) in which feedbacks between human social systems and ecological functions are modeled explicitly. Of course, human well-being and environmental quality are tightly linked (Pickett and Cadenasso 2008), and this is where the opportunities for integration are clear. By addressing both ecological function and the need for beauty in built environments, habitat fragments of varying sizes, designed with insect conservation in mind, can create linked systems along the urban–suburban–exurban gradient worldwide.

Endangered arthropods are species in Arthropoda, whose eradication is likely within a reasonable time frame. Assessing the quantity of arthropod imperiled species is troublesome, on the grounds that countless are yet to be named or portrayed. Autonomous assessments demonstrate that there are several undocumented arthropods on Earth. Arthropods have been effective as a gathering on this planet. However, the growing human population has prompted extirpation of several arthropod species through deforestation, ordinary cultivation, slice-and-copy strategies in the tropics, natural surrounding discontinuity by means of urban improvement, and exorbitant and injudicious use of pesticides.

The role of arthropod in the ecology of planet Earth is vital to the survival of a large number of animals that prey upon arthropods. This includes mammals, avifauna, fishes, reptiles, and amphibians. In addition, arthropods constitute a bulk of faunal pollinators, so that the survival of crops as well as millions of natural flora species depends on robust and biologically diverse arthropod populations. The survival of diverse arthropods is essential for the survival of higher animals in the food chain and food webs. Extinction of arthropod species threatens hundreds of thousands of mammals, birds, amphibians, and reptiles and plants too (Tables 18.3 and 18.4).

Agriculture results in a monoculture that cannot support the biodiversity nurtured by the natural systems. Arthropods represent the largest number of species that is displaced by farming. In tropical regions the major threat is slash-and-burn agricultural techniques adopted by indigenous people. Pesticides use is also a major threat to arthropod species survival. Habitat fragmentation has special methods of endangerment.

Table 18.3 Summary of recorded species extinction in insects

Order	Number of species	Island	Continent
Ephemeroptera	2	–	2
Orthoptera	1	–	1
Phasmoptera	1	1	–
Dermaptera	1	1	–
Plecoptera	1	–	1
Homoptera	2	2	–
Coleoptera	10	10	–
Diptera	3	2	1
Trichoptera	4	1	4
Lepidoptera	33	32	1
Hymenoptera	3	3	–
Total	61	51	10

Table 18.4 List of well-known extinct insects

Sl No.	Insects	Place
1	Tobias caddisfly (*Hydropsyche tobiasi*)	USA
2	Xerces blue butterfly (Glaucopsyche xerces)	North America
3	Antioch katydid (Neduba extincta)	California (USA)
4	Rocky mountain grass hopper (*Melanoplus spretus*)	USA
5	Large blue butterfly (*Maculinea arion*)	UK
6	Black-veined white butterfly (*Aporia crataegi*)	UK

There is just one taxonomist for every 425 species at present. If there are ten million insect species, the ratio becomes 1:5669. This taxonomic impediment emphasizes the urgent need for more taxonomists to inventory our biodiversity

18.9 The Following Is a Small Fraction of the Potentially Hundreds of Thousands of Endangered Arthropods

- California freshwater shrimp (*Syncaris pacifica*)
- Delhi Sands flower-loving fly (*Rhaphiomidas terminatus abdominalis*), due to severely limited range of habitat and development
- South African black millipedes (*Doratogonus* spp.), due to habitat destruction
- Kentucky cave shrimp (*Palaemonias ganteri*)
- Salt Creek tiger beetle (*Cicindela nevadica lincolniana*)
- Smith's blue butterfly (*Euphilotes enoptes smithi*), due to human overpopulation of coastal dune areas and associated highway and land development
- Spruce-fir moss spider (*Microhexura montivaga*)
- Tooth cave spider (*Neoleptoneta myopica*)
- White-clawed crayfish (*Austropotamobius pallipes*)

References

Blair RB, Launer AE (1997) Butterfly diversity and human land use: species assemblages along an urban gradient. Biol Conserv 80:113–125

Breuste JH (2004) Decision making, planning and design for the conservation of indigenous vegetation within urban development. Landsc Urban Plan 68:439–452

Cappiella K (2001) Land use/impervious cover relationships in the Chesapeake Bay. Watershed Prot Tech 3:835–840

Carpaneto GM, Mazziotta A, Piattella E (2005) Changes in food resources and conservation of scarab beetles: from sheep to dog dung in a green urban area of Rome (Coleoptera, Scarabaeoidea). Biol Conserv 123:547–556

Cohen AN, Carlton JT (1995) Biological report on non indigenous aquatic species in a United States Estuary: a case study of the biological invasions of the San Francisco Bay and Delta. U.S. Fish and Wildlife Service, Washington, DC

Collinge SK (1998) Spatial arrangement of habitat patches and corridors: clues from ecological field experiments. Landsc Urban Plan 42:157–168

Cook WM, Faeth SH (2006) Irrigation and land use drive ground arthropod community patterns in an urban desert. Environ Entomol 35:1532–1540

Cushman JH (1993) The mission blue, Plebejus icarioides missionensis Hovanitz. In: TR New (ed) Conservation biology of Lycaenidae (butterflies). Occasional paper of the IUCN species survival commission, No. 8 Gland, pp 139–140

Debach P (1974) In: Rosen D (ed) Biological control by natural enemies. Cambridge University Press, London, pp 1–113

Dunn RR, Gavin MC, Sanchez MC, Solomon JN (2006) The pigeon paradox: dependence of global conservation on urban nature. Conserv Biol 20:1814–1816

Forman RTT (1995) Land mosaics: the ecology of landscapes and regions. Cambridge University Press, Cambridge

Grimm NB, Grove JM, Pickett STA, Redman CL (2000) Integrated approaches to long-term studies of urban ecological systems. Bioscience 50:571–584

Hafernik JE, Reinhard H (1995) Butterflies of the bay: winners and losers in San Francisco's urban jungle. Am Butterflies 3:4–11

Hardy PB, Dennis RLH (1999) The impact of urban development on butterflies within a city region. Biodivers Conserv 8:1261–1269

Kozlov M (1996) Patterns of forest insect distribution within a large city: Lepidoptera in St. Petersburg, Russia. J Biogeogr 23:95–103

Kuchlein JH, Ellis WN (1997) Climate-induced changes in the microlepidoptera fauna of the Netherland and the implications for nature conservation. J Insect Conserv 1:73–80

McGeoch MA (2002) Insect conservation in South Africa: an overview. Afr Entomol 10:1–10

Nuckols MS, Connor EF (1995) Do trees in urban or ornamental plantings receive more damage by insects than trees in natural forests? Ecol Entomol 20:253–260

Oksanen J, Holopainen JK, Nerg A, Holopainen T (1996) Levels of damage to Scots pine and Norway spruce caused by needle miners along a SO2 gradient. Ecography 19:229–236

Opler PA (1979) insects and American chestnut: possible importance and conservation concern. In: Mac Donald WL, Cech FC, Luchok J, Smith C (eds) Proceeding of the American chestnut symposium, Morganton, W. Va, United state of America 1978, West Virginia University Press Morganton, pp 83–85

Pickett STA, Cadenasso ML (2006) Advancing urban ecological studies: frameworks, concepts, and results from the Baltimore Ecosystem Study. Austral Ecol 31:114–125

Pickett STA, Cadenasso ML (2008) Linking ecological and built components of urban mosaics: an open cycle of ecological design. J Ecol 96:8–12

Pickett STA, Cadenasso ML, Grove JM, Nilon CH, Pouyat RV, Zipperer WC, Costanza R (2001) Urban ecological systems: linking terrestrial ecological, physical, and socioeconomic components of metropolitan areas. Ann Rev Ecol Syst 32:127–157

Powell JA, Parker MW (1993) Lange's metalmark: *Apodemia mormo langei* comstock. In: TR New (ed) Conservation biology of Lycaenidae (butterflies). Occasional paper of the IUCN species survival commission, No. 8 Gland, pp 116–119

Pullin AS (1995) Ecology and conservation of butterflies. Chapman & Hall, London

Randall JM, Rejmanek M, Hunter JC (1998) Characteristics of the exotic flora of California. Fremontia 26:3–12

Rickman JK, Connor EF (2003) The effect of urbanization on the quality of remnant habitats for leaf-mining Lepidoptera of Quercus agrifolia. Ecography 26(6):777–787

Ryan R (2006) The role of place attachment in sustaining urban parks. In: Platt RH (ed) The humane metropolis: people and nature in the 21st-century city. University of Massachusetts, Amherst, pp 61–74

Samways MJ (2005) Insect diversity conservation. Cambridge University Press, Cambridge

Shapiro A (2002) The Californian urban butterfly fauna is dependent on alien plants. Divers Distrib 8:31–40

Shochat E, Stefanov WL, Whitehouse MEA, Faeth SH (2004) Urbanization and spider diversity: influences of human modification of habitat structure and productivity. Ecol Appl 14:268–280

Speight MR, Hails RS, Gilbert M, Foggo A (1998) Horse chestnut scale (*Pulvinaria regalis*) (Homoptera: Coccidae) and urban host tree environment. Ecology 79:1503–1513

Waite TA, Chhangani AK, Campbell LG, Rajpurohit LS, Mohnot SM (2007) Sanctuary in the city: urban monkeys buffered against catastrophic die-off during ENSO-related drought. Ecohealth 4:278–286

Weiss SB (1993) The San Bruno Elfin, Incisalia mossii bayensis (Brown). In: TR New (ed) Conservation biology of Lycaenidae (butterflies). Occasional paper of the IUCN species survival commission, No. 8 Gland, pp 141–143

Wood BC, Pullin AS (2002) Persistence of species in a fragmented urban landscape: the importance of dispersal ability and habitat availability for grassland butterflies. Biodivers Conserv 11:1451–1468

Giant Moths and Their Conservation

19

P.P. Mary, Vasudev Kammar, A.T. Rani, and K.R. Yathish

Abstract

Giant moths were found in North Eastern Regional Institute of Science and Technology campus, Papum Pare, Arunachal Pradesh, India. Both species of moths had bipectinate antennae. The largest moth of this area was identified as Atlas moth, *Attacus atlas* (Saturniidae: series Heterocera: Lepidoptera). It measured 23 cm in breadth from wing to wing. This species is brown with flask-shaped transparent windows on the fore and hind wings. Another moth was identified as luna moth (*Actias selene*), Saturniidae. It measured 19 cm in breadth from wing to wing. It is yellow with oval windows on all wings. Posterior end of hind wings in this species is greatly elongated and looks like a pair of tails. Information on the status of giant moths in Europe, Malaysia, and other countries is provided. The causes of decline in giant moth populations across different habitats are enumerated. Such magnificent and beautiful moths of the Indian subcontinent require conservation to preserve them for posterity.

Keywords

Atlas moth • Luna moth • Arunachal Pradesh • Conservation • Diversity • Moth size

P.P. Mary (✉)
Department of Forestry, North Eastern Regional Institute of Science and Technology (NERIST) Deemed University, Nerjuli 791109, Arunachal Pradesh, India
e-mail: purakat@yahoo.co.in

V. Kammar • A.T. Rani
Department of Entomology, University of Agricultural Sciences (UAS), Gandhi Krishi Vignana Kendra (GKVK), Bengaluru 560065, Karnataka, India
e-mail: 4547vk@gmail.com

K.R. Yathish
Division of Entomology and Nematology, Indian Institute of Horticultural Research, Bengaluru 560089, India

A.K. Chakravarthy, S. Sridhara (eds.), *Arthropod Diversity and Conservation in the Tropics and Sub-tropics*, DOI 10.1007/978-981-10-1518-2_19

19.1 Introduction

Among the lepidopteran insects, series Heterocera (moths) are less colorful compared to Rhopalocera (butterflies). Adult moths are mainly nocturnal and are useful as pollinators. However, caterpillars of moths cause large-scale destruction like the notorious teak defoliator (*Hyblaea puera*) and poplar wood borer (*Trochilium omnatiaeforme*). Moths of India belong to 41 families, while the valuable silk-producing caterpillars/moths belong to mainly two families. Commercially reared mulberry silk moths belong to semidomesticated tussar silk moth Saturniidae. Most of the wild silk moths in Arunachal Pradesh, Northeast India, are poorly studied. Hence the present study to effect conservation of large moths in northeast and other parts of India.

Atlas moth (*Attacus atlas*) was reported from Western Ghats (Mathew 2004), Nagaland (Thangavelu et al. 2002), Assam (Bhattacharya et al. 2004), and Tripura (Majumder et al. 2011). Life history of *Attacus mcmulleni* which closely resembles *Attacus atlas* was studied by Veenakumari et al. (1994), and they found that its larvae were multivoltine. *Attacus taprobanis* is another related species found in South India, Thailand, and Sumatra. Nazar (1990) found that egg, larval, pupal, and adult stages of Atlas moth lasted 6–12, 50–58, 27–31, and 7–10 days, respectively. It was listed among forest insects in Thailand (Hutacharern and Tubtim 1995). A related species, *Actias maenas*, was reported as polyvoltine exhibiting sexual dimorphism in adult stage. Males are bright yellow with brown markings and females are light green (Naessing and Peigler 1984). *A. selene* is also found in Nagaland, India (Kakati and Chutia 2009), and as a pest of *Populous alba* in Himachal Pradesh and Jammu and Kashmir, India (Kakati and Chutia 2009). Outside India, in Britain with over 2,500 moth species recorded, moths comprise an important part of biodiversity with considerable and scientific interest by naturalists. Results of Rothamsted height trap network data have revealed significant declines of larger moths. Similar patterns of decline among moths have also been reported from Finland and Netherlands. Sixty-one species of larger moth declined by 75 % or more over the last 40 years.

19.2 Materials and Methods

Moths from Arunachal Pradesh were collected and photographed at the Department of Forestry, NERIST campus, and observations on biology were recorded. Moreover, contacts with lepidopterists working on giant moths were established to document latest information particularly on the status of giant moths in different localities across different countries. Major sources of information on giant moths worldwide are also provided in this review, as systematic, comprehensive studies on giant moths, particularly in tropical countries, are lacking.

19.3 Observations and Discussion

Moths were photographed in July 2012 and identified as Atlas moth (*Attacus atlas*) (Fig. 19.1) belonging to Saturniidae (Mani 1968). *Attacus atlas* is the largest moth found in Arunachal Pradesh and its wingspan is 23 cm. There is a white and vermilion line dividing both forewings and hind wings into basal half and apical half. Basal half is chocolate brown with flask-shaped transparent windows in each wing. Apical half is chocolate vermilion with black wavy lines in the border of wings except the apex. Apex is whitish yellow, elongated curved with chocolate line in the middle and a black spot near the end of coastal margin. Similar line passes between the thorax and abdomen extending almost 2 cm to the forewings. Antennae are bipectinate. The fingerlike projections are long enough to give a feather-like appearance to the antennae. Atlas moth is related to eri silk moth, *Ricinus communis* (*Attacus ricini*), which feeds on leaves of castor plants. Eri silk is wooly white or off-white staple fiber which is spun into yarns. Its fiber cannot be reeled because it is not continuous and is used mostly for preparation of shawl, jacket, blanket, etc.

Atlas moth is the largest moth in the world in terms of total wing surface area reaching 400 cm^2. Females are larger and heavier than the males. The moths are found in the tropical and subtropical forests of Southeast Asia and common in the Malay archipelago. They are named after the Titan of Greek mythology. In Hong

Fig. 19.1 Atlas moth, *Attacus atlas*

Kong, its Cantonese name translates as snake's head moth, referring to the apical extension of the forewing. Japan has only the *Attacus atlas ryukyuensis* subspecies native to the Yaeyama Islands. A unique specimen of *Attacus atlas* from Java island measured 262 mm. Atlas moth cocoons have been utilized to make purses in Taiwan and other countries.

At night female moths emit pheromones. Males detect the pheromones with large, feather-like antennae and flutter to females. After mating, females lay up to 100 eggs, usually on leaves. Adults die soon after mating. A week later, eggs hatch into caterpillars. Larvae molt four to six times to spin cocoons. Some use strands of silk to pull leaves around themselves. In some species, cocoons are nested inside this protective curtain. Others burrow underground to form cocoons.

Another giant moth was photographed in August 2012 inside the Department of Forestry, NERIST campus, Papum Pare, Arunachal Pradesh. It was identified as luna moth (*Attacus selene*) (Fig. 19.2), Saturniidae. It is an enormous insect measuring 19 cm in wingspan. Posterior end of hind wings in this species is greatly elongated looking like a pair of "tails." It is bright lemon yellow with transparent oval windows near the middle of coastal margin of forewing and middle of apical margin of hind wings. A black line crosses the head extending up to half-length of coastal margin of forewings. Black line is present on the apical margin up to the beginning of the "tail" in hind wings. Two black wavy lines divide the wings into three portions. Prominent "eyespot" is present in the middle portion of each wing. Thick wooly scales are present in the anal margin of hind wings touching the abdomen. Antennae have feathery appearance. Excepting the "tails" this moth has external resemblance with the semidomesticated tussar silk moth (*Antheraea mylitta*) of Assam.

Fig. 19.2 Luna moth (*Actias luna*)

At Kolhapur, Western Ghats, South India, a team of naturalists have found an entire family of the gigantic and rare *Attacus atlas*, the world's largest moth measuring up to 30 cm from wing tip to wing tip (Jadhav 2014). The finding of this species with a wingspan of about 25–30 cm was spotted in the evergreen tropical forests of Chandgad in Western Ghats, South India, which is one of the global biodiversity hot spots. A male and female adult moths with cocoons were spotted on green twigs in the deep woods.

The moth is brown with woollike silk which is considered durable and is farmed as Fagara silk in North India. Tussar silk and wild silk of *Atlas mylitta* are produced on a commercial scale in Bhandara, Gadchiroli, Chandrapur, and Gondiya districts in the East Maharashtra. Recently, the sericulture section of the Shivaji University has successfully mated and fertilized around 150–200 eggs of a female moth, and efforts are on to rear the caterpillars on preferred host plants for their mass rearing in nature (Jadhav 2014).

Atlas moths in India include several species, viz., *Actias selene*, *Antheraea proylei* , *Samia canningi*, *Cricula trifenestrata*, *Atlas mylitta*, *Philosamia ricini*, and *Antheraea assamensis*. These moths are low, weak fliers; the females are homely and localized and do not go far from the habitat but are capable of attracting males from afar. They feed voraciously on the foliage of certain evergreen plants, and after they attain 11.50 cm, they pupate within papery cocoons interwoven into desiccated leaves, eventually transforming and emerging as beautiful moths after a month.

19.3.1 Giant Moths in Malaysia

People in Malaysia are frequently sighting the giant moths. The British Broadcasting Company reported that a large number of the goliath animals have found in urban communities in Malaysia, These moths have disturbed a football match too! As indicated by biologist Anuj Jain, the bugs are pulled in to light-filled urban communities: "their propensity to emigrate looking for new uneaten host plants draw in these moths to light in urban city territories." The moths are innocuous; however, asthma sufferers could be touchy to hairs on their wings, yet their nearness is still sort of dreadful. They are often found dead in the structures of buildings in the morning hours.

19.3.2 The State of Larger Moths in Europe

The giant peacock moth also known as the great peacock moth (*Saturnia pyri*) is the largest moth in Europe. An adult flies from mid-April to June and is distributed in Iberian Peninsula, Southern France, Italy, the Balkans, and North Africa. The Emperor moth is the main individual from the Saturniidae found in the UK. The family is named after the ringed planet Saturn, in view of the large ringed spots found on the wings of these species. The Saturniidae are generally known as wild silk moths in light of the fact that most species build elaborate silk casings.

There are five major *Saturnia* species in Europe which include the Emperor *Saturnia pavonia* and the comparable however much bigger great peacock moth, *Saturnia pyri*. There are additionally a few individuals from other genera including the orange-winged Tau Emperor, *Aglia tau*; the Autumn Emperor, *Perisomena caecigena*; and the Spanish moon moth, *Graellsia isabellae*. Furthermore, there are three presented species – *Antheraea yamamai*, *Antheraea pernyi*, and *Samia cynthia*. The *Saturnia* species are heathland and moorland species; however, they can be discovered additionally in forests and inferior fields. In Britain they are swamp species, but can be found at 1800 m in the Alps.

Around 40 species of migrant moths have appeared in the UK for the first time in the last 15 years with a small number such as the black-spotted chestnut getting established. Other species considered occasional migrants have become established in the UK in recent years, such as the tree-lichen beauty, oak rustic, sombre brocade, Blair's mocha, flame brocade, and Clifden nonpareil.

19.3.3 Moths Are Important

Giant moths are important because of their unique size, shape, and color patterns. They form an important part of natural heritage, and currently they are threatened as their populations at the places of occurrence are fast declining. For many species, action is urgently required to conserve. Recording and monitoring schemes should be in place to assess their current status.

These moths make up a significant part of the indigenous biodiversity (e.g., 2500 species in Britain and 10,000 species in the Indian subcontinent). Many small animals, viz., bats, birds, and other invertebrate animals, utilize moths as prey.

19.3.4 Moths Are Threatened

As many as 62 moth species have become extinct in Britain during the twentieth century, and many more species are at the brink of extinction. The UK Biodiversity Action Plan (BAP) has listed 53 moths as nationally important and has taken steps for their conservation, survey work, scientific research, and area-wide management. However, eight species are considered to be on the verge of extinction.

Conservation efforts and research in the future are needed to execute the existing action plans for their conservation. Additionally, many threatened and scarce species are now known to meet the criteria for inclusion in the UK BAP, and therefore, the number of priority moths may increase after the 2006 review.

19.3.5 Moths Are Declining (Table 19.1)

The aggregate number of moths in Rothamsted trap tests has decimated by one third since 1968. Populace patterns were produced for 337 moth species. Sixty-six

Table 19.1 Larger moth species declined by 75 % or more over 40 years (1968–2007) in Europe

Sl. no.	Common name	Species	% change over 40 years	IUCN category
1	Dusky thorn	*Ennomos fuscantaria*	−98	Endangered
2	V-moth	*Macaria wauaria*	−97	Endangered
3	Garden dart	*Euxoa nigricans*	−97	Endangered
4	Double dart	*Graphiphora augur*	−97	Endangered
5	Hedge rustic	*Tholera cespitis*	−97	Endangered
6	Grass rivulet	*Perizoma albulata*	−96	Endangered
7	Figure of eight	*Diloba caeruleocephala*	−95	Endangered
8	Spinach	*Eulithis mellinata*	−95	Endangered
9	Dark spinach	*Pelurga comitata*	−95	Endangered
10	Anomalous	*Stilbia anomala*	−93	Endangered
11	Dusky-lemon sallow	*Xanthia gilvago*	−92	Endangered
12	White-line dart	*Euxoa tritici*	−92	Endangered
13	Autumnal rustic	*Eugnorisma glareosa*	−92	Endangered
14	Dark-barred twin-spot carpet	*Xanthorhoe ferrugata*	−92	Endangered
15	September thorn	*Ennomos erosaria*	−91	Endangered

percent (226 species) demonstrated a diminishing populace pattern over the 35 years of the study. Such boundless decay is liable to have adverse thump on impacts on different life forms. The decrease of regular moths, alongside decays of butterflies and other untamed life, exhibit a far reaching and serious emergency for common legacy of a country.

In addition, the use of global (International Union for Conservation of Nature) criteria demonstrates that 71 species (21 %) of these "normal" moths are undermined: 15 are delegated imperiled and 56 as powerless. None of these moths have been recorded as Red Data Book species earlier in Britain, and none were thought to warrant any protection need in the UK.

Despite the fact that the major share of populace patterns for bigger moths from the Rothamsted system are negative, a few species have done stupendously well over late decades. Forty-six species have dramatically increased their populace levels (i.e., expanded by no less than 100 %), and a further 23 animal types have expanded by more than half over the 35 years time span.

More species have declined in Southern Britain (75 %) than in Northern Britain (55 %). South East Britain is especially gravely influenced. Some huge connections exist between populace patterns and biological qualities. Moths whose caterpillars feast upon lichens/green growth and conifers have done well, as have species that rest amid the winter months.

19.4 Causes for Decline in Giant Moths

19.4.1 Habitat Loss

Throughout the past 50 years, horticultural escalation, business ranger service, and urban improvement have resulted in disastrous changes in the British scene. The impact on semi-regular living spaces, for example, antiquated forest, heathland and chalk field, and the natural life that possesses them, has been decimating. The broad demolition, adjustment, and discontinuity of these living spaces have severely affected numerous moths, butterflies, and different animals and plants. Large number of life forms have been recorded in Red Data Book and in the UK Biodiversity Action Plan.

There is insufficient evidence at present to determine the level of impact of habitat changes on the decline of common moths. However, the substantial loss of hedgerows, destruction of field margins, reseeding and fertilization of pastures, etc. are likely to have been major factors. A recent comparison of organic and conventional mixed farms in Southern Britain found significantly more moths in the organic farms.

19.4.2 Pesticides

Another element that may significantly affect moth populaces is the enormous increment in pesticide (bug spray and herbicide) use. The aggregate amount of pesticides sold for use in residential patio nurseries expanded by 70 % somewhere around 1992 and 1997 alone. This may have hurried the decrease of patio nursery experts, for example, V-moth and spinach, and may have brought about generous decay of moths. These days, bug spray use might have less impact, as present-day chemicals are less diligent than past ones. Be that as it may, herbicides have empowered agriculturists to lessen weed populaces enormously, which are profoundly affecting the accessibility of nourishment plants for some common moths.

19.4.3 Pollution

Eutrophication (expanded richness of soil and water) is adjusting the plant arrangement and vegetation structure of numerous natural surroundings, with obscure effects on herbivores, for example, moths. The expansions among lichen-sustaining moths, for example, marbled beauty, might be connected to increments in some lichen species, which are thus ascribed to air contamination which results in corrosive downpours.

19.4.4 Vehicles

Moth recorders have suggested that mortality of moths due to strikes against moving vehicles is a major reason for mortality. Nonetheless, there is little proof at present to propose that this reason for mortality has contributed fundamentally to moth populace patterns. Air contamination produced by vehicles, especially the outflow of greenhouse gasses, may be a more important problem in the longer term. On the positive side, vehicles might also assist the geographical expansion of species when moths get hitchhiked on the vehicles.

19.4.5 Light Pollution

Expanding light contamination is another potential reason for change in moth populaces. Outside lighting can have numerous problematic consequences for moth conduct and exposure to predators. In any case, such impacts appear to fluctuate between species, populaces, and even people and with the kind of lighting utilized. All the more essentially, it is hard to isolate out the immediate effect of light contamination on moth populaces from different effects of urbanization and advancement that more often than not go with an expansion in lighting levels. This major issue requires urgent attention of scientists.

19.4.6 Light Pollution and Moth Trapping

The engaging quality of moth trap lights depends incompletely upon the level of foundation lighting and also the sort of globule utilized. For instance, moths gets by and large decay when there is a full moon on a starry evening. Expanded light contamination over late decades may in this way have diminished the productivity of Rothamsted light traps bringing about spurious populace patterns.

19.4.7 Climate Change

Of late several studies on early or late moths and change in moth phenology (e.g., single-brooded species producing partial second broods) were always consistent with climatic changes. It appears to be exceptionally likely that environmental change is likewise applying a significant impact on giant moths. This requires further research.

The precise causes of the decline or increase of giant moths are unknown. However, it seems likely that agricultural intensification and other land use changes have significant impact on many moths as well as climate change. Further research is expected to decide the hidden reasons for the across the board decreases of bigger moths to propose methods to alter the course.

Table 19.2 Larger moth species increasing in Europe

Common name	Species	% change over 40 years
Least carpet	*Idaea rustica*	74,684
Blair's shoulder knot	*Lithophane leutieri*	7878
Treble brown spot	*Idaea trigeminata*	4312
Buff footman	*Eilema depressa*	3884
Scare footman	*Eilema complana*	3590
Satin beauty	*Deileptenia ribeata*	2928
Peacock moth	*Macaria notata*	2409

Select species of larger moths in Europe exploited the changing conditions of the landscape and the environment so well that their numbers gradually increased. A sample of seven such moths is presented in Table 19.2. Over 40 years, the numbers of these moths have dramatically changed.

19.5 Implications

Changes in the degree and nature of appropriate environments are among the prime elements in the decrease of some once-basic moths. Pesticide use, eutrophication, and light contamination may contribute in a few or numerous cases. Environmental change additionally is by all accounts influencing moth circulation, wealth, and phenology and has been ensnared in the moth populace decrease that has been examined in subtle element so far (the garden tiger *Arctia caja*).

- The dangerous position of such a large number of moths demonstrate that activity is required desperately to preserve them at a scene scale and inside new cultivating and forest schemes.
- Results showed that it is critical for government organizations to keep on investing in long haul recording and checking plans, for example, the Rothamsted light-trap system. The improvement of a complete national recording plan covering every single bigger moth is indispensable to survey patterns and execute powerful protection over this imperative gathering of creepy-crawlies.

References

Bhattacharya A, Singh BK, Das PK (2004) Biodiversity of wild silkmoths in Assam (North East India). Ann For 12:208–216

Hutacharern C, Tubtim N (1995) Checklist of forest insects in Thailand. Office of Environmental Policy and Planning, Bangkok, p 392

Jadhav AD (2014) Indian naturalists find giant moth in Western Ghats. IANS 7(14):47
Kakati LN, Chutia BC (2009) Diversity and ecology of wild sericigenous insects in Nagaland, India. Trop Ecol 50(1):137–146
Majumder J, Lodh R, Agarwala BK (2011) *Attacus atlas* L. (Saturniidae): a new distribution recorded from Tripura, Northeast India. NeBIO 2(2):27–28
Mani MS (1968) General entomology. IBH Publishing Co, Oxford, p 597
Mathew G (2004) Insect diversity of India and its conservation. In: Proceedings of National Seminar- cum- workshop on butterfly diversity of Western Ghats, 8–9 October 2005, Malappuram, pp 9–19
Naessing, Peigler (1984) The life history of *Actias maenas* (Saturniidae). J Lepidopterists' Soc 38(2):114–123
Nazar A (1990) The biology of Attacus atlas Linn, a pest of clove leaves. Pemberitaan Penititian Tanaman Ind 16(1):35–37
Thangavelu K, Rao KVS, Pandey VK (2002) Wild silk moth's diversity and conservation. Int J Wild Silk Moths Silk 7:87–93
Veenakumari K, Mohanraj P, Peigler RS (1994) Life history of *Attacus mcmulleni* (Saturniidae) from the Andaman Islands, India. J Res Lepidoptera 313(4):169–179

Biological characteristics of Arrhenotokous and Thelytokous *Trichogramma pretiosum* Riley

20

T. Prabhulinga, S.K. Jalali, K.P. Kumar, and B. Doddabasappa

Abstract

Biological attributes of arrhenotokous (sexual) and thelytokous (asexual) of *Trichogramma pretiosum* Riley were studied on *Corcyra cephalonica* in laboratory. The adult longevity in *Trichogramma pretiosum* arrhenotokous and thelytokous *T. pretiosum* were 10.4 days and 10 days, respectively. The total female progeny produced during entire life span was 147.08 in thelytokous *T. pretiosum* and 90 in arrhenotokous *T. pretiosum*. The sex ratio of arrhenotokous *T. pretiosum* was 1:1 while for thelytokous it was 5:95 (M/F). The intrinsic rate of increase was 0.42 in arrhenotokous *T. pretiosum* and 0.48 in thelytokous *T. pretiosum*. These results revealed that in the net reproductive rate, weekly multiplication rate and the number of estimates in F2 generation are more in thelytokous *T. pretiosum* compared to arrhenotokous *T. pretiosum*. These results confirmed that biological attributes of arrhenotokous and thelytokous *T. pretiosum* are different. Thelytokous *T. pretiosum* have more practical implications in the biological control of lepidopterous pests.

T. Prabhulinga (✉)
Division of Entomology, Central Institute for Cotton Research (CICR), Nagpur 440010, India
e-mail: prabhut07@gmail.com

S.K. Jalali
Division of Entomology, National Bureau of Agricultural Insect Resources (NBAIR), Hebbel, Bengaluru 560024, Karnataka, India

K.P. Kumar
Department of Entomology, University of Agricultural Sciences, Bengaluru 560065, Karnataka, India

B. Doddabasappa
Department of Entomology, College of Horticulture, Kolar, Karnataka, India

A.K. Chakravarthy, S. Sridhara (eds.), *Arthropod Diversity and Conservation in the Tropics and Sub-tropics*, DOI 10.1007/978-981-10-1518-2_20

Keywords
Arrhenotokous • Life table • Thelytokous • *Trichogramma pretiosum*

20.1 Introduction

Trichogramma (Hymenoptera: Trichogrammatidae) are naturally occurring biocontrol agents recorded in almost every habitat, terrestrial or aquatic. Members of the Trichogrammatidae family range in size from 0.2 to 1.5 mm. *Trichogramma* have a wide range of insect hosts, especially among the Lepidoptera. They parasitize some of the most important lepidopteran pests of field crops, forests, fruits and vegetable crops. Trichogrammatids can therefore play a vital role in integrated pest management programmes by destroying the early stage of the pest, thereby reducing or curtailing the use of pesticides (Jalali et al. 2006). They encourage populations of beneficial organisms like parasites, predators and pollinators in cultivated ecosystems. The commercially successful use of *Trichogramma* to control the European corn borer in Europe has demonstrated the potential of this approach (Knutson 1998).

Trichogramma species are the most widely used insect natural enemies in the world, partly because they are easy to mass rear and attack important crop insect pests. A large number of species, subspecies and strains of trichogrammatids (over 200) are distributed throughout the world parasitising eggs of over 200 insect species belonging to 70 families in diverse habitats. In India, about 36 trichogrammatids are recorded (Singh and Jalali 1994). Long-term use of trichogrammatids not only contributes to improvement of agro-biodiversity but also helps in sustaining productivity and crop yields in the cultivated ecosystems.

The present study was conducted using *T. pretiosum*, which attacks a wide range of pest populations, viz. *Helicoverpa armigera* (Hubner), *Conogethes punctiferalis* (Guenee), *Cryptoblabes adoceta* (Turner), *Plutella xylostella* (Linnaeus), *Thysanoplusia orichalcea*, *Phthorimaea operculella* (Fabricius), etc., on many important agricultural and horticultural crops.

The construction of a number of life tables is vital to the description and understanding of the population dynamics of the species, because many insects have discrete generations and their populations are not stationary. The age-specific life table is widely acceptable than the time-specific life table. In this regard an attempt was made to critically assess the biological potential of arrhenotokous and thelytokous *Trichogramma pretiosum* under laboratory conditions. The study will enable use of effective strain of *Trichogramma* for the biocontrol of crop pests.

20.2 Materials and Methods

20.2.1 Insect Culture

1. *Corcyra* culture
2. *Trichogramma* culture

Cultures of arrhenotokous and thelytokous *T. pretiosum* were collected from insectary of the National Bureau of Agricultural Insect Resources, Bangalore, and maintained on the eggs of *C. cephalonica*. The freshly laid, cleaned, UV-irradiated *Corcyra* eggs were taken and glued on a strip of card sheet (6 × 2 cm) in a single layer using gum, and these cards were exposed to both arrhenotokous and thelytokous *T. pretiosum* separately for maintenance of the cultures. Fine streaks of honey and water (1:1) were provided as adult food. After 24 h, the parasitized cards were transferred to fresh glass tubes (15 × 2.5 cm). The parasitoids that emerged from the egg cards were used for further studies. These arrhenotokous and thelytokous *T. pretiosum* were selected for life table studies.

In each replication 20 freshly emerged and mated (for arrhenotokous) females were taken, and the egg card bit containing 1000 eggs was placed in glass vials (15 × 2.5 cm size) daily, and fine streak of 10 % honey was provided as adult food. The procedure was repeated till all females died. After 24 h of exposure, the parasitized egg cards were separated and kept in another fresh glass tube (15 × 2.5 cm) to record the number of females emerged/day. Every day the number of females alive and dead was recorded. This experiment was replicated five times in each arrhenotokous and thelytokous *T. pretiosum*.

The age-specific survival (*lx*) and age-specific fecundity (*mx*) at each pivotal age (*x*) were worked out daily for entire reproductive period to prepare fertility life table as outlined by Southwood (1976). The life table statistics was constructed as detailed by Andrewartha and Birch (1954) and Southwood (1976).

$$\text{The net reproductive rate}(R_0) = \sum l_x m_{x.}\, where,$$

$$\text{Appropriate duration of generation}(T_c) = \sum x l_x m_{x.}$$

$$\text{Appropriate intrinsic rate of increase}(r_c) = \log_e R_0 / T_c$$

$$\text{Precise intrinsic rate of increase}(r_m) = e^{-\text{rm}} x l_x m_x = 1$$

$$\text{Precise generation time}(T) = \log_e R_0 / \text{rm}$$

$$\text{Finite rate of increase}(\lambda) = \text{antilog}_e r_m$$

$$\text{Weekly multiplication rate} = (\lambda)^7$$

$$\text{Number of estimated females in } F_2 = (R_0)^2$$

$$\text{Doubling time}(DT) = \log_e 2 / r_m$$

The age-specific longevity (*lx*) and age-specific fecundity (*mx*) were plotted against age (*x*), and the distribution of mortality was determined by the shape of curve as described by Slobodokin (1962).

20.3 Results and Discussion

The adult longevity in arrhenotokous and thelytokous *T. pretiosum* were 10.4 days and 10 days, respectively (Tables 20.1 and 20.2). The total female progeny produced during entire life span was 147.08 in thelytokous *T. pretiosum* and 90 in arrhenotokous *T. pretiosum*. The sex ratio of arrhenotokous *T. pretiosum* was 1:1 while for thelytokous it was 5:95 (M/F). Data on the parasitism rate of *T. pretiosum* on *C. cephalonica* eggs, based on container sizes, are presented in Tables 20.3 and 20.4. The intrinsic rate of increase was 0.42 in arrhenotokous *T. pretiosum* and 0.48 in thelytokous *T. pretiosum*. These results revealed that in the net reproductive

Table 20.1 Comparative life table statistics of arrhenotokous and thelytokous *T. pretiosum*

Parameters	Formula	Arrhenotokous	Thelytokous
Net reproductive rate (*Ro*)	$\sum l_x m_x$	90.00	147.08
Approximate duration of generation (*Tc*)	$\sum x l_x m_x$	09.68	10.02
Approximate intrinsic rate of increase (*rc*)	loge *Ro*/*Tc*	00.45	00.48
Precise intrinsic rate of increase (*rm*)	$e^{-rm}\ x l_x m_x = 1$	00.42	00.48
Precise generation time (*T*)	loge *Ro*/*rm*	10.40	10.00
Finite rate of increase (λ)	antiloge *rm*	01.52	01.63
Weekly multiplication rate	$(\lambda)^7$	20.40	31.48
No. of estimated females in F_2	$(Ro)^2$	9083.35	24171.42
Doubling time	loge 2/*rm*	01.64	01.41

Table 20.2 Days to adult emergence of arrhenotokous and thelytokous *T. pretiosum* (at 50 female parasitoids/2000 *Corcyra* eggs)

Containers size (sq cm area)	Days to adult emergence *T. pretiosum*		Mean no. of days to adult emergence
	Arrhenotokous	Thelytokous	
37.5	10.0a	10.0	10.0
70	10.0a	10.0	10.0
128	09.0b	09.0	09.0
27,000	09.9a	10.4	10.1
182,250	10.3a	11.9	11.1
Mean	09.8b	10.3	A × B factor
Parameters	A factor	B factor	
F test	*	NS	NS
SEm (±)	0.4	0.6	0.8
CD ($P = 0.01$)	1.0	1.6	2.3

NS Non-significant
*Highly significant

Table 20.3 Parasitism of *Corcyra cephalonica* eggs by arrhenotokous and thelytokous *T. pretiosum* (at 100 female parasitoids/2000 *Corcyra* eggs)

	Parasitism (%) of *T. pretiosum*		
Containers size (sq cm area)	Arrhenotokous	Thelytokous	Mean % parasitism
37.5	78.9 (62.8)c	90.0 (71.7)a	84.5 (67.2)a
70	65.4 (53.8)d	85.9 (68.1)b	75.6 (60.9)b
128	60.7 (51.2)d	77.7 (61.9)c	69.2 (56.5)c
27,000	27.1 (31.2)g	40.7 (39.6)f	33.9 (35.4)e
182,250	7.0 (14.8)i	16.7 (24.1)h	11.8 (19.4)f
Mean	47.8 (42.8)b	62.2 (53.1)a	A × B factor
Parameters	A factor	B factor	
F test	(*)	(*)	NS
SEm (±)	0.7	1.1	1.7
CD ($P = 0.01$)	2.1	3.3	4.7
r value	−0.847	−0.872	−0.862

Figures in parentheses are arc sine transformed value
*Highly significant

Table 20.4 Parasitism of *Corcyra cephalonica* eggs by arrhenotokous and thelytokous *T. pretiosum* (at 50 female parasitoids/2000 *Corcyra* eggs)

	Parasitism (%) of *T. pretiosum*		
Containers size (sq cm area)	Arrhenotokous	Thelytokous	Mean % parasitism
37.5	48.1 (43.9)c	76.3 (61.1)a	62.2 (52.5)a
70	44.6 (41.8)c	66.3 (54.7)b	55.4 (48.2)b
128	26.3 (30.8)d	60.9 (51.4)b	43.6 (41.1)c
27,000	2.1 (7.9)d	18.2 (25.1)e	10.1 (16.5)d
182,250	0.3 (2.2)g	4.7 (11.7)f	2.5 (6.9)e
Mean	24.2 (25.3)b	45.3 (40.8)a	A × B factor
Parameters	A factor	B factor	
F test	(*)	(*)	NS
SEm (±)	0.6	0.9	1.3
CD ($P = 0.01$)	1.7	2.7	3.8
r value	−0.691	−0.807	−0.675

Figures in parentheses are arc sine transformed value
*Highly significant

rate, weekly multiplication rate and the number of estimates in F2 generation are more in thelytokous *T. pretiosum* compared to arrhenotokous *T. pretiosum* (Table 20.1). Similarly Botto et al. (2004) showed that in *Trichogramma nerudai*, the main biological attributes of two Neotropical egg parasitoids, the arrhenotokous *Trichogramma nerudai* Pintureau and the thelytokous *Trichogramma* sp., *T. nerudai*, had a faster developmental rate than *Trichogramma* sp. Both species showed similar life table statistics; *rm* was 0.222 and 0.225 for *T. nerudai* and *Trichogramma* sp., respectively.

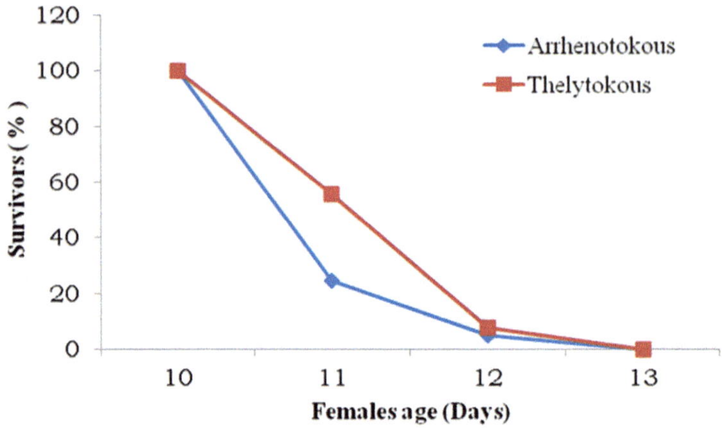

Fig. 20.1 Survivorship curve for females of arrhenotokous and thelytokous *T. Pretiosum*

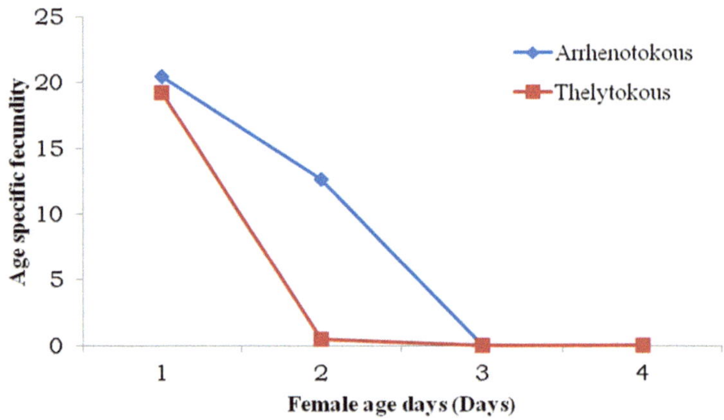

Fig. 20.2 Age-specific fecundity of arrhenotokous and thelytokous *T. pretiosum*

Therefore, from the observations on age-specific fecundity, it can be suggested that release of parasitoids on the first 3 days of emergence would be more beneficial in the biological control programmes as more viable progenies can be sustained (Figs. 20.1 and 20.2). Jalali et al. (2003) reported the age-specific fecundity of *T. bactrae* was always at a maximum on the first day after emergence (Fig. 20.1). The same was revealed in the findings of Nagarkatti and Nagaraja (1977) and Singh and Jalali (1994).

The experimental observations on age-specific fecundity suggested that release of parasitoids on the first 3 days of emergence would be more beneficial in the biological control programmes as more viable progenies can be sustained. Therefore, from the observations on survivorship, it can be suggested that the number of parasitoid emergence was more on the first day of emergence, and it gradually decreased as the days advanced. Therefore, maximum egg parasitisation can be achieved if parasitoids are released as soon as they emerge. The results also confirmed that

Table 20.5 Emergence or parasitism viability of arrhenotokous and thelytokous *T. pretiosum* (at 50 female parasitoids/2000 *Corcyra* eggs)

Containers size (sq cm area)	Emergence or parasitism viability of *T. pretiosum*		Mean % parasitism
	Arrhenotokous	Thelytokous	
37.5	96.3 (80.4)	98.1 (83.0)	97.2 (81.7)
70	97.6 (82.6)	97.5 (82.1)	97.6 (82.3)
128	96.8 (80.4)	98.0 (83.0)	97.4 (81.7)
27,000	98.2 (83.5)	97.4 (82.0)	97.8 (82.8)
182,250	96.6 (79.8)	97.3 (81.87)	96.9 (80.8)
Mean	97.1 (81.3)	97.7 (82.4)	A × B factor
Parameters	A factor	B factor	
F test	NS	NS	NS
SEM (±)	0.7	1.0	1.5
CD ($P = 0.01$)	1.9	2.9	4.1

biological attributes of arrhenotokous and thelytokous *T. pretiosum* are different. Thelytokous *T. pretiosum* has more practical implications in the field of biological control because of high net reproductive rate (Table 20.5).

Acknowledgments The authors wish to express sincere thanks to the Director of the National Bureau of Agricultural Insect Resources, Bangalore, and the University of Agricultural Sciences, GKVK, Bangaluru.

References

Andrewartha HG, Birch LC (1954) The distribution and abundance of animals. University of Chicago Press, Chicago, pp 782

Botto EN, Horny C, Klasmer P, Gerding M (2004) Biological studies on two neotropical egg parasitoid species: *Trichogramma nerudai* and *Trichogramma* sp. (Hymenoptera: Trichogrammatidae). Biocontrol Sci Technol 14(5):449–457

Jalali SK, Singh SP (1993) Superior strain selection of the egg parasitoid *Trichogramma chilonis* Ishii – biological parameters. J Biol Control 7:57–60

Jalali SK, Singh SP, Venkatesan T (2003) Life table characteristics of *Trichogramma toideabactrae* Nagaraja at different temperatures on eggs of *Plutella xylostella* (Linnaeus). Pest Manag Hortic Ecosyst 9:13–17

Jalali SK, Singh SP, Venkatesan T, Murthy KS, Lalitha Y (2006) Development of Endosulfan tolerant strain of egg parasitoid *Trichogramma pretiosum* Ishii (Hymenoptera: Trichogrammatidae). Indian J Exp Biol 44:584–590

Knutson A (1998) The trichogramma manual, Texas Agricultural extension service. The Texas A and M university system, USA. B-6071: 5–98, pp 44

Nagarkatti S, Nagaraja H (1977) Biosystematics of Trichogrammatoidea species. Annu Rev Entomol 22:157–176

Singh SP, Jalali SK (1994) Trichogrammatid. Project Directorate of Biological Control, Bangalore, Tech bull 7:95

Slobodokin LB (1962) Growth and regulation of animal populations. Holt, Rinehart & Winston, New York, pp 104

Southwood TRE (1976) Ecological methods with particular reference to the study of insect populations. Chapman and Hall, London, pp 391

Faunal Composition of Scarab Beetles and Their Hosts in Assam

21

Mrinmoy Das, Badal Bhattacharyya, D. Pujari, and Gautam Handique

Abstract

Field investigations were conducted in the Instructional cum Research Farm and Horticultural Orchard of Assam Agricultural University, Jorhat, to monitor the population of foliage-feeding scarab beetles on eight different host plants during 2010–2012. The highest numbers of beetles were collected from *sthal-padma* (*Hibiscus mutabilis*) (1579 numbers) followed by *agar* (*Aquilaria malaccensis*) (1526 numbers) and guava (*Psidium guajava*) (1348 numbers). The lowest number was found on jute (*Corchorus* sp.) (370 numbers). The host plants, viz., rose (*Rosa chinensis*), ber (*Ziziphus jujuba*), *Silikha* (*Terminalia chebula*), and *Soalu* (*Litsaea polyantha*) recorded 1143, 1143, 1065, and 930 beetles, respectively. The present investigation also revealed 13 numbers of new host plants of scarab beetles, *viz.*, devil tree (*Alstonia scholaris*), albizia (*Albizia odoratissima*), dalchini (*Cinnamomum zeylanicum*), cocoa (*Theobroma cacao*), indigo (*Indigofera zollingeriana*), mango (*Mangifera indica*), Assam lemon (*Citrus lemon*), green gram (*Vigna radiata*), jamun (*Syzygium cumini*), rain tree (*Samanea saman*), mast tree (*Polyalthia longifolia*), jackfruit (*Artocarpus heterophyllus*), and drumstick (*Moringa oleifera*). Through scouting 20 species of scarab beetles were recorded, of which, *Sophrops iridipennis*, *Anomala perplexa*, *Anomala dussumieri*, and *Adoretus renardi* are new records from Assam.

Keywords

Assam • Faunal composition • Host trees • Scarab beetles

M. Das (✉) • B. Bhattacharyya • D. Pujari • G. Handique
All India Network Project on Soil Arthropod Pests, Department of Entomology, Assam Agricultural University, Jorhat 785013, Assam, India
e-mail: mrinmoy.aau@gmail.com

A.K. Chakravarthy, S. Sridhara (eds.), *Arthropod Diversity and Conservation in the Tropics and Sub-tropics*, DOI 10.1007/978-981-10-1518-2_21

21.1 Introduction

'White grub' or 'root grub' is applied to the immature stage of scarab beetles popularly known as cockchafers, chafer beetles, May beetles or June beetles. They belong to the family Scarabaeidae consisting of both useful (coprophagous) as well as harmful (phytophagous) beetles. Coprophagous ones play a significant role in nutrient cycling and phytophagous are known pests of many useful crops. The grubs which feed on the living roots belong to the subfamily Melolonthinae and partly to Rutelinae and Dynastinae. They feed on the roots of many plants, killing many seedlings and sometimes older plants, reducing drought tolerance and thereby affecting final yield (Monck and Pearce 2008). In India, white grubs have become a serious constrain for production of almost all crops grown particularly during kharif season (Mathur and Bhatnagar 2001).They are distributed from high altitude of Himalayas to low altitude of coastal Kerala, from arid and semi-arid tracts of Rajasthan to high rainfall, humid regions of North-East hill region and West coastal peninsula. The white grub has been recognized as the pest of National Importance. During the recent years, white grubs have emerged as important insect pests of potato, jute, sugarcane, pulses and vegetables in Assam (Bhuyan et al. 2009). Therefore, an attempt was made to study the faunal composition of foliage feeding scarab beetles by monitoring them on host trees.

21.2 Materials and Methods

Field investigations were conducted in the Instructional cum Research Farm and Horticultural Orchard of Assam Agricultural University, Jorhat (26″46′ N lat. and 94″12′E long.), during 2011–2012 to monitor population of foliage-feeding scarab beetles. Eight different host plants, *viz*., *sthal-padma* (*Hibiscus mutabilis*), agar (*Aquilaria malaccensis*), ber (*Ziziphus jujuba*), guava (*Psidium guajava*), *Silikha* (*Terminalia chebula*), jute (*Corchorus* sp.), rose (*Rosa chinensis*), and *Soalu* (*Litsaea polyantha*) were selected. The scouting was conducted from March to September 2011. For each host plant, there were five replications, and from each replication, one branch was marked with an aluminum tag before conducting the scouting. The weekly scouting was carried out from 18.30 to 20.30 h, and the observations were recorded as beetle catch per branch. The beetles collected from different host plants at various locations were brought to the laboratory in perforated plastic containers and were sorted out to species level. Attempts were also made to survey host trees of foliage-feeding scarab beetles in the Assam Agricultural University, Jorhat, during April to June 2012 using a powerful flash light (220v) and also following shaking and beating method for the beetle collection. The scarab beetles collected were identified from the Division of Entomology, Indian Agricultural Research Institute, New Delhi, India.

21.3 Results and Discussion

21.3.1 Population Monitoring of Scarab Beetles by Scouting

While monitoring the population of scarab beetles on eight different host plants, the highest numbers of beetles were collected on *Sthal-padma* (1579 numbers) followed by *agar* (1526 numbers) and guava (1348 numbers) (Table 21.1). The lowest number was found on jute (370 numbers). Rose, ber, *Silikha*, and *Soalu* recorded 1143, 1143, 1065, and 930 numbers of beetles, respectively. The maximum numbers of species were collected on *agar* (eight) followed by *sthal-padma* and ber (six) and rose, *Silikha*, and guava (Five). Jute and Soalu recorded four and three numbers of species, respectively. Twelve species belonging to five genera (*Apogonia, Adoretus, Holotrichia, Anomala*, and *Sophrops*) were recorded (Table 21.1).

Thirteen host plants of scarab beetles (other than the previous record of eight host plants (see Table 21.1)), viz., devil tree (*Alstonia scholaris*), albizia (*Albizia odoratissima*), dalchini (*Cinnamomum zeylanicum*), cocoa (*Theobroma cacao*), indigo (*Indigofera zollingeriana*), mango (*Mangifera indica*), Assam lemon (*Citrus lemon*), green gram (*Vigna radiata*), jamun (*Syzygium cumini*), rain tree (*Samanea saman*), mast tree (*Polyalthia longifolia*), jackfruit (*Artocarpus heterophyllus*), and drumstick (*Moringa oleifera*) (Table 21.2 and Fig. 21.1), were recorded from the Assam Agricultural University during 2012. Through scouting 20 species of scarab beetles were recorded; four species, *viz*., *Sophrops iridipennis, Anomala perplexa, Anomala dussumieri*, and *Adoretus renardi*, were new records from Assam.

The variations in the species complex on different host plants may be due to the species under investigation, the preferred host and location. Availability of host plants in specific localities attracts more number of beetles. Similar observations were also reported earlier (John et al. 1998; Held et al. 2001; Rodrigo et al. 2003; Diagne et al. 2006; Sanjun and Pomper 2008; Bhuyan et al. 2009). Light is another influencing factor for the distribution of scarab beetles (Pal 1977). Furthermore, quality and quantity of certain substances present in the food have been reported to influence the development, reproduction, resistance to disease and pigmentation of insects (Bogawat and Pandey 1967). However, no realistic information are available on host plant preference of the foliage-feeding scarab beetles in relation to morpho-physiological and biochemical characteristics of the foliage of preferred host plants. This aspect needs further study.

Table 21.1 Population monitoring of scarab beetles by scouting on select hosts in Assam (2011)

Sl. no.	Host plants	Mean monthly population of scarab beetles collected by scouting*							Total collection	Major species recorded
		March	April	May	June	July	Aug.	Sept.		
1.	Rose (*Rosa chinensis*)	6.80	8.15	9.15	10.10	11.00	6.40	5.50	1143	*Adoretus bicolor*, *Apogonia* sp., *Adoretus aerial*, *Apogonia ferruginea*, *Holotrichia* sp.
2.	*Sthal-Padma* (*Hibiscus mutabilis*)	9.95	11.40	12.45	13.45	14.55	9.55	7.70	1579	*Holotrichia consanguinea*, *Adoretus bicolor*, *Apogonia ferruginea*, *Apogonia* sp., *Holotrichia* sp., *Anomala* sp.
3.	Guava (*Psidium guajava*)	8.25	9.35	10.40	11.40	12.50	8.60	7.00	1348	*Apogonia* sp., *Sophrops iridipennis*, *Holotrichia consanguinea*, *Apogonia ferruginea*, *Adoretus aerial*
4.	Jute (*Corchorus* sp.)	–	–	1.35	7.70	8.35	1.10	–	370	*Adoretus bicolor*, *Apogonia* sp., *Apogonia ferruginea*, *Apogonia aerea*
5.	Ber (*Ziziphus jujube*)	6.95	8.30	9.30	10.30	11.30	6.20	5.10	1143	*Holotrichia consanguinea*, *Adoretus bicolor*, *Apogonia* sp., *Adoretus bomblinota*, *Adoretus* sp., *Holotrichia* sp.
6.	*Agar* (*Aquilaria malaccensis*)	9.45	11.05	11.95	12.95	13.90	9.15	7.80	1526	*Holotrichia consanguinea*, *Adoretus bicolor*, *Apogonia* sp., *Adoretus aerial*, *Adoretus* sp., *Sophrops iridipennis*, *Holotrichia serrata*, *Holotrichia* sp.
7.	*Silikha* (*Terminalia chebula*)	6.50	7.50	8.50	9.50	10.50	6.05	4.70	1065	*Sophrops* sp., *Holotrichia consanguinea*, *Apogonia* sp., *Holotrichia serrata*, *Holotrichia* sp.
8.	*Soalu* (*Litsaea polyantha*)	5.65	6.60	7.50	8.40	9.55	4.90	3.90	930	*Adoretus bicolor*, *Apogonia* sp., *Holotrichia* sp.

*Mean of 4 weeks

Table 21.2 Host plants of scarab beetles in Assam (2012)

SL No.	Common name	Botanical name	Family	Species recoded
1	Albizia	*Albizia odoratissima*	Leguminosae	*Holotrichia consanguinea*, *Apogonia* sp., *Anomala dussumieri*, *Maladera insanabilis*
2	Mango	*Mangifera indica*	Anacardiaceae	*Apogonia* sp., *Sophrops iridipennis*
3	Indigo	*Indigofera zollingeriana*	Fabaceae	*Apogonia* sp., *Holotrichia serrata*, *Anomala perplexa*, *Anomala dussumieri*, *Apogonia blanchardi*
4	Jamun	*Syzygium cumini*	Myrtaceae	*Holotrichia consanguinea*, *Adoretus bicolor*, *Holotrichia serrata*, *Onthophagus* sp.
5	Drumstick	*Moringa oleifera*	Moringaceae	*Holotrichia consanguinea*, *Anomala perplexa*
6	Assam lemon	*Citrus Lemon*	Rutaceae	*Brahmina coriacea*, *Sophrops iridipennis*, *Holotrichia* sp.
7	Green gram	*Vigna radiate*	Leguminosae	*Onthophagus* sp., *Apogonia* sp., *Anomala* sp., *Apogonia blanchardi*
8	Mast tree	*Polyalthia longifolia*	Annonaceae	*Anomala* sp., *Apogonia* sp., *Apogonia ferruginea*
9	Rain tree	*Samanea saman*	Fabaceae	*Adoretus* sp., *Anomala perplexa*
10	Devil tree	*Alstonia scholaris*	Apocynaceae	*Adoretus bicolor*, *Apogonia* sp., *Sophrops iridipennis*
11	Dalchini	*Cinnamomum zeylanicum*	Lauraceae	*Sophrops iridipennis*, *Holotrichia consanguinea*, *Anomala perplexa*
12	Coco	*Theobroma cacao*	Malvaceae	*Sophrops iridipennis*, *Apogonia* sp.
13	Jackfruit	*Artocarpus heterophyllus*	Moraceae	*Adoretus aerial*, *Sophrops iridipennis*, *Holotrichia serrata*

21.4 Studies from Other Areas

Many investigations have been carried out to find the role of dung beetles in regenerating dung and to find the way of using them as agents of decomposition in the dung decaying process and in the control of dung breeding arthropods, predominantly flies (Ridsdill-Smith 1981; Anderson et al. 1984; Roth et al. 1988). This study was aimed to determine dung beetle diversity, abundance, and species richness and present data on biodiversity and population of cattle dung beetles collected during a 4-month period in five areas of Ardebil Province. Another study on dung beetles has come from Azerbaijan, Eastern Europe. The collected dung beetle specimens belonged to nine genera and 15 species. They were collected from cow dung

Anomala dussumieri *Adoretus renardi* *Anomala pellucida* *Sophrops irridipennis*

Fig. 21.1 New records of four species of scarabs beetles from Assam

pads in different parts of the province. Out of 231 collected dung beetles, *Onthophagus taurus* (34.19 %) were identified as most prevalent species, while *O. amyntas* (0.86 %), *O. speculifer* (0.86 %), and *O. furcatus* (0.86 %) were the least prevalent (Fig. 21.1). The Meshkinshahr and Sarein areas had the most and the least densities, i.e., 45.02 % and 3.89 %, respectively. *Copris lunaris* were present in high numbers in Namin while *Aphodius lugens* were numerous in Neer. Overall, Meshkinshahr and Sarein had the highest and lowest diversities, respectively. Eight species were collected in Meshkinshahr including *Sisyphus schaefferi*, *O. taurus*, *C. lunaris*, *Chironitis pamphilus*, *O. amyntas*, *Caccobius schreberi*, *Aphodius lugens*, and *Geotrupes stercorarius.*

21.4.1 Species Diversity and Abundance Pattern

During the current research, the adults from five different transects were collected and Shannon indices were calculated as an index of diversity within the habitat. The Shannon diversity index indicated that transect I was relatively diverse (1.32) followed by transect II (1.25), transect V (1.02), transect IV (0.88), and lastly transect III (0.73). The Simpson and Shannon J (evenness) indices also gave almost the same results. The other studies in these lines are as follows:

(a) Thakare et al. (2011) studied the species composition of dung beetles from Maharashtra, India, and recorded the occurrence of 26 beetle species from eight subfamilies. Of the family Scarabaeidae, the Scarabaeinae was the dominant subfamily in terms of species richness (15 species) and abundance, followed by Melolonthinae (two species), Rutelinae (two species), and Cetoniinae (two species), and the subfamilies Geotrupinae, Hybosorinae, Orphninae, and Dynastinae were represented by only one species in the surveyed area. Eight species could be identified at the species level. The *Onthophagus* was the most species-rich genera with six species followed by *Onitis* (three species), *Copris* (two species), *Anomala* (two species), and *Catharsius* (two species).

(b) Species composition of scarabaeid beetles in rose gardens in Bangalore, Karnataka (India), revealed 13 species of scarabaeid beetles belonging to nine genera representing three subfamilies: Melolonthinae, Rutelinae, and

Cetoniinae. Maximum of five species belong to subfamily Melolonthinae and three species each belonged to Rutelinae and Cetoniinae. Among the species identified, *Holotrichia serrata*, *Schizonycha ruficollis*, *Anomala bengalensis*, and *Adoretus versutus* were found to be the most dominant leaf feeders, and *Maladera* sp. and *Apogonia ferruginea* were documented damaging and preferring on rose flowers over others. The scarabaeid adults' emergence began after the first rain in April, and it continued up to the last week of September. Peak numbers of beetles were found between 19.00 and 19.30 h, and subsequently no emergence was found from each species of scarabaeids. The extent of leaf damage incurred by *H. serrata*, *S. ruficollis*, and *A. versutus* was significantly more on rose leaf despite the presence or absence of flowers. The outcome of the study may be helpful in pest management of rose pests (Kumar et al. 2009).

Dung beetles belong to the Coleoptera, family Scarabaeidae, and consist approximately 5000 species worldwide including 12 tribes (González-Maya and Mata-Lorenzen 2008). However, there may be many more dung beetles because thousands of existing species have not yet been described or discovered. Dung beetles constitute a group of high ecological and economic importance due to the several ecosystem services they offer in. The removal of animal dung and waste reduces micro- and macro invertebrate parasites, facilitates nutrient cycling, contributes to soil conditioning and aeration, and assists in seed spread (Spector 2006). None have addressed the ecology, community structure, and succession pattern of the dung beetles associated with the herbivorous mammals in Ardabil Province of Iran. The occurrence of suitable habitats and favorable climate conditions in Iran benefits the maintenance of scarabaeids dung beetles in nature. In a study carried out by Mowlavi et al. (2009), they were able to find larval stages of *Gongylonema* spp. and *Rhabditis* spp. inside dissected beetles. In a similar survey, the number of other free living nematodes such as *Rhabditis* spp. as well as *Pelodera* spp. has been observed externally on the body surface of the beetles. Overall, Meshkinshahr and Sarein areas had the highest and lowest diversities, respectively. Eight species were collected in Meshkinshahr (45.02 %). One factor that might explain a higher overall diversity of dung beetles in Meshkinshahr and Neer areas is mammal diversity as well as moderate climate conditions. Low diversity and abundance in Sarein may be related with lack of enough mammals. Mammalian fauna in Ardabil Province includes a wide variety of wild and domestic species. The main mammal in the study area is cow. Dung beetles are to a large extent adapted to man-domesticated habitats. The practice of maintaining large herbivores such as cows and horses in pastures has raised the abundance of several species substantially. The "suitability" of dung is also critical because several species can have varied preferences for dung which are guided by their life history strategies. For example, the moisture content of dung is an important factor for some species (Gittings and Giller 1997). Some species of dung beetles prefer specific habitats (grassland or forest) or certain soil types (sand or clay). Some species feed on dung of only one species of animal, while others are not so limited.

Some of the beetles in the current study were reported from other provinces in Iran including *Caccobius schreberi* (L.) (from Golestan and Tehran), *Onthophagus furcatus* Fab. (from Golestan, Kordestan, and Tehran), *Onthophagus speculifer* Sols (from Azerbaijan Gharbi and Fars), *Oniticellus fulvus* (Goez.) (from Fars, Gilan, Golestan, and Kordestan), *Chironitis pamphilus* (Men.) (from Azerbaijan Gharbi, Chaharmahal and Bakhtiari, Kohgiluyeh and Boyer-Ahmad, Kordestan, Mazandaran, and Tehran), and *Geotrupes spiniger* Marsh (from Gilan) (Barari 2001).

The diversity of regional dung beetle communities is primarily affected by vegetation cover, soil type and moisture, and resource (dung) availability (Doube 1987; Davis 2002). Since dung beetles exhibit varying preference for dung pads of different mammals based on texture, the structure of dung beetle communities is influenced by the local mammalian fauna. Many studies in tropical South America and Southeast Asia have demonstrated that deforestation caused by human activities (e.g., logging, pasture clearing, and plantation cutting) reduces the diversity of dung beetle communities. Deforestation has always decreased the number of mammals producing dung for beetles (Estrada and Coates-Estrada 2002). This suggests that although vegetation cover may be most important for determining the structure of dung beetle communities, mammalian fauna may substantially influence the diversity of dung beetle communities.

There are several reports of biodiversity of dung beetles in the world (Kanda et al. 2005; Durães et al. 2005). In a study in India, the abundance of small rollers (*Sisyphus*) and the low presence of large rollers (*Gymnopleurus*) in pitfall traps and further studies indicated the dominance of smaller rollers in elephant dung of the region (Sabu et al. 2006). In conclusion, dung beetles have an important function in farmland ecosystems. Their role in dung breakdown and as valuable prey for vertebrate predators should not be underestimated. Because of their economic importance, their use as biodiversity indicators is an important ecological and conservation factor. A complete round the year survey considering the differences in fauna of the beetles in relation to mammal fauna and environmental factors may provide more information on the community ecology of the dung beetles. Additionally, further sampling effort is needed in the area to obtain a comprehensive pool and distribution of potential species. The present finding represents an important advance in the knowledge of this important group of indicator species in the region. Scientists assume that it is crucial to sustain the benefits of controlling livestock pests by using veterinary chemicals; the adverse side effects of these on dung fauna are of significant importance.

Acknowledgments The authors are indebted to Dr. V.V. Ramamurthy, Division of Entomology, Indian Agricultural Research Institute, New Delhi, 110012. Authors are also thankful to Dr. Swaroop Singh, Ex Network Coordinator, All India Network Project on Soil Arthropod Pests, Agricultural Research Station (Durgapura), Rajasthan, India.

References

Anderson JR, Merrit RW, Loomis EC (1984) The insect-free cattle dung fouling of range land pastures. J Econ Entomol 77:133–141

Barari H (2001) Insects of Iran. Plant Pests and Diseases Research Institute, Tehran, p 64

Bhuyan U, Bhattacharyya B, Pujari D, Borah AALH, Saud R, Goswami J (2009) Host plants of scarab beetles in Assam. National seminar on frontiers in plant physiology towards sustainable agriculture, 5–7 November, Assam Agricultural University, Jorhat 785013, Assam, pp 96

Bogawat JK, Pandey SN (1967) Food preferences in *Aula cophora* sp. Indian J Entomol 29(4):349–352

Davis ALV (2002) Dung beetle diversity in South Africa: influential factors, conservation status, data inadequacies and survey design. Afr Entomol 10:53–65

Diagne A, Richard NS, Hammond MA (2006) Adult *Phyllophaga ephilida* host plant feeding preference. Fla Entomol 89(3):391–395

Doube BM (1987) Spatial and temporal organization in communities associated with dung pats and carcasses. In: Gee JHR, Giller PS (eds) Organization of communities: past and present. Blackwell, Oxford, pp 255–280

Durães R, Martins WP, Vaz-De-Mello FZ (2005) Dung Beetle (Coleoptera: Scarabaeidae) assemblages across a natural forest-Cerrado ecotone in Minas Gerais, Brazil. Neotropical Entomol 34(5):721–731

Eberhard ML, Busillo C (1999) Human Gongylonema infection in a resident of New York City. Am J Trop Med Hyg 61:51–52

Estrada A, Coates-Estrada R (2002) Dung beetles in continuous forest, forest fragments and an agricultural mosaic habitat island at Los Tuxtlas, Mexico. Biodivers Conserv 11:1903–1918

Gittings T, Giller PS (1997) Life history traits and resource utilisation in an assemblage of north temperate *Aphodius* dung beetles (Coleoptera: Scarabaeidae). Ecography 20:55–66

González-Maya JF, Mata-Lorenzen J (2008) Dung-beetles (Coleoptera: Scarabeidae) from the Zona Protectora Las Tablas, Talamanca, Costa Rica. Check List 4(4):458–463

Held DW, Eaton T, Potter DA (2001) Potential for habituation to a neem based feeding deterrent to Japanese beetles, *Popillia japonica*. Entomol Exp Appl 101(1):25–32

John H, Daniel A, Thomas R (1998) Attraction of Japanese beetles to host plant volatiles in field trapping experiments. Ann Rev Entomol 27(2):395–400

Kanda N, Yokota T, Shibata E, Sato H (2005) Diversity of dung-beetle community in declining Japanese sub alpine forest caused by an increasing sika deer population. Ecol Res 20:135–141

Kumar S, Sankar M, Sethuraman V, Musthak A (2009) Population dynamics of white grubs (Coleoptera: Scarabaeidae) in the rose environment of Northern Bangalore, India. Indian J Sci Technol 2(1):46–52

Mathur YS, Bhatnagar A (2001) White grubs and its management in ground nut cropping system. In: Sharma, Mathur, Gupta (eds) Indian phytophagous scarabs and their management. pp 48–61

Molavi G, Massoud J, Gutierrez Y (2006) Human Gongylonema infection in Iran. J Helminthol 80(4):425–428

Mowlavi G, Mikaeili E, Mobedi I, Kia E, Masoomi L, Vatandoost H (2009) A survey of dung beetles infected with larval nematodes with particular note on *Copris lunaris* beetles as a vector for *Gongylonema* sp. in Iran. Korean J Parasitol 47(1):13–17

Pal SK (1977) Relative abundance of Scarabaeid beetles on light traps. Indian J Entomol 39:197–200

Ridsdill-Smith TJ (1981) Some effects of three species of dung beetle (Coleoptera: Scarabaeidae) in South-Western Australia on the survival of the bush Fly, *Musca autumnalis* Walker (Diptera: Muscidae) in dung pads. Bull Entomol Res 71:425–433

Rodrigo J, Mercader L, Isaacs R (2003) Damage potential of rose chafer and Japanese beetle in Michigan vineyards. Great Lakes Entomol 36(3 and 4):34–41

Roth JP, Macqueen A, Bay DE (1988) Scarab activity and predation as mortality factors of the buffalo fly *Haematobia irritans exigu* a, in Central Queensland. Southwest Entomol 13:119–125

Sabu TK, Vinod KV, Vineesh PJ (2006) Guild structure, diversity and succession of dung beetles associated with Indian elephant dung in South Western Ghats forests. J Insect Sci 17:1–12

Sanjun G, Pomper KW (2008) Grape cultivar feeding preference of adult Japanese beetles. Hortscience 43(1):196–199

Spector S (2006) Scarabaeine dung beetles (Coleoptera: Scarabaeidae: Scarabaeinae): an invertebrate focal taxon for biodiversity research and conservation. Coleopt Bull 60(5):71–83

Thakare VG, Zade VS, Chandra K (2011) Diversity and abundance of Scarab Beetles (Coleoptera: Scarabaeidae) in Kolkas Region of Melghat Tiger Reserve (MTR), District Amravati, Maharashtra, India. World J Zool 6(1):73–79

Honeybees and Beekeeping: The Global Scenario

22

D.P. Abrol, Uma Shankar, K.S. Nitin, and G. Basana Gowda

Abstract

Diversity of geographical features particularly in the tropics and subtropics plays a key role in determining the topography, climate, and plant species of the region. Such regions provide abundant opportunities for both migratory and non migratory beekeeping. Current agricultural transformation, once linked to apicultural operations, offers much scope for income generation from beekeeping. Till now only 10 % of the existing potential has been utilized. For instance, India has a potential to keep about 120 million bee colonies that can provide self-employment to over 6 million rural and tribal families. In terms of production, these bee colonies can produce over 1.2 million tonnes of honey and about 15,000 tonnes of beeswax. Organized collection of forest honey and beeswax using improved methods can result in an additional production of at least 120,000 tonnes and 10,000 tonnes of honey and beeswax, respectively. Thus, it is expected to generate income of worth satisfying needs of five million tribal families. The present global status as well as future strategies for conservation of beekeeping is discussed in detail.

Keywords

Beekeeping • Beeswax • Diversity • Honey

D.P. Abrol (✉) • U. Shankar
Division of Entomology, Sher-e-Kashmir University of Agricultural Sciences & Technology
FOA, Chatha, 18009 Jammu, Jammu & Kashmir, India
e-mail: dharam_abrol@rediffmail.com; umashankar.ento@rediffmail.com

K.S. Nitin
Division of Entomology and Nematology, Indian Institute of Horticultural Research (IIHR), Hessarghatta Lake Post, Bengaluru 560089, Karnataka, India

G.B. Gowda
ICAR-National Rice Research Institute, Cuttack 753 006, Odisha, India
e-mail: basanagowda4099@gmail.com

A.K. Chakravarthy, S. Sridhara (eds.), *Arthropod Diversity and Conservation in the Tropics and Sub-tropics*, DOI 10.1007/978-981-10-1518-2_22

22.1 Introduction

Recent disappearance of honeybees across the globe with the diagnosis of colony collapse disorder has raised concerns about the threats posed to food security and ecosystems (Barrionuevo 2007; UNEP 2010; Mathews 2010). Debates about honeybees loss and its causes continue to rage. In part, the controversy stems from the combined complexity of pollinator relations and suspected multiple, interrelated causes such as chemicals, pests, disease, vegetation changes, weather, etc. The complexity also results in inconsistent manifestations of disappearances from year to year and region to region, leading to further contestations and confusions. The problem and solution(s) continue to haunt and provoke (Thompson et al. 2003).

Beekeeping has turned into an inexorably prominent practice to get tasty nectar as a wellspring of occupation and nourishment security through pollination (Walker and Crane 2000; Agrawal 2014). Honey and honeybees have a long history in India. Honey bees developed in this area a large number of years prior and delivered nectar from nectars of blooming plants that possessed large amounts of rich backwoods surviving everywhere throughout the nation. Nectar was the main sweet nourishment tasted by the ancient Indians meandering in these woodlands. "Honey hunting – is an antiquated practice in India as appeared, for instance, in hole compositions going back to 11,000 BC found in Madhya Pradesh (Suryanarayana 2002) and in Ancient Egyptian drawings and works of art (Crane 1999). The historical backdrop of beekeeping is attached in and connected to nectar chasing and related practices. Assembling wild nectar is still a typical practice in numerous parts of the world; in India it is evaluated that 22,000 tonnes of wild nectar is gathered by nectar seekers yearly – doubling the measure of nectar delivered by the overseen beekeeping area (Wakhle and Pal 2000).

India has properly been known as a "place that is known for nectar." The nation has a portion of the most established records of nectar industry as ancient works of art delineating hives and the soonest endeavors to reap nectar. India has for quite some time been surely understood for its transcendent custom of beekeeping and nectar chasing. Beekeeping is presently settled and thriving as a beneficial agro-woodland-based industry. Beekeeping assumes a vital part in the present connection of commercialization of agribusiness and liberalization of economy. It covers the whole extent of bumble bee assets, honeybee items, beekeeping rehearses, fertilization administrations, and their interface with business frameworks and environment trustworthiness.

22.2 Evolution

Bees are flying insects closely related to wasps and ants. Bees are a monophyletic family within the superfamily Apoidea, now thought as a covered *Anthophila*. There are nearly 20,000 known species of bees in seven to nine recognized families, though many are undescribed and the actual number is probably higher. The majority of species are single/alone laying eggs in tunnels they dig out themselves. In some species small numbers of females may share a single-tunnel system, and in others there may be a semiorganization/social organization involving an order among the females (related to certain bees being ranked above or below other bees). Bees provide a supply of food (honey and pollen) for the larvae, but there is no progressive feeding of the larvae by the adult bees. Bees are found on every continent except Antarctica, in every home/place where something lives, and on the planet that contains insect (brought pollen from one plant to another)-flowering plants. Bees are changed/ready for feeding on nectar and pollen.

Honeybees have a long proboscis (an unpredictable "tongue") that empowers them to get the nectar from blooms. They have radio wires generally comprised of 13 fragments in males and 12 in females, as is common for the superfamily. Minor stingless honeybee species exist whose laborers are under 2 mm (0.079″) long. The biggest honeybee on the planet is *Megachile pluto*, a leaf-cutter honeybee whose females can achieve a length of 39 mm (1.5″). Individuals from the family Halictidae, or sweat honeybees, are the most widely recognized kind of honeybee in the Northern Hemisphere; however, they are little and regularly mixed up for wasps or flies.

It is imagined that honeybees initially advanced from wasps – Crabronidae – in this way predators of different creepy crawlies. Fossil proof is scanty; however, honeybees most likely showed up on the planet about the same time as blooming plants in the cretaceous period, 146–74 million years back. The most seasoned known fossil honeybee, stingless honeybee named *Trigona prisca*, was found in the upper cretaceous of New Jersey, USA, and dates from 96 to 74 million years prior. It is undefined from cutting-edge *Trigona*.

The change from bug prey to dust may have come about because of the utilization of prey creepy crawlies which were blossom guests and were incompletely secured with dust when they were bolstered to the wasp hatchlings. This same developmental situation has additionally happened inside the vespoid wasps, where the gathering known as "dust wasps" likewise advanced from savage precursors. As of not long ago, the most established nonpressure honeybee fossil had been *Cretotrigona prisca* in New Jersey and of Cretaceous age, a meliponine. As of a late reported honeybee fossil, of the class Melittosphex, is viewed as "a wiped out genealogy of dust gathering Apoidea sister to the present day honey bees" and dates from the early Cretaceous (~100 mya). Determined elements of its morphology ("apomorphies") is placed unmistakably inside the honeybees; however, it holds two unmodified genealogical attributes ("plesiomorphies") of the legs (two midtibial goads and a thin rear basitarsus), demonstrative of its transitional status.

The oddity is that honeybees are particular as fertilization specialists, with behavioral and physical adjustments that particularly improve fertilization and are by and large more productive at the undertaking than other pollinating bugs, for example, scarabs, flies, butterflies, and dust wasps. The presence of such flower masters is accepted to have driven the versatile radiation of the angiosperms and, thus, the honeybees themselves.

22.2.1 Development of Subspecies

Despite the fact that it has for quite some time been realized that there are numerous sorts of bumble bees, just as of late has a far-reaching order been endeavored considering the contrasts in physical characters among subspecies and their present topographical appropriation, the land proof indicating their causes, and the course of their resulting development and circulation.

Like the stingless honeybees, bumble bees initially developed in tropical conditions. The fossil record appears at times when Europe had a tropical atmosphere. As the atmosphere got cooler, the open settling sorts would not have possessed the capacity to make due with the exception of relocating to the tropical locale of Southern Asia. For most of the tertiary time, Africa was confined from Europe via ocean, and no tertiary sorts of bumble bee reached Africa even after an area scaffold was built up. It is likely that the improvement of cutting-edge warm homeostasis in bumble bees which allowed the control of cool calm zones in this manner happened in Southern Asia, potentially in the Himalayan area. Once settled, the hole-settling cerana-mellifera sort would spread east and west, in the long run involving both tropic and cool and mild zones.

In spite of the fact that cerana honeybees more likely than not imparted a typical progenitor to mellifera, they have developed into isolated species. It is impractical to cross cerana with mellifera notwithstanding utilizing instrumental insemination, on the grounds that the two species are presently hereditarily contrary and feasible eggs don't come about because of the cross preparation. Different contrasts incorporate their varying responses to illnesses, infestations, and predators.

How far mellifera honeybees infiltrated into Northern and Western Europe amid the warm interims between glaciations of the Pleistocene time frame is a matter of guess; what is sure is that no bumble bees could have existed north of the Mediterranean district, the Iberian promontory, and Southwestern France at Ice Age.

A honeybee taking after *Apis dorsata* yet much littler was available in the upper Miocene. It is felt that *Apis florea* and *A. dorsata* may have existed as particular species as ahead of schedule as the Oligocene time frame. It has not been conceivable to gauge when honeybees of mellifera/cerana sort initially showed up on earth. Mellifera and cerana more likely than not procured separate characters amid the last part of the tertiary time.

Among living honeybee aggregates, the "short-tongued" honeybee family Colletidae has generally been viewed as the most "primitive" and sister taxon to the rest of the honeybees. In the twenty-first century, nonetheless, a few scientists have guaranteed that the *Dasypodaidae* is the basal gatherer, the short, wasp-like mouthparts of colletids being the consequence of concurrent development, as opposed to characteristic of a plesiomorphic condition. This subject is still under open deliberation, and the phylogenetic connections among honeybee families are ineffectively caught on. In some species, groups of cohabiting females may be sisters, and if there is a division of labor within the group, then they are considered semisocial. If, in addition to a division of labor, the group consists of a mother and her daughters, then the group is called eusocial. The mother is considered the queen and the daughters are workers. These castes may be purely behavioral alternatives, in which case the system is considered "primitively eusocial" (similar to many paper wasps), and if the castes are morphologically discrete, then the system is "highly eusocial." Highly eusocial bees live in colonies. Each colony has a single queen, many workers, and, at certain stages in the colony cycle, drones. Honeybee hives can contain up to 40,000 bees at their annual peak, which occurs in the spring (October–February). The true honeybees (genus *Apis*) have arguably have the most complex social behavior among the bees. *A. mellifera* is the best-known bee species and one of the best known of all insects. There are 29 subspecies of *A. mellifera*, native to Europe, the Middle East, and Africa.

22.2.2 Diversity

Insects evolved 350 million years ago. Some changes in insect fauna were noticed in the Permian, Mesozoic, Triassic, and Jurassic periods. When flowering plants became established in the cretaceous period, many insects were found associated with them. Of the several insect orders, Hymenoptera is the largest group of insects which includes at least a quarter million species including all social insects such as bees, wasps, and ants (except termites). Honeybees have settled almost all over the planet. They live both in regions with cold climates and long severe winters and in the tropics where winters never occur and the summer temperatures are higher. Bee's adaptability to different climates and environments has proved to be genuinely amazing. As a result of specific climatic conditions and peculiarities of nectariferous flora, the group developed different breeds of honeybees during evolutionary history.

22.2.3 Origin and Distribution of Apis

Bees as a gatherer seem to have their focal point of root in Southeast Asia (counting the Philippines), as everything except one of the surviving animal categories is local to that area, including the most primitive living species (*Apis florea* and *A. andreniformis*). The main *Apis* honeybees show up in the fossil record in stores dating around 40 million years back amid the Eocene time frame; that these fossils are from Europe does not as a matter of course demonstrate that Europe is the place the sort began, as the probability of fossils being found in Southeast Asia is little, regardless of the fact that that is the genuine beginning. At around 30 million years before present, they seem to have created social conduct and fundamentally are for all intents and purposes indistinguishable with advanced bumble bees.

The nine major *Apis* species are *Apis mellifera*, *A. cerana*, *A. koschevnikovi*, *A. nigrocincta*, *A. nuluensis*, *A. dorsata*, *Apis florea*, *A. laboriosa*, and *A. andreniformis*. Out of these nine species, five initial species nest in cavities which have a number of combs. The last four that nest in the open have a single comb. *Apis* species are divided into three lineages: the cavity nesting bees, *Apis mellifera*, *A. cerana*, *A. koschevnikovi*, *A. nigrocincta*, and *A. nuluensis*; open nesting, the dwarf bees, *A. florea* and *A. andreniformis*; and giant bees, *A. dorsata* and *A. laboriosa*. Of the nine species, only *A. mellifera* and *A. cerana* have been "domesticated" for a long time (Koeniger 1976). *A. mellifera* is the most studied and economically exploited species. All *Apis* species, except for *A. mellifera*, are native to Southeast Asia. This region is a center of *Apis* diversity and makes scientists pay great attention to the recently recognized species such as *A. nigrocincta* and *A. nuluensis*.

There are nine different species of bees that make honey. The most commonly recognized honeybee species, *A. mellifera*, is native to Africa and Europe and subdivided into about 24 subspecies.

In India, European bees were successfully introduced in 1965. There had been several attempts to bring the exotic bees into different parts of the country through the past eight decades. One can assume that all these stocks perished because no reports on their performances exist. Successful establishment of the European bees was possible due to a combination of several factors. Important among these are selection of appropriate strain of the bee that could adapt to the hot agriculture climate, change in the cropping pattern that helped in ensuring a continuity in bee forage, and scientific methods of introduction of the exotic strain in an alien climate.

22.3 Honeybee Species

India has five types of *Apis*, i.e., the genuine bees – *Apis florea*, *A. cerana*, *A. dorsata*, and *A. laboriosa* Smith and *Apis mellifera*, a European species – and a few types of stingless honeybees of the variety *Trigona* (Fig. 22.1a–f). There are signs that no less than an extra species near *A. florea* exists. Of the *A. cerana*, three subspecies have so far been distinguished. Kshirsagar (1983) and Verma (1992) showed that an aggregate of seven races under three perceived subspecies exists. Notwithstanding this assortment, all the *Apis* species have an amazingly comparable life cycle. The science of the stingless honeybees is likewise practically comparative. *Apis* honeybees have a very much created social association and have comparable standing separation, division of work, searching, guard, and conceptive practices. Of the few bumble bee species, *Apis cerana* and *A. mellifera* are tamed species and fill the need of business beekeeping.

Fig. 22.1 Different types of honeybee species in India: (**a**) *Apis indica*, (**b**) *Apis mellifera*, (**c**) *Apis dorsata*, (**d**) *Apis florea*, (**e**) *Trigona bengalensis*, (**f**) *Trigona* (*Tetragonula*) *iridipennis*

22.3.1 Dwarf Honeybee

As its name suggests, *Apis florea* is the littlest of the genuine bumble bees, both in body size of specialists and home. A home of *A. florea* comprises of a solitary brush, whose upper part extends to frame a peak that encompasses the branch or other article from which the brush is suspended. Smaller person bumble bees home in the open, however not without disguise: most homes are swung from slim branches of trees or bushes secured with moderately thick foliage, for the most part from 1 to 8 m over the ground. The honeybee is for the most part found in fields or swamps in tropics and subtropics. It is once in a while found in heights above 1500 m. The homes are inherent shrubs, thickly leaved little trees in greenhouses and plantations, holes of structures or shielded boxes or divider specialties in urban ranges, and on firmly put stalks of harvest plants like *Sorghum*. The midget honeybee can make due in extremely hot and dry atmospheres with surrounding temperatures achieving 50 °C or more. The brush design is like that in different *Apis* species, aside from the nectar stockpiling parcel that is particular in *A. florea* brush. Additionally, in nectar generation, *A. florea* which is an essential pollinator of yields in dry and semi-dry areas can likewise be used for pollination.

22.3.2 Giant or Rock Bee

The giant honeybees (*Apis dorsata*) are predominantly found in or near forests, although at times nests may be observed in towns near forest areas. The bee shares the open-air, single-comb nesting habits of *Apis florea*, suspending its nest from the undersurface of its support, such as a tree limb or cliff. In general, *A. dorsata* tends to nest high in the air, usually from 3 to 25 m above ground. In tropical forests in Thailand, many nests are suspended in *Dipterocarpus* trees from 12 to 25 m high: this tree is probably preferred as a relatively safe nesting site because its smooth bark and its trunk rising for 4–5 m before branching out make it difficult of access to terrestrial predators. Nonetheless, about three-quarters of the worker population of a colony of giant honeybees is engaged in colony defense. While birds are common predators of *A. dorsata*, the workers' large body size protects them reasonably well against ant invasion, so that the sticky bands of propolis characterizing the nests of the dwarf honeybee are not found surrounding the nests of *A. dorsata*, nor are the nests hidden by dense foliage. Nests of *A. dorsata* may exist singly or in groups; it is not uncommon to find 10–20 nests in a single tall tree, known locally as a "bee tree." In Southeast Asian countries like Thailand and Vietnam, tree harboring more than 100 nests is occasionally seen in or near the tropical forests.

Rock honeybees scavenge notwithstanding amid moonlit evenings (Diwan and Salvi 1965). Its flight reach is more than 5 km (Koeniger and Vorwohl 1979). In the ordinary rummage conditions, they have been seen to visit sources 2–3 km far from the home. The honeybees have a normal tongue length of 6.683 mm (Ruttner 1988). These elements give a vast rummaging range, both in territory and assortment of plant species to the stone honeybees. A yearly generation of around 2500 tonnes of honeybees' wax from the wild shake honeybees was accounted for in 1969 (Phadke et al. 1969). Rock honeybee is a critical pollinator of a few products.

22.3.3 Eastern Honeybee

Apis cerana or the Asian bumble bee (or the eastern bumble bee) are little bumble bees of Southern and Southeastern Asia. In the wild, the oriental bumble bees build their different brush homes in dim-walled areas, for example, caverns, rock pits, and empty tree trunks. The typical settling site is, as a rule, near the ground, not more than 4–5 m high. The honeybees' propensity for settling oblivious empowers man to keep them in uniquely built vessels, and for a great many years, *A. cerana* has been kept in various types of hives, i.e., dirt pots, logs, boxes, divider openings, and so

Table 22.1 Comparative morphometric, behavioral, and economic characteristics of *A. mellifera* and *A. cerana*

Characteristics	*A. mellifera*	*A. cerana*
Body weight (mg)	90–120	50–70
Tongue length (mm)	5.7–7.2	4.39–5.53
Nectar load (mg)	40–80	30–40
Pollen load (mg)	12–29	7–14
Flight range (km)	2–5	0.8–2
Egg laying capacity of queen per day	800–1800	300–800
Colony buildup at honey flow	40,000–60,000	25,000–30,000
Swarming	Little	High tendency
Absconding	Very little	Very high tendency
Aggressiveness	Usually calm	Mostly furious
Yield under Indian conditions (kg/colony)	25–30	4–5

Source: Chahal et al. (1995)

on. In spite of the moderately late presentation of mobile edge hives, provinces of *A. cerana* kept in conventional hives are still a typical sight in the towns of most Asian nations. Subsequently, the wild homes of the oriental bumble bee in tropical Asia support less setbacks in being chased by man than those of the smaller person and goliath bumble bees. The honeybee dwelling in an extensive variety of topographical territories from tropical beachfront to the mild Himalayan regions reaches at around 3000 m. Provinces are found in backwoods or agrarian regions in the fields and even in urban regions with great vegetation. The nest consists of several parallel combs with a uniform distance between them. The nests have usually six to eight combs. A wide variation occurs in number depending upon the period of stay of the nest in the location, space available in the nest site, and its shape. Sometimes only three to four combs that are narrow but about a meter long are found. In natural nests that lived for over 2 years, up to 15 normal-sized combs were found that yielded over 10 kg honey (Table 22.1).

The Indian hive bee does not use propolis as the European bees do. This may be an adaptation to tropical climate, where hive ventilation assumes importance. The cracks in the floorboard or gaps in the hive walls or frame joints are not sealed. This may attract pests like wax moth. One of the characteristic features of the hive bee is fanning used for ventilation of the hive. During nectar flow, large quantities of water have to be removed from the dilute honey in combs, before it ripens. In their experiments on introducing queens of the European bee into the colonies of the Indian bee, Dhaliwal and Atwal (1970) observed the workers of both the species fanning side by side, but heads oriented in opposite directions.

A. cerana has some of the special qualities such as:

(a) It is gentle to handle, industrious, and well adapted to the ecological conditions of South and Southeast Asia.
(b) It is less susceptible than *A. mellifera* to nosema disease, is not seriously affected by *Varroa*, and is less prone to the attack of predatory wasps.
(c) It can effectively defend against diseases, parasites, and predators and does not require chemical treatments as compared to *A. mellifera*.
(d) Its variety of geographical races/populations that exists in South and Southeast Asia provides excellent opportunities for the genetic improvement of this native species through selective breeding.
(e) Through genetic engineering techniques, it may be possible to introduce desirable genes from *A. cerana* into *A. mellifera*.
(f) It is sympatric in distribution and coexists with the two other species of Asiatic honeybees, *A. dorsata* and *A. florea*, without any adverse ecological consequences.
(g) For pollination purposes, it is superior to *A. mellifera* as it is more suitable for cross-pollinating entomophilous crops grown in the small holdings because of shorter flight range and longer foraging hours.

Use of bee hives for pollination of agricultural and horticultural crops is another field gaining importance in recent years. There is an increasing demand for bee colonies by orchardists growing apples, litchi, and citrus fruits; by producers of cucumber and other cucurbits, seeds, cole crops, spices, onion and vegetable crops, and flower seeds; and also by farmers growing sunflower. Payment by the farmer to the beekeeper for pollination service is also becoming a common practice. Bee colonies migrated to farm and orchard areas, to tide over adverse periods, can be utilized for crop pollination. Such migrations can be doubly beneficial to the beekeeper.

22.3.4 The European Honeybee

The Western honeybee or European honeybee (*Apis mellifera*, Linnaeus, 1758) is a species of honeybee comprised of several subspecies or races. European bees were successfully introduced in India in 1965. The European bee is similar to the Indian hive bee in biology, nesting, foraging, colony defense, and other behavior features, with minor differences. According to Chahal et al. (1995), the European bee possesses definite superiority over the indigenous bee in Punjab. It can colonize areas where the indigenous bee is not present or cannot do well. It can yield four to five times as much honey as the Indian bee (Table 22.1). However, the performance of *A. mellifera* may not be the same in other agroclimatic conditions. For example, *A. mellifera* doing well in Punjab may not survive the new climatic and vegetation conditions in other regions (Table 22.1).

22.3.5 Stingless Bees

Normally known as dammar honeybees (*Trigona* spp.), these are very little in size and look like little mosquitoes or flies. They are circulated in tropics, subtropics, and mild locales. The honeybees construct homes in dull fenced in areas like pits in branches or trunks of trees, ant colonies, termite burrows in the ground, divider fissure, or any deserted repository like logs, pots, and tins. The homes of *Trigona* are bunches of little, uniform, globular cells of wax. These pots are the cells in which the young are raised. The pots are firmly stacked touching each other or isolated; every cell or group of cells being associated with others by braces or mainstays of wax. Dust and nectar are put away in prominently huge oval cells built near the brood cell groups or at their outskirts well separated from them. The brood cells of *Trigona* [*Tetragona* (now called *Tetragonula*) *iridipennis* Smith] from Castle Rock, Karnataka (Kshirsagar and Chauhan 1977), were overall 3.12 mm expansive and 4.09 mm high. The nectar pots were around 6.64 mm wide and about the same in stature. The dust cells were like nectar cells in expansiveness yet were 6.78 mm in stature. In the wake of selecting a reasonable settling site, another swarm shuts all splits, openings, and hole in the home with paste like sticky propolis. The home is connected with the outside air just through a little opening. The opening is regularly made into a passage utilizing wax and propolis. The passage opening is closed by propolis each night after all the outside work is finished for the day. It is opened each morning empowering the foragers to continue their field obligations. Taking care of the dammar honeybees is moderately simple. Not at all like the genuine bumble bees, these honeybees don't sting as the resistance component, yet built up a similarly viable gnawing conduct, in shielding their home. At the point of exasperation, the honeybees assault the adversary in huge numbers, more often than not selecting delicate organs like eyes, the nose, and ears as their objective. Gnawing with the mandibles is very bothering for a few of the targets of dammar honeybees yet can be effortlessly ensured against by man.

Honey of the dammar bee is dark amber in color and shows a highly positive polarization. In the five samples of dammar bee, honeys from Castle Rock, Karnataka, and Indore, Madhya Pradesh (Phadke (1968)), were found containing levulose about 32 % and dextrose only 20 %. However, dextrins were nearly 6 %. The proportion of ash was also quite high, indicating a high mineral content. Neto (1949) emphasizing the need for taxonomic work on the dammar bees reports the remarks of H. S. Schwarz, a world specialist, thus "the number …. (of the species of the dammar bees) … when the size of India is taken into consideration is notably small." In Malaysia 29 species of Meliponinae (stingless bees) were recorded. India which has a larger geographic area and variety of vegetations and climates should have a larger number of species. Rasmussen (2013) in a recent study on the diversity and taxonomy of stingless bees of Indian subcontinent reported nine species of stingless bees which include *Lepidotrigona arcifera* (Cockerell), *Lisotrigona cacciae* (Nurse), *Lisotrigona mohandasi* Jobiraj and Narendran, *Tetragonula* aff. *laeviceps* (Smith), *Tetragonula bengalensis* (Cameron), *Tetragonula gressitti* (Sakagami), *Tetragonula iridipennis* (Smith), *Tetragonula praeterita* (Walker), and *Tetragonula ruficornis* (Smith). Vijayakumar and Jeyaraaj (2014) also reported *Tetragonula iridipennis* belonging to *Tetragonula* subgenus in Nellithurai Village, Tamil Nadu, India.

Pugh (1947) from Assam gave a short account of the species of stingless bees. He distinguished these as types of *Melipona*. These are most likely *Trigona*, since no *Melipona* is known to exist in India. A fascinating element of all the three species is their similitude in size and appearance to the diminutive person honeybee. Ngap siwor, recognized by Pugh as *Melipona tunneli*, is unique in relation to the next two in having yellowish spots on the stomach area. *Melipona khasiana*, privately known as ngap hamang, is somewhat littler than *M. tunneli* and does not frame the long passage at the home passageway. Yield from these honeybees from a couple of accumulations is around one kilogram. Both these species are kept in earthen pots set in the open cells of houses in towns. Ngap khyndew, *M. terrestris*, is the third species that lives in underground depressions made by termites. Nectar is likewise gathered a few times from their homes and somewhat over a kilogram of nectar is got in a year. Dammar honeybees are tough and work in harsh climate conditions. Disregarding their short scrounging range, they can misuse a few little blossomed plants like weeds and create nectar. They can be kept in any container that gives satisfactory space to home building. They withstand unpleasant circumstance, taking care of and repairing rapidly the home that is harmed. They recoup quickly after unfavorable conditions that drain the state quality. They require little care by the beekeeper. The settlements can be separated by basic division of the brood group. The part of the bunch without the ruler raises promptly another ruler and an undeniable new province gets set up. These components make dammar honeybees perfect for keeping for nectar generation, or for fertilization of agrarian and agricultural crops that have little blossoms, and can't be helpful to other bigger honeybees. Dammar honeybee nectar is much sought after for its high therapeutic worth.

Stingless honeybees assume an incredible part in the economy of the nation. Notwithstanding when a few types of *Apis* are utilized for nectar generation, the stingless honeybees can be kept by ranchers for nectar creation and for enhanced yields of their harvests. The honeybees require little consideration. They are safe, not at all like the bumble bees, whose sting is entirely excruciating. The hive for stingless honeybees is very straightforward and does not request any incredible craftsmanship. Consequently at any rate, raising of stingless honeybees should be supported.

22.4 Beekeeping Industry

Honey industry involves honeybee, flowering plants that provide food to the bees, and beekeepers who manipulate bees according to the climate and vegetation for their own benefit. With the evolution and advancement of human societies, the association between man and honeybees gradually changed from one of hunting and killing to that of an organized industry, in which honeybees were nurtured and made to produce honey and other valuable, nutritive, medicinal, and industrial materials (Crane 1990).

Nectar is the sweet fluid delivered by bumble bees from nectars and other sweet substances on plants by expanding their digestive proteins and by aging the resultant blend. Nectar is along these lines a result of collaboration of bumble bees, the exceedingly propelled creepy crawly gatherers, and plants. Honeybees get nectar, the sugar sustenance from blossoms, and utilize it for their vitality necessity. Honeybees additionally gather dust grains, the male conceptive units of blossoming plants that comprise of life-supporting proteins and amino acids. Dust gives the honeybees all prerequisites of proteins for egg creation, larval development, and general quality of the apiary. Honeybees rely on plants for their presence and development. Nectar industry includes bumble bees, blooming plants, and man.

Wild bee states have been abused for gathering of nectar in India and in a couple of other Southeast Asian nations. The normal bumble bee settlers that have been struck are those of the monster bumble bee, *A. dorsata*, and the oriental hive honeybee, *A. cerana*. Tribal populaces and woodland tenants in a few sections of India have been nectar gathering from wild honey bee homes as their conventional calling. The wild honey bee has been the regular and real wellspring of nectar. Maybe as a result of this, most references in antiquated Indian writing, likewise in Ayurveda, identify this honeybee. The strategies for gathering of nectar and beeswax from these homes have changed just marginally throughout the centuries. Planned insurance of honeybee homes in stamped tree depressions and occasional collecting of nectar was the initial phase in beekeeping. The following stride may have been effective to draw in honeybees into logs that had been emptied out, so that a cavity was shaped to oblige the apiary, and to keep such log hives in secured honeybee gardens.

Swarms of bees were attracted to receptacles made of clay, logs of wood, or niches in the outer walls of dwellings. At the end of the flowering seasons, honey was harvested by driving away the bees with smoke and squeezing out honey from the combs. No effort was made to prevent bees from leaving the hive in the process. In some remote inaccessible forest areas, beekeeping may have changed but little from that practiced centuries ago.

Among the hive items are nectar, beeswax, honeybee-gathered dust, illustrious jam, honeybee venom, and propolis. Today a few nations in Asia, Africa, the USA, and Australia have very much created apicultural industry with an extensive variety of honeybee items including dietary, helpful, and restorative arrangements from honeybee-gathered dust, regal jam, propolis, and honeybee venom, other than nectar and beeswax.

Apiaries neither interest extra land space nor do they contend with farming or creature cultivation for any information. The beekeeper needs just to save a couple of hours in a week to take care of his honeybee provinces. Beekeeping is along these lines preferably suited to him as low-maintenance occupation. Beekeeping constitutes an asset of manageable pay era to the country and tribal agriculturists. It gives them important nourishment as nectar, protein-rich dust, and brood. Honeybee items additionally constitute critical elements of society and conventional drug. Apiculture has been connected with farming and agriculture and truth be told contributes, not inconsequential, to the environmental equalization of the area including the neighborhood developed and regular verdure.

More recently, beekeeping industry in India has got major blow due to spread of *Varroa destructor* which has annihilated more than 80 % of the bee stocks of *Apis mellifera* (Abrol 2009a). Honey industry in India is at present in utter chaos. Production of honey has already been dwindling in the past few decades due to indiscriminate deforestation and consequent depletion of sources of food to the honeybees. Similar situation is now found in other Asian and African countries.

Honeybees suffer from various viruses, bacteria, protozoa, parasitic mites, wax moths, predatory Arachnida, wasps, birds, and mammals. In most instances, the appearance of disease is abrupt and instantaneous which plays havoc with honeybee colonies in apiaries.

All the diseases and enemies of honeybees present in other countries have been recorded from India also (Table 22.2) (Singh 1962, 1979; Chhuneja et al. 2002; Abrol 1997, 2007, 2009b; Abrol and Kakroo 2003; Abrol and Sharma 2007, 2009). The two viral diseases *Apis iridescent virus* and *Thai sacbrood virus* have played havoc with Indian honeybee *Apis cerana* during the last decade.

Table 22.2 Brood and adult diseases of honeybees, their causative agents, and occurrence in India

Disease	Causative agent	Bee species affected	Symptoms/color of the brood	Locations
Brood diseases				
American foulbrood (AFB)	Bacteria, *Bacillus* larvae	*Apis cerana indica*	Dull-white dead brood becoming brown to white	Nainital (Uttar Pradesh)
European foulbrood (EFB)	Bacteria, *Melissococcus plutonius*	*A. cerana indica*	Dull-white dead brood turning yellow to dark brown	Mahabaleshwar Maharashtra Karnataka (around Castel rock near Goa border) Jammu, Punjab, Himachal
Sacbrood disease (SBV)	Virus, *Morator aetatulus*	*A. mellifera*	Grayish or straw colored becoming brown, grayish black, or black or blackhead or blackhead end darker	Not recorded from India so far
Thai sacbrood virus (TSBV) disease	Virus, *Morator aetatulus* (Thai strain)	*A. cerana indica*	Grayish or straw colored becoming brown, grayish black, or black or blackhead end darker	Whole India except Andhra Pradesh and Orissa

(continued)

Table 22.2 (continued)

Disease	Causative agent	Bee species affected	Symptoms/color of the brood	Locations
Chalkbrood disease (CB)	Fungus, *Ascosphaera apis*	*A. cerana indica*	White chalklike mass sometimes referred to as mummy	Not recorded from India so far
Store brood (SB)	Fungus, *Aspergillus flavus*	*A. cerana indica* *A. mellifera*	The fungus forms a characteristic whitish-yellow color like ring near the head end of the infected larvae after death. Infected larvae become hardened and difficult to crush, hence called "stone brood"	Not recorded from India so far
Adult diseases				
Nosema disease	Protozoan, *Nosema apis*	*A. cerana*	Shining swollen abdomen	Uttar Pradesh, Himachal Pradesh, Jammu and Kashmir Punjab, Assam Orissa, Nagaland
	Nosema ceranae	*A. mellifera*	Shining swollen abdomen	Not recorded from India so far
Amoeba disease	Protozoan, *Malpighamoeba mellificae*	*A. mellifera*	Cysts in Malpighian tubules	Not recorded from India so far
Bee paralysis	Filterable virus	*A. mellifera*	Black hairless shiny bees	Not recorded from India so far
Septicemia	Bacterium *Pseudomonas apiseptica*	*A. mellifera*	Destruction of connective tissue of legs, wings, and the antennae	Not recorded from India so far
Clustering disease	Iridescent bee virus	*A. cerana indica*	Bees leave combs and form clusters on the wall of hive or outside the hive become sluggish; queen stops egg laying and drawlers appear around the colony	Jammu and Kashmir, Himachal Pradesh, Punjab Maharashtra

Abrol (2009a, b)

22.5 Brood Diseases

A number of mite species have also been reported associated with honeybees in India which include parasitic mites, phoretic mites, and stored provision mites (Abrol and Kakroo 1997, Table 22.3). Of all these mites, *Varroa destructor* is most serious which has wiped out more than 90 % of the stocks wherever it was detected. The other predators and enemies which attack honeybees are listed in Table 22.4. Of the various pests and predators, wax moths and wasps cause considerable damage in the Orient and Africa.

Studies clearly establish that beekeeping industry is at great risk due to the attack of several diseases and enemies which call for regular monitoring to manage these diseases well in time to save the beekeeping industry for food security. European foulbrood brings on microorganism *Melissococcus pluton* which has been changed to *Melissococcus plutonius*. Another strain of Nosema sickness *Nosema ceranae* (microsporidian protozoan) that causes an ailment of the eastern honey bee, *Apis cerana*, has been recognized connected with spring diminishing, "vanishing infection, fall breakdown, honey bee elimination disorder, or Province Collapse Disorder (CCD)" in *A. mellifera*. Obviously this pathogen has hopped host from *Apis cerana* to *Apis mellifera* in the past 10 years and is spreading quickly. *Nosema ceranae* has a pathology that is not quite the same as *Nosema apis*. The causative specialist for Nosema in the USA is *Nosema apis*, which seldom causes real misfortunes in contaminated provinces. Nonetheless, high death rate exhibits that *N. ceranae* is exceptionally pathogenic to *Apis mellifera*. Honeybees succumb inside eight days after introduction to *N. ceranae* which is quicker than honeybees presented to *N. apis*. Some new infections and microorganisms have been found. Evidently, regular

Table 22.3 Occurrence of various species of mites associated with honeybees in India

Mite species	Bee host	Occurrence in India
Acarapis woodi (endoparasitic mite)	*A. cerana*	Present
Varroa jacobsoni (ectoparasitic mite)	*A. cerana*	Present
Varroa destructor (ectoparasitic mite)	*A. mellifera*	Present
Euvarroa sinhai (ectoparasitic mite)	*A. florea*	Present
Tropilaelaps clareae (ectoparasitic mite)	*A. cerana*	Present
	A. mellifera	
	A. dorsata	
Tropilaelaps clareae (ectoparasitic mite)	*A. cerana*	Present
	A. mellifera	
	A. dorsata	
Tyrophagus longior (provision mite)	*Apis cerana*	Present
	A.mellifera	
Neocypholaelaps indica (phoretic mite)	*A.cerana*	Present
	A.mellifera	
	A.dorsata	

Abrol (2009a, b)

Table 22.4 Occurrence of various pests, predators and enemies of honeybees in India

Category enemy/ animal	Disease/enemy	Causative agent	Bee species affected
Moths	*Achroia grisella*	Lesser wax moth	*A.cerana*
	Galleria mellonella	Greater wax moth	*A.florea*
			A.cerana
			A.florea, *A. mellifera*
			A.dorsata
	Acherontia styx	Hawk moth	*A.cerana*
			A.mellifera
Pseudoscorpion	*Ellingsenius indicus*	False scorpion	*A.cerana*
			A. mellifera
Wasps	*Vespa orientalis*, *V. cincta*, *V. velutina*, *V. magnifica*, *V. mandarinia*, *V. tropica*, *V. ducalis*, *V. analis*, *V. asiatica*, *V. austriaca Vespa auraria*, *V. basalis*, *V. nursei*, *V. flaviceps*, *V. structur*, *Vespula vulgaris*, *V. germanica*	All are predatory wasps	All the honeybee species
Birds	*Merops orientalis*	Predatory bird	All the honeybee species
	Merops apiaster	Predatory bird	All the honeybee species
	M. superciliosus	Predatory bird	All the honeybee species

Abrol (2009a, b)

surveillance of honeybee diseases is needed to save the beekeeping industry so that diagnosis of diseases is initiated well in time before any catastrophe occurs because care and management of diseases is easier in early stages of infection or attack (Abrol 2001).

22.5.1 World Honey Production Scenario

China is the biggest maker (40 %) and exporter (35 %) of nectar on the planet followed by Mexico, Argentina, Ukraine and Mexico supplying 20 for each penny though and Argentina supplying 15–20 for every penny. Germany is the world's biggest purchaser, importing 90,000 tonnes of nectar items every year with per capita utilization 1.5 kg/individual when contrasted with India 3–5 g (Table 22.5).

In India honey production is about 70,000 tonnes/year and exports 25,000–27,000 tonnes/year to >42 countries, e.g., the European Union, the Middle East, and

Table 22.5 Comparative honey consumption pattern in different countries of the world

Sl. No.	Country	Honey consumption per person per year
1	India	3–5 g
2	Greece	1.8 kg
3	Australia	1.6 kg
4	Germany	1.5–2.0 kg
5	EU countries like Italy, Spain, France, and Hungary	0.6–0.9 kg
6	UK	0.4 kg
7	China	100 g per year
8	Russia	500 g per year
9	Serbia, Ukraine	600 g per year
10	USA	800–1000 g per year
11	Japan	300 g per year
12	Poland	400 g per year
13	Switzerland	2.0 kg
14	Nepal	3 g per year
15	Turkey	850 g
16	Croatia	400 g
17	Netherlands	300–400 g
18	Indonesia	300 g
19	Bulgaria	400–500 g
20	Australia	1.6 kg

the USA. The data in Table 22.6 and Fig. 22.2 show that China is the largest exporter of honey in the world accounting for 11.8 % of the world share as compared to India which accounts for 3.7 % only. Argentina is the largest importer of honey with 18.5 % share of world honey imports (Table 22.7). The honey industry in India is still far behind to achieve its goal as India has also to import 15.8 % of the honey from other countries.

In India major honey-producing states include Punjab, Haryana, Uttar Pradesh, Bihar, and West Bengal. The foreign exchange earned amounts to Rs 200 crore (2002–2003).

Furthermore in India, the sale price of honey is very less from Rs 25 to Rs 45 per kg compared to Rs 55 to Rs 80 a kg in the USA, Argentina, and Brazil. Organized collection of rock bee and other forest honey and beeswax using improved methods can result in an additional production of at least 50,000 tonnes of honey and 5000 tonnes of beeswax. This can generate income to about half a million tribal families.

22.6 Declining Bee Populations

The population of honeybees and, what's more, different species of pollinators that fertilize plants is declining at a disturbing rate which has debilitated the presence of vegetation, and this descending pattern could harm many financially critical

Table 22.6 Top exporters of honey in the world

Top exporters	2009 value	Share (%)
China	$284,064,882	11.80
Germany	$212,519,898	8.90
Mexico	$164,486,793	6.90
Brazil	$134,944,059	5.60
New Zealand	$117,387,647	4.90
Spain	$116,734,881	4.90
Hungary	$114,880,765	4.80
India	$87,560,291	3.70
Canada	$81,491,036	3.40
Vietnam	$74,327,554	3.10
All others	–	42 %

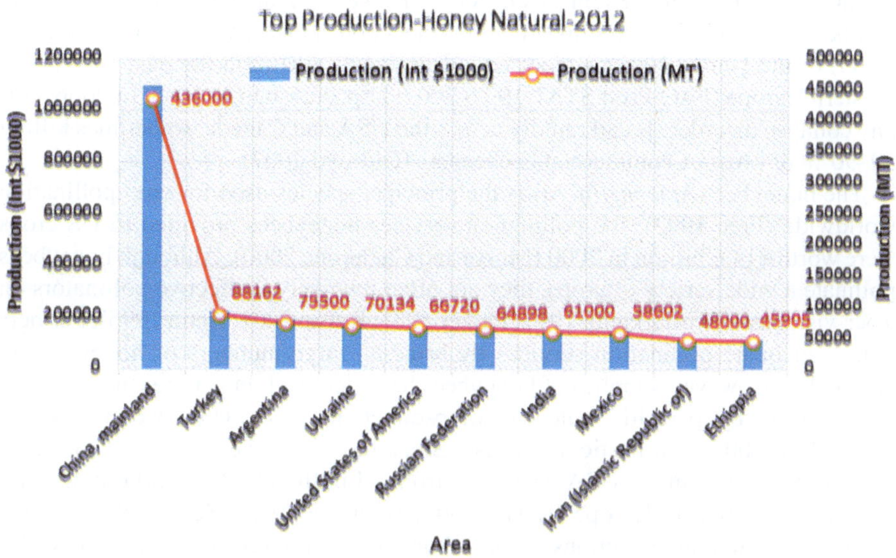

Fig. 22.2 Top production of honey during 2012 (Source: http://www.beekeeping.com)

products. A decrease in pollinator populaces is one type of worldwide change that really can possibly modify the shape and structure of physical and biological communities. The decrease in pollinator populace and differences presents genuine risk to farming generation and protection and upkeep of biodiversity of florea in numerous parts of the world. One pointer of the decrease in common arthropod pollinators is diminishing product yields and quality regardless of essential agronomic inputs. Insect pollination is essential for production of good-quality marketable fruits. Though a number of species of bees are managed for pollination (Strickler and Cane 2003), the honeybees (*Apis mellifera* L.) are the most important

Table 22.7 Major importers of honey in the world

Imports from country	2010 value	Share (%)
Argentina	$54,227,583	18.50
Vietnam	$46,946,545	16.00
India	$46,356,898	15.80
Canada	$39,729,080	13.60
Brazil	$28,958,207	9.90
Malaysia	$24,915,368	8.50
Indonesia	$13,653,764	4.70
Mexico	$10,012,673	3.40
New Zealand	$4,185,579	1.40
Thailand	$3,866,587	1.30
All others	–	6.90

pollinators (Free 1993; Klein et al. 2007). However, managed honeybee colony numbers are declining in some regions; the Food and Agriculture Organization (FAO) of the United Nations reported a decrease of >40 % in the past 10 years in Western Europe (FAO Prod STAT 1997–2007; http://faostat.fao.org/). In 2006, colony collapse disorder spread rapidly across the USA and Canada, sometimes killing 50–90 % of hives of commercial beekeepers (Oldroyd 2007).

The honeybee (*Apis mellifera*) is the principal species used for crop pollination worldwide (Free 1993); the pollination services honeybees provided to US crops were worth $14.6 billion in 2000 (Morse and Calderone 2000). Although honeybees pollinate a wide variety of crops, they are often relatively ineffective pollinators on a per-visit basis (Westerkamp 1991; Parker 1981; Kevan 2001; Batra 1995). Farmers obtain adequate pollination services by bringing large numbers of honeybees to crop fields. However, supplies of honeybees have declined, in part because of problems caused by parasitic mites (Thompson et al. 2003; Underwood and van Engelsdorp 2007) and pesticide misuse. Since the 1970s the number of managed honeybee colonies in the USA decreased from >4 million to 2.41 million; declines were also reported in Europe (Kevan 2001; Williams 2002). Several documented examples show that reductions in bee abundance can cause reduced crop yields (Kevan 1977; Ricketts et al. 2004).

22.6.1 Pollinator Decline

In India, 50 million hectares is under entomophilous crops, cross-pollinated by different abiotic and biotic agents. Considering the fact that there are indications of differing pollinator decline impacts between developing and developed world, the same requires closer scrutiny for developing countries like India, Bangladesh, Sri Lanka, the Philippines, Vietnam, and others which have undergone over four decades of pesticide-dependent agriculture and are undergoing changes in land use due to urbanization and industrialization. Evidently, the deliberate misuse of pesticides has posed a major threat to pollinators and pesticide-induced declines in bee abundance in many countries of the world (Johansen 1977). The decline in pollinator population and diversity presents a serious threat to agricultural production and conservation and maintenance of biodiversity in several parts of the world (Buchmann and Nabhan 1996; Johansen and Mayer 1990). One indicator of the decline in natural insect pollinators is decreasing crop yields and quality despite necessary agronomic inputs. Examples can be found in Himachal Pradesh in northwest India, where despite all agronomic inputs, production and quality of fruit crops such as apples, almonds, cherries, and pears is declining. Another example is the declining rice yields in the Philippines, Vietnam, Indonesia, and other Southeast Asian countries. Extreme negative impact of declining pollinator populations can be seen in other areas, for example, several farmers in Jammu and Kashmir were disappointed with lower yields and quality of apples as a result of poor pollination of apple trees (Partap 2003).

In spite of its monetary value, biodiversity, the Asian hive honeybee, *A. cerana* is enduring sharp decrease and is undermined with elimination in its whole local environment. For instance, in Japan, beekeeping with this local honeybee species has been totally supplanted by European bumble bee, *A. mellifera*, and just a couple of beekeepers and exploration establishments are raising *A. cerana* (Sakai 1992). In China, out of more than 8.5 million settlements of honeybees kept in current hives, 70 % are intriguing to *A. mellifera* (Zhen-Ming et al. 1992). Additionally, in South Korea, just 16 % beekeeping is with local *A. cerana*, and the remaining has been supplanted by extraordinary *A. mellifera* (Choi 1984). This pollinator emergency appears to be more intense in scenes ruled by yearly products (e.g., grains and oil seeds) as these seriously oversaw and most exceptionally bothered monocultures don't give natural chances to useful arthropods. Large amounts of unsettling hamper the foundation of pollinator populaces, and this could diminish harvests, for example, of canola, flax, safflower, sunflower, tomatoes, peppers, strawberries, and cucurbits. Some cole crops require bug fertilization for seed generation. Steps required to defeat the issue of pollinator decay are:

(a) Teach the general population on the significance of pollinators.
(b) Bring issues to light of the fertilization emergency.
(c) Prepare the up-and-coming era of specialists and taxonomists.
(d) Bolster national arrangements for the preservation of honeybees and expansion of the consciousness of governments, industry, and people in general.
(e) Embrace research on option pollinator.
(f) Agriculturists have uncovered that more than 90 % of the ranchers had no information of fertilization.
(g) There is a requirement for national approach on pollinators. Fertilization has not just been underplayed by the organizers, government powers, and the agriculturists who have overlooked through and through.
(h) There is requirement for zonation for European and Indian hive honeybees.
(i) There is a requirement for protection of pollinators (Allsopp et al. 2008) by giving settling destinations, great rummage, and toxic substance-free environment.

Similarly, managing flowering weeds at desirable levels to provide alternative forage to pollinators within crop fields has been a totally neglected habitat management tactic for encouraging pollinators. Many advances, however, have been achieved in biological control where entomologists and agroecologists continually manipulate weeds and other floral diversity to enhance predators and parasitoids of pests (Altieri and Nicholls 2004). The same principles that apply in conservation biological control can be applied to enhance pollinator services, thus simultaneously achieving plant protection and pollination.

22.6.2 Pesticides

Pesticides are also responsible for pollinator population decline. Neonicotinoids are highly likely to be responsible for triggering "colony collapse disorder" in honeybee hives that were healthy prior to the arrival of winter. The loss of honeybees in many countries in the last decade has caused widespread concern because about three-quarters of the world's food crops require pollination. In December 2013, the EU banned the use of three neonicotinoids for 2 years. Many believe that the increasing use of chemical pesticides and herbicides, which honeybees ingest during their daily pollination rounds, is largely to blame. Commercial beehives are also subjected to direct chemical fumigation at regular intervals to ward off destructive mites. Another leading suspect is genetically modified crops, which may generate pollen with compromised nutritional value.

It may be that the buildup of both synthetic chemicals and genetically modified crop pollen has reached a "tipping point," stressing bee populations to the point of collapse.

22.6.3 Radiation

Bee populations seem to be vulnerable to many other factors, such as the recent increase in atmospheric electromagnetic radiation as a result of increased use of cell phones and wireless communication towers. The nonionizing radiation given off by such devices may interfere with bees' ability to navigate. A small study at Germany's Landau University found that bees would not return to their hives when mobile phones were placed nearby. Further research is currently underway in the USA to determine the extent of such radiation-related phenomena on bees and other insect populations.

22.6.4 Global Warming

Biologists also wonder if global warming may be exaggerating the growth rates of pathogens such as the mites, viruses, and fungi that are known to take their toll on bee colonies. The unusual hot-and-cold winter weather fluctuations in recent years, also blamed on global warming, may also be wreaking havoc on bee populations accustomed to more consistent seasonal weather patterns.

22.7 Future Action Plan

22.7.1 At Industry Level

22.7.1.1 Extension

1. Manufacturing, stocking, and distributing of bee boxes, honey extractors, and other beekeeping equipment
2. Qualitative improvement in worked out areas
3. Survey of new potential areas
4. Establishment of demonstration cum training centers
5. Establishment of bee nursery units
6. Registration of beekeepers

22.7.1.2 Research

1. Standardization of beekeeping equipment.
2. Standardization of bee management methods for different zones of India.
3. Protection of bees from harmful chemicals like insecticides, miticides and fungicides.
4. Maintenance and conservation of all the species of honeybees and assessing their utility as pollinators.
5. Improving the strains of honeybees by selection from local stocks, imbibing in them disease resistance and high honey-yielding qualities.
6. Assessment of crops and other flowering plants for their degree of utility to honeybees.
7. Preparation list of multiple tree species region wise having staggered flowering and useful to honeybees during their flowering.
8. Coordinating with agroforestry, farm forestry, and social forestry department for introduction of multipurpose plant species useful to honeybees.
9. Quarantine organizations may be developed.
10. Standardization and preservation of other bee products may be taken up.

22.7.1.3 Training and Education

1. Syllabi of schools and colleges must include topics on apiculture as graded course.
2. Trainings must be organized for unemployed youths and hobbyists.

22.7.1.4 Marketing

1. Marketing of other bee products like beeswax, propolis, bee-collected pollen, etc., to be encouraged.
2. Quality control of bee products should be assured.
3. Export market for bee products be explored.
4. Marketing organization and marketing channels may be established to avoid excessive stocking of honey in any region.

22.7.1.5 At Pollinator End

1. Biodiversity, characterization and taxonomy, multiplication, conservation, and domiciliation of pollinators (Abrol 1993).
2. Protocol for multiplication of pollen vectors
3. Prepare bee maps in the country with respect to the distribution and availability of different bee species in different crops, i.e., scenario or status of bees in time and space.
4. A database on pollinators, their abundance, behavior, floral rewards, and pollinator-plant interaction needs to be generated.
5. The role of honeybees in crop pollination needs to be given due importance.
6. Non-*Apis* bees, pollinators such as stingless bees, carpenter bees, bumble bees, megachilids, and many other bees need to be promoted for crop pollination.
7. In addition to cross-pollination, the role of insects in self-pollinated crops needs to be worked out.
8. Role of insects in increasing crop yields qualitatively and quantitatively.
9. Use of pollinators for seed/hybrid seed production using CMS lines and its economics.
10. Isolation distances for pure-seed production assume greater significance while using pollinators for seed production to reduce contamination and increase the purity of the seed.
11. Use of pollinators for garden and greenhouse crops.
12. Bee lure, attractants, and repellents to be used to increase the attractiveness to the crop or repel the bees from insecticidal treated crop.
13. Safety of pollinators from pesticides, diseases, parasites, and predators and from adverse climatic conditions.
14. Fast and safe multiplication/domiciliation of pollinators both for greenhouse and open fields.
15. Need to extend technology through demonstration, farmers participatory trials, and human resource development (HRD) by organizing on different aspects of insect pollinators for scientists.
16. Need for development of infrastructural facilities.

22.7.1.6 Biotechnological Interventions

1. Biotechnology-related research with honeybees is being conducted at US Department of Agriculture (USDA), USA. The goal of this research is to develop the technology to locate, manipulate, and transfer genes in honeybees. The first gene to transfer will be the one which can be identified easily so that verification of a successful transfer is simplified. Once a reliable transfer method is developed, other genes that will result in an improved stock can be transformed.
2. One class of genes that is being evaluated as the most promising for honeybees is the genes coding for proteins known as cecropins. These genes are from insects and produce proteins that have very strong bactericidal and fungicidal effects, even at low concentrations. Thus, by transferring a single gene into

honeybees, resistance to American foulbrood (AFB), European foulbrood (EFB), and chalkbrood might all be obtained simultaneously. It is possible that honeybees contain cecropins, but in some cases, they are either induced or turned on late when the larva is already infected and fails to provide acceptable level of resistance. Therefore, by examining honeybee DNA to determine the sequence of cecropin proteins, it might be possible to modify the expression of this gene in honeybee larvae so that the cecropin protein is present to respond to the challenge of various diseases equally in *Apis mellifera* and *Apis cerana*.

3. Another area of biotechnology-related research emerging is that of resistance to mites. *A. cerana* is the natural host of *Varroa jacobsoni*, but in some ways, it is resistant to the effects of this mite to keep the levels very low. As an alternative to chemical control methods, it would be desirable to have the same résistance in *A. mellifera* as is present in *A. cerana*. Since these two species do not hybridize, the use of recombinant DNA methods is the only solution to transfer this resistance.
4. The genetic diversity of *A. mellifera* is well established. In *A. cerana*, only three species have been recognized from India. All these subspecies seem to have different geographical populations/subgroups or ecotypes (Verma 1992; Abrol and Khan 2002) which can be used for improvement through selective breeding.
5. Similarly, in *Apis cerana*, the transformation for the cecropin gene is an obvious choice. An important problem for *A. cerana* is a virus disease, Thai sacbrood. A disease caused by virus would not be affected by the cecropin protein. Identification of a gene in *A. mellifera* which conferred resistance to this sacbrood virus would provide a gene which would be important to *A. cerana* and would only be available through recombinant DNA technology.

22.7.1.7 Future Prospects

Honey industry in the country can well become a major foreign exchange earner if international standards are met. Beekeeping is an age-old tradition in India but it is considered a no-investment profit giving venture in most areas. Of late it has been recognized that it has the potential to develop as a prime agri-horticultural- and forest-based industry. Honey production is a lucrative business and it generates employment. The informal sector is providing up to 70 % of the honey and beeswax market in India. Indian honey has a good export market. With the use of modern collection, storage, beekeeping equipment, honey-processing plants, and bottling technologies, the potential export market can be tapped. One problem is the quality of honey production. From a buyer's point of view, quality honey is essential. But India is lacking on that front. There is a need to look specifically at how to promote quality production and develop an export market. For tapping its potential, there is a need to chalk out suitable export strategy. Some of the points which merit attention of the policy makers in this respect include:

Application of advanced technology for collection, and processing of honey, adhering strictly to the quality standards including health regulations, laid down by markets such as the EU, Japan, and the USA.

- Recognition of beekeeping as agro-industry
- Priority allocation and concessions to be made applicable for material needed for beekeeping, like wood for bee boxes, sugar for supplementing feeds to bees, and medicines for bees' diseases
- Campaigning abroad about the quality of our honey
- Developing an efficient export marketing network to optimize the production and exports
- Creating an Indian logo as a joint effort of exporters, APEDA, and the Ministry of Commerce and Industry, government of India. The brand equity thus created can be better marketed for higher sales realization. Timely implementation of the above steps is likely to pave the way for a quantum jump in the export of honey from the country in the coming years.

22.7.1.8 Need for Certification

- Certification is an indicator that the honey-manufacturing company is competent and this has been established through a third-party analysis, done professionally, using the most recent and up-to-date technologies, procedures, and equipments.
- India is becoming an increasingly important supplier of high-quality, mild honey with versatility for the North American market. India has been identified as a potential point of transhipment for Chinese honey en route to the USA.
- Certification will help identify and certify honey that has met the expectations of government, industry, and consumers: Fully traceable, high-quality honey of pure Indian origin.
- Certification would help to preserve the image of Indian honey and that of legitimate Indian exporters and validate the livelihood of approximately 250,000 beekeepers.

References

Abrol DP (1993) Honeybee – virus infection and immunity. Korean J Apic 8(2):116–118

Abrol DP (1997) Bees and beekeeping in India. Kalyani Publishers, Ludhiana, p 450

Abrol DP (2001) Import – export policy of honey bees and other pollinators. Indian Bee J 63(3 & 4):55–58

Abrol DP (2007) Honeybee and rapeseed – a pollinator plant interaction. In: Dr Gupta SK (ed) Advances in botanical research: rapeseed breeding, vol 45. Elsevier, London, pp 339–367

Abrol DP (2009a) Bees and beekeeping in India, 2nd edn. Kalyani Publishers, Ludhiana, p 720

Abrol DP (2009b) Discovery of a bacterial and nosema disease in honeybee *Apis cerana colonies* in Jammu, India. In: 2nd international beekeeping congress August 19–21, 2008, Thimphu, Bhutan

Abrol DP, Kakroo SK (1997) Observations on concurrent parasitism by mites on four honeybees species in India. Trop Agric 74(2):150–156

Abrol DP, Kakroo SK (2003) Pests, predators and pathogens of honeybee *Apis mellifera* L.in Jammu, India. In: Conservation and management of bees for sustainable development and APIEXPO-2003 held at Bangalore University, Bangalore from 13–18, October, 2003

Abrol DP, Khan AA (2002) Characterization of ecotypes of Indian honeybee *Apis cerana* F in the northwest Himalayas. Korean J Apic 17(1):13–20

Abrol DP, Sharma D (2007) Morphogenic and toxic effects of some pesticides on honeybees, *Apis mellifera* L brood. J Res SKUAST J 6(2):144–159

Abrol DP, Sharma D (2009) Honeybee mites and their management, 1st edn. Kalyani Publishers, Ludhiana, p 200

Agrawal TJ (2014) Beekeeping industry in India: future potential. Int J Res Appl Nat Soc Sci 2(7):133–140

Allsopp MH, de Lange WJ, Veldtman R (2008) Valuing insect pollination services with cost of replacement. PLoS One 3:e3128

Altieri MA, Nicholls CI (2004) Biodiversity and pest management in agroecosystems, 2nd edn. The Haworth Press, New York

Barrionuevo A (2007) Bees vanish, and scientists race for reasons. The New York Times. April, 24

Batra SWT (1995) Solitary bees for blueberry pollination. In: Proceedings of the 7th southeastern blueberry conference, Savannah, USA, pp 69–72

Buchmann SL, Nabhan GP (1996) The forgotten pollinators. Island Press, Washington, DC

Chahal BS, Brar HS, Gatoria GS, Jhajj HS (1995) Italian honeybees and their management. Punjab agricultural university, Ludhiana, p 116

Chhuneja PK, Gatoria GS, Aulakh RK, Singh J, Garcha S (2002) Surveillance of *Apis. mellifera* Linn. colonies for bee diseases and enemies in the Punjab (India). Paper presented at 6th International Conference and World Apiexpo-2002, pp 133–134

Choi SY (1984) Brief report on the status of Korean beekeeping. In: Proceedings of the expert consultation beekeep. *Apis mellifera*. Tropical Sub-tropical Asia. FAO, Rome, pp 170–190

Crane E (1990) Bees and beekeeping: science, practice and world resources. Heinemann, Oxford

Crane E (1999) The world history of beekeeping and honey hunting. Routledge, London

Dhaliwal GS, Atwal AS (1970) Interspecific relations between *Apis cerana indica* and *Apis mellifera*. J Apic Res 9:53–59

Diwan VV, Salvi SR (1965) Some interesting behavioural features of *Apis dorsata* Fab. Indian Bee J 27(1):52

FAO Prod STAT (1997–2007) http://faostat.fao.org

Free JB (1993) Insect pollination of crops. Academic, San Diego

Johansen CA (1977) Pesticides and pollination. Annu Rev Entomol 22:177–192

Johansen CA, Mayer DF (1990) Pollinator protection: a bee and pesticide handbook. Wicwas Press, Cheshire

Kevan PG (1977) Blueberry crops in Nova Scotia and New Brunswick – pesticides and crop reductions. Can J Agric Econ 25:61–64

Kevan PG (2001) In: Stubbs CS, Drummond FA (eds) Bees and crop pollination: crisis, crossroads, conservation. Entomological Society of America, Lanham, pp 7–68

Klein AM, Vaissiere BE, Cane JH, Steffan-Dewenter I, Cunningham SA, Kremen C, Tscharntke T (2007) Importance of pollinators in changing landscapes for world crops. Proc R Soc B Biol Sci 274:303–313

Koeniger N (1976) Interspecific competition between *Apis florea* and *Apis mellifera* in the tropics. Bee World 57:110–112

Koeniger N, Vorwohl G (1979) Competition for food among four sympatric species of Apini in Sri Lanka (*Apis dorsata*, *Apis cerana*, and *Trigona iridipennis*). J Apic Res 18:95–109

Kshirsagar KK (1983) Morphometric studies on the Indian hive bee Apis cerana indica F. I-morphometric characters useful in identification of intraspecific taxa. In: Proceedings of the 2nd international conference on apiculture in tropical climates, New Delhi, pp 254–261

Kshirsagar KK, Chauhan RM (1977) Preliminary observations on oviposition behaviour in *Trigona* (*Tetragona*) *iridipennis* Smith. Indian Bee J 39:22–23

Mathews F (2010) Planetary collapse disorder: the honeybee as portent of the limits of the ethical. Environ Ethics 32(4):351–364

Morse RA, Calderone NW (2000) The value of honey bees as pollinators of U.S. crops in 2000. Bee Cult 128:2–15

Neto NP (1949) Notas bionomicas sobre meliponineos. II. Sobre a pilhagem. Papeis Avulsos do depto. De Zoologia da Secr Agr Do Est De São Paulo, pp 13–32

Oldroyd BP (2007) What's killing American honey bees? Publ Libr Sci Biol 5:e168. doi:10.1371/journal.pbio.0050168

Parker FD (1981) How efficient are bees in pollinating sunflowers? J Kansas Entomol Soc 54(1):61–67

Partap U (2003) Cash crop farming in the Himalayas: the importance of pollinator management and managed pollination. Case Study No.:10. In: Biodiversity and the ecosystem approach in agriculture, forestry. http://fao.org///docrep/005/y4586e11.htm

Phadke RP (1968) Studies on Indian honeys. Proximate composition and physico-chemical characteristics of honeys from the wild honeybees, *Apis dorsata*, *Apis florea* and *Trigona*. Indian Bee J 30:3–8

Phadke RP, Nair KS, Nandekar KU (1969) Indian beeswaxes. I Their physico-chemical constants. Indian Bee J 31:52–55

Pugh BM (1947) Varieties of bees found in Assam (Extract form the report of the Assistant Agriculture Officer, North East Frontier Agency, Assam on "Agriculture in the Mynner (Jirang) state, Khasi Hills, Assam"). Ind Bee J 9:62

Rasmussen C (2013) Stingless bees (Hymenoptera: Apidae: Meliponini) of the Indian subcontinent: diversity, taxonomy and current status of knowledge. Zootaxa 3647(3):401–428

Ricketts TH, Daily GC, Ehrlich PR, Michener CD (2004) Economic value of tropical forest to coffee production. Proc Natl Acad Sci U S A 101(34):12579–12582

Ruttner F (1988) Biogeography and taxonomy of honeybees. Springer, Berlin

Sakai T (1992) Apis cerana beekeeping in Japan. In: Verma LR (ed) Honeybees in mountain agriculture. Oxford & IBH Publishing Co. Pvt. Ltd., New Delhi, pp 235–238

Singh S (1962) Beekeeping in India. Indian Council of Agricultural Research, New Delhi

Singh Y (1979) Pollination activity on strawberry at Jeolikote. (Distt. Nainital, India). Indian Bee J 41:17–19

Strickler K, Cane JH (2003) For non-native crops, whence pollinators of the future? Thomas Say Publications in Entomology. Entomological Society of America, Lanham

Suryanarayan MC (2002) Bees of India. http://www.manadasofpune5.homestead.com/BeesIndia.html. Accessed Dec 2002

Thompson H, Ball R, Brown M, Bew M (2003) Varroa destructor resistance to pyrethroid treatments in the United Kingdom. Bull Insectology 56:175–181

Underwood RM, van Engelsdorp D (2007) Colony collapse disorder: have we seen this before? Bee Cult 135:13–15

UNEP (2010) UNEP emerging issues: global honey bee colony disorder and other threats to insect pollinators. United Nations Environment Programme, Nairobi

Verma LR (1992) Honeybees in mountain agriculture. Oxford and IBH Publishing Co., New Delhi, p 274

Vijayakumar K, Jeyaraaj R (2014) Taxonomic notes on *Trigona* (*Tetragonula*) *iridipennis* Smith (Hymenoptera: Apidae) from India. J Threatened Taxa 6(11):6480–6484

Wakhle DM, Pal N (2000) Honey and hive products in India – present status. In: Proceedings of the 7th international conference on tropical bees: management and diversity and 5th AAA conference, held 19–25 March 2000 in Chiang Mai, Thailand. International Bee Research Association, Cardiff

Walker P, Crane E (2000) The history of beekeeping in English Gardens. Garden Hist 28(2):231–261

Westerkamp C (1991) Honeybees are poor pollinators – why? P I Syst Evol 177:71–75

Williams IH (2002) The EU regulatory framework for GM crops in relation to bees. Bee World 83:24–35

Zhen-Ming J, Guanhang Y, Shuangxiu H, Shikui L, Zaijin R (1992) The advancement of apicultural science and technology in China. In: Verma LR (ed) Honeybees in mountain agriculture. Oxford/IBH Publishing Co Pvt Ltd, New Delhi, pp 133–147

Nominating the Bee Trees of Nandagudi/Ramagovindapura as a World Heritage Site

23

Stephen Petersen and Muniswamyreddy Shankar Reddy

Abstract

Nandagudi/Ramagovindapura, a township in the north Bangalore city, holds a high and unique value as a living pollinator exhibit, demonstrating with spectacular force the value and importance of pollinators especially *Apis* spp. and the need for their protection. The road to being nominated on the list of potential World Heritage Sites is a long and arduous journey requiring a significant contribution of time and support. With the proximity to Bangalore University and its concurrent reputation as "excellence in apiculture and agriculture," it offers an opportunity for field research on *Apis dorsata* Fab. in a convenient and proximal location. The authors are requesting resolutions to be passed or letters of support to be submitted by the International Insect Science Congress in favor of establishing the bee trees of Nandagudi/Ramagovindapura as a property for consideration for nomination as a World Heritage Site. Resolutions of support are also being sought from the Indian Pollinator Initiative (IPI), the Asian Apicultural Association (AAA), the Apimondia, the Global Conference on Entomology, other entomological-oriented associations, departments of apiculture and agriculture at national, state, and university levels worldwide, as well as environmental action groups. Individuals interested in contributing to these efforts are urged to contact either or both of the authors.

Keywords

Apis • Pollinator • South India • World Heritage Site

S. Petersen (✉)
Toklat Apiaries, 1153 Donna Dr., Fairbanks, AK 99712, USA
e-mail: akbeeman2000@yahoo.com

M. Shankar Reddy
Department of Zoology, Bangalore University, Jnana Bharathi, Bengaluru 560 056, India
e-mail: jenureddy@gmail.com

A.K. Chakravarthy, S. Sridhara (eds.), *Arthropod Diversity and Conservation in the Tropics and Sub-tropics*, DOI 10.1007/978-981-10-1518-2_23

23.1 Introduction

The giant Asian honey bee (*Apis dorsata*) is distributed across most of Southeast Asia and the Indian subcontinent with geographically isolated subspecies confined to population pockets (e.g., *A. d. breviligula* in the Philippines and *A. d. binghami* on Sulawesi in Indonesia). A closely related species, *Apis laboriosa*, is restricted to cliff nesting in higher elevations across the Himalayas to as far East as Vietnam (Trung et al. 1996). Frequently *dorsata* exhibits aggregate nesting (many colonies nesting in close proximity but not necessarily related) on trees, cliffs, and man-made structures such as water towers, billboards, and tall buildings with scores of nests frequently reported in one location. In an agricultural area, 38 km north of Bangalore city (Karnataka, India), a high concentration of nesting *A. dorsata* (as many as 2000 colonies) in 11 trees within 5–7 km radius (see Table 23.1) is found, setting the record in the concentration of nesting *dorsata* bees.

The rural agricultural setting where the bees are nesting has, in the past, fended off urban sprawl and governmental development plans including a special economic zone (SEZ) initiative and an attempt by the Bangalore Metropolitan Region (BMR)

Table 23.1 Number of bee colonies recorded during the last 15 years (1998–2012) and the income generated through honey harvesting

Year	Number of *Apis dorsata* bee colonies[a]	Number of colonies harvested[b]	Income generation[c](INR)	Remarks
1998	252	70	Rs. 12,000.00	During the study period, farmers report that there has been a tremendous increase in the productivity of agriculture, horticulture, and the forest produce in the study area
1999	310	110	Rs. 21,000.00	
2000	370	150	Rs. 32,000.00	
2001	410	150	Rs. 32,000.00	
2002	432	170	Rs. 34,000.00	
2003	410	160	Rs. 30,000.00	
2004	442	165	Rs. 32,000.00	
2005	475	150	Rs. 30,000.00	
2006	566	160	Rs. 34,000.00	
2007	570	180	Rs. 36,000.00	
2008	625	200	Rs. 38,000.00	
2009	617	*	*With a view to conserving populations of *Apis dorsata*, the harvesting of honey has been stopped	
2010	630	*		
2011	620	*		
2012	622 (average number of nests over 15 years = 424)	*		

[a]Number of colonies counted in one tree only in the village of Ramagovindapura
[b]Average number of colonies harvested from one tree (Ramagovindapura) over 11 years ≈ 148
[c]Average income for village (11 years) = Rs.30, 100 (≈ $548 USD)

to develop a "satellite city" in the region. Although the need for rational expansion of the rapidly growing city of Bangalore cannot be dismissed, little consideration has been given nor input taken from the villagers. The effects of urbanization on the *A. dorsata* bee populations would be understandably negative, primarily because of loss of forage potential.

The authors, with the support of the International Insect Science Congress (IISC), the Apimondia (World Congress on Beekeeping), the Asian Apiculture Association (AAA), the Indian Pollinator Initiative (IPI), and various apicultural, agricultural, and environmental organizations, are hoping to have the area surrounding the bee trees nominated as World Heritage Site (WHS) because of the ecological, natural, and economical (benefits of pollinators) aspects of the site. There is a long road ahead to accomplish this goal (Fig. 23.1).

For more than 15 years, the banyan and peepal (*Ficus benghalensis* and *Ficus religiosa*: Moraceae) trees of Ramagovindapura have hosted hundreds of colonies of the giant Asian honey bee (*Apis dorsata*). The pollinating efforts of these bees have (according to the villagers) increased the crop yields of the surrounding area, their defecation flights have enhanced soil fertility, and they have put the village on the map as a tourist attraction less than 40 km from Bangalore. An opportunity exists to set aside and protect these valuable natural treasures as a **World Heritage Site (WHS)**, a long and burdensome project. At one point in 2008, there were plans to turn the areas into special economic zones slated for development; local beekeepers and ecologists managed to organize and thwart the plan. However, as long as the area exists unprotected, there will be threats such as the new integrated townships in the Bangalore Metropolitan Region proposal which may threaten the area. These plans, which are opposed by the agricultural residents of the Nandagudi township

Fig. 23.1 Right – January, 2010, some of the 630 colonies of *Apis dorsata* nesting in a banyan tree (*Ficus benghalensis*) in the village of Ramagovindapura, near Nandagudi, 36 km from Bangalore on Chintamani Road (Photo by Dr. M. S. Reddy)

and the village of Ramagovindapura, are not in the best interests of local farmers and the bees.

As part of their agenda, the recently formed organization, "Indian Pollinator Initiative" (IPI), has decided to pursue the nomination of this important area for recognition as a World Heritage Site. IPI is eliciting support from entomological societies, apicultural associations, nature groups, and ecology advocates. This paper documents the current status, the imminent threats, and the way forward to nomination.

23.1.1 The City and the Countryside

Bangalore (metropolitan population 8.7 million – 2011; *Wikipedia*) is the fifth most populous metropolitan in India. It is well known as the hub of information technology in India; it is among the top ten preferred entrepreneurial locations in the world (Economic Times of India 2012), and is the second fastest-growing major metropolitan city in India, but the expansion of Bangalore confronts substantial pollution along with other logistical and socioeconomic problems. Not far into the countryside, rural residents farm as they have for many generations growing grains, fruits, and vegetables and are normally self-sufficient. To date the locals have been successful in warding off metropolitan expansion, but as land values, encouraged by speculators, skyrocket, it becomes harder to resist the temptation of a one-time cash sale for immediate gain and perhaps with little regard for the security of future generations.

The villagers are well aware of the pollination efforts and benefits of the honey bees in the area (an attitude to be disseminated); they claim they can get five harvests per year (as opposed to four) all attributable to the presence of the bees. Though not on the "must-see" list of tourists (except perhaps bee enthusiasts) visiting Bangalore, the area has potential as an ecotourism destination, a source for organically produced vegetables and crops, and a quiet break away from the cacophony of Bangalore, all within a short drive from the city center. Designation as a World Heritage Site (WHS) will provide the necessary environmental protection so the area would be ecologically protected for generations to come.

The village of Ramagovindapura is located in the Nandagudi township situated 38 km northwest of Bangalore and is primarily an agricultural zone. There are several granite quarries in the vicinity and a few Hindu temples, and the area is near the locally famous Nandi Hills, a large granitic pluton. Some of the *Apis dorsata* bees are present at the location on a year-round basis, but the majority of the colonies migrate to parts unknown during part of the year. Typically the large majority of the bees begin to arrive in October of each year, the population swells to thousands by late January, honey is sometimes harvested in March, and the bees will depart by April leaving a few colonies behind.

Other *Apis* species existing in the area including *Apis florea* and some managed, as well as feral colonies, of *Apis cerana*. Currently, to our knowledge, there are no colonies of *Apis mellifera* (the imported European honey bee) in the area. If the area was successful in obtaining designation as a WHS, the importation of nonnative *Apis mellifera*, with its potential of disease importations, should be banned.

23.2 Social and Economic Implications of WHS Designation

As this paper is the first step in the long process of obtaining WHS nomination (and future designation), there have not been any formal studies of the economic and social impact that this will have on the potential area; simple and informal conversations have been held with a handful of village leaders – their response, to date, in favor has been 100 % positive.

23.2.1 Agricultural Aspects

Many agriculturists and residents in the Nandagudi/Ramagovindapura area are well aware of the benefits of pollination and the need to protect indigenous species of pollinators. A report from Africa concludes that 20–30 colonies of *Apis dorsata* can deposit 800–1200 kg of nitrogen fertilizer spread in the defecation area per year (Annamalai 2012). If substantiated, this is a significant source of "free" fertilizer, adding to the organic image desired for the area.

One of the project goals which will be proposed is to turn the WHS zone into a model organic farming area with ongoing training sessions for residents on integrated pest management (IPM), organic farming methods, vermiculture, composting, and other bio-friendly techniques. Casual conversations with Bangalore residents indicate a willingness to travel to the area with the assurance of procuring high-quality, organically produced food items. Bangalore, concurrent with its exponential growth, has a high percentage of upper-middle-class residents willing to spend money for quality foods and honey. This will benefit the farmers and village residents by providing good prices for organically grown food items as well as the opportunities for medium to small business enterprises.

23.2.2 Economic Aspects and Opportunities

Beside the enhanced opportunities in the agricultural sector, other economic opportunities exist. These can take the form of agritourism (spend a couple of days on a rural farm in India), homestays, souvenir sales, handicrafts, local tour guide services (encouraging school children to pursue English language and biodiversity studies), sale of services (food and snacks), local transportation (auto, bicycle rentals, horse carts, etc.), and farm gate sales of produce and honey.

23.2.3 Educational Aspects and Opportunities

If nominated and hopefully selected as a WHS, the Nandagudi/Ramagovindapura bee trees will be unique as the world's first honey bee WHS. On the western slope of the Southern Ural Mountain in Russia (Bashkortostan region), a current WHS, the "Bashkir Ural," occupies a territory of about 450 km^2 which encompasses a state entomological wildlife reserve "AltynSolok" (which means "golden bee tree") designated to conserve and insure the traditional methods and techniques of beekeeping with local strains of *Apis mellifera* (Bashkortostan Republic info 2007). Its designation as a WHS includes other cultural and historical factors and thus cannot be considered solely entomological.

The general public is woefully ignorant of the role that honey bees and their pollination efforts play in our lives and food security. By establishing the WHS at Ramagovindapura, it will help raise awareness of the vital role of pollinators and the need to protect them. Information booklets and brochures should be made available as part of the outreach education effort. Schools could have field trips to the area, and it could become a "tourist destination" especially for the environmentally aware tourist segment.

23.3 Steps to Nomination as a WHS (UNESCO 2011)

When the authors first proposed the concept, it seemed like a pretty straightforward task, but after a little checking, we realized that it will be an onerous multiyear project with no guarantee of success. Consulting the web resources, we found several booklets outlining the process, Preparing World Heritage Site nominations manual (138 pages), World Heritage Information Kit (32 pages), Operational Guidelines for the Implementation WHS Convention (175 pages), plus sites offering international assistance – from which we plan to seek assistance. To quote from the WHS nominations manual:

> *Along with increasingly comprehensive requirements, the preparation of nominations has become an important, but rather complex process which requires a good understanding of the various requirements. The participation of local people in the nomination process is also essential to enable them to have a shared responsibility with the State Party in the maintenance of the property, and has to be strongly encouraged.* (UNESCO 2011)

Quoting further from the manual:

> *Lack of preparation time is the biggest enemy of successful nominations. Far too many are prepared against unrealistically short timeframes. It can take at least a year to set up appropriate support mechanisms and gather material, and a further year to write the nomination text and consult stakeholders. When research is needed, protection has to be achieved, and new management systems put in place and documented, so the process might take much longer. If the aim is a successful nomination that leads to inscription on the World Heritage List and long-term conservation and presentation of the property, a realistic timeframe should be allowed. Too often, lack of adequate preparation time leads to*

> *deferred or referred nominations, which is frustrating for States Parties, the World Heritage Committee and the Advisory Bodies.*
>
> *Sometimes political commitments are made which set an unrealistic timeframe for preparing a nomination, resulting in a nomination dossier which is inadequate and not ready for evaluation.* (*Ibid*)

The concept of "outstanding universal value" is paramount and underpins the World Heritage Convention concept. The basic purpose of nominations is to describe what a property consists of, why it demonstrates potential outstanding universal value, and how this value will be sustained, protected, conserved, managed, monitored, and communicated.

23.3.1 Tentative List

The first step a country must take is to make an "inventory" of its important natural and cultural heritage sites located within its boundaries which are considered to be of cultural and/or natural heritage reflecting potential **outstanding universal values** and therefore suitable for inscription on the World Heritage List. It is then placed on the list for the particular country to submit for consideration as a WHS.

By preparing a tentative list and selecting properties from it, a State Party can plan when to present each nomination dossier for a particular property. The World Heritage Centre offers advice and assistance to the State Party in preparing this dossier, which needs to be as comprehensive as possible, making sure the necessary documentation and maps are included. The nomination is submitted to the World Heritage Centre to check that it is complete. Once a nomination file is completed, the World Heritage Centre sends it to the appropriate advisory body/bodies for evaluation.

23.3.2 Short (But Not All Inclusive) Task List

As the process moves forward, there are several baseline tasks which must be done which will form a foundation for the proposal; they include but are not limited to:

- Defining the boundaries of the proposed area and the all-important buffer area
- Participatory rural appraisal discussions with village stakeholders detailing their inputs
- Socioeconomic impact studies
- Negotiations with road commissions to enhance accessibility
- Negotiations with village stakeholders as to their desires for an organic-based agricultural zone and community enforcement (if practical)
- Promotion of the concept and garner of support from organizations, state bodies, and individuals

- Scientific studies including (perfect venue for university student research) migration patterns of the *dorsata*, light range and forage patterns, and value of pollination and fertility contribution
- Any other ideas

23.3.3 Selection Criteria

To be included on the World Heritage List, sites must be of *outstanding universal value* and meet at least one of ten criteria, as well as the relevant conditions of integrity and authenticity and requirements for protection and management.

23.4 The Bee Trees as a Natural World Heritage Site (UNESCO 2011)

The World Heritage Convention defines natural heritage as:

- Natural features consisting of physical and biological formations or groups of such formations, which are of outstanding universal value from the aesthetic or scientific point of view.
- Geological and physiographical formations and precisely delineated areas which constitute the habitat of threatened species of animals and plants of outstanding universal value from the point of view of science or conservation.
- Natural sites or precisely delineated natural areas of outstanding universal value from the point of view of science, conservation, or natural beauty.
- The requirements outlined above are mostly met by Nandagudi/Ramagovindapura rural complex in the opinion of the authors.
- Briefly there are ten criteria for the selection of cultural or natural heritage sites, three of which in the opinion of the authors are definitely met by Nandagudi/Ramagovindapura, and they are as mentioned below:

> *Criterion (vii): contain superlative natural phenomena or areas of exceptional natural beauty and aesthetic importance.*
>
> *Criterion (ix): be outstanding examples representing significant ongoing ecological and biological processes in the evolution and development of terrestrial, fresh water, coastal and marine ecosystems and communities of plants and animals.*
>
> *Criterion (x): contain the most important and significant natural habitats for* in situ *conservation of biological diversity, including those containing threatened species of Outstanding Universal Value from the point of view of science or conservation.* (UNESCO 2011)

References

Annamalai M (2012) Defecation Behavior in Insects. Power point presentation in partial fulfillment of Ph. D. requirement at Tamil Nadu Agricultural University, Dr. M. Muthuraman Chairman, Coimbatore, India, slide # 84 of 86 data attributed to Dr. Garth Cambray. http://www.scienceinafrica.co.za/2007/april/yellowrain.htm. Accessed 24 Dec 2012

Bashkortostan Republic info (2007) Shulgan-Tash National Nature Preserve. http://eng.bashkortostan450.ru/nature/natural-monuments/preserve_132.html. Accessed 24 Dec 2012

Economic Times of India (2012) Bangalore among the top 10 preferred entrepreneurial locations. http://economictimes.indiatimes.com.12 April 2012. Accessed 24 Dec 2012

Trung LQ, Dung PX, Ngan TX (1996) A scientific note on first report of Apis laboriosa F Smith, 1871 in Vietnam. Apidologie 27(6):487–488

UNESCO (2011) Preparing World Heritage Nominations (Second edition, 2011), Published in November 2011 by the United Nations Educational, Scientific and Cultural Organization

Requirements for Sustainable Management of *Apis dorsata* Fab. with Rafter Method

24

Stephen Petersen and Muniswamyreddy Shankar Reddy

Abstract

The manufacture of purpose-made nesting sites to attract migrating swarms of the giant Asian honeybee/Bambara bee (*Apis dorsata* Fab.) is termed "rafter beekeeping" and is practiced at several locations across Southeast Asia. Raftering to be successful, several criteria must be met, some structural; others are environmental and cultural. This paper examines these criteria and their applications. This method of harvesting honey can be introduced into potential environments in parts of tropics and subtropics.

Keywords

Apis dorsata • Implementation criteria • Rafter beekeeping • Rafter potential

24.1 Introduction

Rafter beekeeping is so named because purpose-made nesting sites are placed in the proper environment at an angle (≈30°) resembling the rafters (structural roof components) of a building. As they are placed typically near the ground, they allow easy access (little to no climbing involved) for honey harvest. Because of the angle at which they are secured, the bees store the majority of the honey toward the top of

S. Petersen (✉)
Toklat Apiaries, 1153 Donna Dr., Fairbanks, AK 99712, USA
e-mail: akbeeman2000@yahoo.com

M. Shankar Reddy
Department of Zoology, Bangalore University,
Jnana Bharathi, Bengaluru 560 056, India
e-mail: jenureddy@gmail.com

A.K. Chakravarthy, S. Sridhara (eds.), *Arthropod Diversity and Conservation in the Tropics and Sub-tropics*, DOI 10.1007/978-981-10-1518-2_24

the rafter resulting in a "honey head" that may be sustainably harvested with no detriment to the brood.

The oft quoted phrase "build it and they will come" may hold true for golf courses and theme parks, but it is not always the case for rafter beekeeping. There are four essential criteria that must be met before this type of management can be implemented for *Apis dorsata*:

- A lack of "natural" nesting sites (tall trees, water towers, cliffs) which will force the bees to nest on the purpose-made rafters
- An abundance of bee-friendly floral resources available at least on a seasonal basis
- A widely recognized tenure or ownership of the rafters by individuals enforced by community laws and culture
- Locally available, long-lasting, sustainable materials from which to construct the rafters and, of course, the presence of bees themselves

24.2 Current Status of Rafter Beekeeping

Rafter beekeeping has been practiced in select areas for many hundreds of years, but its origins are difficult to precisely state (Petersen 2010a; Tấn et al. 1997). It is our belief that the origins lay with some very astute observations – "the bees like flowers and high smooth branches; I have lots of flowers in my area which the bees will like but no tall trees as suitable nesting sites…..perhaps if I put up a "branch" they would nest there."

This surmise comes from my personal observations (Petersen 2010b) in Danau Sentarum National Park in West Kalimantan, Indonesia, where, during the rainy season in the seasonally inundated Kapuas Lake system, there is a tremendous bee forage potential but all the trees have small diameter branches and don't extend more than five to six m above the high waters (a lack of "natural" nesting sites). A very similar situation can be found in the Bay of Kampong Som (south coast of Cambodia) where the smaller branches of seaside mangroves and a bit further back from the ocean. The shaggy bark and almost-vertical orientation of *Melaleuca leucadendron* limbs are not conducive to nesting but provide excellent forage.

Rafter beekeeping is (and has been) practiced in several locations in Southeast Asia (Map 24.1) and is referred to by several local names (Table 24.1). It can be divided into two main groups based on the season when bees produce honey in the area – dry season (e.g., area around Siem Reap, Cambodia) and wet season (e.g., Kapuas/Danau Sentarum, West Kalimantan, Indonesia).

Dry season rafters are less intricate than their rainy season counterparts. Most frequently dry season rafters are simple round logs (without any shaggy bark) supported at either end by a forked pole driven into the ground or a natural fork in a small tree. The rainy season rafters, on the other hand, are hewn to shape being concave on the upper surface (much like a rain gutter) and convex (simulating a smooth branch) on the lower surface.

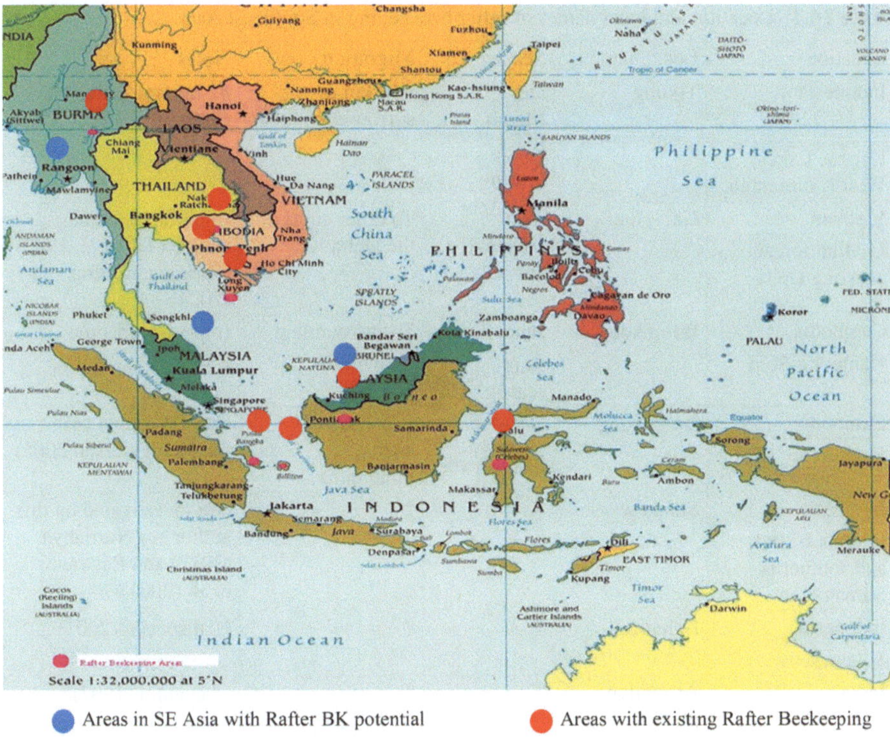

Map 24.1 Distribution of rafter beekeeping across SE Asia

A similar method to the aforementioned rafters was employed by Mahendra (1997) where "attraction planks" were coated with beeswax and suspended in locations formerly occupied by migrating swarms of *Apis dorsata*. After the planks had been occupied and comb production initiated, the planks would be lowered on one end forming an angle where the bees would store honey on the uppermost apex. The plank could be lower for ease of harvesting (Mahendra 1997).

24.2.1 Some Important Criteria for Establishing Rafter Beekeeping

For many years in the early 1900s Westerners attempted to "teach the bees" to become cavity nesters as managed in the West, but all such attempts failed. Rather than trying to adapt the bees to human management, efforts with rafter beekeeping center around adapting management schemes to fit the bees.

24.2.1.1 Lack of Natural Nesting Sites

Apis dorsata dorsata is a single-comb, open-nesting species of honeybee frequently found in aggregate nesting sites; other subspecies include *A. d. binghami* (found in Sulawesi and some neighboring islands) and *A. d. breviligula* found in the

Table 24.1 Location and local names of rafter beekeeping in Southeast Asia

Location	Local name	Season	Migration	Notes
Indonesia	Tikung	Dec–March	<50 km to surrounding area	#2 income activity (Petersen 2010a)
Kapuas Lakes, West Kalimantan		Rainy season		
Vietnam	Gac-glue	Two seasonal harvests	To other areas of the Delta	Important income source (Tan et al. 1997) Waring and Jump (2004)
U Minh forest Mekong Delta				
Cambodia	Bong kong	June–Sept	To surrounding mountains <50 km	Important income source
Kampong Som		Rainy season		
Cambodia	Bong kong	Dec–April	<25 km to nearby mountains	For some, sole cash income (Petersen 2005)
Siem Reap		Dry season		
Indonesia	Sunggau	?	?	Not investigated by this author see Nurtjahya (2012) and Purwanto et al. (2001)
Bangka Island and southern Sumatra				
Indonesia	Enhance	?	?	Hadisoesilo (2001)
Sulawesi				
India	Attraction planks	?	?	Mahendra (1997)

Philippines. A close relative of *dorsata* is the Himalayan rock bee (*Apis laboriosa*), found living above 1200 m only on cliff faces. The use of *dorsata* (sensu lato) bees in raftering is common but not prevalent in all environments. *A. d. binghami* is utilized in Central Sulawesi (Hadisoesilo 2001); the use of *A. d. breviligula* has not been reported. The fact that the latter two subspecies are not aggregate nesters may facilitate attraction to single, man-made rafters.

This lack of "natural" nesting sites is a commonality in rafter-friendly environments; it may be the small branches that are found in the flooded forests (Kapuas Lake region, West Kalimantan, Indonesia), the unsuitable shaggy bark texture of predominant trees (e.g., *Melaleuca leucadendron*), the vertical nature of the branches (e.g., *Acacia mangium*), or simply the small secondary growth in a degraded forest (Siem Reap, Cambodia, several locations in Indonesia).

24.2.1.2 Abundance of Bee Forage in the Area

It is important to establish the honey potential of an area by observations of the phenology (bloom time) of the bee-friendly plants. Local people typically have a reasonable understanding of which flowers are most attractive to bees and their respective blooming season; they are also aware of the seasonal migrations in response to blooms that are typical of *dorsata* (*s.l.*). Although *dorsata* will fly several kilometers on foraging trips, they are much more efficient at close range.

An example may be seen in the coastal-strip environment of Terengganu on the east coast of Peninsular Malaysia; the area may be generally characterized as a band of mangroves on the seaward side, a band of *Melaleuca* further inland, a short agricultural plain, and a series of forested hills where an emergent tree, tualang (*Koompassia excelsa*), is found. The tualang is the preferred nesting site for *dorsata*, tall, almost inaccessible to predators, smooth barked, and having horizontal branches.

Left – Aggregate nesting is common with *Apis dorsata dorsata* here pictured on a tualang (*Koompassia excelsa*) tree near Terengganu, on the east coast of West Malaysia. Not only does the height deter easy harvesting but also the number of nests – if one nest is disturbed, the other may attack en masse.

Left – Substantial areas of bee-friendly forage (*Barringtonia aculeatum*) must be present even if on a seasonal basis. If there is a single predominate species, a monofloral honey may be obtained; several attractive species blooming at the same time ensure a more reliable honey harvest.

Left – *Acacia mangium* is frequently grown in plantations for use as wood/pulp (paper). It has some blooms almost all year but does have a peak period; it is a poor pollen producer, so there must be a good pollen source (e.g., *Mimosa pudica*) to provide a protein source for good brood rearing. *Eucalyptus* (depending on species may be a good bee plant) is also grown in plantations.

24.2.2 Tenure or System of Ownership

Typically wild bees are seen as a common, "free" resource to be exploited by whoever discovers them. For rafter beekeeping to be successful and sustainable, there must be a sense of "ownership" (note the small sign in the photo to the left designating who owns that particular rafter) which is locally enforced by surrounding communities. From a project development standpoint, this is most frequently the biggest obstacle to surmount. The best success story comes from the Asosiasi Periau Danau Sentarum (APDS), a community-based organization which deals with the collection, processing, certification, and distribution of "*dorsata* honey," a certified organic product. To utlilize the benefits of this organization (*periau*), each participant must adhere to the following criteria:

- They must own and operate at least 25 "tikung" (local name for rafters).
- They must fall within his local village administration unit (see Map 24.2).
- Their location must be mapped and recorded.
- Harvest data and other records must be kept.
- Agreed-upon organic practices must be used at all times.
- Infringement of rules is grounds for dismissal from the *periau*; this may be outside of the realm of beekeeping, e.g., cutting timber without permit or fishing in restricted areas.
- Adhere to the rules and regulations enforced by traditional and community/peer pressure ("*adat*").

In areas of Cambodia, community forest associations (CFAs) provide the framework for ownership. The problem is that unless the CFA is a widely recognized entity by *all* of the communities in the area, "outsiders," with no respect for the ownership, will frequently harvest the honey often in an unsustainable manner (i.e., taking the brood or killing the bees to obtain the honey). This theft is a major obstacle to attempts at establishing a rafter beekeeping enterprise as, even though it is

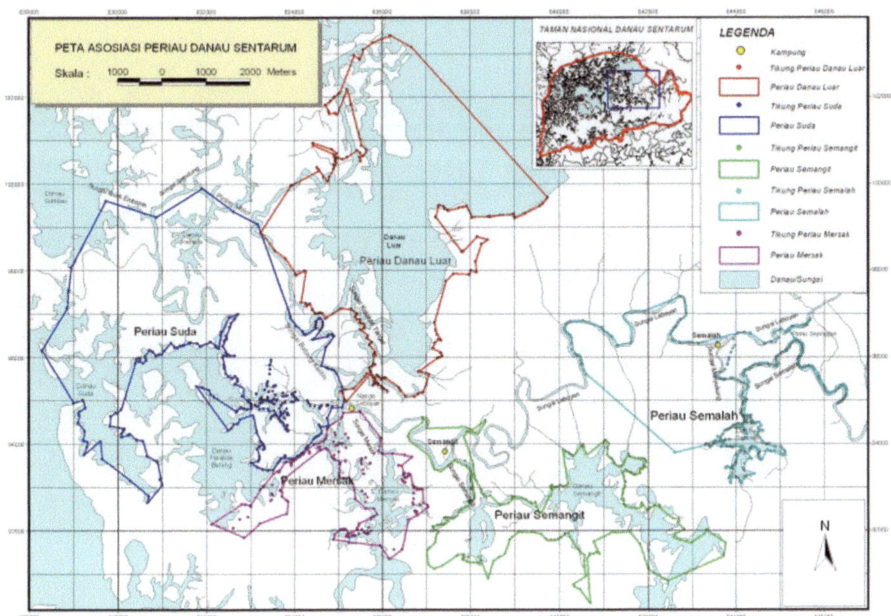

Map 24.2 Showing the location of rafters and the harvest district boundaries (Courtesy of APDS, West Kalimantan)

obvious that the rafters are man-made for the purpose, the bees themselves are seen as a "wild resource" to be harvested on a first-come basis. It does take some special skills to harvest honey without massive stinging attacks, but determined thieves can circumvent this problem (e.g., using insecticides).

The APDS mentioned above in West Kalimantan has the advantages that all communities in the area respect the *periau* traditions ("*adat*") and access for harvesting is only by boat reducing the number of people "dropping by" to harvest. The map on the left shows how the rafters are mapped and harvest whereas boundaries are established through consensus.

24.2.3 Effect of Rafters Utilizing Suitable Materials

Placement of rafters, types of wood used, angle of repose, orientation, and protection have been documented by several authors (see Table 24.1 for references). They can be grossly divided into rainy season and dry season depending on the season the bees migrate to the local area. In both cases support of the rafter must be secure and hold the rafter at an angle (normally about 30° from the horizontal). This allows the bees to store honey at the upper end of the rafter facilitating a sustainable harvest by taking only the "honey head" and leaving the brood intact. The bees, depending on the floral resources, will frequently rebuild the honey section allowing for further harvest(s).

	Left – One end of the rafter is partially dug into the ground, the other balanced (precariously) on a vertical post forming the requisite angle of ≈30° (near Kampong Som, Cambodia).	

Right – A more substantial rafter placement is supported by two forked poles taken from the surroundings and dug into the ground. In both cases note the very small structure of the surrounding vegetation; there is ample bee forage (***Melaleuca***) but no natural nesting sites.

Right – Using hand tools, a rafter is hewn from tembusu wood (*Fagraea fragrans*), a long-lasting (40+ years) local wood. The upper surface (above left) is concave, as in a rain gutter, allowing water to readily run off; the lower surface (above right) is rounded and smoothed to mimic a large branch. The notched ends with pegs facilitate securing the rafter in place. Photos from Kapuas Lake region, West Kalimantan, Indonesia.

An angle of about 30° with the horizontal seems to be an effective slope; even in situations where one end of the rafter is buried in the ground on a steep slope (30–40°), the angle with the horizontal still remains at 30° ("*tingku*" see Hadisoesilo 2001).

In India, Mahendra (1997) describes how, using his planks and a series of slings, he first secures the wax-coated planks to former nesting sites (from previous migrations) and then, after they are occupied and comb building initiated, one end of the plank is lowered slightly and secured at the 30° slope position. The bees then store the honey on the upper portions, and it is subsequently harvested by lowering the whole nest to the ground using a series of pulleys. This may be an appropriate technology in situations where there are large numbers of aggregate nesting *dorsata*, but we would imagine it would have to be closely monitored to prevent theft.

A flyway or tunnel through the thick vegetation maintained in front of the rafter's upper end allows the bees an easy access to the rafter colony. In Vietnam the beekeepers say that a clearing or opening directly in front of the rafter is beneficial. Several rafters can be placed near one another so long as they face the clearing (compass orientation appears to have little effect).

In areas where vegetation may be patchy, beekeepers often provide "shade" by cutting or bending small local branches over the rafter in a tentlike fashion. This is more prevalent in secondary growth areas and on hillsides (Hadisoesilo 2001; Purwanto et al. 2001; Nurtjahya 2012).

Material for the construction of rafters should be locally available and not pose a threat to the environment by its utilization. Woods that last a long time in the particular environment, are not repellant to bees (local knowledge), and are structurally sound are preferred. Frequently suitable material is available just meters away from the desired rafter site location (Siem Reap and Kampong Som areas in Cambodia, U Minh Forest in Vietnam). In other situations (especially in the manufacture of rainy season rafters), material must be selected, shaped, and then transported to the rafter site (Danau Sentarum, West Kalimantan). In Vietnam rafters are frequently made by selecting the appropriate-sized tree (*Melaleuca* spp.) (≈20–25 cm diameter), splitting it lengthwise and then peeling the shaggy bark from the outer portions. This results in two rafters (2–2.5 m long) that are suspended from two forked poles at a 30° angle.

24.2.4 Harvesting Honey from Rafters

One of the major goals of implementing rafter beekeeping (besides the obvious near-the-ground ease of access) is the ability to harvest a major portion of the stored and ripe (i.e., capped over) honey with minimal disturbance to the bees. The upward sloping angle of the rafter makes this possible. Honeybees of all species (*Apis*) store pollen and honey above the brood nest. With cavity-nesting bees, man has devised ways of removing this honey using honey supers (i.e., superimposed above the brood nest). With single-comb, open-nesting species such as *dorsata*, this is not always as easy as the honey, stored along the upper edge of the comb, also serves as an attachment for the whole comb.

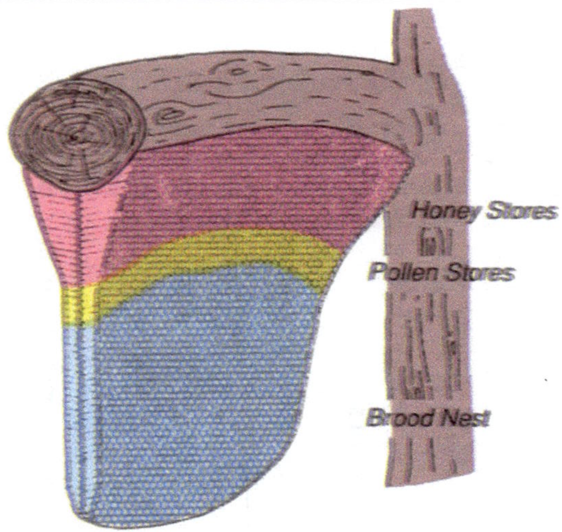

Above left – Honey (pink) is stored above the pollen (yellow band) and brood (blue) making it difficult to harvest honey while leaving the suspensional integrity of the comb intact.	Right – The "honey head" outlined by the red arc may be taken with little loss of suspension strength. The honey is concentrated in the uppermost portion of the comb.

To drive the bees from the comb during harvest, only cool-white smoke (preferably from a natural source, no oily rags or plastic) or a natural bee repellant, e.g., *Amomum aculeatum*, should be used. In no instance should the bees be chased off using open flame or insecticides. Smoke, properly used, is a very effective deterrent to the bees and should protect the beekeepers from massive stinging incidents – *provided they stay in the smoke*!

Certain plants, e.g., *Amomum aculeatum*, have a repellant effect on *dorsata* bees and are used to deter stings by the aboriginal peoples of the Andaman Islands while harvesting (Petersen 2005; Dutta et al. 1983). These plants (and others) warrant further investigation.

24.3 Processing and Marketing Rafter Honey

Quality control in the value chain/marketing aspects of rafter beekeeping is important and begins in the field during harvesting. Rubber gloves and clean harvesting implements (knives and buckets) should be used when harvesting honey; care should be exercised so that only the "ripe" (capped over) portion of the honey head is taken and no pollen, unripe honey (i.e., high moisture), or brood should contaminate the food product. After harvesting, steps should be taken to insure that no contamination or atmospheric moisture will spoil the product. High moisture (>22 %) is a common problem with all tropical honeys, and it is not uncommon to have fermentation problems.

With assurances of quality control and traceability, rafter honey can demand a higher price in niche markets due to its organic and unadulterated nature. Instead of competing with "off-the-grocery-store-shelf" prices, producers should strive to market a signature brand with label and quality characteristics reflecting its uniqueness.

Rafter beekeeping can provide a means of sustainably harvesting an eco-friendly product provided the criteria listed above are met. There are many areas which may hold potential for this activity in South and Southeast Asia.

Acknowledgments Many thanks to the rafter beekeepers we have interacted over the last 13–14 years and the NGOs who have, in some fashion or another, supported our efforts.

References

Dutta T, Ahmed R, Abbas R, Rao MKV (1983) Discovery of a plant in the Andaman Islands that Tranquilizes *Apis dorsata*. Bee World 64:158–163

Hadisoesilo S (2001) Tingku, a traditional management technique for *Apis dorsata binghami* in Central Sulawesi. In: IBRA (eds) Proceedings of the 7th international conference on tropical bees management and diversity and the 5th Asian Apiculture Association conference, 19–25 March 2000, Chiang Mai Thailand, International Bee Research Association, Cardiff, pp 309–312

Mahendra DB (1997) Improved methods of honey harvest from rock bee colonies. Indian Bee J 59(2):95–98

Nurtjahya K (2012) The Architecture of Sunggau in Banka Island, Indonesia – An Artificial wild Honeybee Nest. Paper presented at 11th Asian Apicultural Association conference, Terengganu, West Malaysia. 28th Sept–2nd Oct 2012

Petersen S (2005) Adventures in beekeeping – rafter beekeeping in Cambodia. Am Bee J 145(3):222–224

Petersen S (2010a) Production of natural bee repellants from indigenous SE Asian plant species, private publication available on request from author

Petersen S (2010b) Rafter Beekeeping in Borneo. Am Bee J 150(5):479–483

Purwanto DB, Hadisoesilo S, KasnoKoeniger N, Lunderstedt J (2001) Sunggau system: a sustainable method of honey production from Indonesia with the giant Asian Honeybee *Apis dorsata.* In: IBRA (eds) Proceedings of the 7th international conference on tropical bees management and diversity and the 5th Asian Apiculture Association conference, 19–25 March 2000, Chiang Mai, Thailand, International Bee Research Association, Cardiff

Tan N, Chinh PH, Thai PH, Mulder V (1997) Rafter bee keeping with *Apis dorsata*: some factors affecting the occupation of rafters by bees. J Apic Res 36:49–54

Waring C, Jump D (2004) Rafter beekeeping in Cambodia with *Apis dorsata*. Bee World 85(1):14–18

RNA Interference: History, Mechanisms, and Applications in Pest Management

25

Subba Reddy Palli

Abstract

RNA interference (RNAi) is emerging as a new technology for use in arthropod pest management. Unique features of RNAi-based methods including the specificity to the target pests and ability to knock down genes associated with insecticide resistance make this technology superior to the currently used pest management methods. The RNAi works best in coleopteran insects and also works well in some hemipteran insects. The RNAi does not work well in dipterans and lepidopteron insects. The RNAi could be used to study biology of insects, identify new target sites, fight insecticide resistance, regulate metabolism, produce sterile male insects, and suppress insect pest's populations. In the near future, further research to improve RNAi in insects belonging to Hemiptera, Lepidoptera, and other orders will lead to development of efficient and cheaper methods for production and delivery of double-stranded RNA, understanding potential for resistance development and effects on nontarget organisms. This will help in moving RNAi technology to field applications.

Keywords

Beetle • dsRNA • Pest management • RNAi

S.R. Palli (✉)
Department of Entomology, College of Agriculture, University of Kentucky,
Lexington, KY 40546, USA
e-mail: rpalli@uky.edu

A.K. Chakravarthy, S. Sridhara (eds.), *Arthropod Diversity and Conservation in the Tropics and Sub-tropics*, DOI 10.1007/978-981-10-1518-2_25

25.1 Introduction

Ribonucleic acids (RNAs) including microRNAs and small interference RNAs (siRNAs) are known to control gene expression (Fire et al. 1998). MicroRNAs are produced endogenously, but the siRNAs are processed products of double-stranded RNAs (dsRNA). The target-specific interference of siRNA on mRNA transcription, stability, and translation results in a reduction in the amount of protein produced from the target gene due to RNA interference (RNAi) (Fire et al. 1998). The RNAi has been extensively used in basic research, and many applications based on RNAi have been developed to solve problems in human and plant health as well as in many other fields (Perrimon et al. 2010). The journal *Science* declared RNAi as the technology of the year in 2002. Nobel Prize in medicine for the year 2006 was awarded to discoverers of RNAi – Fire and Mello. In this chapter, discussion is focused on RNAi technology and its applications with an ultimate goal of designing target-specific management methods for insect pests.

25.2 Historical Perspective

Early hints of gene silencing were observed in plants (Napoli et al. 1990), and this phenomenon was described as co-suppression, and the exact molecular mechanism was not determined (Jorgensen 1992). Gene silencing was also reported in viral resistance in plants and animals. For example, insertion of a La Crosse virus gene into an infectious Sindbis virus expression vector and its ability to induce interference was reported (Powers et al. 1994). Gene silencing approaches were employed to engineer resistance to California serogroup virus replication in mosquito cells and mosquitoes (Powers et al. 1996). Gene silencing was also reported in fungus *Neurospora crassa* (Cogoni et al. 1996) and nematodes (Fire et al. 1998). Soon after discovery, functioning of RNAi was demonstrated in nematodes, insects, plants, and trypanosomes (Bucher et al. 2002; Marie et al. 2000; Misquitta and Paterson 1999; Montgomery et al. 1998; Moss 2001). In 2001, the first successful RNAi after exposing mammalian cells to siRNA was reported (Hope 2001). During the next few years, improvements in design, synthesis, and delivery of dsRNA or siRNA led to the development and testing of RNAi-based methods to treat human diseases as well as to control pests and diseases that attack crops and trees (Brown et al. 2003; Ding 2005; Hoa et al. 2003; Means et al. 2003; Schroder 2003; Tomoyasu and Denell 2004; Worby et al. 2001). In the fruit fly, *Drosophila melanogaster*, RNAi methods were employed to determine functions of genes involved in growth, development, reproduction, and behavior (Perrimon et al. 2010). High-throughput RNAi-based screening assays were developed in cell lines as well as in flies, and these screening assays were used to learn about functions of a number of genes involved in regulation of almost every aspect of insect life (Agaisse et al. 2005; Baeg et al. 2005; Bai et al. 2009; Cherry et al. 2005; DasGupta et al. 2005, 2007; Echeverri and Perrimon 2006; Eggert et al. 2004; Flockhart et al. 2006; Lu et al. 2007; Ni et al. 2009; Perrimon and Mathey-Prevot 2007a, b; Sepp et al. 2008; Wang et al. 2009).

25.3 The Mechanism of RNA Interference

The dsRNA introduced into cells by viruses or introduced from outside triggers RNAi response. The enzyme dicer digests dsRNA into small interference RNAs (siRNAs, 18–21 base pairs in length) (Dykxhoorn et al. 2003; Hannon 2002; McCaffrey et al. 2002). The siRNAs are recruited to the RNA-induced silencing complex (RISC) containing a number of proteins including Argonautes (Elbashir et al. 2001a; Nykanen et al. 2001). The RISC complex facilitates siRNA hybridization to the complimentary DNA/mRNA resulting in a block in replication, transcription, and/or translation (Elbashir et al. 2001b) resulting in knockdown or knockout of the target gene. Exogenously applied dsRNAs induce immune response through interferon in vertebrates. Therefore, only siRNA can be used to trigger RNAi. In contrast, in insects and other invertebrates, dsRNA does not induce immune response. Therefore, long dsRNAs could be used to trigger RNAi.

In general, 200–400 bp fragments of target gene are PCR amplified using primers containing bacterial T7 promoter and either genomic DNA or cDNA from the target insect as a template. Then the PCR product is used to synthesize RNA from both the strands of DNA using T7 RNA polymerase. The RNAs produced from both the strands hybridize to each other producing dsRNA. The DNA in the reaction is digested with DNase I and the dsRNA is purified. The dsRNAs are delivered to cells by adding to the growth medium or by transfection using lipids. Either injection or feeding is used to deliver dsRNA to animals. The dsRNAs delivered to insect cells are digested into multiple siRNA by dicer-2 enzyme present in these cells. The siRNAs are then recruited to the RISC complex, which facilitates hybridization of siRNAs to target DNA/RNA resulting in knockdown of target gene. Unlike in vertebrates, optimization of siRNAs for each target gene is not normally required for efficient silencing of genes in insects. This advantage is one of the major factors contributing to widespread use of dsRNA to trigger RNAi in insects.

The efficiency of RNAi among insects belonging to different orders varies quite a bit. The RNAi works the best in coleopteran insects. The RNAi seems to work well in some hemipteran insects (including hoppers and bugs). The RNAi does not work well in dipteran and lepidopteran insects. Many factors including degradation by dsRNases; inhibition of RNAi machinery by microorganism present in insects; inefficiency of uptake, transport, and endosomal escape; and difference in expression and composition of factors involved in RNAi have been suggested as the reasons for differential efficiency of RNAi among insect species. Differences in RNAi efficiency depending on the method of application of dsRNA have also been reported. In the locust, injection of dsRNA efficiently silences target genes, but dsRNA feeding does not cause silencing of target genes (Luo et al. 2013). Also, tissue-specific RNAi response to dsRNA has been observed in both desert and migratory locusts (Ren et al. 2014; Wynant et al. 2012). RNAi works very well in the fat body and other tissues but ovaries are insensitive to RNAi. Differences in the stability and uptake of dsRNA among these tissues are likely responsible for differential effects of dsRNA among tissues tested (Ren et al. 2014). RNases responsible for degradation of dsRNA and proteins involved in transport of dsRNA into cells have been identified in desert and migratory locusts (Luo et al. 2012, 2013; Wynant et al. 2014). Recent studies in the pea aphid, *Acyrthosiphon pisum*, showed that dsRNA degradation is one of the major factors responsible for lack of efficient RNAi response in these insects (Christiaens et al. 2014).

25.4 RNAi in the Red Flour Beetle, *Tribolium castaneum*

RNA interference is highly efficient and systemic in beetles. RNAi has been used extensively in the red flour beetle, *Tribolium castaneum*, a pest on stored products. Injection of dsRNA into the hemocoel of the pupae resulted in knockdown in the expression of zygotic genes in the embryos developed from the eggs laid by adult females developed from the dsRNA injected pupae (Bucher et al. 2002). Injection of dsRNA into larvae causes knockdown of target genes in the larvae, pupae, and adults (Tomoyasu and Denell 2004). RNAi has been extensively used to determine functions of a number of genes involved in growth, development, reproduction, and insecticide resistance in these beetles (Aikins et al. 2008; Arakane et al. 2005, 2008; Bitra et al. 2009; Cerny et al. 2008; Choe et al. 2006; Farzana and Brown 2008; Minakuchi et al. 2008; Niu et al. 2008; Ober and Jockusch 2006; Parthasarathy and Palli 2008, 2011; Parthasarathy et al. 2008, 2010a, b; Tan and Palli 2008; van der Zee et al. 2005; Xu et al. 2010; Zhu et al. 2010a). Although gene silencing has been reported after feeding dsRNA to *T. castaneum* larvae (Whyard et al. 2009), most of the studies reported to date used injection method. The RNAi effect in *T. castaneum* appears to be highly effective because injection of minute quantities of dsRNA causes efficient knockdown in gene expression suggesting an amplification of injected dsRNA. The RNAi effect in *T. castaneum* is also systemic and is not cell autonomous (Fig. 25.1).

25.5 Applications of RNAi in Basic Science

RNAi has played an important role in advances made in understanding physiological processes in insects. Prior to discovery of RNAi, gene knockout was possible only in a few model insects such as the fruit fly. RNAi enabled gene knockdown in non-model insects especially that are major pests. We used RNAi to gain insights into juvenile hormone action and hormonal regulation of reproduction in the red flour beetle, *T. castaneum*.

JH Receptor Methoprene tolerant (Met) has been proposed as a JH receptor, but the scientific community did not accept the proposal readily because Met null mutants are viable due to the presence of a second member of bHLH-PAS family, germ cell expresser (gce) (Ashok et al. 1998; Wilson and Ashok 1998; Wilson and Fabian 1986). Tribolium genome contains only one homolog of both Met and gce. Injection of Met dsRNA into the penultimate instar larvae caused premature metamorphosis, and injection of Met dsRNA into the final instar larvae caused premature

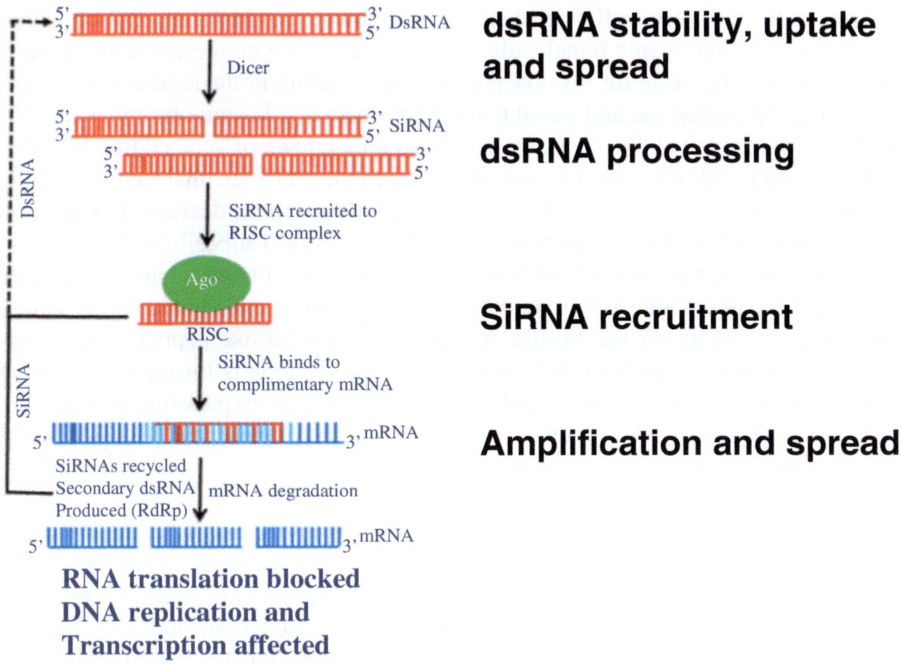

Fig. 25.1 A schematic diagram of pathway for functioning of RNAi in cells. Ago, Argonaute; RISC, RNA-induced silencing complex

development of adult structures, and the larvae tried to skip pupal stages and developed adult structures (Konopova and Jindra 2007; Parthasarathy et al. 2008). These are typical phenotypes expected from an animal deficient in JH action. These RNAi studies helped to establish Met as a JH receptor. RNAi also helped to identify steroid receptor coactivator (SRC) as the protein required for JH action (Zhang et al. 2011). Knockdown in the expression of all members of bHLH-PAS family members by injecting dsRNA targeting gene coding for these proteins into final instar larvae showed that besides Met, SRC is also required for JH induction of target genes. RNAi also played an important role in establishing kr-h1 as an important transcription factor involved in juvenile hormone action (Kayukawa et al. 2013).

Hormonal Regulation of Female and Male Reproduction RNAi was employed to identify key players involved in regulation of female and male reproduction. Knockdown in expression of gene coding for female proteins involved in juvenile hormone, ecdysone, and insulin-like peptide action showed that juvenile hormone regulates vitellogenin production, ecdysone regulates oocyte maturation, and insulin signaling is required for both vitellogenin production and oocyte maturation. Knockdown of the same genes in males revealed that juvenile hormone and insulin-like peptides regulate production of accessory gland proteins and insulin signaling is required for sperm production.

Knockdown of insulin/IGF (IIS) signaling pathway terminal transcription factor, FOXO, in previtellogenic female adults increased vitellogenin mRNA and protein levels suggesting a role for insulin-like peptide signaling in the regulation of vitellogenin gene expression and possible cross talk between juvenile hormone and IIS pathways (Sheng et al. 2011). RNAi-mediated knockdown of gene coding for juvenile hormone acid methyltransferase, an enzyme that catalyzes final step in juvenile hormone biosynthesis or methoprene tolerant, a JH receptor, decreased expression of gene coding for insulin-like peptides influenced FOXO subcellular localization, resulting in a decrease in both mRNA and protein levels of vitellogenin. Exogenous juvenile hormone application to previtellogenic female beetles induced the expression of gene coding for insulin-like peptide 2 and insulin-like peptide 3 and vitellogenin. These data showed that juvenile hormone functions through insulin-like peptide signaling pathway and regulates vitellogenin gene expression, thus settling a long-standing controversy on whether juvenile hormone directly or indirectly regulate vitellogenin gene expression.

Juvenile Hormone Regulation of Metabolism Insulin/IGF-1 signaling and juvenile hormone signaling cross talk in regulation of resistance to starvation. In *Tribolium* RNAi aided knockdown of juvenile hormone acid methyltransferase, methoprene tolerant or insulin-like peptide 2, decreased lipid and carbohydrate metabolism and extended the survival of starved beetles (Xu et al. 2013). This effect on extension of life span could be restored by injection of bovine insulin into juvenile hormone acid methyltransferase RNAi beetles but not by application of JH III to insulin-like peptide 2 RNAi beetles suggesting that juvenile hormone controls starvation resistance by regulating synthesis of insulin-like peptide. Juvenile hormone regulation of starvation resistance is achieved through maintaining trehalose homeostasis by regulating trehalose transport and metabolism as well as utilization of stored nutrients in starved adults.

25.6 Applications of RNAi to Identify Target Sites

There is a constant fight between humans and insects. Humans discover and develop new insecticides to control insects that destroy crops and trees and transmit deadly diseases. The insects on the other hand develop resistance to overcome the new challenges and survive. Insects developed resistance to almost all classes of chemicals introduced for their control. Majority of insecticides developed to date target nervous system making it easy for insects to develop resistance against these compounds. **Therefore, identification and characterization of new non-neuronal target sites are urgently needed. RNAi could help in this respect.**

Nuclear Receptors Nuclear receptors are a group of proteins characterized by the presence of classified, a well-conserved DNA-binding domain and a less conserved ligand-binding domain. Nuclear receptors function as regulators of diverse signaling and metabolic pathways. Nuclear receptors are switched on and off by small molecule ligands with properties desirable in an insecticide. Therefore, the nuclear receptors are attractive targets for new insecticides. Nineteen canonical and two Knirps family nuclear receptors were identified in the genome of *Tribolium*. RNAi-aided knockdown of all 21 nuclear receptors and showed that 10 out of the 19 nuclear receptors (TcE75, TcHR3, TcHR4, TcEcR, TcUSP, TcFTZ-F1, TcHR51, SVP, TcHR38, TcHR39) play key roles in regulation of metamorphosis (Tan and Palli 2008). In the absence of these nuclear receptors, the *Tribolium* larvae were unable to complete metamorphosis successfully. Silencing of five other nuclear receptors (TcTll, TcDsf, TcHNF4, and TcHR78) resulted in defects in offspring production. In addition, knockdown on TcHNF4, TcHR78, TCHR51, and TcDsf influenced egg production. Knockdown of TcTll blocked embryonic development. Knirps-like receptor knockdown affected adults and caused a reduction in egg production. Knockdown of TcEagle, TcE78, TcHR83, TcHR96, TcPNR-like, and TcERR did not show much effect. Thus, by employing RNAi, it was possible to identify functions of 15 out of 21 nuclear receptors in *Tribolium* within a short time.

RNAi studies also showed that E75, HR3, EcR, USP, SVP, FTZ-F1, and HR4 are required for vitellogenesis and oogenesis (Xu et al. 2010). Knockdown of these seven nuclear receptors resulted in a reduction in the levels of vitellogenin in mRNAs and affected the oocyte maturation. Knockdown of HR96, HR51, HR38, HR39, TTll, Dsf, and Knirps-like blocked embryogenesis. These studies showed that at least 17 out of the 21 nuclear receptors identified play key roles in female reproduction and embryogenesis. In male *Tribolium*, knockdown of 11 nuclear receptors (E75, E78, FTZ-F1, HR38, HR4, Knirps-like, HNF4, Tailless, HR51, Dsf, and HR39) caused more than 50 % reduction in the eggs laid by the female beetles mated with RNAi male beetles (Xu et al. 2012). While E78 and HR39 are required for sperm production, E75 and HR38 are required for production of male accessory gland proteins.

The Basic Helix-Loop-Helix Transcription Factors The basic helix-loop-helix transcription factors control cellular proliferation, tissue differentiation, development, and detoxification. Insect genomes contain 50 helix-loop-helix transcription factors. We identified 53 bHLH genes in *Tribolium* genome. Phylogenetic analyses classified these 53 genes into ten families: PAS, HES, Myc/USF, Hand, Mesp, Shout, p48, NeuroD/Neurogenin, Atonal, and AS-C. Knockdown of seven members of the PAS and HES families affected the growth and development of *Tribolium* (Bitra et al. 2009). The steroid receptor coactivator showed the most severe phenotypes. Knockdown of four bHLH transcription factors (TcSRC, TcSim1, TcAsh, and TcDaughterless) killed female beetles. Knockdown of 16 bHLH genes affected oogenesis, and knockdown of 13 bHLH transcription factors affected embryogenesis. TcSide1 and TcSpineless knockdown affected both oogenesis and embryogenesis (Bitra and Palli 2010). Thirty-one bHLH transcription factors are required for survival of female beetles, reproduction, and embryogenesis.

The G Protein-Coupled Receptors The G protein-coupled receptors regulate behavior, development, and reproduction. The G protein-coupled receptors are the therapeutic targets for about 40 % of prescription drugs. However, there is no commercial insecticide targeting insect G protein-coupled receptor. We identified 111 non-sensory GPCRs in *Tribolium* genome (Bai et al. 2011). Knockdown of eight G protein-coupled receptors caused severe developmental arrest and ecdysis failure. The G protein-coupled receptors include dopamine-2 like receptor (TC007490/D2R) and latrophilin receptor (TC001872/Cirl). The majority of larvae injected with TC007490/D2R dsRNA died during larval stage prior to entering pupal stage, suggesting that this GPCR is essential for larval growth and development. These identified G protein-coupled receptors could serve as excellent targets for developing screening assays to identify new chemicals for insect control.

25.7 Applications of RNAi to Fight Insecticide Resistance

Approximate cost of discovery, development, and commercialization of an insecticide is about 300 million dollars. Insects developed resistance to most of the insecticides introduced for their control. If the insects develop resistance to insecticides introduced for their control quickly, the companies investing large funds to develop insecticides may not be able to make profits. This is one of the main reasons for significant drop in introduction of new insecticides in recent past. RNAi could help fight this problem. Insects develop resistance by altering target site or by altering levels or function of proteins involved in metabolic, penetration, and behavioral resistance. RNAi could help fight every other form of resistance except target site resistance.

25.7.1 RNAi for Fighting Metabolic Resistance in Tribolium

Cytochrome P450s play important role in insecticide resistance. We identified a P450 gene responsible for deltamethrin resistance observed in the QTC279 strain of *Tribolium*. CYP6BQ9 is expressed 200-fold higher in the deltamethrin-resistant QTC279 strain when compared with its expression in the deltamethrin-susceptible Lab-S strain. Knockdown of CYP6BQ9 reduced deltamethrin resistance by more than 50 %. Expression of CYP6BQ9 in *Drosophila* made flies deltamethrin resistant. CYP6BQ9 enzyme expressed in baculovirus metabolized deltamethrin to 4-hydroxy deltamethrin. These studies showed that it is possible to use RNAi to fight insecticide resistance (Zhu et al. 2010b).

25.7.2 RNAi for Fighting Insecticide Resistance in Bed Bugs

Of late bed bug *Cimex lectularius* is becoming a nuisance pest in most parts of the world. We identified a number of P450 and ABC transporter genes involved in pyrethroid resistance in bed bugs. We exploited the traumatic insemination mechanism of bed bugs to inject dsRNA into bed bug females. Knockdown of resistance-associated genes in bed bugs showed the relative contribution of each mechanism toward overall resistance development. Knockdown of P450s in bed bugs reduced deltamethrin resistance in resistant insects (Zhu et al. 2013). Similarly, knockdown of ABC transporters reduced resistance in resistant populations. Knockdown of NADPH-cytochrome P450 reductase increased susceptibility to deltamethrin in resistant populations but not in the susceptible population of bed bugs (Zhu et al. 2012a).

25.8 Applications of RNAi to Produce Sterile Male Insects

Sterile male technologies are being used to eradicate insect pests. Often, it is time consuming and expensive to isolate males from females and to sterilize males using methods such as irradiation. RNAi could be employed to overcome these problems. We identified genes responsible for sex determination in *Tribolium* and used them to produce all male population. Sex in insects is determined by a series of regulators ultimately controlling sex-specific splicing of a transcription factor, doublesex. We identified doublesex and its regulator transformer in *Tribolium* (Shukla and Palli 2012a, b, 2013, 2014). RNA interference-aided knockdown of transformer in pupa or adults caused a change in sex from females to males by diverting the splicing of doublesex pre-mRNA to male-specific isoform. All the pupae and adults developed from transformer dsRNA injected final instar larvae were males. Knockdown of transformer could be used for elimination of females from the progeny resulting in the production of all male progeny. Similarly, RNAi can be used to knock down gene coding for important regulators of spermatogenesis to sterilize males. Therefore, it is possible to produce sterile males by knocking down both transformer and spermatogenesis regulator.

25.9 Applications of RNAi to Control Insect Pests

Delivery of dsRNA Through Plants Soon after discovery of RNAi, dsRNAs targeting insect genes were expressed in corn and cotton plants. The expression of target genes in insects feeding on transgenic plants containing dsRNA was knocked down resulting in the death of target insects (Baum et al. 2007; Mao et al. 2007). These initial reports caused an excitement in application of RNAi for controlling insect pests. DsRNA-expressing cotton plants have been produced, and the bollworm larvae reared on these transgenic plants expressing dsRNA targeting insect gene coding for P450 enzyme showed retarded growth and caused less damage (Mao et al. 2011). The expression of dsRNA targeting three genes highly expressed in the midgut and coding for hexose transporter, carboxypeptidase, and trypsin-like serine protease gene in the rice plants caused a decrease in expression of these target genes in nymphs of brown plant hopper feeding on transgenic rice plants, but the survival of these nymphs was not affected (Zha et al. 2011). The aphid, *Myzus persicae*, fed on *Arabidopsis thaliana* plants producing *MpC002*, and *Rack-1* dsRNAs showed a reduction in mRNA levels of target genes and the aphids feeding on these plants produced less progeny (Pitino et al. 2011). *Helicoverpa armigera* larvae that fed on transgenic tobacco plants expressing ecdysone receptor (EcR) dsRNA showed molting defects and larval lethality. Interestingly, these transgenic tobacco plants expressing *H. armigera* EcR dsRNA also showed resistance against the beet armyworm, *Spodoptera exigua*. This may be due to the longer and conserved region of EcR gene fragment used to prepare dsRNA. The dsRNA expressed in plants would have produced enough siRNAs that matched 100 % with the regions of *S. exigua* EcR mRNAs (Zhu et al. 2012b). *H. armigera* hormone receptor 3 dsRNA expressed in tobacco plants also disrupted development of larvae feeding on the transgenic plants (Xiong et al. 2013). Whitefly *Bemisia tabaci* feeding on transgenic tobacco plants expressing *v-ATPaseA* dsRNA showed reduced levels of v-ATPaseA mRNA and increased mortality (Thakur et al. 2014). These studies demonstrate the feasibility of using plant-expressed dsRNA to control insect pests.

Expression of dsRNA in *E. coli* The nematode *Caenorhabditis elegans* that fed on dsRNA expressed in *E. coli* showed successful knockdown of target genes, and this approach has been used to knock down all the genes in this worm to learn about their function (Kamath and Ahringer 2003; Kamath et al. 2003). This could be an alternative method of delivering dsRNA to control insects. It is possible to express dsRNA targeting critical genes in insects and spray heat-killed bacteria on the crops to control insect pests. A bacterial strain that is deficient in RNase III, an enzyme that degrades dsRNAs, has been developed and used in production of dsRNA. Feeding heat-killed bacteria that produced dsRNA targeting five genes efficiently knocked down all five target genes tested in Colorado potato beetle. Loss of function of these target genes caused larval mortality and decrease in feeding and insect growth (Zhu et al. 2010a). Similar results have been reported for tephritid fruit fly (Li et al. 2011) and beet armyworm (Tian et al. 2009).

Feeding AdoHcy hydrolase (SAHase) dsRNA expressed in *E. coli* decreased SAHase and Kr-h1 mRNA levels, reduced juvenile hormone titer, and caused the death of larvae and pupae and blocked adult emergence in Colorado potato beetle (Zhou et al. 2013). Feeding ryanodine receptor dsRNA caused knockdown of RyR in the larvae and adults and resulted in a decrease in chlorantraniliprole-induced mortality in Colorado potato beetle (Wan et al. 2014). Knockdown of P450 enzyme, shade, was achieved by feeding shade dsRNA resulting in a reduction in the hydroxylation of ecdysone, delay in development, and death of Colorado potato beetle larvae and pupae (Kong et al. 2014). Prohibitin, an essential protein for Colorado potato beetle viability, has been identified as Cry3Aa binding protein (Ochoa-Campuzano et al. 2013). This protein is also involved in cellular processes including mitochondrial function, cell proliferation, and development (Mishra et al. 2010). Feeding prohibitin dsRNA and treatment with Cry3Aa enhanced Cry3Aa toxin-induced mortality by threefold and the time to kill was reduced, and as a result 100 % mortality was achieved in 5 days after treatment in Colorado potato beetle. This is an interesting example of using a combination of RNAi and Bt toxin to achieve efficient control of insect pests.

The cost of bacteria expressed dsRNA could be much lower when compared to that of in vitro synthesized dsRNA. In most RNAi studies, often a knockdown but not a knockout of gene expression is achieved. Therefore, the silencing of gene and its effect is generally transient. To avoid the risk of resistance development, multiple applications of dsRNA are necessary. Because of lower costs and the ease in manufacturing large quantities, bacteria expressed dsRNA could be sprayed on the crop plants whenever necessary. Recent reports suggest that dsRNA effects are highly species specific even when targeting highly conserved genes (Whyard et al. 2009). To conquer the insecticide resistance, dsRNA-based insecticide exhibits advantages when compared to the chemical insecticides. If an insect pest develops resistance to one particular target site, it would be easier to switch to another region of the same target or to an entirely different target gene.

In vitro Synthesis of dsRNA The use of in vitro synthesized dsRNA for controlling insect pests and pathogens that attack insects has also been proposed. Although it is relatively straightforward to synthesize large quantities of dsRNA in vitro, cost of production is the major factor that needs to be considered in these dsRNA-based insecticides to compete with chemical insecticides.

25.10 Conclusion and Prospective

The contribution of RNAi to the basic understanding of insect biology is irrefutable. However, the contribution of RNAi for insect pest management is at the infancy. RNAi holds a great promise for developing target-specific environmentally friendly pest management methods. The initial attempts on the use of RNAi in insect pest management are encouraging. However, the development and commercialization of RNAi-based products for pest management is rather slow. Availability of highly effective BT transgenic plants to control insect pests, lack of reliable methods to knock down genes in sucking pests and lack of efficient and cost-effective methods for production and delivery, concerns about potential for resistance development by insects could be among the reasons for slow progress. Although transgenic plants are the preferred delivery method, it has its own limitations such as public hesitance in acceptance and unavailability of methods to produce transgenic plants for some of the crop pests. Production of dsRNA in bacteria and spraying heat-killed bacteria using existing methods for formulation and application of natural BT may work well in controlling pests such as Colorado potato beetle. Spraying of in vitro synthesized and formulated dsRNA using existing machinery used for insecticide applications may be another option.

Further research to improve RNAi in other insects, especially sucking pests, for identification of target genes to improve speed of kill and increase specificity, to improve production and delivery methods, to understand potential resistance development, and to better understand effects on nontarget organisms in the ecosystem will help in moving this technology from lab to field. One day, insecticides based on RNAi will be used in the field for pest management, but how soon this will occur depends on how well research community responds to the existing limitations and develops methods to overcome these limitations.

Acknowledgments The research in Palli laboratory was supported by the National Science Foundation (IBN 0421856), the National Institutes of Health (GM070559-07), and the National Research Initiative of the USDA-NIFA (2011-67013-30143). This report is contribution number 15-08-075 from the Kentucky Agricultural Experimental Station.

References

Agaisse H, Burrack LS, Philips JA, Rubin EJ, Perrimon N, Higgins DE (2005) Genome-wide RNAi screen for host factors required for intracellular bacterial infection. Science 309:1248–1251

Aikins MJ, Schooley DA, Begum K, Detheux M, Beeman RW, Park Y (2008) Vasopressin-like peptide and its receptor function in an indirect diuretic signaling pathway in the red flour beetle. Insect Biochem Mol Biol 38:740–748

Arakane Y, Muthukrishnan S, Kramer KJ, Specht CA, Tomoyasu Y, Lorenzen MD, Kanost M, Beeman RW (2005) The Tribolium chitin synthase genes TcCHS1 and TcCHS2 are specialized for synthesis of epidermal cuticle and midgut peritrophic matrix. Insect Mol Biol 14:453–463

Arakane Y, Specht CA, Kramer KJ, Muthukrishnan S, Beeman RW (2008) Chitin synthases are required for survival, fecundity and egg hatch in the red flour beetle, *Tribolium castaneum*. Insect Biochem Mol Biol 38:959–962

Ashok M, Turner C, Wilson TG (1998) Insect juvenile hormone resistance gene homology with the bHLH-PAS family of transcriptional regulators. Proc Natl Acad Sci U S A 95:2761–2766

Baeg GH, Zhou R, Perrimon N (2005) Genome-wide RNAi analysis of JAK/STAT signaling components in *Drosophila*. Genes Dev 19:1861–1870

Bai JW, Sepp KJ, Perrimon N (2009) Culture of Drosophila primary cells dissociated from gastrula embryos and their use in RNAi screening. Nat Protoc 4:1502–1512

Bai H, Zhu F, Shah K, Palli SR (2011) Large-scale RNAi screen of G protein-coupled receptors involved in larval growth, molting and metamorphosis in the red flour beetle. BMC Genomics 12:388

Baum JA, Bogaert T, Clinton W, Heck GR, Feldmann P, Ilagan O, Johnson S, Plaetinck G, Munyikwa T, Pleau M, Vaughn T, Roberts J (2007) Control of coleopteran insect pests through RNA interference. Nat Biotechnol 25:1322–1326

Bitra K, Palli SR (2010) The members of bHLH transcription factor superfamily are required for female reproduction in the red flour beetle, *Tribolium castaneum*. J Insect Physiol 56:1481–1489

Bitra K, Tan A, Dowling A, Palli SR (2009) Functional characterization of PAS and HES family bHLH transcription factors during the metamorphosis of the red flour beetle, *Tribolium castaneum*. Gene 448:74–87

Brown AE, Bugeon L, Crisanti A, Catteruccia F (2003) Stable and heritable gene silencing in the male vector *Anopheles stephensi*. Nucleic Acids Res 31:1–6

Bucher G, Scholten J, Klingler M (2002) Parental RNAi in Tribolium (Coleoptera). Curr Biol 12:R85–R86

Cerny AC, Grossmann D, Bucher G, Klingler M (2008) The Tribolium ortholog of knirps and knirps-related is crucial for head segmentation but plays a minor role during abdominal patterning. Dev Biol 321:284–294

Cherry S, Doukas T, Armknecht S, Whelan S, Wang H, Sarnow P, Perrimon N (2005) Genome-wide RNAi screen reveals a specific sensitivity of IRES-containing RNA viruses to host translation inhibition. Genes Dev 19:445–452

Choe CP, Miller SC, Brown SJ (2006) A pair-rule gene circuit defines segments sequentially in the short-germ insect *Tribolium castaneum*. Proc Natl Acad Sci U S A 103:6560–6564

Christiaens O, Swevers L, Smagghe G (2014) DsRNA degradation in the pea aphid (*Acyrthosiphon pisum*) associated with lack of response in RNAi feeding and injection assay. Peptides 53:307–314

Cogoni C, Irelan JT, Schumacher M, Schmidhauser TJ, Selker EU, Macino G (1996) Transgene silencing of the al-1 gene in vegetative cells of Neurospora is mediated by a cytoplasmic effector and does not depend on DNA-DNA interactions or DNA methylation. EMBO J 15:3153–3163

DasGupta R, Kaykas A, Moon RT, Perrimon N (2005) Functional genomic analysis of the Wnt-Wingless signaling pathway. Science 308:826–833

DasGupta R, Nybakken K, Booker M, Mathey-Prevot B, Gonsalves F, Changkakoty B, Perrimon N (2007) A case study of the reproducibility of transcriptional reporter cell-based RNAi screens in Drosophila. Genome Biol 8(9):R203

Ding SW (2005) RNAi: mechanisms, biology and applications. FEBS Lett 579:5821–5821

Dykxhoorn DM, Novina CD, Sharp PA (2003) Killing the messenger: short RNAs that silence gene expression. Nat Rev Mol Cell Biol 4:457–467

Echeverri CJ, Perrimon N (2006) High-throughput RNAi screening in cultured cells: a user's guide. Nat Rev Genet 7:373–384

Eggert US, Kiger AA, Richter C, Perlman ZE, Perrimon N, Mitchison TJ, Field CM (2004) Parallel chemical genetic and genome-wide RNAi screens identify cytokinesis inhibitors and targets. PLoS Biol 2:2135–2143

Elbashir SM, Lendeckel W, Tuschl T (2001a) RNA interference is mediated by 21-and 22-nucleotide RNAs. Genes Dev 15:188–200

Elbashir SM, Martinez J, Patkaniowska A, Lendeckel W, Tuschl T (2001b) Functional anatomy of siRNAs for mediating efficient RNAi in *Drosophila melanogaster* embryo lysate. EMBO J 20:6877–6888

Farzana L, Brown SJ (2008) Hedgehog signaling pathway function conserved in Tribolium segmentation. Dev Genes Evol 218:181–192

Fire A, Xu S, Montgomery MK, Kostas SA, Driver SE, Mello CC (1998) Potent and specific genetic interference by double-stranded RNA in *Caenorhabditis elegans*. Nature 391:806–811

Flockhart I, Booker M, Kiger A, Boutros M, Armknecht S, Ramadan N, Richardson K, Xu A, Perrimon N, Mathey-Prevot B (2006) FlyRNAi: the Drosophila RNAi screening center database. Nucleic Acids Res 34:D489–D494

Hannon GJ (2002) RNA interference. Nature 418:244–251

Hoa NT, Keene KM, Olson KE, Zheng L (2003) Characterization of RNA interference in an *Anopheles gambiae* cell line. Insect Biochem Mol Biol 33:949–957

Hope IA (2001) RNAi surges on: application to cultured mammalian cells. Trends Genet 17:440

Jorgensen R (1992) Silencing of plant genes by homologous transgenes. AgBiotech News Inform 4:265N–273N

Kamath RS, Ahringer J (2003) Genome-wide RNAi screening in *Caenorhabditis elegans*. Methods 30:313–321

Kamath RS, Fraser AG, Dong Y, Poulin G, Durbin R, Gotta M, Kanapin A, Le Bot N, Moreno S, Sohrmann M, Welchman DP, Zipperlen P, Ahringer J (2003) Systematic functional analysis of the *Caenorhabditis elegans* genome using RNAi. Nature 421:231–237

Kayukawa T, Tateishi K, Shinoda T (2013) Establishment of a versatile cell line for juvenile hormone signaling analysis in *Tribolium castaneum*. Sci Rep 3:1570
Kong Y, Liu XP, Wan PJ, Shi XQ, Guo WC, Li GQ (2014) The P450 enzyme Shade mediates the hydroxylation of ecdysone to 20-hydroxyecdysone in the Colorado potato beetle, *Leptinotarsa decemlineata*. Insect Mol Biol 21(2):147–58
Konopova B, Jindra M (2007) Juvenile hormone resistance gene Methoprene-tolerant controls entry into metamorphosis in the beetle *Tribolium castaneum*. Proc Natl Acad Sci U S A 104:10488–10493
Li XX, Zhang MY, Zhang HY (2011) RNA interference of four genes in adult *Bactrocera dorsalis* by feeding their dsRNAs. PLoS One 66:e17788
Lu JR, Ruhf ML, Perrimon N, Leder P (2007) A genome-wide RNA interference screen identifies putative chromatin regulators essential for E2F repression. Proc Natl Acad Sci U S A 104:9381–9386
Luo Y, Wang XH, Yu D, Kang L (2012) The SID-1 double-stranded RNA transporter is not required for systemic RNAi in the migratory locust. RNA Biol 9:663–671
Luo Y, Wang X, Wang X, Yu D, Chen B, Kang L (2013) Differential responses of migratory locusts to systemic RNA interference via double-stranded RNA injection and feeding. Insect Mol Biol 22:574–583
Mao YB, Cai WJ, Wang JW, Hong GJ, Tao XY, Wang LJ, Huang YP, Chen XY (2007) Silencing a cotton bollworm P450 monooxygenase gene by plant-mediated RNAi impairs larval tolerance of gossypol. Nat Biotechnol 25:1307–1313
Mao YB, Tao XY, Xue XY, Wang LJ, Chen XY (2011) Cotton plants expressing CYP6AE14 double-stranded RNA show enhanced resistance to bollworms. Transgenic Res 20:665–673
Marie B, Bacon JP, Blagburn JM (2000) Double-stranded RNA interference shows that Engrailed controls the synaptic specificity of identified sensory neurons. Curr Biol 10:289–292
McCaffrey AP, Meuse L, Pham TT, Conklin DS, Hannon GJ, Kay MA (2002) RNA interference in adult mice. Nature 418:38–39
Means JC, Muro I, Clem RJ (2003) Silencing of the baculovirus Op-iap3 gene by RNA interference reveals that it is required for prevention of apoptosis during *Orgyia pseudotsugata* M nucleopolyhedrovirus infection of Ld652Y cells. J Virol 77:4481–4488
Minakuchi C, Namiki T, Yoshiyama M, Shinoda T (2008) RNAi-mediated knockdown of juvenile hormone acid O-methyltransferase gene causes precocious metamorphosis in the red flour beetle *Tribolium castaneum*. FEBS J 275:2919–2931
Mishra S, Ande SR, Nyomba BLG (2010) The role of prohibitin in cell signaling. FEBS J 277:3937–3946
Misquitta L, Paterson BM (1999) Targeted disruption of gene function in Drosophila by RNA interference (RNA-i): a role for nautilus in embryonic somatic muscle formation. Proc Natl Acad Sci U S A 96:1451–1456
Montgomery MK, Xu S, Fire A (1998) RNA as a target of double-stranded RNA-mediated genetic interference in *Caenorhabditis elegans*. Proc Natl Acad Sci U S A 95:15502–15507
Moss EG (2001) RNA interference: it's a small RNA world. Curr Biol 11:R772–R775
Napoli C, Lemieux C, Jorgensen R (1990) Introduction of a chimeric chalcone synthase gene into petunia results in reversible co-suppression of homologous genes in trans. Plant Cell 2:279–289
Ni JQ, Liu LP, Binari R, Hardy R, Shim HS, Cavallaro A, Booker M, Pfeiffer BD, Markstein M, Wang H, Villalta C, Laverty TR, Perkins LA, Perrimon N (2009) A drosophila resource of transgenic RNAi lines for neurogenetics. Genetics 182:1089–1100
Niu BL, Shen WF, Liu Y, Weng HB, He LH, Mu JJ, Wu ZL, Jiang P, Tao YZ, Meng ZQ (2008) Cloning and RNAi-mediated functional characterization of MaLac2 of the pine sawyer, *Monochamus alternatus*. Insect Mol Biol 17:303–312
Nykanen A, Haley B, Zamore PD (2001) ATP requirements and small interfering RNA structure in the RNA interference pathway. Cell 107:309–321

Ober KA, Jockusch EL (2006) The roles of wingless and decapentaplegic in axis and appendage development in the red flour beetle, *Tribolium castaneum*. Dev Biol 294:391–405

Ochoa-Campuzano C, Martinez-Ramirez AC, Contreras E, Rausell C, Real MD (2013) Prohibitin, an essential protein for Colorado potato beetle larval viability, is relevant to Bacillus thuringiensis Cry3Aa toxicity. Pestic Biochem Physiol 107:299–308

Parthasarathy R, Palli SR (2008) Proliferation and differentiation of intestinal stem cells during metamorphosis of the red flour beetle, *Tribolium castaneum*. Dev Dyn 237:893–908

Parthasarathy R, Palli SR (2011) Molecular analysis of nutritional and hormonal regulation of female reproduction in the red flour beetle, *Tribolium castaneum*. Insect Biochem Mol Biol 41:294–305

Parthasarathy R, Tan A, Palli SR (2008) bHLH-PAS family transcription factor methoprene-tolerant plays a key role in JH action in preventing the premature development of adult structures during larval-pupal metamorphosis. Mech Dev 125:601–616

Parthasarathy R, Sheng Z, Sun Z, Palli SR (2010a) Ecdysteroid regulation of ovarian growth and oocyte maturation in the red flour beetle, *Tribolium castaneum*. Insect Biochem Mol Biol 40:429–439

Parthasarathy R, Sun Z, Bai H, Palli SR (2010b) Juvenile hormone regulation of vitellogenin synthesis in the red flour beetle, *Tribolium castaneum*. Insect Biochem Mol Biol 40:405–414

Perrimon N, Mathey-Prevot B (2007a) Applications of high-throughput RNA interference screens to problems in cell and developmental biology. Genetics 175:7–16

Perrimon N, Mathey-Prevot B (2007b) Matter arising – off targets and genome scale RNAi screens in Drosophila. Fly 1:1–5

Perrimon N, Ni JQ, Perkins L (2010) In vivo RNAi: today and tomorrow. Cold Spring Harb Perspect Biol 2(8):a003640

Pitino M, Coleman AD, Maffei ME, Ridout CJ, Hogenhout SA (2011) Silencing of aphid genes by dsRNA feeding from plants. PLoS One 6:e25709

Powers AM, Higgs S, Olson KE, Carlson JO, Beaty BJ (1994) Interference to La Crosse virus superinfection by an antisense RNA generated from an infectious Sindbis virus vector. Virus Res 32:57–67

Powers AM, Kamrud KI, Olson KE, Higgs S, Carlson JO, Beaty BJ (1996) Molecularly engineered resistance to California serogroup virus replication in mosquito cells and mosquitoes. Proc Natl Acad Sci U S A 93:4187–4191

Ren D, Cai Z, Song J, Wu Z, Zhou S (2014) dsRNA uptake and persistence account for tissue-dependent susceptibility to RNA interference in the migratory locust, *Locusta migratoria*. Insect Mol Biol 23:175–184

Schroder R (2003) The genes orthodenticle and hunchback substitute for bicoid in the beetle Tribolium. Nature 422:621–625

Sepp KJ, Hong P, Lizarraga SB, Liu JS, Mejia LA, Walsh CA, Perrimon N (2008) Identification of neural outgrowth genes using genome-wide RNAi. PLoS Genet 4:e1000111

Sheng Z, Xu J, Bai H, Zhu F, Palli SR (2011) Juvenile hormone regulates vitellogenin gene expression through insulin-like peptide signaling pathway in the red flour beetle, *Tribolium castaneum*. J Biol Chem 286:41924–41936

Shukla JN, Palli SR (2012a) Doublesex target genes in the red flour beetle, *Tribolium castaneum*. Sci Rep 2:948

Shukla JN, Palli SR (2012b) Sex determination in beetles: production of all male progeny by parental RNAi knockdown of transformer. Sci Rep 2:602

Shukla JN, Palli SR (2013) *Tribolium castaneum* Transformer-2 regulates sex determination and development in both males and females. Insect Biochem Mol Biol 43:1125–1132

Shukla JN, Palli SR (2014) Production of all female progeny: evidence for the presence of the male sex determination factor on the Y chromosome. J Exp Biol 217:1653–1655

Tan A, Palli SR (2008) Identification and characterization of nuclear receptors from the red flour beetle, *Tribolium castaneum*. Insect Biochem Mol Biol 38:430–439

Thakur N, Upadhyay SK, Verma PC, Chandrashekar K, Tuli R, Singh PK (2014) Enhanced Whitefly Resistance in Transgenic Tobacco Plants Expressing Double Stranded RNA of v-ATPase A Gene. PLoS One 9:e87235

Tian HG, Peng H, Yao Q, Chen HX, Xie Q, Tang B, Zhang WQ (2009) Developmental control of a lepidopteran pest Spodoptera exigua by ingestion of bacteria expressing dsRNA of a non-midgut gene. PLoS One 4(7):e6225

Tomoyasu Y, Denell RE (2004) Larval RNAi in Tribolium (Coleoptera) for analyzing adult development. Dev Genes Evol 214:575–578

Van der Zee M, Berns N, Roth S (2005) Distinct functions of the Tribolium zerknullt genes in serosa specification and dorsal closure. Curr Biol 15:624–636

Wan PJ, Guo WY, Yang Y, Lu FG, Lu WP, Li GQ (2014) RNAi suppression of the ryanodine receptor gene results in decreased susceptibility to chlorantraniliprole in Colorado potato beetle, *Leptinotarsa decemlineata*. J Insect Physiol 63:48–55

Wang J, Zhou XB, Li FH, Bradley PL, Chang SF, Perrimon N, Wong STC (2009) An image score inference system for RNAi genome-wide screening based on fuzzy mixture regression modeling. J Biomed Inform 42:32–40

Whyard S, Singh AD, Wong S (2009) Ingested double-stranded RNAs can act as species-specific insecticides. Insect Biochem Mol Biol 39:824–832

Wilson TG, Ashok M (1998) Insecticide resistance resulting from an absence of target-site gene product. Proc Natl Acad Sci U S A 95:14040–14044

Wilson TG, Fabian J (1986) A *Drosophila melanogaster* mutant resistant to a chemical analog of juvenile hormone. Dev Biol 118:190–201

Worby CA, Simonson-Leff N, Dixon JE (2001) RNA interference of gene expression (RNAi) in cultured Drosophila cells. Sci STKE 2001(95):pl1

Wynant N, Verlinden H, Breugelmans B, Simonet G, Vanden Broeck J (2012) Tissue-dependence and sensitivity of the systemic RNA interference response in the desert locust, *Schistocerca gregaria*. Insect Biochem Mol Biol 42:911–917

Wynant N, Santos D, Verdonck R, Spit J, Van Wielendaele P, Vanden Broeck J (2014) Identification, functional characterization and phylogenetic analysis of double stranded RNA degrading enzymes present in the gut of the desert locust, Schistocerca gregaria. Insect Biochem Mol Biol 46:1–8

Xiong YH, Zeng HM, Zhang YL, Xu DW, Qiu DW (2013) Silencing the HaHR3 gene by transgenic plant-mediated RNAi to disrupt *Helicoverpa armigera* development. Int J Biol Sci 9:370–381

Xu J, Tan A, Palli SR (2010) The function of nuclear receptors in regulation of female reproduction and embryogenesis in the red flour beetle, *Tribolium castaneum*. J Insect Physiol 56:1471–1480

Xu J, Raman C, Zhu F, Tan A, Palli SR (2012) Identification of nuclear receptors involved in regulation of male reproduction in the red flour beetle, *Tribolium castaneum*. J Insect Physiol 58:710–717

Xu J, Sheng Z, Palli SR (2013) Juvenile hormone and insulin regulate trehalose homeostasis in the red flour beetle, *Tribolium castaneum*. PLoS Genet 9:e1003535

Zha WJ, Peng XX, Chen RZ, Du B, Zhu LL, He GC (2011) Knockdown of midgut genes by dsRNA-transgenic plant-mediated RNA interference in the hemipteran insect *Nilaparvata lugens*. PLoS One 6(5):e20504

Zhang Z, Xu J, Sheng Z, Sui Y, Palli SR (2011) Steroid receptor co-activator is required for juvenile hormone signal transduction through a bHLH-PAS transcription factor, methoprene tolerant. J Biol Chem 286:8437–8447

Zhou LT, Jia S, Wan PJ, Kong Y, Guo WC, Ahmat T, Li GQ (2013) RNA interference of a putative S-adenosyl-L-homocysteine hydrolase gene affects larval performance in *Leptinotarsa decemlineata* (Say). J Insect Physiol 59:1049–1056

Zhu F, Parthasarathy R, Bai H, Woithe K, Kaussmann M, Nauen R, Harrison DA, Palli SR (2010a) A brain-specific cytochrome P450 responsible for the majority of deltamethrin resistance in the QTC279 strain of *Tribolium castaneum*. Proc Natl Acad Sci U S A 107:8557–8562

Zhu F, Xu JJ, Palli R, Ferguson J, Palli SR (2010b) Ingested RNA interference for managing the populations of the Colorado potato beetle, *Leptinotarsa decemlineata*. Pest Manag Sci 67:175–182

Zhu F, Sams S, Moural T, Haynes KF, Potter MF, Palli SR (2012a) RNA interference of NADPH-cytochrome P450 reductase results in reduced insecticide resistance in the bed bug, *Cimex lectularius*. PLoS One 7:e31037

Zhu JQ, Liu S, Ma Y, Zhang JQ, Qi HS, Wei ZJ, Yao Q, Zhang WQ, Li S (2012b) Improvement of pest resistance in transgenic tobacco plants expressing dsRNA of an insect-associated gene EcR. PLoS One 7:e38572

Zhu F, Gujar H, Gordon JR, Haynes KF, Potter MF, Palli SR (2013) Bed bugs evolved unique adaptive strategy to resist pyrethroid insecticides. Sci Rep 3:1456

Optimization of Fermentation Parameters and In Vitro Efficacy of Native *Bacillus thuringiensis* Isolates Against *Spodoptera litura*

26

Rahul Amin, Kuldeep Khatri, Deepak Panpatte, Leena Pathak, Ankit Patel, Noushad Parvez, Harsha Shelat, Janardan Jani, and Rajababu Vyas

Abstract

Insect pests are the major cause for damage to agricultural crops. Current strategies aimed to reduce crop losses primarily rely on chemical pesticides. Microbial pesticides and transgenic crops with intrinsic pest resistance are considered to be a promising alternative developed over the last few decades. *Bacillus thuringiensis* (*Bt*) is a gram-positive, sporeforming, soil bacterium which produces different insecticidal crystal proteins encoded by *cry* genes and is an effective tool to control insects, mites, nematode, etc. Native isolates of *Bt* (CT-1, CT-2, TOB-1) recovered from farms of Anand Agricultural University checked with standard strains of *Bt* var. *kurstaki* and *Bt* HD-73. Bioassay of the *Bt* isolates proved that CT-2 was more effective than all other isolates against the *S. litura*. The LC_{50} value of CT-2 was 4.99×10^6 (4.99 µl) as compared with TOB [1.41×10^7 (14.10 µl)] and *Btk* and HD-73 [3.85×10^7 (38.50 µl)] at 96 h. It was observed that the percent larval mortality increased with a consequent increase in dose as well as time up to 168 h. Mass production of native *Bt* isolates showed that combination of molasses and corn steep liquor was found best followed by molasses alone and combination of molasses, corn steep liquor, rice bran, and wheat bran.

Keywords

Bacillus thuringiensis • In vitro • Optimization • *Spodoptera litura*

R. Amin • K. Khatri (✉) • D. Panpatte • L. Pathak • A. Patel
N. Parvez • H. Shelat • J. Jani • R. Vyas
Department of Microbiology, Biological Control Research Laboratory, Anand Agricultural University, Anand 388110, Gujarat, India
e-mail: kuldeepkhatri33@gmail.com

A.K. Chakravarthy, S. Sridhara (eds.), *Arthropod Diversity and Conservation in the Tropics and Sub-tropics*, DOI 10.1007/978-981-10-1518-2_26

26.1 Introduction

Microbial pesticides and transgenic crops with intrinsic pest resistance are considered to be a promising alternative to pesticides developed over last few decades. The first generation of insect-resistant transgenic plants is based on insecticidal Cry proteins (δ-endotoxin) from *Bacillus thuringiensis* (*Bt*). The second generation of insect-resistant plants under development will include both *Bt* and non-*Bt* toxins with novel modes of action and have a wide spectrum of activity against insect pests. Among various microbes reported worldwide, bacterial pathogen, *B. thuringiensis* Berliner, is found as most effective agent for suppressing lepidopteran insects. The members of genus *Spodoptera* (Lepidoptera: Noctuidae) are the prime agricultural pests across the world. Castor leaf-eating caterpillar, *Spodoptera litura* Fabricius, is a well-known, ubiquitous, polyphagous, multivoltine lepidopteran which attacks 112 cultivated crops all over the world (Moussa et al. 1960).

One of the desirable characteristics of *B. thuringiensis* is production of toxins, which have shown activity against not only Lepidoptera, Diptera, Hymenoptera, Isoptera, Orthoptera, and Coleoptera but also showed activity against nematodes, mites, lice, aphids, and ants (Lacey and Goettel 1995). Scientists have identified at least 29 different crystals and δ-endotoxins effective against pests; thus *B. thuringiensis* can produce one or more types of *Cry* proteins (Entwistle 1993). The insecticidal crystal proteins (ICPs) of *B. thuringiensis* were originally classified as CryI, CryII, CryIII, and CryIV proteins based on their toxic activities. More precisely, related proteins within each family are further assigned divisional letters such as CryIA, CryIB, CryIC, etc. Even more closely related proteins within each division are given names such as CryIC1, CryIC2, etc. (Höfte and Whiteley 1989).

In India, transgenic crops have resulted in substantial economic benefits amounting to US$78.40 billion 1996–2010 and $14 billion for 2010 alone. The future of transgenic crops appears to be encouraging with the number of countries adopting these crops expected to grow and their global area and the number of farmers planting transgenic crops expected to increase. In India, a significant 39 % reduction in the number of insecticide sprays has led to an 88 % increase in profitability. One of the major public concerns about transgenic crops is their effect on nontarget organism. The result of a number of studies has demonstrated that the effects of transgenic crops on nontarget organism including natural enemies and other arthropods are likely to be much less severe than those of the broad-spectrum insecticides (Dhaliwal and Koul 2007, 2010). Therefore, a study was initiated on native isolates of *Bt* isolated from soils collected at Anand, Gujarat. The isolates of *Bt* were evaluated for their efficacy against the polyphagous, defoliator pest, *S. litura*. Experiments were also conducted on mass production of *Bacillus thuringiensis* on agro-wastes.

26.2 Materials and Methods

26.2.1 Native and Standard Strains Used

Native isolates of *B. thuringiensis* which were isolated from soil collected from farms at Anand Agricultural University, Anand, were tested against different crops pests. Standard strains of *B. thuringiensis* var. *kurstaki* HD-1 and HD-73 were obtained from the *Bacillus* Genetic Stock Center, Department of Biochemistry, Ohio State University, Ohio.

26.2.2 Pilot-Scale Mass Production of *Bacillus thuringiensis* (*Bt*) on Agro-wastes

To devise small-scale production technique for local entrepreneurs and educated farmers to meet their own requirements with less capital investment and to generate local employment, attempts were made to mass produce *B. thuringiensis* in laboratory following shake flask method.

Composition of different media tested for mass production of *Bt*:

1. Luria broth + 5 % glucose
2. 30 % coconut water + 70 % distilled water
3. 30 % rice waste water + 70 % distilled water
4. 30 % molasses + 70 % distilled water
5. 30 % corn steep liquor + 70 % distilled water
6. 30 % rice bran + 5 % glucose + 70 % distilled water
7. 30 % wheat bran + 5 % glucose + 70 % distilled water
8. 15 % rice bran + 15 % wheat bran + 5 % glucose + 70 % distilled water
9. 20 % molasses + 10 % corn steep liquor + 70 % distilled water
10. 20 % rice waste water + 10 % corn steep liquor +70 % distilled water
11. 10 % coconut water + 10 % rice waste water + 10 % corn steep liquor + 70 % distilled water
12. 10 % coconut water + 10 % rice waste water + 10 % molasses + 70 % distilled water
13. 5 % molasses + 5 % corn steep liquor + 10 % rice bran + 10 % wheat bran + 70 % distilled water
14. 5 % rice waste water + 5 % corn steep liquor + 10 % rice bran + 10 % wheat bran + 70 % distilled water
15. 5 % coconut water + 5 % rice waste water + 5 % molasses + 7.5 % rice bran + 7.5 % wheat bran + 70 % distilled water

26.2.3 Shake Flask Procedure (Obeta and Okafor 1984)

Native *B. thuringiensis* isolates grown on nutrient agar slants at 34 °C for 72 h and stored at 4 °C were inoculated into treatment flasks (1–15) containing respective test media at 10^5 spores/ml/flask and incubated at 30 °C on rotator shaker at 150 rpm. Cultures were examined at regular intervals of 24 h by counting the spores using Neubauer's hemocytometer under phase-contrast microscope up to 15 days. The number of bacterial spores were recorded to know which medium gave best growth of *B. thuringiensis*.

$$\text{Spore count} = \frac{\text{No. of spores}}{\text{No. of square}} \times \text{No. of total} \times \text{dilution} \times \text{depth} \times 1000\ \text{Squares factor Counted}$$

Bacterial growth was plotted in a graphically manner to identify the best media which gave highest growth of *B. thuringiensis*. For colony-forming unit (CFU), serial dilution from 10^{-1} to 10^{-12} of all culture grown in different media was prepared, and 100 μl of 10^{-8} to 10^{-12} dilutions was spread on selective agar media (MY agar with polymyxin β-sulfate) and incubated for 48 h at 30 °C. After incubation, colonies were considered on plates showing colonies between 30 and 300 per 0.1 ml of diluted culture to determine the CFU and calculation numbers using the formula mentioned below:

$$\text{CFU} = \frac{\text{Colony formed} \times \text{Dilution factor}}{\text{Aliquot taken}}$$

26.2.3.1 Bioassay of δ-Endotoxin of *B. thuringiensis* Isolates Against Neonate Larvae of *S. Litura* Fab.

Rearing of Target Insect A large number of egg masses were collected from castor and kept in clean plastic bowls for hatching to start and maintain the *S. litura* Fab. in the laboratory. Galvanized iron cages and tubes previously cleaned with 10 % formaldehyde to avoid the contamination of other microorganisms were used for collection and rearing. The larvae were reared on fresh castor leaves in galvanized iron cages (25 cm D) covered with a piece of black cotton cloth held in position with the help of string and rubber rings. Fresh and clean castor leaves were provided as food to the larvae every day. The larvae in later stages near pupation were transferred to a cage containing moist pad of sterilized soil for pupation. The pupae thus obtained were transferred to a plastic bowl having moist blotting paper at the bottom and kept in a wooden oviposition cage with glass walls (30 × 30 × 30 cm) and allowed to emerge as adults. On emergence, the adult male and female in the ratio of 3:1 were transferred to oviposition cages and fed with 10 % honey. Fresh tender leaves of castor as well as plastic stripes were hanged in the cage for egg laying. The eggs recovered from oviposition cages were transferred to chemically sterile plastic bowls (13 cm D × 6 cm H) for hatching. Neonate larvae were used for laboratory studies.

Table 26.1 Artificial diet for neonate larvae of *S. litura*

Diet composition		
Sr. no.	Ingredients	gm/100 ml
1.	Chick pea flour + milk powder (6:1)	16.0
2.	Vitamins (B-complex solution)	10.5
3.	Cholesterol	0.05
4.	Choline chloride	0.09
5.	Ascorbic acid	0.1
6.	Cellulose powder	1.67
7.	Agar	1.67
8.	Methyl-p-hydroxybenzoate	0.2
9.	Ascorbic acid	0.03

Table 26.2 Ascorbic treatments

T_1	*Btk* HD-1
T_2	*Bt* CT-1
T_3	*Bt* CT-2
T_4	*Bt* TOB
T_5	*Btk* HD-73
T_6	Control (sterile distilled water)

Preparation of Diet (Table 26.1) Agar powder and methyl-p-hydroxybenzoate were mixed in half the quantity of required water and dissolved by autoclaving. The other ingredients No.1, 2, 3, 4, 6, and 9 were mixed with other half quantity of water in blender. The hot agar solution and nutrient homogenate solution were then mixed together, and ascorbic acid was added after cooling the diet at 50 °C (Table 26.2). Warm diet was poured in a clean diet dispenser for equal distribution in insect-rearing trays.

26.2.4 Bioassay Procedure (Navon and Klein 1990)

Different doses (2 μl, 4 μl, 8 μl, 16 μl, 32 μl, 64 μl ,128 μl, and 256 μl) of crystal protein extract were spread on artificial diet previously solidified in wells (2 g diet/well) of bioassay trays (45 × 20 cm) having eight compartments and each compartment having 16 wells, with each well having 2 cm diameter and 1.5 cm depth.

Different doses (2 μl, 4 μl, 8 μl, 16 μl, 32 μl, 64 μl, 128 μl, and 256 μl) of each respective test solutions from stock solution having 1×10^9 crystals were poured by Gilson micropipette in designated compartments of the trays and were spread on diet surface by mechanical shaking. Simultaneously, 50 μl sterile distilled water was spread in control wells. After 10 min, each well was filled with healthy, laboratory-reared neonate larva of *S. litura*. Plastic transparent stickers with micropores were pasted to cover each compartment to prevent the escape of larvae from the wells.

Complete procedure was carried out in a laminar flow chamber. Trays were kept at 29 °C for 7 days in BOD incubator. Thus, a total of six bioassays following eight different doses including control, replicating each treatment for three times, were carried out in six different trays for three *B. thuringiensis* isolates including standard *Btk* HD-1 and *Btk* HD-73 against larvae under laboratory conditions. Dead larvae of *S. litura* recovered from trays were primarily examined using magnifying lens for disease symptoms. *B. thuringiensis*-induced mortality was further confirmed taking random samples from the bioassay trays following staining and microscopy (100×).

26.2.5 Statistical Analysis

The cumulative mortality of the insect on the days following treatment was transformed to percent mortality values. Periodic observations were analyzed following the analysis of variance (ANOVA) method (Steel and Torries 1980), and means were compared by Duncan's new multiple range test (DNMRT) (Duncan 1955). Significance levels were within confidence limits of 0.05 or less.

26.3 Results and Discussion

Results of the studies on pilot-scale mass production of *Bt* isolates on agro-wastes (by-products) and bio-efficacy of δ-endotoxin of *Bt* against *S. litura* (Fab.) under laboratory conditions are presented and discussed hereafter.

26.3.1 Pilot-Scale Mass Production of *Bt* on Agro-wastes

Mass production of native *B. thuringiensis* was performed by shake flask methods to develop a technique for small-scale production at cheaper cost by utilizing various agro-waste products listed in Table 26.3.

Table 26.3 Content of agro-wastes (standard average composition) tried for mass production of *B. thuringiensis*

Sr. no.	Waste products	Specific gravity (gm/ml)	Total carbon (%)	Total nitrogen (%)	Average oil content (%)	C/N ratio
1.	Corn steep liquor	1.19	13.03	3.30	13.05	3.95
2.	Molasses	1.38	33.85	1.14	33.94	29.7
3.	Coconut water	0.93	131.25	17.5	10.00	7.50
4.	Rice bran	0.72	48.28	2.44	17.03	19.79
5.	Wheat bran	0.79	105.60	22.00	10–12	4.80

26.3.1.1 Spore Count of Native *B. thuringiensis* Multiplied on Different Agro-wastes

From Table 26.4, it is evident that among individual agro-waste test media, liquid media T_5 (molasses + distilled water) gave highest spore count of CT-1 (4.7×10^{11}), CT-2 (4.6×10^{11}), TOB (5.0×10^{11}), and *Btk* HD-73 (5.2×10^{11}), and in combination of semisolid medium T_{13} (molasses + corn steep liquor + rice bran + wheat bran + distilled water), spore count was next best CT-1 (5.7×10^{12}), CT-2 (4.7×10^{12}), TOB (6.1×10^{12}), and *Btk* HD-73 (5.8×10^{12}) followed by liquid media T_9 (molasses + corn steep liquor + distilled water) spore count of CT-1 (6.0×10^{12}), CT-2 (6.0×10^{12}), TOB (6.8×10^{12}), and *Btk* HD-73 (6.8×10^{12}) for different *B. thuringiensis* cultures.

Table 26.4 Spore count of *B. thuringiensis* isolates at harvest

Treatments		Count/ml			
		CT-1	CT-2	TOB	*Btk* HD-73
T_1	LB broth with 5 % glucose	3.25×10^{9h}	5.0×10^{9hi}	5.25×10^{9f}	4.5×10^{9f}
T_2	Coconut water + distilled water	5.25×10^{8h}	4.8×10^{8i}	6.2×10^{8f}	5.0×10^{8f}
T_3	Rice waste water + distilled water	3.8×10^{8h}	4.3×10^{8i}	5.0×10^{8f}	4.2×10^{8f}
T_4	Corn steep liquor + distilled water	5.1×10^{12d}	3.8×10^{12d}	5.2×10^{12d}	4.6×10^{12d}
T_5	Molasses + distilled water	5.5×10^{12c}	4.3×10^{12c}	5.6×10^{12c}	5.5×10^{12c}
T_6	Rice bran with 5 % glucose + distilled water	4.7×10^{11e}	4.6×10^{11e}	5.0×10^{11e}	5.2×10^{11e}
T_7	Wheat bran with 5 % glucose + distilled water	3.8×10^{9h}	5.6×10^{9hi}	5.6×10^{9f}	4.5×10^{9f}
T_8	Rice bran + wheat bran + 5 % glucose + distilled water	4.3×10^{10g}	4.8×10^{10gh}	5.3×10^{10f}	4.5×10^{10f}
T_9	Molasses + corn steep liquor + distilled water	6.0×10^{12a}	6.0×10^{12a}	6.8×10^{12a}	6.8×10^{12a}
T_{10}	Rice waste water + corn steep liquor + distilled water	4.2×10^{9h}	6.5×10^{9hi}	7.0×10^{9f}	5.8×10^{9f}
T_{11}	Coconut water + rice waste water + corn steep liquor + distilled water	3.5×10^{10h}	4.6×10^{10hi}	5.0×10^{10f}	4.2×10^{10f}
T_{12}	Coconut water + rice waste water + molasses + distilled water	5.5×10^{10g}	6.2×10^{10g}	8.5×10^{10f}	5.6×10^{10f}
T_{13}	Molasses + corn steep liquor + rice bran + wheat bran + distilled water	5.7×10^{12b}	4.7×10^{12b}	6.1×10^{12b}	5.8×10^{12b}
T_{14}	Rice waste water + corn steep liquor + rice bran + wheat bran + distilled water	4.3×10^{11f}	4.8×10^{11e}	5.8×10^{11e}	5.6×10^{11e}
T_{15}	Coconut water + rice waste water + molasses + rice bran + wheat bran + distilled water	4.8×10^{11f}	5.6×10^{11e}	6.5×10^{11e}	6.1×10^{11e}
SEm		1.18	1.37	8.55	5.61
CD 5 %		3.56	4.13	2.57	1.69
CV %		10.50	14.05	7.02	4.81

**Figures indicating common letters do not differ significantly from each other at 5 % level of significance according to DNMRT

Table 26.5 Colony count of native *B. thuringiensis* isolates

Treatments		CFU/ml			
		CT-1	CT-2	TOB	*Btk* HD-73
T_1	LB broth with 5 % glucose	3.0×10^9	4.6×10^9	4.90×10^9	4.2×10^9
T_2	Coconut water + distilled water	5.0×10^8	4.5×10^8	6.0×10^8	4.8×10^8
T_3	Rice waste water + distilled water	3.4×10^8	4.0×10^8	4.8×10^8	4.0×10^8
T_4	Corn steep liquor + distilled water	4.8×10^{12}	3.5×10^{12}	5.0×10^{12}	4.4×10^{12}
T_5	Molasses + distilled water	5.3×10^{12}	4.2×10^{12}	5.3×10^{12}	5.2×10^{12}
T_6	Rice bran with 5 % glucose + distilled water	4.5×10^{11}	4.6×10^{11} e	5.0×10^{11}	5.4×10^{11}
T_7	Wheat bran with 5 % glucose + distilled water	3.5×10^9	5.3×10^9	5.5×10^9	4.3×10^9
T_8	Rice bran + wheat bran + 5 % glucose + distilled water	4.2×10^{10}	4.5×10^{10}	5.2×10^{10}	4.5×10^{10}
T_9	Molasses + corn steep liquor + distilled water	5.9×10^{12}	6.0×10^{12}	6.5×10^{12}	6.7×10^{12}
T_{10}	Rice waste water + corn steep liquor + distilled water	4.2×10^9	6.3×10^9	6.5×10^9	5.6×10^9
T_{11}	Coconut water + rice waste water + corn steep liquor + distilled water	3.2×10^{10}	4.4×10^{10}	4.8×10^{10}	4.0×10^{10}
T_{12}	Coconut water + rice waste water + molasses + distilled water	5.2×10^{10}	6.0×10^{10}	8.3×10^{10}	5.5×10^{10}
T_{13}	Molasses + corn steep liquor + rice bran + wheat bran + distilled water	5.4×10^{12}	4.5×10^{12}	6.0×10^{12}	5.5×10^{12}
T_{14}	Rice waste water + corn steep liquor + rice bran + wheat bran + distilled water	4.2×10^{11}	4.6×10^{11}	5.5×10^{11}	5.6×10^{11}
T_{15}	Coconut water + rice waste water + molasses + rice bran + wheat bran + distilled water	4.8×10^{11}	5.3×10^{11}	6.4×10^{11}	6.1×10^{11}

Thus, media T_9, T_{13}, and T_5 were found suitable for small-scale mass production compared to check T_1 (LB broth with 5 % glucose).

Colony-forming unit observation of agro-waste (by-products) media showed that in individual substrates, liquid media T_5 gave best CFU for different cultures, viz., CT-1 (5.5×10^{12}), CT-2 (4.3×10^{12}), TOB (5.6×10^{12}), and *Btk* HD-73 (5.5×10^{12}) (Table 26.5). In combination treatments media, T_{13} gave CFU, CT-1 (5.4×10^{12}), CT-2 (4.5×10^{12}), TOB (6.0×10^{12}), and *Btk* HD-73 (5.5×10^{12}) followed by media T_9 highest where CFU were in CT-1 (5.9×10^{12}), CT-2 (6.0×10^{12}), TOB (6.5×10^{12}), and *Btk* HD-73 (6.7×10^{12}) and remain at par. Though coconut water contains highest carbon (131.25), it gave low spore count, and colony count was also lowest for all *Bt* cultures, which indicates that there may be some compound present in coconut water which has negative impact on growth of *B. thuringiensis*.

Overall (Fig. 26.1), results indicated that the combination of molasses and corn steep liquor was found best followed by molasses alone, and combination of molasses, corn steep liquor, wheat bran, and rice bran is cheaper for mass production of native *B. thuringiensis* by semisolid/liquid pilot-scale methods.

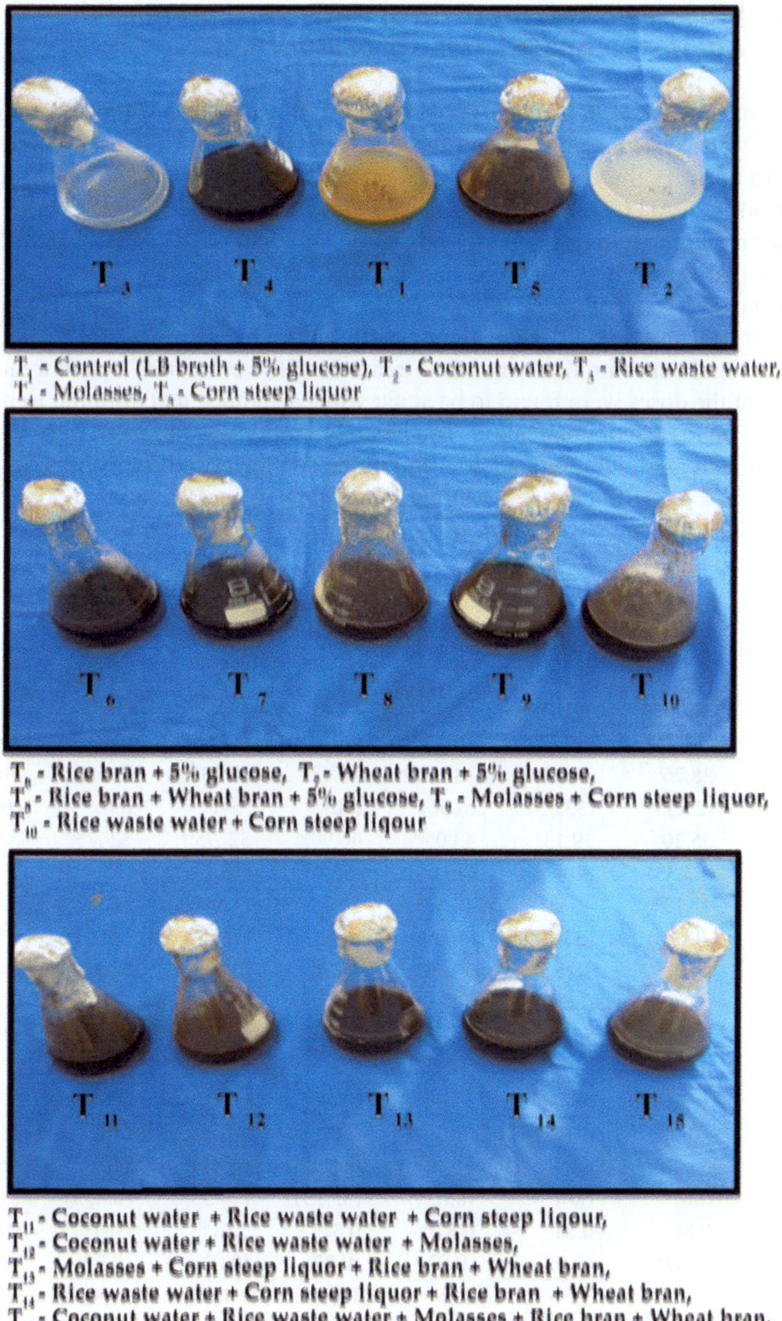

Fig. 26.1 Mass production of *Bt* on different agro-wastes

26.4 Bioassay of δ-Endotoxin of Native *B. thuringiensis* Against Neonate Larvae of *S. litura* Fab.

The promising native *Bt* isolates, CT-1, CT-2, TOB, and standard *Btk* HD-1 and *Btk* HD-73, were systematically tested through bioassay against neonate larvae of *S. litura* to quantify the optimum bacterial dose for effective suppression of insects. The result of GAU CT-2 bioassay against *S. litura* is presented in Table 26.6.

The crystal mixture of *Bt* was tested at 2 μl, 4 μl, 8 μl, 16 μl, 32 μl, 64 μl, 128 μl, and 256 μl different doses of 10^9 concentrations, and the larval mortality was examined from 24 h onward of the treatment. The results revealed that the CT-2 (256 μl of 10^9 crystal mixture/ml) gave significantly higher mortality (83.33 %) as compared to other concentrations at 96 h of treatment. However, after 24 h of treatment, the rest of the doses were found to be at par with each other, and a similar trend was observed up to 96 h of application. The larval mortality was observed to be increasing with increased duration and concentration. The lowest mortality was observed (40.0 %) at 2 μl of 10^9 crystal mixture/ml. In untreated control, larval mortality was not noticed.

Table 26.6 Bioassay of native CT-2 isolate against neonate larvae of *S. litura* Fab.

Treatment	% mortality after hrs						
doses	24	48	72	96	120	144	168
2 μl	21.13^{g**}	30.77^{fg}	32.98^{g}	39.13^{f}	48.82^{d}	48.82^{d}	48.82^{d}
	(13.33)*	(26.66)	(30)	(40)	(56.66)	(56.66)	(56.66)
4 μl	28.76^{f}	35.20^{ef}	337.20^{fg}	44.98^{ef}	50.83^{d}	50.83^{d}	50.83^{d}
	(23.33)	(33.33)	(6.66)	(50)	(60)	(60)	(60)
8 μl	35.20^{ef}	39.13^{de}	43.05^{ef}	46.90^{de}	52 .75^{cd}	52 .75^{cd}	52 .75^{cd}
	(33.33)	(40)	(46.66)	(53.33)	(53.33)	(53.33)	(53.33)
16 μl	41.13^{de}	46.90^{cd}	48.82^{de}	52.75^{cd}	58.98^{bc}	58.98^{bc}	58.98^{bc}
	(43.33)	(53.33)	(56.66)	(63.33)	(73.33)	(73.33)	(73.33)
32 μl	44.98^{cd}	48.82^{c}	50.74^{cd}	56.76^{bc}	61.19^{b}	61.19^{b}	61.19^{b}
	(50)	(56.66)	(60)	(70)	(76.66)	(76.66)	(76.66)
64 μl	50.74^{bc}	54 .76^{bc}	56.76^{bc}	58.98^{bc}	58.98^{bc}	58.98^{bc}	58.98^{bc}
	(60)	(66.66)	(70)	(73.33)	(73.33)	(73.33)	(73.33)
128 μl	56.76^{ab}	58.98^{ab}	63.40^{ab}	63.40^{b}	63.40^{b}	63.40^{b}	63.40^{b}
	(70)	(73.33)	(80)	(80)	(80)	(80)	(80)
256 μl	63.40^{a}	63.40^{a}	66.11^{ab}	66.11^{ab}	66.11^{ab}	66.11^{ab}	66.11^{ab}
	(80)	(80)	(83.33)	(83.33)	(83.33)	(83.33)	(83.33)
Control	0.81^{h}	0.81^{hi}	0.81^{gh}	0.81^{h}	0.81^{h}	0.81^{hi}	0.81^{gh}
	(0.02)	(0.02)	(0.02)	(0.02)	(0.02)	(0.02)	(0.02)
SEm±	2.28	2.51	2.09	1.98	2.14	2.14	2.14
CV%	10.01	9.76	7.66	6.69	6.62	6.62	6.62

1 μl = 10^6 crystal/ml from stock solution 10^9 crystal/ml

*Figures in parentheses are the retransformed values

**Figures indicating common letters do not differ significantly from each other at 5 % level of significance according to DNMRT

Table 26.7 Bioassay of native TOB isolate against neonate larvae of *S. litura* Fab.

Treatments	% mortality after hrs						
doses	24	48	72	96	120	144	168
2 μl	10.76^{f}**	23.35^{d}	28.76^{fg}	32.98^{f}	41.13^{f}	41.13^{f}	41.13^{f}
	(6.66) *	(16.66)	(23.33)	(30)	(43.33)	(43.33)	(43.33)
4 μl	21.13^{ef}	26.05^{d}	30.98^{f}	37.20^{ef}	43.05^{ef}	43.05^{ef}	43.05^{ef}
	(13.33)	(20)	(26.66)	(36.66)	(46.66)	(46.66)	(46.66)
8 μl	28.76^{de}	32.98^{cd}	37.20^{e}	41.13^{de}	46.90^{de}	46.90^{de}	46.90^{de}
	(23.33)	(30)	(36.66)	(43.33)	(53.33)	(53.33)	(53.33)
16 μl	35.20^{cd}	39.13^{bc}	43.05^{de}	46.90^{cd}	52.75^{cd}	52.75^{cd}	52.75^{cd}
	(33.33)	(40)	(46.66)	(53.33)	(63.33)	(63.33)	(63.33)
32 μl	39.21^{bcd}	43.05^{bc}	44.98^{cd}	50.74^{bc}	54.76^{c}	54.76^{c}	54.76^{c}
	(40)	(46.66)	(50)	(60)	(66.66)	(66.66)	(66.66)
64 μl	44.98^{bc}	48.80^{ab}	50.74^{bc}	52.75^{bc}	56.76^{bc}	56.76^{bc}	56.76^{bc}
	(50)	(56.66)	(60)	(63.33)	(70)	(70)	(70)
128 μl	50.74^{ab}	54.76^{a}	56.76^{ab}	56.76^{b}	61.19^{ab}	61.19^{ab}	61.19^{ab}
	(60)	(66.66)	(70)	(70)	(76.66)	(76.66)	(76.66)
256 μl	56.76^{a}	58.98^{a}	61.19^{a}	63.40^{a}	63.40^{a}	63.40^{a}	63.04^{a}
	(70)	(73.33)	(76.66)	(80)	(80)	(80)	(80)
Control	0.81^{g}	0.81^{gh}	0.81^{gh}	0.81^{g}	0.81^{gh}	0.81^{g}	0.81^{g}
	(0.02)	(0.02)	(0.02)	(0.02)	(0.02)	(0.02)	(0.02)
SEm±	3.42	3.18	1.81	1.78	1.79	1.79	1.79
CV%	17.74	14.12	7.50	6.82	6.19	6.19	6.19

1 μl = 10^6 crystal/ml from stock solution 10^9 crystal/ml
*Figures in parentheses are the retransformed values
**Figures indicating common letters do not differ significantly from each other at 5 % level of significance according to DNMRT

In TOB bioassay, the *S. litura* mortality was 80.0 % at highest dose tested (256 μl of 10^9 crystal mixture/ml) at 96 h (Table 26.7). Significantly low mortality (30.0 %) was recorded in 2 μl dose, and the mortality rate was observed to be increased with increase in duration as well as concentrations. Larval mortality was not noticed in untreated control.

In standard *Btk* HD-1 bioassay, the *S. litura* mortality was 70.0 % at highest dose tested (256 μl of 10^9 crystal mixture/ml) at 120 h (Table 26.8). Significantly low mortality (33.3 %) was recorded in 2 μl dose, and *Btk* HD-73 bioassay *S. litura* mortality was 70.0 % at highest dose tested (256 μl of 10^9 crystal mixture/ml) at 96 h, and mortality was not noticed in untreated control (Table 26.9). Significantly low mortality (23.3 %) was recorded in 2 μl dose as compared to CT-1 bioassay. *S. litura* mortality was 66.66 % at highest dose tested (256 μl of 10^9 crystal mixture/ml) at 120 h, and in untreated control, larval mortality was not noticed. Significantly low mortality (23.3 %) was recorded in 2 μl dose (Table 26.10).

Similarly, Whitlock et al. (1991) also isolated two native isolates of *Bt* (K-2074 and K-2178) in Taiwan through an active screening program and found that isolates

Table 26.8 Bioassay of native *Btk* HD-1 isolate against neonate larvae of *S. litura* Fab.

Treatments	% mortality after hrs						
doses	24	48	72	96	120	144	168
2 µl	15.94^{ef**}	23.84^{d}	28.76^{de}	30.98^{de}	35.20^{de}	35.20^{de}	35.20^{de}
	(7.6)*	(16.3)	(23.3)	(26.3)	(33.3)	(33.3)	(33.3)
4 µl	26.05^{de}	30.77^{cd}	34.91^{cd}	37.12^{cd}	43.05^{cd}	43.05^{de}	43.05^{de}
	(19.3)	(26.3)	(33.3)	(36.3)	(46.3)	(46.3)	(46.3)
8 µl	28.76^{cd}	34.99^{bc}	41.05^{abc}	41.05^{c}	42.97^{cd}	42.97^{cd}	42.97^{cd}
	(23)	(33.3)	(43.3)	(43.3)	(46.3)	(46.3)	(46.3)
16 µl	33.19^{cd}	37.20^{bc}	39.13^{bc}	43.05^{c}	46.90^{bc}	48.82^{bcd}	48.82^{bcd}
	(30)	(36.3)	(40)	(46.3)	(53.3)	(56.3)	(56.3)
32 µl	34.99^{bcd}	37.20^{bc}	43.05^{abc}	43.05^{c}	44.98^{bc}	44.98^{bc}	44.98^{bc}
	(33)	(36.3)	(46.3)	(46.3)	(50)	(50)	(50)
64 µl	41.13^{abc}	41.13^{ab}	43.05^{abc}	44.98^{bc}	50.74^{abc}	50.74^{abc}	50.74^{abc}
	(43)	(43.3)	(46.3)	(50)	(60)	(60)	(60)
128 µl	46.90^{ab}	46.90^{a}	48.82^{ab}	52.75^{ab}	52.75^{ab}	52.75^{ab}	52.75^{ab}
	(53)	(53.3)	(56.3)	(63.3)	(63.3)	(63.3)	(63.3)
256 µl	50.83^{a}	48.82^{a}	50.74^{a}	54.76^{a}	56.76^{a}	56.76^{a}	56.76^{a}
	(60)	(56.3)	(60)	(66.3)	(70)	(70)	(70)
Control	0.81^{f}	0.81^{f}	0.81^{f}	0.81^{f}	0.81^{f}	0.81^{f}	0.81^{f}
	(0.02)	(0.02)	(0.02)	(0.02)	(0.02)	(0.02)	(0.02)
SEm±	3.87	2.77	2.96	2.38	2.54	2.54	2.54
CV%	20.75	13.33	13.09	9.93	9.79	9.79	9.79

1 µl = 10^6 crystal/ml from stock solution 10^9 crystal/ml
*Figures in parentheses are the retransformed values
**Figures indicating common letters do not differ significantly from each other at 5 % level of significance according to DNMRT

were highly pathogenic against *S. litura* and superior to HD-1 (*Btk*) culture. Another laboratory trial of *Bt* from the USA revealed that the early stages of *S. mauritia* and *Nephantis serinopa* were more susceptible as compared to older ones (Ayyar 1961). In India, different doses of *Bt* var. *kurstaki* formulations Cutlass, Delfin, and Bactec were tested at 10,000, 5000, 2500, 1250, 625, 312.50, 156.25, and 78.13 µg/ml against neonate larvae of *H. armigera* and *S. litura*. The results revealed that, among all three *Btk* products, Cutlass was most toxic against *H. armigera* and Delfin against *S. litura* (Patel 1995). Laboratory evaluation of different *Bt* subspecies showed that *Btk* (NCIM 2514) at 10^8 spore/ml concentration caused more than 85 % mortality of neonate larvae of the *S. litura* (Puntambekar et al. 1997). At higher *Bt* concentrations, increased larval mortality was also observed by Jayanthi and Padmavathamma (1997) who reported maximum mortality at highest concentration tested.

Table 26.9 Bioassay of native *Btk* HD-73 isolate against neonate larvae of *S. litura* Fab.

Treatments doses	% mortality after hrs						
	24	48	72	96	120	144	168
2 μl	8.05^{g**}	21.13^{e}	21.13^{g}	28.76^{f}	28.76^{f}	28.76^{f}	28.76^{f}
	(3.33)*	(13.33)	(13.33)	(23.33)	(23.33)	(23.33)	(23.33)
4 μl	18.42^{ef}	21.13^{e}	28.76^{ef}	30.98^{ef}	35.20d^{ef}	35.20d^{ef}	35.20d^{ef}
	(10)	(13.33)	(23.33)	(26.66)	(33.33)	(33.33)	(33.33)
8 μl	26.55^{de}	28.76^{d}	30.98^{de}	35.20^{e}	37.12^{de}	37.12^{de}	37.12^{de}
	(20)	(23.33)	(26.66)	(33.33)	(36.66)	(36.66)	(36.66)
16 μl	230.98^{ed}	35.20^{c}	37.20^{ed}	41.13^{d}	41.13^{cd}	41.13^{cd}	41.13^{cd}
	(6.66)	(33.33)	(36.66)	(43.33)	(43.33)	(43.33)	(43.33)
32 μl	37.20^{bc}	39.21^{bc}	43.05^{bc}	44.98^{cd}	46.90^{bc}	46.90^{bc}	46.90^{bc}
	(36.66)	(40)	(46.66)	(50)	(53.33)	(53.33)	(53.33)
64 μl	41.13^{ab}	41.13^{bc}	43.05^{bc}	46.90^{bc}	46.90^{bc}	46.90^{bc}	46.90^{bc}
	(43.33)	(43.33)	(46.66)	(53.33)	(53.33)	(53.33)	(53.33)
128 μl	44.98^{ab}	44.98^{ab}	46.90^{ab}	50.74^{b}	50.74^{ab}	50.74^{ab}	50.74^{ab}
	(50)	(50)	(53.33)	(60)	(60)	(60)	(60)
256 μl	48.82^{a}	50.74^{a}	50.74^{a}	56.76^{a}	56.76^{a}	56.76^{a}	56.76^{a}
	(66.6)	(60)	(60)	(70)	(70)	(70)	(70)
Control	0.81^{fg}	0.81^{de}	0.81^{fg}	0.81^{f}	0.81^{ef}	0.81^{g}	0.81^{f}
	(0.02)	(0.02)	(0.02)	(0.02)	(0.02)	(0.02)	(0.02)
SEm±	2.79	1.96	2.09	1.53()	2.11	2.11	2.11
CV%	16.15	9.98	10.03	6.61	8.81	8.81	8.81

1 μl = 10^6 crystal/ml from stock solution 10^9 crystal/ml

*Figures in parentheses are the retransformed values

**Figures indicating common letters do not differ significantly from each other at 5 % level of significance according to DNMRT

The dosage-mortality response of *S. litura* at different concentrations of *Bt* isolates, CT-1, CT-2, and TOB, and standards *Btk* HD-1 and *Btk* HD-73, was subjected to probit analysis, and the respective LC_{50} value for 5 *Bt* culture against *S. litura* was calculated (Table 26.11). The LC_{50} value CT-2 isolate against *S. litura* was found to be 4.99 × 10^6 crystal/ml (4.99 μl) followed by TOB isolate 1.41 × 10^7 crystal/ml (14.10 μl) at 96 h of 50 % mortality. Thus, among native *Bt* isolates, CT-2 was found to be more pathogenic and potential compared to TOB and CT-1 isolates (Fig. 26.2; Table 26.7). The results are in conformity to Maghodiya (2002) who tested crude spore-crystal mixture of these isolates and reported similar trend. Moreover, CT-2 is also reported as highly toxic to *S. litura* and other lepidopteran pests. The LC_{50} values of CT-2 against *S. litura* were 15.69 and 12.96 times higher compared to standard checks *Btk* HD-1 and *Btk* HD-73. While LC_{50} value of TOB against *S. litura* was 44.33 and 36.62 times higher as compared to checks *Btk* HD-1 and *Btk* HD-73.

Table 26.10 Bioassay of native CT-1 isolate against neonate larvae of *S. litura* Fab.

Treatments doses	% mortality after hrs						
	24	48	72	96	120	144	168
2 μl	23.84[cd]**	28.76[cd]	37.12[bc]	39.13[b]	43.05b[cd]	43.05[b]	43.05[b]
	(16.66)*	(23.33)	(36.66)	(40)	(46.66)	(46.66)	(46.66)
4 μl	26.55[bc]	28.76[cd]	33.19[c]	37.20[b]	39.21[de]	39.21[de]	39.21[de]
	(20)	(23.33)	(30)	(36.66)	(40)	(40)	(40)
8 μl	30.98[bc]	34.99[bc]	44.05[abc]	41.05[ab]	42.27[bcd]	42.27[bcd]	42.27[bcd]
	(26.66)	(33.33)	(43.33)	(43.33)	(46.66)	(46.66)	(46.66)
16 μl	28.27[bc]	32.98[bcd]	32.98[c]	37.20[b]	41.05[cd]	41.05[cd]	41.05[cd]
	(23.33)	(30)	(30)	(36.66)	(43.33)	(43.33)	(43.33)
32 μl	34.99[abc]	34.99[bc]	39.13[bc]	41.05[ab]	43.05[bcd]	43.05[b]	43.05[b]
	(33.33)	(33.33)	(40)	(43.33)	(46.66)	(46.66)	(46.66)
64 μl	39.13[ab]	39.13[ab]	41.13[abc]	44.98[ab]	48.82[abc]	48.82[ab]	48.82[ab]
	(40)	(40)	(43.33)	(50)	(56.66)	(56.66)	(56.66)
128 μl	44.98[a]	43.05[ab]	44.98[ab]	46.09[ab]	50.74[ab]	50.74[ab]	50.74[ab]
	(50)	(46.66)	(50.00)	(53.33)	(60)	(60)	(60)
256 μl	46.98[a]	48.82[a]	48.82[a]	50.83[a]	54.76[a]	54.76[a]	54.76[a]
	(53.33)	(56.66)	(56.66)	(60)	(66.66)	(66.66)	(66.66)
Control	0.81[f]	0.81[e]	0.81[d]	0.81[f]	0.81[e]	0.81[f]	0.81[d]
	(0.02)	(0.02)	(0.02)	(0.02)	(0.02)	(0.02)	(0.02)
SEm±	3.80	3.07	2.77	2.92	2.85	2.85	2.85
CV%	20.54	15.19	12.74	12.48	11.26	11.26	11.26

1 μl = 10^6 crystal/ml from stock solution 10^9 crystal/ml

*Figures in parentheses are the retransformed values

**Figures indicating common letters do not differ significantly from each other at 5 % level of significance according to DNMRT

Table 26.11 Probit analysis of bioassay of native and standard *B. thuringiensis* isolates against *S. litura* Fab.

Isolates	Chi²[a]	Slope	LC_{50}	Fiducial limit (95 %) Lower	Upper	Regression equation
Btk HD-1	3.39	0.44	3.18×10^{7}[b] (31.80 μl)	5.54	9.46	$Y = 1.67 + 0.44x$
CT-1	6.16	0.27	1.88×10^{7}[c] (18.80 μl)	3.19	10.46	$Y = 3.00 + 0.27x$
CT-2	2.16	0.64	4.99×10^{6}[b] (4.99 μl)	5.91	12.60	$Y = 0.66 + 0.64x$
TOB	0.95	0.61	1.41×10^{7}[b] (14.10 μl)	2.36	11.92	$Y = 0.61 + 0.61x$
Btk HD-73	0.95	0.58	3.85×10^{7}[b] (38.50 μl)	2.46	12.53	$Y = 0.54 + 0.58x$

[a]Table value of chi² $[P_{(0.05)} = 12.59]$

[b]Observations of 96 h when 50 % and above mortality was achieved

[c]Observations of 120 h when 50 % and above mortality was achieved

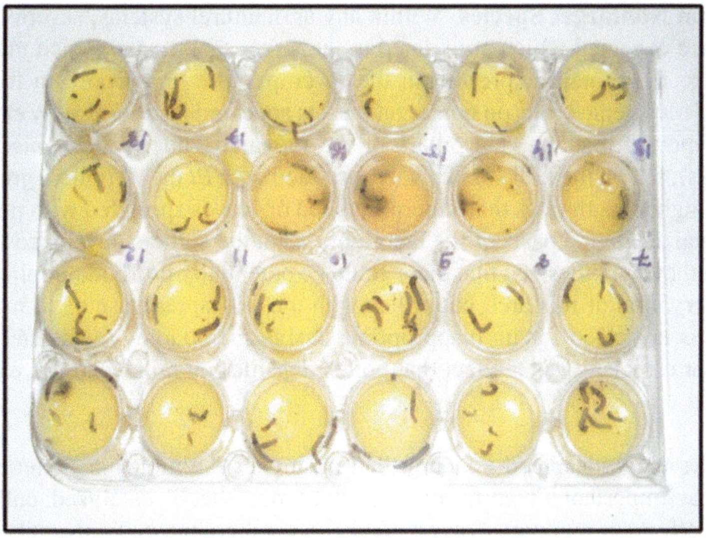

Fig. 26.2 Dead larvae of *S. litura* on artificial diet treated with different δ-endotoxin doses of test *B. thuringiensis* isolates on fifth day

26.5 Discussion

For mass production of native *Bt* isolates, combination of 20 % molasses and 10 % corn steep liquor was found best followed by 30 % molasses alone. Combinations of 5 % molasses, 5 % corn steep liquor, 10 % wheat bran, and 10 % rice bran are cheaper sources for in vitro pilot-scale mass production of native *B. thuringiensis* cultures following shake flask technique.

Bioassay of the *B. thuringiensis* isolate proved that CT-2 is more effective than all other isolates against the *S. litura*. The LC_{50} value of CT-2 was 4.99×10^6 (4.99 μl) as compared with TOB, 1.41×10^7 (14.10 μl), and *Btk* HD-73, 3.85×10^7 (38.50 μl) at 96 h. It was observed that the percent larval mortality increased with a consequent increase in dose as well as time up to 168 h.

Agricultural wastes and by-products like molasses, corn steep liquor, wheat bran, and rice bran are prospective for mass production of native *B. thuringiensis* isolates on a pilot scale by shake flask technique. Toxicity test carried out for confirmation of *B. thuringiensis* potency indicated an effective native *B. thuringiensis* isolate CT-2 for control of noxious insect, *S. litura*. In Gujarat, yield losses due to *S. litura* range from 12 to 25 %. For management of this pest, several chemicals are in use since the last four to five decades, but looking to the environmental pollution and residue problems, there is an urgent need to employ suitable microbial agents as alternatives (Thanki 1999). *B. thuringiensis* has the special property of producing toxic parasporal crystal protein (*Cry* protein) effective against many insect pests (Lecadet et al. 1999). Hence, efforts to isolate potential strains definitely help to create a healthy agroecosystem.

Impact on Nontarget Species Within any agricultural systems, several nontarget species are expected which are related to the target species and which may be susceptible to the protein expressed in the *Bt* crops. The first report on the adverse effects of transgenic crops on nontarget insects appeared in 1999 (Losey et al. 1999). It was reported that the caterpillars of the monarch butterfly, *Danaus plexippus* (Linnaeus), fed on the milkweed plants sprinkled with *Bt* maize pollen, grew slower, and had higher death rates than caterpillars fed on leaves sprinkled with pollen from non-*Bt* maize. On the basis of the laboratory data, the authors developed a scenario in which they hypothesized that there could be potentially profound implications for the conservation of monarch butterflies with the widespread use of *Bt* maize. This report was criticized for its inappropriate design methodology and interpretation. The major criticism was the fact that pollen densities on milkweed leaves were not quantified.

Despite the widespread concern that arose over the monarch butterfly controversy, one important outcome of this study was that it catalyzed collaborative research efforts of numerous government, nongovernment, and university scientists on this issue. Laboratory research carried out and concluded that monarch butterfly populations that developed in fields where maize hybrids were planted remained largely unaffected from exposure to *Bt* maize pollen (Sears et al. 2001). In other studies, there was no relationship between pollen deposition from transgenic maize and mortality of the black swallowtail butterfly, *Papilio polyxenes* Fabricius, and the milkweed tiger moth, *Euchaetes egle* Drury. There was no adverse effect on overall development of nontarget herbivore, *Mamestra brassicae* (Linnaeus), when reared on *Plutella xylostella* (Linnaeus)-resistant *Bt* Chinese cabbage (Kim et al. 2007).

Predators The effects of transgenic plants on the activity and abundance of predators vary across crops, insect species, and the transgenes in question. In a recent study (Wei et al. 2008), Cry1Ac was detected in *Chrysoperla carnea* (Stephens) larvae fed on resistant *Plutella xylostella* (Linnaeus) larvae reared on *Bt* oilseed rape. However, no Cry1Ac could be detected in *C. carnea* larvae when the lacewings were transferred to *Plutella xylostella* larvae reared on non-transgenic plants, indicating that *C. carnea* is able to metabolize plant-produced Cry1Ac. There was no effect on premaginal development or mortality of *C. carnea* when reared on *Rhopalosiphum padi* (Linnaeus) fed on *Bt* maize. Similarly, survival, aphid consumption, development, and reproduction of *Hippodamia convergens* (Guérin-Méneville) are not influenced when fed on *Myzus persicae* (Sulzer) reared on potatoes expressing §-endotoxin. Feeding *C. carnea* on *Tetranychus urticae* Koch (which ingested *Bt* toxin from the transgenic plants) or *R. padi* (which did not ingest the *Bt* toxin) did not affect survival or development of the predator. Field surveys have shown little impact of *Bt* maize on predator species numbers or densities. In tritrophic studies with the hemipteran predator, *Orius insidiosus* (Say), there was no effect when feeding on *Bt*-intoxicated European corn borers. In this case, the results were confirmed with direct feeding studies on *Bt* corn silks and observations of populations in *Bt* and non-*Bt* maize fields (Dhaliwal and Koul 2010).

The movement of Cry1Ac toxin from *Bt* transgenic cotton was investigated through specific predator-prey pairings, viz., *Spodoptera exigua* (Huber) and *Podisus maculiventris* (Say); *Tetranychus urticae* Koch and *Geocoris punctipes* (Say); *T. urticae* and *Nabis roseipennis* Reuter; and *Frankliniella occidentalis* (Pergande) and *O. insidiosus*. *T. urticae* exhibited 16.8 times more toxin in their bodies than that expressed in *Bt* cotton plant, followed by *S. exigua* (1.05 times) and *F. occidentalis* immature and adults (0.63 and 0.73 times, respectively). Of the toxin in the respective herbivorous prey, 4 %, 40 %, 17 %, and 14 % of that amount were measured in the predators, *G. punctipes*, *P. maculiventris*, *O. insidiosus*, *and N. rosiepennis*, respectively. However, there was no developmental time, survival, longevity, and fecundity of *P. maculiventris* and other predators (Torres and Ruberson 2006). Similarly, no detrimental effects were observed on development, fecundity, longevity, and egg viability of a piercing and sucking predators, *Orius sauteri* (Poppius) when fed with *Aphis gossypii* Glover reared on *Bt* cotton (Zhang et al. 2008).

A significant increase in the mortality and delay in the development of *C. carnea* were observed when fed on *Spodoptera littoralis* (Boisduval) and *Ostrinia nubilalis* (Huber) which had ingested *Bt* toxin from transgenic corn. The mean total immature mortality for *C. carnea* raised on *Bt*-fed prey was 62 %, as compared with 37 % on *Bt*-free prey. However, experimental design in this study did not make distinction between a direct effect due to the *Bt* protein on the predator and indirect effect of consuming a suboptimal diet consisting of sick or dying prey that had succumbed to the *Bt* toxin. Thus, the effects observed appear to reflect the poor nutritional quality of *Bt*-susceptible prey rather than any toxic effect of the *Bt* protein on lacewings. In another study, high concentrations of Cry1Ab (100 µg of Cry1Ab/ml of artificial diet) fed directly to lacewings larvae were toxic. A concern has been expressed about the methods used in this study. The dose is at least 30 times higher than that found in most maize tissues, and in the field setting, the lacewing larvae have a choice of other insects or eggs to feed on, so field exposures would be intermittent, rather than continuous. Direct exposure of the coccinellid predators, *Cheilomenes sexmaculatus* (Fabricius) larvae to *Bt* toxin, resulted in reduced larval survival and adult emergence. However, there were no adverse effects of *Bt* toxin on *C. sexmaculatus* when the larvae were reared on *Aphis craccivora* Koch fed on different concentrations of Cry1Ab or Cry1Ac in artificial diet (Dhaliwal and Koul 2007, 2010; Sharma 2012).

Parasitoids In general transgenic plants seem to have little or no effect on parasitoids of insect pest. In fact, increased level of parasitism by *Campoletis sonorensis* (Cameron) on *H. virescens* has been observed on transgenic tobacco as compared to non-transgenic plants. Physiological mechanism was put forth to support this phenomenon, i.e., toxic plants generally caused larvae to grow more slowly which may increase the duration of attack by natural enemies. In another study, activity of *Cardiochiles nigriceps* Viereck on *H. virescens* was not influenced by transgenic plants, which may alter their chances of encountering by parasitoids. Similarly, transgenic corn was observed to have chances of encounter by parasitoids and had no adverse effects on the parasitization of *O. nubilalis* by *Eriborus terebrans* (Gravenhorst) and *Macrocentrus grandii* Goidanich. However, the larval development and mortality of the parasitoids, *Parallorho gaspyralophagus* (Marsh), were adversely affected, when reared on *Bt*-susceptible insects that had fed on *Bt* maize, but the fitness of the emerging adults was not impacted (Bernal et al. 2007).

Schuler et al. (2004) found no adverse effect on diamondback moth parasitoid, *Cotesia plutellae* Kurdjumov, by feeding on Cry1Ac-resistant larvae. No Cry1Ac protein was detected in newly emerged larvae of the parasitic wasp, *Cotesia vestalis* Haliday, fed on diamondback moth larvae which had fed on *Bt* oilseed rape (Wei et al. 2008). Similarly, no significant changes were observed in the parasitization rate, larval period, pupal period, cocoon weight, or adult emergence rate when the parasitoid, *Microplitis mediator* (Haliday), was reared on the *M. brassicae* larvae fed with *Bt* transgenic Chinese cabbage (Kim et al. 2008). Studies on the impact of *Bt* broccoli plants on *Pteromalus puparum* (Linnaeus), an endoparasitoid of *Pieris rapae* (Linnaeus), indicated that there was an adverse effect on parasitism rate, developmental time, total number, and longevity of *P. puparum*. However, no Cry1C toxin was detected in newly emerged *P. puparum* adults developing in *Bt*-fed hosts. Moreover, no negative effect was found on the progeny of *P. puparum* developing from the *Bt* plant-fed host when subsequently supplied with a healthy host (*P. rapae* pupae). The negative impact of *Bt* broccoli on *P. puparum* resulted from poor quality of the host rather than direct effects of the *Bt* toxin (Chen et al. 2008).

There was a significant reduction in cocoon formation and adult emergence of the ichneumonid parasitoid, *Campoletis chlorideae* Uchida, reared on *Helicoverpa armigera* (Huber) larvae fed on the leaves of transgenic cotton before and after parasitization. However, no *Bt* toxins were detected in *H. armigera* larvae and parasitoid cocoons. Reduction in cocoon formation was because of early mortality of *H. armigera* larvae due to *Bt* toxin in the leaves of transgenic cotton. Thus, the effects of transgenic plants on the activity and abundance of parasitoids vary across crops and the insect species involved. In general, no major adverse effects have been observed on host-specific parasitoids. These are largely due to early mortality of host larvae or poor nutritional quality of the insect host, rather than direct toxicity of the insecticidal proteins. Such effects are common for all pest control interventions and are not regarded as risk (Sharma 2012).

Pollinators Pollination is another factor that must be considered in terms of possible effects of transience products on beneficial insects. Some reports indicate that transgenic plants seem to have low or no harmful effects on the life span and behavior of honeybees. Trypsin inhibitors, wheat germ agglutinin, serine protease inhibitor from soybean, cysteine protease inhibitor from rice, chicken egg white cystatin, and Bowman-Birk-type SBTI do not produce harmful effects on honeybees at the concentrations expressed in transgenic plants. The chitinase transgene in genetically modified oilseed rape did not affect leading performance of honeybees; beta-1,3-glucanase affected the level of conditioned responses and CpTI-induced marked effects in both conditioning and test phase, especially at high concentrations.

However, these experiments do not reflect the situation in plants, since the levels of protein employed in feeding assays are orders of magnitude higher than expressed by pollen. The diversity and density of the foraging insects have been found to be similar on the transgenic and non-transgenic cultivars. There were also no differences in nectar and pollen as well as foraging activity between transgenic and non-transgenic plants. Thus, it can be assumed that transgenic crops do not pose a major threat to the activity and abundance of pollinators (Sharma 2012).

Secondary Insect Pests The large-scale cultivation of transgenic crops with resistance to certain insect pests may result in secondary insect pest problems becoming a serious constraint in crop production. It may, therefore, become necessary to resort to spraying in order to control the secondary insect pests, which would adversely affect the natural enemies. Most field crops are attacked by several insect pests, and in the absence of competition from the major insects pests, the secondary insects may assume the status of major pests, e.g., leafhoppers, mired bugs, root feeders, etc., and this may offset some of the advantages of the insect-resistant transgenic crops. The transgenic *Bt* cotton, which is resistant to bollworms, is susceptible to sucking pests like the jassids, *Amrasca biguttula* (Ishida); whitefly, *Bemisia tabaci* (Gennadius); and mealybug, *Phenacoccus solenopsis* Tinsley and less effective against *Spodoptera litura* (Fabricius). There are also no differences in the susceptibility of transgenic and non-transgenic cotton varieties to boll weevil and aphids. The impact of transgenic *Bt* cotton on cotton aphid, *Aphis gossypii* Glover, was not significant after reared for one and two generations, with even some positive effects shown in survival rate and fecundity in the third generation. However, there was significant short-term impact of transgenic Bt + CpTI cotton, as the survival, lifetime fecundity, longevity, and feeding behavior of the aphids were negatively influenced in the first two generations, but the aphids apparently developed fitness in the third generation. In recent study, *Bt* cotton had no observed detrimental effects on *A. gossypii* (Zhang et al. 2008). Effective and timely control measures should be adopted for the control of secondary pests on transgenic crops. There is a need to deploy protease inhibitor and lectin genes that are effective against sucking pests, along with the *Bt* genes, to make genetically modified plants to be more effective against insect pest for sustainable crop protection (Sharma 2012).

References

Ayyar GR (1961) Preliminary trails with an entomogenous bacterium, *B. thuringiensis* Berliner new to India. Curr Sci 30:29–30

Bernal XE, Rand AS, Ryan MJ (2007) Sexual differences in responses to non-conspecific advertisement calls: receiver permissiveness in male and female tungara frogs. Anim Behav 73:955–964

Chen M, Zhao J-Z, Collins HL, Earle ED, Cao J, Anthony MS (2008) A critical assessment of the effects of *Bt* transgenic plants on parasitoids. PLoS One 3(5), e2284

Dhaliwal GS, Koul O (2007) Biopesticides and pest management: conventional and biotechnological approaches. Kalyani Publishers, New Delhi, p 455
Dhaliwal GS, Koul O (2010) Quest for pest management: from green revolution to gene revolution. Kalyani Publishers, New Delhi, p 386
Duncan GB (1955) Multiple range & multiple tests. Biometrics 42:1–42
Entwistle PF (1993) *Bacillus thuringiensis*, an environmental biopesticide: theory and practice. Wiley, New York, pp 171–191
Höfte H, Whiteley HR (1989) Insecticidal crystal proteins of *Bacillus thuringiensis*. Microbiol Rev 53:242–255
Jayanthi PDK, Padmavathamma K (1997) Laboratory evaluation of toxicity of *Bacillus thuringiensis* subsp. *kurstaki* to larvae of mulberry silkworm, Bombyx mori L. J Entomol Res NewDelhi 21(1):45–50
Kim JK, Kraemer, Neefus GP, Chung CD, Yarish IK (2007) The effects of temperature and ammonium on growth, pigment production and nitrogen uptake in four species of Porphyra native to the coast of New England. J Appl Phycol 19:431–440
Lacey LA, Goettel MS (1995) Current developments in microbial control of insect pests and prospects for the early 21st century. Entomophage 40:3–27
Lecadet MM, Franchon E, CosmaoDumanoir C, Ripoteau H, Hamon S, Laurent P, Thiery I (1999) Updating the H-antigen classification of *Bacillus thuringiensis*. J Appl Mocrobiol 86:660–672
Losey JE, Rayor LS, Carter ME (1999) Transgenic pollen harms monarch larvae. Nature 399:214
Maghodiya A (2002) Isolation of native *Bacillus thuringiensis* and its bioefficacy against leaf eating caterpillars of castor, *Spodoptera litura* Fabricius and *Achaea janata* Linnaeus. Unpublished M.Sc. thesis submitted to Gujarat Agricultural University, Anand
Moussa AM, Zather MA, Kothy F (1960) Abundance of cotton leaf worm *Prodenia litura* F. in relation to host plants. Host plants and their effect on biology (Lepidopteran: Agrotidae – Zanobiinae). Bull Soc Ent Egypte 44:241–251
Navon A, Klein M (1990) *Bacillus thuringiensis* potency bioassay against *Heliothis armigera*, *Earias insulana* and *Spodoptera litoralis* larvae based on standardized diets. J Invert Path 55:387–393
Obeta JAN, Okafor N (1984) Medium for the production of primary powder of Bacillus thuringiensis subsp. israelensis. Appl Environ Microbiol 47(4):863–867
Patel MC (1995) Boiefficacy of some *Bacillus thuringiensis* and neem (Azadirachtin) fomulations against cotton bollworms. M.Sc. thesis (Agril. Entomology). Gujarat Agricultural University, Sant Kabir Nagar
Puntambekar US, Mukherjee SN, Ranjekar PK (1997) Laboratory screening of different *Bacillus thuringiensis* strains against certain lepidopteron pests and subsequent field evaluation on the pod boring pest complex of Pigeonpea (*Cajanuscajan*). Antonie-Van-Leeuwenhoek 71:319–323
Schuler TH, Denholm I, Clark SJ, Stewart CN, Poppy GM (2004) Effects of Bt plants on the development and survival of the parasitoid *Cotesia plutellae* (Hymenoptera: Braconidae) in susceptible and Bt-resistant larvae of the diamondback moth, *Plutella xylostella* (Lepidoptera: Plutellidae). J Insect Physiol 50:435–443
Sears MK, Hellmich RL, Stanley-Horn DE, Oberhauser KS, Pleasants JM, Mattila HR, Siegfried BD, Dively GP (2001) Impact of Bt corn pollen on monarch butterfly populations: a risk assessment. Proc Natl Acad Sci U S A 98:11937–11942
Sharma HC (2012) Climate change effects on activity and abundance of insects: implications for crop protection and food security. In: Kang MS, Banga SS (eds) Combating climate change: an agriculturel perspective. Taylor and Francis, Boca Raton
Steel RGD, Torrie JH (1980) Principal and procedure of statistics. A biometrical approach, 2nd edn. Mc-Growhillcogakusha Ltd, New Delhi, p 137
Thanki KV (1999) Ph.D. thesis (Agril. Entomology). Gujarat Agricultural University, Sant Kabir Nagar, Gujarat

Torres JB, Ruberson JR (2006) Interactions of Bt-cotton and the omnivorous big-eyed bug *Geocoris punctipes* (Say), a key predator in cotton fields. Biol Control 39:47–57
Wei HC, Rollins, Fabian J, Hayes L, Polevoy M, Bazinet G, Brill JA (2008) Depletion of plasma membrane PtdIns (4,5) P2 reveals essential roles for phosphoinositides in flagellar biogenesis. J Cell Sci 121(7):1076–F1084
Whitlock VH, Lo MC, Kuo MH, Soong TS (1991) Two new isolates of Bacillus thuringiensis pathogenic to Spodoptera litura. J Invert Pathol 58:33–39
Zhang L, Jacob DJ, Boersma KF, Jaffe DA, Olson JR, Bowman KW, Worden JR, Thompson AM, Avery MA, Cohen RC, Dibb JE, Flock FM, Fuelberg HE, Huey LG, McMillan WW, Singh HB, Weinheimer AJ (2008) Transpacific transport of ozone pollution and the effect of recent Asian emission increases on air quality in North America: an integrated analysis using satellite, aircraft, ozonesonde, and surface observations. Atmos Chem Phys 8:6117–6136

MIX
Papier aus verantwortungsvollen Quellen
Paper from responsible sources
FSC® C105338

If you have any concerns about our products,
you can contact us on
ProductSafety@springernature.com

In case Publisher is established outside the EU,
the EU authorized representative is:
Springer Nature Customer Service Center GmbH
Europaplatz 3, 69115 Heidelberg, Germany

Printed by Libri Plureos GmbH
in Hamburg, Germany